Advances in Experimental Medicine and Biology

Volume 1149

Advances in Microbiology, Infectious Diseases and Public Health

This book series focuses on current progress in the broad field of medical microbiology, and covers both basic and applied topics related to the study of microbes, their interactions with human and animals, and emerging issues relevant for public health. Original research and review articles present and discuss multidisciplinary findings and developments on various aspects of microbiology, infectious diseases, and their diagnosis, treatment and prevention.

The book series publishes review and original research contributions, short reports as well as guest edited thematic book volumes. All contributions will be published online first and collected in book volumes. There are no publication costs.

Advances in Microbiology, Infectious Diseases and Public Health is a subseries of *Advances in Experimental Medicine and Biology*, which has been publishing significant contributions in the field for over 30 years and is indexed in Medline, Scopus, EMBASE, BIOSIS, Biological Abstracts, CSA, Biological Sciences and Living Resources (ASFA-1), and Biological Sciences. 2018 Impact Factor: 2.126.

More information about this subseries at http://www.springer.com/series/13513

Shigeru Kamiya • Steffen Backert
Editors

Helicobacter pylori in Human Diseases

Advances in Microbiology, Infectious Diseases and Public Health Volume 11

Springer

Editors
Shigeru Kamiya
Faculty of Health Sciences
Kyorin University
Mitaka, Tokyo, Japan

Steffen Backert
Division of Microbiology;
Dept of Biology
University of Erlangen-Nuremberg
Erlangen, Bayern, Germany

ISSN 0065-2598 ISSN 2214-8019 (electronic)
Advances in Experimental Medicine and Biology
ISSN 2365-2675 ISSN 2365-2683 (electronic)
Advances in Microbiology, Infectious Diseases and Public Health
ISBN 978-3-030-21915-4 ISBN 978-3-030-21916-1 (eBook)
https://doi.org/10.1007/978-3-030-21916-1

This Springer imprint is published by the registered company Springer Nature Switzerland AG.
The registered company address is: Gewerbestrasse 11, 6330 Cham, Switzerland

Preface

Helicobacter pylori is a spiral-shaped, microaerophilic Gram-negative bacterium that colonizes the human gastrointestinal tract, primarily the stomach. *H. pylori* is one of the most common causes of human infection, especially in developing countries, where the incidence is quite high (80–90%). *H. pylori* infection can commonly persist throughout life. The pathogen has been identified as an etiological agent of chronic acute/chronic gastritis, peptic ulcer disease, gastric cancer and gastric MALT (mucosa-associated lymphoid tissue) lymphoma. In addition, *H. pylori* has been considered to be a class-I (definite) carcinogen in humans according to the Working Group of the World Health Organization International Agency for Research on Cancer concluded in 1994.

H. pylori infection causes not only various gastric disorders as described above, but also extragastric diseases such as immunogenic (idiopathic) thrombocytopenic purpura (ITP) and iron-deficiency anemia (IDA). Although a number of virulence factors such as the vacuolating cytotoxin VacA, *cag* pathogenicity island with the CagA effector protein, adhesins, serine protease HtrA and the urease enzyme are known to be involved in the virulence of by this microorganism, exact mechanisms by which *H. pylori* infection causes pathogenic effects on host are still not fully understood.

A promising vaccine candidate against *H. pylori* is currently not available. The infection can be eradicated by a combination therapy using proton pump inhibitors and antimicrobial drugs including clarithromycin (CLR) and amoxicillin (AMX), and consequently decrease the burden of above diseases. However, as the resistance rates of *H. pylori* strains against CLR and AMX have been increased substantially, alternative therapies using novel proton pump inhibitors or different antimicrobial agents need to be developed in near future.

In this eBook, genomic structure and virulence factors of *H. pylori* as well as detailed pathogenic mechanisms by *H. pylori* infection are reviewed by worldwide expert contributors. In addition, from the clinical aspects, epidemiology, diagnosis, treatment by several drugs, the use of probiotics and prophylaxis by new vaccination strategies are introduced with updated literatures. The eBook will be useful for not only basic microbiologists, but also researchers in the fields of pathology, biochemistry and genomics as well as medical students/scientists.

Tokyo, Japan Shigeru Kamiya
Erlangen, Germany Steffen Backert
March 2019

Contents

Abbreviations

Abl	Abelson kinase
ADP-Hep	ADP-β-D-manno-heptose
ACE	Angiotensin-converting enzyme
ACRG	Asian cancer research group
ADP	Adenosine diphosphate
ADR	Alternate DNaseI resistant
AID	Activation-induced cytidine deaminase
AJ	Adherens junction
AKAP	A-kinase-anchoring protein
AKT/ PKB	Protein kinase B
AlpA	Adherence-associated lipoprotein A
ALPK1	Alpha protein kinase 1
Am-1	Amerindian-I
AmiE	Amidase
AmiF	Formidase
AMX	Amoxicillin
AP	Apurinic/apyrimidinic
AP-1	Activator protein-1
APC	Antigen presenting cell
APC	Adenomatous polyposis coli protein
APE	AP endonuclease
APOBEC	Apolipoprotein B mRNA-editing enzyme, catalytic polypeptide-like
ARF	Alternate reading frame protein product of the CDKN2A locus
ARF-BP1	ARF-binding protein 1
Arg2	Arginase II
ARHGAP26	Gene name for: Rho GTPase activating protein 26
ARHGAP6	Gene name for: Rho GTPase activating protein 6
ARID1A	AT-rich interactive domain-containing protein 1A
ASA	Acetylsalicylic acid
ASC	Adult stem cell
ASPP2	Apoptosis-stimulating protein of p53 2
ATF-2	Activating transcription factor 2
ATL	Adult T cell leukemia
ATM	Ataxia telangiectasia mutated kinase
ATP	Adenosine triphosphate

B2M	Beta-2-microglobulin
B3GNT5	Beta-1, 3-N-acetylglucosaminyltransferase
BabA	Blood group antigen binding adhesin
BAC	Bacterial artificial chromosome
Barx1	BarH-like homeobox 1
Bcl-2	B-cell leukemia/lymphoma 2 protein
Bcl-3	B-cell lymphoma encoded protein 3
Bcl-xL	B-cell lymphoma-extra large
BE	Barrett´s esophagus
BECN1	Beclin 1
BER	Base excision repair
BMDC	Bone marrow derived cell
BMP	Bone morphogenetic protein
BrdU	Bromodeoxyuridine
cag	Cytotoxin-associated genes
cagA	Cytotoxin-associated gene A
CagA	Protein encoded by *cagA*
cagL	Cytotoxin-associated gene L
*cag*PAI	*cag* pathogenicity island
1CAMP	Cationic antimicrobial peptides
cAMP	Cyclic adenosine monophosphate
Cas9	CRISPR associated protein 9
CASP 1	Caspase 1
CEACAM	Carcinoembryonic antigen-related cell adhesion molecule
CCKBR	Cholecystokinin B receptor (also called CCK_2)
CCND1	Gene name for: Cyclin D1
CCNE1	Gene name for: Cyclin E1
CcrM	*Caulobacter crescentus* DNA methyltransferase
CD	Cluster of differentiation
Cdc6	Cell division control protein 6
CDH1	Gene name for: E-cadherin
CDKN	Cyclin-dependent kinase inhibitor gene
Cdx2	Caudal type homeobox 2
CEACAM	Carcinoembryonic antigen-related cell adhesion molecule
CFU	Colony forming unit
CGD	Chronic granulomatous disease
CGRP	Calcitonin gene-related peptide
CGT	Cholesterol alpha glycosyl-transferase
CHK2	Checkpoint kinase 2
CI	Confidence interval
CIMP	CpG island methylator phenotype
CIN	Chromosomal instability
cITP	Chronic immune thrombocytopenic purpura
CK1	Casein kinase 1
CLD18	Gene name for: Claudin 18
CLR	Clarithromycin

CLR	C-type lectin receptor
CM	CagA multimerization sequence
CM^E	East Asian CagA-specific CM
CM^W	Western CagA-specific CM
c-Met	Tyrosine kinase (also called hepatocyte growth factor receptor)
comB	*H. pylori* competence gene B
comH	*H. pylori* competence gene H
Cox2	Cyclooxygenase 2
CpG	Cytosine-guanine repeats
CREB	cAMP responsive element binding protein
CRISPR	Clustered regularly interspaced short palindromic repeats
CRPIA	Conserved repeat responsible for phosphorylation independent activity
Csk	Carboxy-terminal Src kinase
CSMD	CUB and sushi multiple domains
CT	Cytosine-thymine
CT	Concomitant therapy
CTNNA1	Gene name for: α-1-Catenin
CTNNB1	Gene name for: β-1-Catenin
CTNND1	Gene name for: δ-1-Catenin
CXCL8	C-X-C motif chemokine ligand (also called IL-8)
CYP2C19	Cytochrome P450 2C9 gene
DC	Dendritic cell
DC-SIGN	Dendritic cell-specific HIV-1 receptor (also called CD209)
DGC	Diffuse gastric cancer
DKK	Dickkopf 1
DLBCL	Diffuse large B cell lymphoma
dmLT	Double-mutant *E. coli* Heat-labile toxin
DNA	Deoxyribonucleic acid
DSB	Double-strand DNA breaks
dsDNA	Double-stranded DNA
DU	Duodenal ulcer
DupA	Duodenal ulcer promoting gene A
EAC	Esophageal adenocarcinoma
EBER	EBV encoded small RNA
EBNA	EBV nuclear antigens
EBNA-LP	EBNA-leader protein
EBV	Epstein-Barr virus
ECL	Enterochromaffin-like
EGF	Epidermal growth factor
EGFR	Epidermal growth factor receptor
EHEC	Enterohemorrhagic *Escherichia coli*
ELK1	ETS domain containing protein
EMT	Epithelial mesenchymal transition
EPIYA	Glu-Pro-Ile-Tyr-Ala sequence motif
EoE	Eosinophilic esophagitis

ER	Endoplasmic reticulum
ERBB2(HER2)	Erb-b2 receptor tyrosine kinase 2
ERCC	Excision repair cross-complementing gene
ERK1/2	Extracellular signal–regulated kinase 1/2
ESC	Embryonic stem cell
ETEC	Enterotoxigenic *E. coli*
ETS2	E26 oncogene homolog 2
FAK	Focal adhesion kinase
FAP	Familial adenomatous polyposis
FasL	Fas ligand
FBXO24	F-box protein 24
FGC	Familial gastric cancer
FGF-10	Fibroblast growth factor-10
FGFR2	Fibroblast growth factor receptor 2
FHIT	Fragile histidine triad
Fic	Filamentation-induced by cAMP
FIGC	Familial intestinal gastric cancer
FlaA	Flagellin A
FoxP3	Forkhead box P3
FtsK	Filamentus temperature-sensitive cell division protein K
FUT3	Fucosyltransferase 3
GAPPS	Gastric adenocarcinoma and proximal polyposis of the stomach
GAPs	GTPase-activating protein
GATA6	GATA binding protein 6
gB	EBV glycoprotein B
GC	Gastric cancer
G-DIF	Diffuse subtype of GC
GERD	Gastro-esophageal reflux disease
GGT	Gamma-glutamyl transpeptidase
gH	EBV glycoprotein H
G-INT	Intestinal subtype of GC
GI	Gastrointestinal
GIN	Gastrointestinal intraepithelial neoplasia
gL	EBV glycoprotein L
GMDS	GDP-mannose 4,6-dehydratase
gp350	EBV glycoprotein 350
gp42	EBV glycoprotein 42
Grb2	Growth factor receptor-bound protein 2
GroEL	Heat shock protein 60
GSH	Glutathione (reduced form)
GSK-3β	Glycogen synthase kinase 3 beta
GU	Gastric ulcer
GyrA	Subunit A of DNA gyrase
GyrB	Subunit B of DNA gyrase
H2AX	Histone H2A variant X
H_2RA	H_2-receptor antagonist

hBD	Human beta-defensin
hBD3	Human-β-defensin 3
HBP	Heptose-1,7-bisphosphate
HBV	Hepatitis B virus
HCV	Hepatitis C virus
HDGC	Hereditary diffuse gastric cancer
HDM2	Human double minute 2
HER2/neu	Human epidermal growth factor receptor 2
HGF	Hepatocyte growth factor
HHLO	*Helicobacter heilmannii*-like organism
HHI	*H. pylori* HtrA inhibitor
HIV	Human immunodeficiency virus
HLA	Human leukocyte antigen class II
HLA-B	Histocompatibility complex class I
HNF4α	Hepatocyte nuclear factorα
HNPCC	Hereditary non-polyposis colorectal cancer
HO-1	Heme oxygensase 1
Hof	*Helicobacter* related
Hom	*Helicobacter* outer membrane
Hop	*H. pylori* outer membrane protein
HopQ	Helicobacter Outer membrane protein Q
HopZ	Helicobacter Outer membrane protein Z
Hor	Hop-related
HpAG	*H. pylori*-associated gastritis
HpSA	*H. pylori* stool antigen test
HPV	Human papillomavirus
HTLV-1	human T-lymphotropic virus type 1
HtrA	High temperature requirement A
HU	Histone-like protein
Hupki	Human TP53 knock-in
IARC	International agency for research on cancer
IBD	Inflammatory bowel disease
IDA	Iron deficiency anemia
IEL	Intraepithelial lymphocytes
IFN	Interferon
IFNγ	Interferon gamma
IGHV	Immunoglobulin heavy chain variable region
IHF	Integration host factor
IKK	IκB kinase
IL	Interleukin
ILC	Innate lymphoid cells
IM	Intestinal metaplasia
IMC	Inner membrane complex
iNOS	Inducible nitric oxide synthase
INS-GAS	Insulin-gastrin
INSR	Insulin receptor
IP_3	Inositol triphosphate

IPD	Interpulse duration
iPSC	Induced pluripotent stem cells
IRAK	IL-1 receptor-associated kinase
IRF	Interferon-regulatory factor
ISGF3	IFN-stimulated gene factor 3
ITT	Intention-to-treat
JAK	Janus kinase
JAM	Junctional adhesion molecule
JUP	Junction plakoglobin
KCNQ1	Potassium voltage-gated channel subfamily Q member 1
KLF5	Krueppel like factor 5
Kras	Kirsten rat sarcoma oncogene
KSHV	Kaposi sarcoma herpesvirus
LacdiNAc	GalNAcβ1-4GlcNAc glycan motif
LabA	lacdiNAc-specific adhesin
LBP	Lipopolysaccharide binding protein
Le^b	Lewis blood group antigen B
Le^x	Lewis X
LES	Lower esophageal sphincter
LFA-1	Lymphocyte function-associated antigen 1
LFS	Li-Fraumeni syndrome
Lgr5	Leucin-rich-repeat-containing G-protein coupled receptor 5
Lig	DNA ligase
LL37	37-residue amphipathic α-helical cathelicidin
LMP	Latent membrane protein
LOH	Loss of heterozygosity
LP	Leader peptide
LPS	Lipopolysaccharide
LR	Low risk
Lrp	Global regulatory protein
LRR	Leucin-rich repeat
LSP1	Lymphocyte specific protein 1
LT	Heat-labile toxin
LTB	Heat labile toxin B subunit
LVFX	Levofloxacin
m4C	N4-methylcytosine
m5C	5-methylcytosine
m6A	N6-methyladenine
MAGI-1	Membrane-associated guanylate kinase with inverted orientation 1
MAIT	Mucosal-associated invariant T cells
MALDI-MS	Matrix-assisted laser desorption ionization-mass spectrometry
MALT	Mucosa-associated lymphoid tissue
MAP	Mitogen-activated protein
MAP	Microtubule-associated protein
MAPK	Mitogen-activated protein kinase
MARK	Microtubule affinity-regulating kinase

MDCK	Madin-Darby canine kidney
MCV	Merkel cell polyomavirus
MHC	Major histocompatibility complex
Mincle	Macrophage inducible C-type lectin
MINCLE	Macrophage inducible C-type lectin
MIC	Minimum inhibitory concentration
Mist1	Basic helix-loop-helix family member a15
MGMT	0-6-Methylguanine-DNA methyltransferase
MKI	MARK kinase inhibitor
MLC	Myosin light chain
MLCK	Myosin light chain kinase
MLH	Human homolog of MMR from *Escherichia coli*
MLN	Mesenteric lymph nodes
MMP	Matrix metalloprotease
MMR	DNA mismatch repair
MPO	Myeloperoxidase
MRN	MRE11-RAD50-NBS1 complex
mRNA	Messenger ribonucleic acid
MS	Multiple sclerosis
MSI	Microsatellite instability
MSI-H	High microsatellite instability
MSI-L	Low microsatellite instability
MSS	Microsatellite stable
Mtase	Methyltransferase
MTHFR	Methylenetetrahydrofolate reductase
mTOR	Mechanistic target of rapamycin
MtrA	Response regulator of a *M. tuberculosis* two-component signal transduction system MtrAB
MtrB	Histidine kinase of a *M. tuberculosis* two-component signal transduction system MtrAB
MTZ	Metronidazole
MUC1	Mucin-1
MUC5AC	Mucin-5AC
MUC5B	Mucin-5B
MUC6	Mucin-6
MUC7	Mucin-7
MukB	SMC homolog
MUPP	Multi-PDZ domain protein
MyD88	Myeloid differentiation primary response gene 88
MZB	Marginal zone B
NAC	N-acetylcysteine
NADPH	Nicotinamide adenine dinucleotide phosphate
NapA	Neutrophil-activating protein A (also called HP-NAP or Dps)
NER	Nucleotide excision repair
NET	Neuroendocrine tumor
NF-κB	Nuclear factor-kappa B
NFAT	Nuclear factor of activated T-cells

NGS	Next Generation Sequencing
NHANES	National Health and Nutrition Examination Survey
NHEJ	Non-homologous end joining
NHPH	Non-*Helicobacter pylori Helicobacter*
NK	Natural killer
NHNN	Non-*H. pylori* and non-NSAID
NLR	Nucleotide binding domain and leucine-rich-repeat-containing-proteins
NLRs	Nod-like receptors
NLRP	Nucleotide-binding domain, leucine-rich-containing family, pyrin domain-containing
NLEP1	NLR pyrin domain containing 1
NLRP3	NOD-like receptor pyrin domain-containing-3
NLRC4	NLR card domain containing 4
NO	Nitric oxide
NOD	Nucleotide binding oligomerization domain
NOG	Noggin
NOS2	Nitric oxide synthase 2
NOX	NAPDH oxidase
NSAID	Non-steroidal anti-inflammatory drug
~P	Phosphate group
NOXA1	NADPH oxidase activator 1
NPPB	Non-specific chloride channel inhibitor
OCT1	Octamer transcription factor1
ODC	Ornithine decarboxylase
ODN	Oligodeoxynucleotide
OGG	Oxoguanine DNA glycosylase
OipA	Outer inflammatory protein A
OLGA	Operative link for gastritis assessment
OLGIM	Operative link on intestinal metaplasia assessment
OMP	Outer membrane protein
OMV	Outer membrane vesicle
$ONOO^-$	Peroxinitrite
OR	Odds ratio
ORC1	Origin recognition complex subunit 1
ORF	Open reading frame
oriC	Origin of chromosome replication
oriT	Origin of transfer
oriV	Origin of vegetative replication
PAI	Pathogenicity island
PAMP	Pathogen-associated molecular pattern
Pap	Pyelonephritis-associated pili
PAR1	Protease-activated receptor 1
PARK	Parkin gene
PARP1	Poly [ADP-ribose] polymerase 1
Par1b	Partitioning-defective 1b kinase
ParA	Chromosome partitioning protein ParA

ParB	Chromosome partitioning protein ParB
parS	Centromere-like sequence
PBMC	Peripheral blood mononuclear cells
P-CAB	Potassium-competitive acid blocker
PCAs	Parietal cell autoantibodies
PCR	Polymerase chain reaction
pDC	Plasmacytoid dendritic cell
PDCD1LG2	Programmed cell death 1 ligand 2
PDE4D	Phosphodiesterase 4D
PD-L1	Programmed cell death ligand 1
PD-L2	Programmed cell death ligand 2; same as PDCD1LG2
PD1-L1	Programmed cell death protein ligand-1
Pdx1	Pancreatic and duodenal homeobox 1
PG	Peptidoglycan
PGC	Pepsinogen C
phasevarion	Phase variable regulon
PhoA-PhoB	Two-component system of the Pho (phosphate) regulon
PhoB	Response regulator of two-component regulatory system PhoA-PhoB
PhoR	Histidine kinase of two-component regulatory system PhoA-PhoB
PI3K	Phosphatidylinositide 3-kinase
PIK3CA	Phosphatidylinositol-4,5-bisphosphate 3-kinase, catalytic subunit alpha
PJS	Peutz-Jeghers syndrome
PKC	Protein kinase C
PLC	Phosphoinositide phospholipase C
PLT	Platelet
PMN	Polymorphonuclear neutrophil
PP	Per-protocol
PPI	Proton pump inhibitor
PRK2	Protein kinase C-related kinase 2
PRR	Pattern recognition receptor
PS	Phosphatidylserine
PSC	Pluripotent stem cells
PSCA	Prostate stem cell antigen
pSLT	*Salmonella* virulence plasmid
PTEN	Phosphatase and tensin homolog
PTPN	Gene encoding non-receptor type protein tyrosine phosphatase
pTyr	Phosphorylated tyrosine
PUD	Peptic ulcer disease
Rac1	Ras-related C3 botulinum toxin substrate 1
RAP	Recurrent abdominal pain
RCT	Randomized controlled trial
REL	Member of the NF-κB family of transcription factors
RGD	Arg-Gly-Asp sequence motif

RGDLXXL	Arg-Gly-Asp-Leu/Met-X-X-Leu/Ile sequence motif
RhoA	Ras homolog gene family A
RHS	RGD helper sequence
RIDA	Regulatory inactivation of DnaA activity
R-M systems	Restriction-modification systems
RNA	Ribonucleic acid
RNF43	Ring finger protein 43
RNI	Reactive nitrogen intermediate
RNS	Reactive nitrogen species
ROC	Receiver-operating characteristic
RocF	Urea-producing arginase
ROCK	Rho-Kinase
ROS	Reactive oxygen species
RPTP	Receptor protein tyrosine phosphatase
RR	Relative risk
rRNA	Ribosomal RNA
RSPO	R-spondin1
RTK	Receptor tyrosine Kinase
RT-PCR	Reverse transcriptase-polymerase chain reaction
RUNX3	Runt-related transcription factor 3
SabA	Sialic acid binding adhesin
SAT	Stool antigen test
SCFA	Short chain fatty acids
SeqA	Sequestration protein A
SFK	Src family kinase
SH2	Src homology 2
SH3	Src homology 3
SHH	Sonic hedgehog
SHP1/2	Src homology region 2 domain-containing phosphatase-1/2
SLB	Single layer antiparallel β-sheet
SLC1A2	Solute carrier family 1 member 2
s-Le^x	sialyl-Lewis X
SLT	Soluble lytic transglycoylase
SMC	Structure maintenance of chromosomes
SMOX	Spermine oxidase
SMRT	Single molecule real-time
SNP	Single-nucleotide polymorphism
Soj	Sporulation initiation inhibitor Soj
Sox9	Sex determining region Y (SRY)-box 2
SPEM	Spasmolytic polypeptide-expressing metaplasia
Spo0J	Chromosome partitioning protein
SPR	Surface plasmon resonance
Src	Sarcoma kinase
SRF	Serum response factor
SSB	Single-strand binding
ssDNA	Single-stranded DNA
SSM	Slipped strand mispairing
STAT3	Signal transducer and activator of transcription factor-3

STR	Streptomycin
T4SS	Type IV secretion system
TBK-1	Serine/Threonine protein kinase-1
TCGA	The Cancer Genome Atlas
TCF/LEF	T cell-specific transcription factor/Lymphoid enhancer binding factor
TCR	T cell receptor
TER	Transepithelial electrical resistance
TFF	Trefoil factor
TET	Tetracyclin
TGF-beta	Transforming growth factor beta
Th	T helper cell
TIFA	TRAF-interacting protein with forkhead-associated domain
TIMP3	Metalloproteinase inhibitor 3
TIR	Toll/IL-1 receptor domain
TIRAP	TIR domain containing receptor protein
TJ	Tight junction
TLR	Toll-like receptor
TLR2	Toll-like receptor 2
Tnfrsf19	Tumor necrosis factor receptor super family 19
TNF-α	Tumor necrosis factor alpha
TP53	Tumor protein p53
tra	Transfer genes
T-RFLP	Terminal restriction fragment length polymorphism
TRAF	TNF receptor associated factor
TRIF	TIR domain containing adaptor inducing interferon-β
TRD	Target recognition domain
TRD1	Target recognition domain 1
TRD2	Target recognition domain 2
Treg	Regulatory T cell
TWIST1	Twist related protein 1
UBT	Urea breath test
UC	Ulcerative colitis
Ure	Urease
UreB	Urease subunit B
UV	Ultraviolet
VacA	Vacuolating cytotoxin A
Vav	Rac-specific nucleotide exchange factor
VEGFA	Vascular endothelial growth factor A
Vil1	Villin-1
WHO	World Health Organization
WPIYA-D	East Asian CagA-specific EPIYA motif
WWOX	WW domain containing oxidoreductase
XRCC	X-ray repair containing oxidoreductase
Y2H	Yeast two hybrid
YY1	Ying Yang binding motif
ZO-1	Zonula Occludens-1

Adv Exp Med Biol - Advances in Microbiology, Infectious Diseases and Public Health (2019) 1149: 1–16
https://doi.org/10.1007/5584_2019_356

Published online: 24 April 2019

The Story of *Helicobacter pylori*: Depicting Human Migrations from the Phylogeography

Langgeng Agung Waskito and Yoshio Yamaoka

Abstract

Helicobacter pylori is a spiral-shaped Gram-negative bacterium, which has infected more than half of the human population. Besides its colonisation capability, the genetic diversity of *H. pylori* is exceptionally well structured and belongs to several distinct genetic populations, depicting various prehistorical human migration events. The evolutionary relationship of *H. pylori* with its host had been started at least ~100,000 years ago. In addition, the discovery of the ancient *H. pylori* genome from a European Copper Age glacier mummy, "The Iceman", gave the idea that the second out of Africa migration resulted in the recombinant population of hpEurope at least about 5300 years ago. The advancement of next-generation genome sequencing discovered the prophage of *H. pylori* and could discriminate the big population of hpEurope into two different subpopulations. In addition, the implementation of the chromopainter/fineSTRUCTURE algorithm to the whole genome analysis of *H. pylori* provides a finer resolution population genetics of *H. pylori*; therefore it could also depict the recent migrations within the past 500 years after colonial expansion. This discovery shows that the genetic recombination of *H. pylori* strains is far more dynamic compared to its human host, but still maintains the similarity to its host, suggesting that *H. pylori* is a handy tool to reconstruct the human migration both in the past and the recent.

Keywords

Helicobacter pylori · Human migration · Phylogeography · Population genetics · Next-generation sequencing

L. A. Waskito
Department of Environmental and Preventive Medicine, Faculty of Medicine, Oita University, Yufu, Japan

Institute of Tropical Disease, Universitas Airlangga, Surabaya, Indonesia
e-mail: langgengaw@gmail.com

Y. Yamaoka (✉)
Department of Environmental and Preventive Medicine, Faculty of Medicine, Oita University, Yufu, Japan

Department of Medicine, Gastroenterology and Hepatology Section, Baylor College of Medicine, Houston, TX, USA
e-mail: yyamaoka@oita-u.ac.jp

1 Introduction

Helicobacter pylori is a spiral Gram-negative bacterium, which demonstrated its roles in gastritis, peptic ulcer and gastric cancer formation in humans (Uemura et al. 2001). Since its original revelation in the early 1980s, many ongoing studies had been conducted and showed that this bacterium infected more than half of the human population (Hooi et al. 2017). The route of infection is still controversial; however, the most reasonable route is oral-oral, which indicates the possibility of transmission within families, mainly from parents to their children. In addition to the vertical transmission, *H. pylori* also could be transmitted between unrelated people living nearby (Kivi et al. 2003; Björkholm et al. 2001). These routes of transmission could become the fundamental evolution of this bacterium, which we are going to discuss later.

The evolutionary process resulting in genetic diversity of the organism involves mutation, recombination, migration, selection and genetic drift. These forces are typically sufficient to explain the genetic differences occurring in the evolutionary process. These forces drive the *H. pylori* to evolve as it had been colonising the highest selective pressure environment in the human stomach. The high selective pressure and long-term colonization, in addition to the nature of *H. pylori,* which has unusually high mutation rates (Björkholm et al. 2001) and homologous recombination rates (Falush et al. 2001; Suerbaum et al. 1998). This resulted in extremely high DNA sequence diversity that is much higher compared to that of other bacteria (Achtman et al. 1999) and even 50-fold higher than its human host (Li and Sadler 1991). These attributes are responsible for making *H. pylori* considerably as the most diverse pathogenic bacterium worldwide (Fischer et al. 2010). Interestingly, the high diversity of *H. pylori* genomes is very well structured, which could be divided into several distinct clusters. The population genetics of *H. pylori* is very consistent with the hosting people on different continents; therefore it reflects the human migration events in the past.

The multilocus sequence typing (MLST) approach has been used for many years to characterise the phylogeographic features of bacteria, including *H. pylori*. The combination of MLST sequence data and the STRUCTURE algorithm have discovered the population genetics of *H. pylori,* which correlates with the human migration events (Falush et al. 2003b; Linz et al. 2007). One of the most important virulence genes, *cagA* (Backert et al. 2010; Posselt et al. 2013), also has been proven to be a geographic predictor alongside with MLST and it could differentiate the isolates from East Asian and the Western isolates (Yamaoka 2010; Achtman et al. 1999). In recent years, the next-generation sequencing (NGS) methodology became less expensive, and thus available to generate more whole genome sequences. NGS analysis can obtain much more sequence data and in a shorter time, compared to the conventional Sanger sequencing approach. Therefore, NGS data can give us even better insights in terms of phylogeographic characterisation of *H. pylori* at the recent recombination point of view (Thorell et al. 2017) as well as the new phylogeographic tools, the bacteriophages (Vale et al. 2017). In addition, the discovery of a 5300-year-old *H. pylori* genome from a European Copper Age glacier mummy, "The Iceman", gave us an idea of the hybridisation between Asia and African populations of *H. pylori* (Maixner et al. 2016). Here, we summarise the current understanding of phylogeographic of *H. pylori* as genomic-based evidence to support the concept of human migration events, characterised by the MLST, specific virulence genes and entire genomes.

2 Characterization the Population Genetics of *H. pylori*

Population genetics characterisation is a useful technique to discover the genetic background of a given microorganism. Understanding the genetic background gives us an idea of the relationship of microorganisms and the nature of infection (discover the origin of the microbe and the way of transmission, for example in a particular outbreak); clinical outcomes (discovery of several virulence factors) and the geographic location of isolated microorganisms. In the case of *H. pylori*, which has an incredible highly and well-structured genetic diversity, the population genetics could reconstruct the evolutionary history.

We can characterise population genetics of bacteria in many ways. To characterise the population genetics of a given bacterium, the MLST was first proposed and applied to *Neisseria meningitidis* to determine the lineage from different invasive diseases and healthy carrier in 1998

(Maiden et al. 1998). For each cluster, the cluster analysis was applied to assign numbers arbitrarily and dendrograms by calculating the pairwise differences in multi-locus allelic profiles. This method resulting in ~470 bp fragments from altogether six genes identified meningococcal lineages associated with invasive diseases of humans (Maiden et al. 1998).

In 1999, Achtman and co-workers applied a similar approach to assign the population genetics to *H. pylori*. Their studies proposed the MLST using seven housekeeping genes (i.e., *atpA, efp, mutY, ppa, trpC, ureI*, and *yphC*) and two virulence-associated genes (*vacA* and *cagA*). This method resulted in a clonal descent, which could be observed, and this separation between one population compared to others reflected the geographical origin of the *H. pylori* isolates (Achtman et al. 1999). Currently, the method using seven housekeeping genes became a standard tool to describe the genetic populations of *H. pylori* using MLST analysis.

One of the most common approaches inferring population genetics based on the MLST data is the STRUCTURE algorithm (Pritchard et al. 2000). This algorithm was built using the Bayesian probability approach followed by Markov Chain Monte Carlo (MCMC), which inferred the posterior probability of the given MLST data into several population genetics. However, as *H. pylori* had a lot of interspecies recombination and mutation events, inferring the population genetics of this bacterium is challenging. Employing this condition, another model is implemented in the STRUCTURE software, known as the *linkage model* (Falush et al. 2003a). This model relies on the admixture linkage disequilibrium that is resulted when a gene flow (migration) occurs between genetically distinct populations. Therefore, it infers a finite number of "ancestral" or precursor number of populations (Falush et al. 2003a). In case of *H. pylori*, with the increasing number of investigated isolates from the original MLST study, this approach succeeded to build the population genetics of *H. pylori* and linear with the geographical origin of the isolates (Falush et al. 2003b).

The advancement of NGS put the genetic information of the microorganism to the next level. Increasing the high-throughput data, which contain high abundance of sequences, needed to consider the effectivity and efficiency of the analysis. One of the drawbacks of the STRUCTURE analysis is the time-consuming property. Therefore, a new approach was introduced, and it is called chromopainter and fineSTRUCTURE, which is less time consuming with the larger dataset. The chromopainter algorithm relies on the consideration that each sequence in a given sample is regarded in turn as a recipient, whose chromosomes are reconstructed using chunks of DNA donated by the other individuals. The result of this algorithm can be summarised as a matrix of co-ancestry, which reveals the critical information directly about the ancestral relationship among individuals (Lawson et al. 2012). Therefore, we can obtain a higher resolution of the genetic relationship between one strain and others at the donor-recipient wise. By applying these main methods, population genetics of *H. pylori* could reflect the major human migration events out of Africa, from the prehistorical migrations to the recent migrations.

3 The Phylogeographic of *H. pylori* from Out of Africa to the Pacific

There is accumulating evidence, which showed that modern humans migrated out of Africa to the Arabian Peninsula approximately 60,000–150,000 years ago (kya), then independently went to Europe and Asia (Cavalli-Sforza et al. 1994). By applying the Bayesian inferring algorithm implemented in the STRUCTURE programme, Linz and co-workers demonstrated the simulation, which indicated that *H. pylori* had spread from East Africa approximately ~58 kya, then the population was spreading through East Asia (Linz et al. 2007). The simulation resulted in several different populations, including hpEurope, hpAsia2, hpEastAsia and hpSahul with a characteristic of decreasing genetic diversity by the increasing distance from Africa, reflecting the genetic drift of each population in the *H. pylori* (Linz et al. 2007).

The European population (hpEurope) is the biggest in terms of number and distribution in the *H. pylori* population genetics. The hpEurope cluster includes almost all *H. pylori* isolated from European countries as far east as Southeast Asian countries (Table 1). In addition, hpEurope also possessed very distinct properties from the other population genetic clusters in global observations. The European genetic properties of *H. pylori* were characterised by great genetic diversity, even higher than in Africa. Observations made with the MLST sequences using *linkage model* (Falush et al. 2003a) implemented in STRUCTURE algorithm showed that the hpEurope was formed by the hybridisation between two ancestral populations, known as Ancestral Europe 1 (AE1) and Ancestral Europe 2 (AE2) (Falush et al. 2003b). Originally, the AE1 was defined as the population found in Ladakh in north India, especially *H. pylori* from Ladakhi Muslim, which showed a distinct ancestral relation to either East Asian or European populations, respectively (Wirth et al. 2004), and currently known as hpAsia2; the AE2 is currently known as hpNEAfrica. The result showing that the current hpEurope population arose by the recombination between AE1 and AE2, suggests that the introduction of *H. pylori* into Europe occurred more than once in history (Moodley et al. 2012). Hence, it opens a question of how and when was the second introduction of *H. pylori* in the Eurasian land, especially in Europe, which needs to be solved in future studies.

In terms of where the second introduction of *H. pylori* in Eurasian land is coming from, implementation of the *linkage model* (Falush et al. 2003a) of STRUCTURE could capture the distribution of AE1 and AE2 of European *H. pylori*.

Table 1 The population genetics of *H. pylori* and geographic localization

Population	Sub-population	Locations	Main references
hpAfrica2	hspNorthSan	Namibia, Angola	Moodley et al. (2012) and Falush et al. 2003b)
	hspSouthSan	South Africa	
hpAfrica1	hspWAfrica	Senegal, the Gambia, Burkina Faso, Morocco, Algeria, Nigeria, Cameroon, South Africa	Nell et al. (2013), Linz et al. (2014), and Falush et al. (2003b)
	hspSAfrica	Namibia, Angola, South Africa, Madagascar	
	hspCAfrica	Cameroon, Namibia	
hpAfrica1/ hpEurope	hspAfrica1Nicaragua	Central America	Thorell et al. (2017)
	hspAfrica1NAmerica	North, Central and South America	
	hspMiscAmerica	Central and South America	
	hspEuropeColombia	South America	
hpNEAfrica	hspCentralNEAfrica	Sudan, Cameroon, Nigeria, Algeria	Linz et al. (2007), and Nell et al. (2013)
	hspEastNEAfrica	Sudan, Ethiopia, Somalia, Algeria	
hpEurope	hspNorthEurope	Europe as far east as Southeast Asia	Thorell et al. (2017) and Falush et al. (2003b)
	hspSouthEurope		
hpAsia2	hspIndia	India, Bangladesh, Malaysia, Thailand, the Philipines, Nepal	Linz et al. (2007), Wirth et al. (2004), and Tay et al. (2009)
	hspLadakh	India (Himalaya)	
hpSahul	hspAustralia	Australia	Moodley et al. (2009)
	hspNGuinea	New Guinea	
hpEastAsia	hspEasia	China, India, Malaysia, Singapore, Taiwan, Thailand, Cambodia, Vietnam, Japan, Korea	Moodley et al. (2009), Falush et al. (2003b), and Linz et al. (2007)
	hspMaori	Taiwan, the Philippines, Japan, Samoa, New Caledonia, Wallis and Futuna, Indonesia, New Zealand	
	hspAmerind	Canada, the USA, Venezuela, Colombia, Peru	

The implementation of the *linkage model* showed a different ratio between AE1/AE2 with the higher proportion of AE1 in northern Europe and a lower proportion of AE1 in southern Europe (Falush et al. 2003b). The utilisation of whole genome sequence data using chromopainter/fineSTRUCTURE algorithm also captured those kinds of information in a finer scale manner. The first study implementing chromopainter/fineSTRUCTURE algorithm to the *H. pylori* genome showed that the European population was divided into two subgroups, Europe_sg1 (mostly strains from Northern Europe) and Europe_sg2 (mostly strains from Southern Europe) (Yahara et al. 2013). Another study implementing chromopainter/ fineSTRUCTURE to *H. pylori* also showed that the co-ancestry level of hpNEAfrica strains was higher in *H. pylori* isolates from Southern Europe than the Northern Europe ones (Maixner et al. 2016). In addition, the new phylogeographic characterization method generated from two prophage genes of *H. pylori* (integrase and holin) showed different European populations, present mainly in Northern Europe and Southern Europe (Vale et al. 2015). These data suggest that the introduction of the second *H. pylori* wave was coming from south of Eurasian continent as a "second out of Africa" event.

The time-wise of the second *H. pylori* introduction in the Eurasian land is somewhat puzzling until today. One assumption time is about 30–40 kya when modern humans settled Europe via two entering routes: from Turkey along the Danube corridor into Eastern Europe, and along the Mediterranean coast with early Neolithic farmers (Correa and Piazuelo 2012). Another assumption is, when after 52 kya based on the split time between hpNEAfrica (as the descendant of AE1) and the hpAfrica1, which was predicted 35–52 kya (Moodley et al. 2012). However, the discovery and characterization of *H. pylori* inside the stomach of "The Iceman", a European copper age mummified human, revealed intriguing evidence of the second introduction time of *H. pylori* in Europe. The MLST characterisation of "The Iceman" was almost pure of AE1 with the minimum admixture of AE2 (6.5%, probability interval 1.5–13.5%) (Maixner et al. 2016). The genome analysis showed a high co-ancestry relation to the Indian hpAsia2 *H. pylori* and with most of the European hpEurope *H. pylori* strains. "The Iceman" is believed to be killed in the Italian Ötztal Alps mountains ~5300 years ago, which is located in the European land. This discovery suggests, if *H. pylori* from the Iceman is representative of the time, the low level of AE2 admixture indicated there was no introduction of AE2 prior to *H. pylori* from the Iceman. Therefore, the AE2 ancestry observed in hpEurope today is a result of AE2 introgression into Europe after the Copper Age or at least later in Central Europe (Ötzal Alps) (Maixner et al. 2016). However, the limitation of this conclusion was due to the data from only one strain.

From all *H. pylori* populations, which have evolved outside Africa, hpAsia2 is probably the most intriguing. This population evolved among the people who either did not follow a southern coastal migration route or who settled in the early phase of migration and later began expanding to the Western and Central Asia (Fig. 1). The distribution of hpAsia2 was widespread through Asians and Europeans prior to the evolution of current hpEurope strains. In addition, hpAsia2 might have accompanied the humans who settled in Europe ~40 kya (Moodley 2016). However, in the modern days the hpAsia2 population was mostly replaced by the hpEurope, especially in western Eurasia, but it still could be found in the South and South East Asian countries, including Nepal, Bangladesh, Malaysia, Myanmar, Thailand and India (Wirth et al. 2004; Aftab et al. 2017; Miftahussurur et al. 2015a; Breurec et al. 2011a; Subsomwong et al. 2017; Tay et al. 2009). The hpAsia2 *H. pylori* can be differentiated into two subpopulations, hspLadakh from the isolated Himalayan region of northern India and hspIndia, which is found among Indians, Malays and Thais (Breurec et al. 2011a; Tay et al. 2009). The presence of hpAsia2 in Finland and Estonia showed the evidence of the remaining of AE1, which initially inhabited the Eurasian land before the introduction of AE2 from the south.

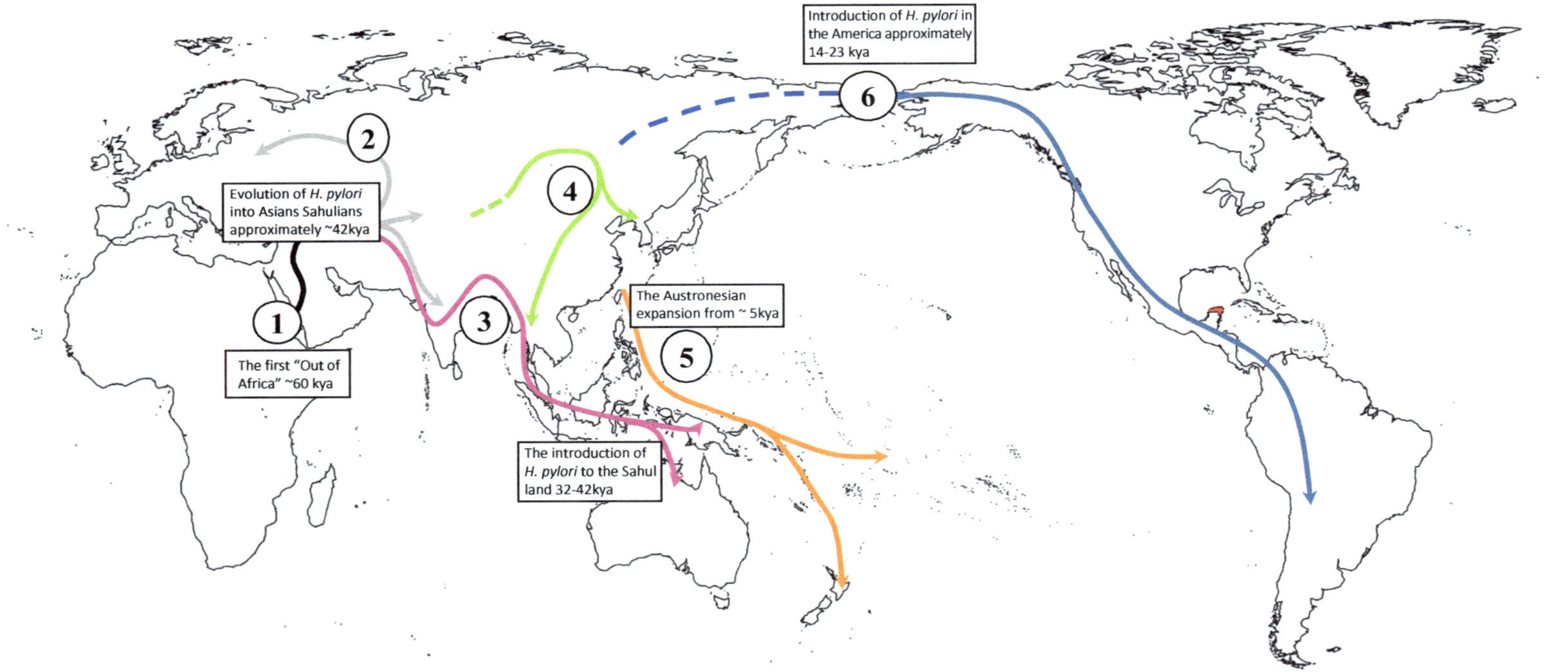

Fig. 1 The global human migration waves from africa to the pacific were constructed based on previous studies (Yamaoka et al. 2002; Falush et al. 2003b; Linz et al. 2007; Moodley et al. 2009). (*1*) The first major "Out of Africa" wave approximately happened ~60 kya; (*2*) The Asian *H. pylori*, especially hpAsia2 population, then occupying almost all western to a middle area of the Eurasian continent until the introduction of AE2, which would make the hybridisation with hpEurope; (*3*) A subsequent human population carries hpSahul and migrated towards southern coastal of Eurasian to the Sahul land; (*4*) The Chinese expansion during Zhou Dynasty in last 3000 years, which also expanding the hspEastAsia to East and South-East Asian countries; (*5*) The Austronesian expansion carrying hspMaori from Taiwan towards the island Southeast Asia and Polynesia; (*6*) The expansion sub-population of hpEastAsia, hspAmerind to America via Bering Strait

hpEastAsia is common among *H. pylori* isolates from East Asia and native American people. hpEastAsia is divided into three different subpopulations; hspMaori, hspAmerind, and hspEAsia and it was simulated to be split from hpAsia2 30–50 kya (Moodley et al. 2009). The hspEAsia is the biggest sub-population among the hpEastAsia population. This subpopulation was mainly observed in most East Asian countries, including China, South Korea, Japan, Thailand and Vietnam (Subsomwong et al. 2017; Breurec et al. 2011a; Linz et al. 2007; Falush et al. 2003a). The homogeneous distribution of the subpopulation of hspEAsia across those countries reflects the Chinese expansion language (family, Sino-Tibetan) during the last 3000 years, but mainly during the expansion of Zhou Dynasty (1100–211 BC) (Moodley 2016). Among three subpopulations of hpEastAsia, the hspAmerind was isolated among the diverse indigenous American population in North and South America, and it has its uniqueness characterised by very low genetic diversity compared to hpEastAsia, suggesting that the formation of Amerindian subpopulation was formed by a small group of people who migrated to America. This fact was different from the assumption of *H. pylori* in America was introduced by Chinese and Japanese people who migrated to America in more recent migration events. Instead, a subgroup was split from the big hpEastAsian population and went to America via crossing the Bering Strait approximately 12,000 years ago (Yamaoka et al. 2002). In addition, our previous data showed that four strains isolated from the Ainu ethnic group, living in Hokkaido, a northern island of Japan, belonged to the hspAmerind population (Gressmann et al. 2005), suggesting that the split between hpEastAsia and hspAmerind happened prior crossing the Bering Strait, possibly by human migrations in boats over the pacific ocean.

hspMaori was first observed in an isolate from Polynesians (Maoris, Tongans, and Samoans) of New Zealand, but was also observed in a small proportion of people in the Philippines and Japan (Linz et al. 2007; Falush et al. 2003b). In 2009, Moodley and co-workers discovered more hspMaori subpopulations among native Taiwanese, New Caledonia, and Torres Straits (an island located between Australia and New Guinea, which has been extensively visited by Polynesians), suggesting that the hspMaori is the marker for the entire Austronesian expansion rather than only for Polynesians (Moodley et al. 2009). Genetic analyses showed that the Taiwanese hspMaori have a significantly higher genetic diversity compared to the Pacific hspMaori ($\pi_{95} = 1.79$–1.82% vs 1.58–1.62%) and the indigenous Taiwanese isolates were isolated from the tribe that speak 5 of 10 subgroups of Austronesian family of languages, whereas the Pacific clades were isolated from individual speaking variants of Malayo-Polynesian, suggesting that the source of Austronesian expansion was in Taiwan. The split between indigenous Taiwanese hspMaori and Pacific hspMaori was predicted 4.9–5.0 kya (Moodley et al. 2009).

3.1 Observation of Asia2 and EastAsian Split from CagA Perspective

The split between hpAsia2 and hpEastAsia is also captured by the *cagA* gene, which encodes a highly immunogenic protein (CagA), located at the 3′ prime end of the *cag* pathogenicity island (PAI). The *cag*PAI encodes a type IV secretion system, through which CagA is delivered into host cells (Backert et al. 2015, Naumann et al. 2017). After delivery into gastric epithelial cells, CagA becomes tyrosine-phosphorylated at Glu-Pro-Ile-Tyr-Ala sequence motif (EPIYA) located in the 3′ region of the *cagA* gene (Backert and Selbach 2008; Zhang et al. 2015). Supporting the idea that *H. pylori* mirrors the human evolution, it was reported that the structure of the 3′ region of the *cagA* varies between strains from East Asian and Western countries (Yamaoka et al. 1998; Yamaoka et al. 1999; Yamaoka et al. 2000; Yamaoka et al. 2002). Isolates from the Western countries (Europe and Americas) mostly possess segments of the so-called EPIYA-ABC type, which are well known as Western-type CagAs, whereas strains from East Asian countries possess segments of the EPIYA-ABD type, which were classified as East Asian-type CagAs. Recent studies in the borderline of Europe and Middle East countries (i.e. Turkey and Iran), reporting that the distribution Western-type CagA was predominant (Kocazeybek et al. 2015; Honarmand-Jahromy et al. 2015). The same pattern was also reported in Nepal (Miftahussurur et al. 2015a). Interestingly, in Bhutan, the country that shares a border with India, almost all strains possessing East Asian-type CagA with more than half having multiple repeats in the 3′ region, which is very rare in other countries (Matsunari et al. 2016). When moving to South-East Asian countries, the predominant CagA genotype was shifted from Western-type to East Asian-type. For example, in Vietnam and Malaysia, East Asian-type CagA is predominant by 96% and 56%, respectively (Uchida et al. 2009; Schmidt et al. 2009). However, Cambodia and Thailand showed a little different pattern with the neighbour countries, with the Western-type CagA being predominant (59% and 54%, respectively) (Breurec et al. 2011b; Chomvarin et al. 2012). The split between Western-type CagA and East Asian-type CagA was a result of sequence rearrangement within the *cagA* gene, including the CagA multimerisation sequence (CM) and the EPIYA-motifs. This rearrangement in the left half of the EPIYA-D segment, characteristic of East Asian CagA, was derived from Western-type EPIYA with Amerind-type EPIYA as intermediate, through recombination of specific sequences within the gene (Furuta et al. 2011; Correa and Piazuelo 2012).

3.2 Peopling of Australians and New Guineans: hpSahul

A subsequent split occurred from the hpAsia2 population, which migrated towards a South-East tip of the Sundaland (i.e., the Malay Peninsula, Sumatra, Java, Borneo, and Bali), and went to a continent called Sahul (i.e. Australia, New Guinea, and Tasmania). Observations on the human parental markers showed that the indigenous Australians are closely related to New Guineans (Hudjashov et al. 2007). However, the observations in *H. pylori* showed that a single population inhabits the indigenous people Australian and New Guinean as hpSahul. Furthermore, detailed analysis showed that hpSahul was divided into two subpopulations, hspAustralia and hspNGuinea (Moodley et al. 2009).

The split between the Asian population and the Sahulian population was somewhat mystifying. The evidence based on Pleistocene human archaeological sites showed that the human colonisation in the Pleistocene Sahul was approximately between 42–48 kya (Pope and Terrell 2008; Allen and O'Connell 2014; Gillespie 2002). This dating was a somewhat before the split dating observed by Moodley and co-workers, which was ~31–37 kya (Moodley et al. 2009). Interestingly, the observation of single phylogeny of hpSahul suggested that the introduction of *H. pylori* into Sahul occurred once, and was followed by the split between hspNGuenia and hspAustralia, because there was no sign of gene flow between those two split sub-populations (Moodley 2016).

Distinct genetic drift was also observed in the *H. pylori* strains isolated from individuals in the Highlander New Guineans, which was captured by the CagA characterisation. The CagA characterisation of highlander New Guinean strain PNG84A (Montano et al. 2015) showed the presence of the unique AB-type EPIYA CagA. Our data also showed the presence of the unique ABB-type CagA, predominantly in strains isolated from Papuan ethnic people (Miftahussurur et al. 2015c). In fact, this unique ABB-type CagA has a very similar B-segment compared to that observed in strain PNG84A. These data suggest a different genetic drift from the one demonstrated by Furuta and co-workers (Furuta et al. 2011)

4 The Migration Inside Africa and the Intimation with the Host

4.1 The Origin of *H. pylori* and Split of hpAfrica2

H. pylori is believed to accompany the humans who migrated out of Africa ~58 kya. This intriguing fact leads us to the assumption that, if the *H. pylori* and modern human migrated together out of Africa, then the initial colonisation of *H. pylori* in the human stomach has occurred far prior to these migration events. Moodley and co-workers demonstrated the possible initial colonisation of *H. pylori* in the human stomach using a rooted, fully resolved and calibrated global clonal phylogeny (Moodley et al. 2012). This approach resulted from the coalescent to single common ancestor, which has occurred approximately 100 kya (range: 88–116 kya, Fig. 2). However, lineage sorting and population bottleneck would wipe out the possible extended older lineage in this time frame. Therefore, it is possible that the association between human and *H. pylori* is older than this estimation (Moodley 2016).

Another intriguing fact about the origin of *H. pylori* is the very distinct relationship between hpAfrica2 and other genetic *H. pylori* populations. The hpAfrica2 population exhibits a great diversity compared to other populations and is exclusively possessed by individuals in the southern part of Africa, including South Africa, Namibia, and southern Angola. The distinct relationship between hpAfrica2 and other *H. pylori* was attributed to the different migration waves once the *H. pylori* successfully colonised modern humans. In addition, the exclusive possession on the specified location in Africa should have an

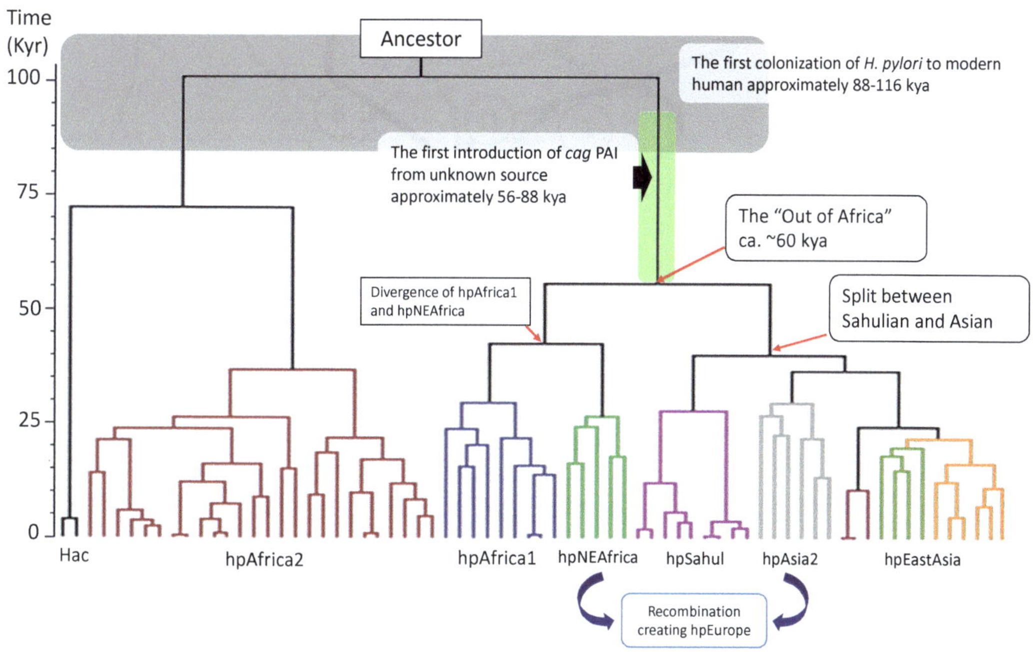

Fig. 2 Global phylogeny of *H. pylori* simulating the time split between populations and the estimated time in which the *cag*PAI was transferred to the *H. pylori* genome constructed based on previous papers (Moodley et al. 2009; Olbermann et al. 2010; Moodley et al. 2012). The housekeeping gene sequence diversity is structured into two super-lineages which coalescent approximately ~88–116 kya. The absence of the *cag*PAI in the hpAfrica2 population was attributed by the introduction of the *cag*PAI after a split into other *H. pylori,* but prior the first "out of Africa" wave approximately ~58–88 kya

association with the San Hunter-Gatherers. Indeed, hpAfrica2 was found in individuals from non-San people in Southern Africa (Falush et al. 2003b); however, there was a distinct relation inside the hpAfrica2 between Northern San and Southern San-Bantu speakers (Fig. 3), suggesting that the original host of hpAfrica2 was the Northern San Hunter-Gatherers and was then split into the Southern San Hunter-Gatherers approximately 32–47 kya (Moodley et al. 2012).

4.2 The hpNEAfrica and the Acquisition of *cag*PAI

Another independent migration wave from the human ancestor to the north, aside from the wave out of Africa, resulted in another genetic *H. pylori* population, called hpNEAfrica. This population was spread along the central Sahel and North Africa and shared its distribution with the population hpAfrica1 (Linz et al. 2007). Its frequency was

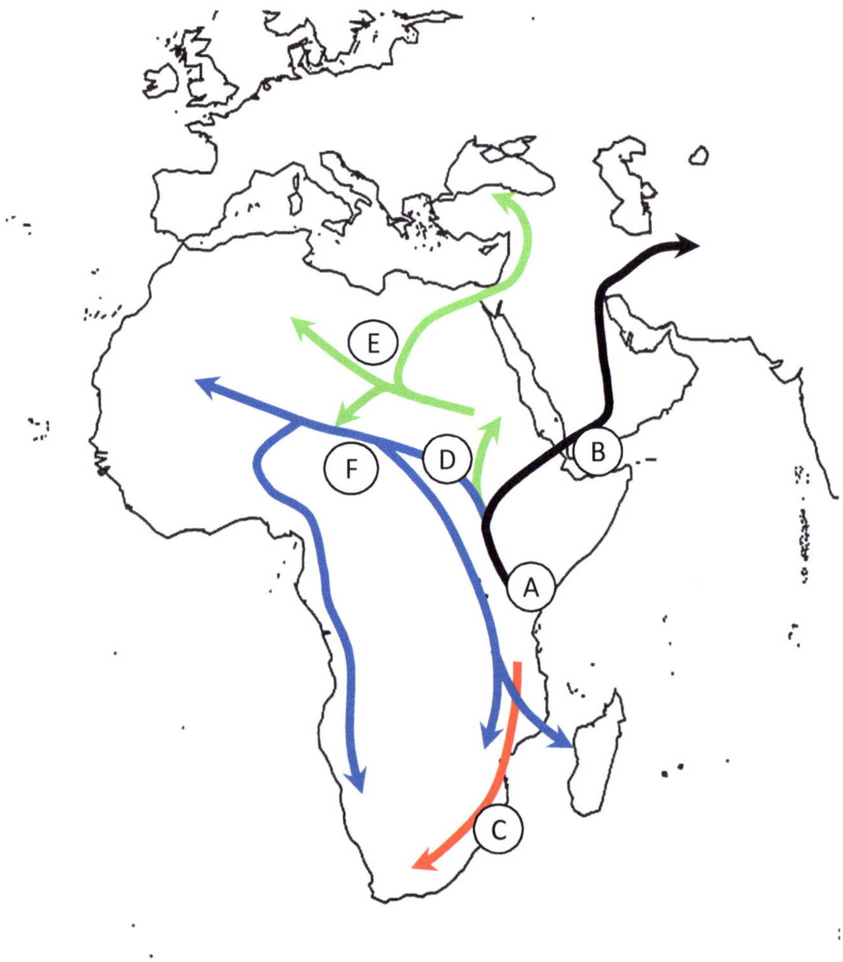

Fig. 3 The reconstruction of migration events and initial coalescent of *H. pylori* in Africa based on previous findings (Nell et al. 2013; Moodley et al. 2012; Linz et al. 2014; Maixner et al. 2016). (*A*) The first coalescent of *H. pylori* to human from the unknown ancestor; (*B*) The first out of Africa; (*C*) The other wave of the ancestor to the south, which evolved to the hpAfrica2 population; (*D*) *H. pylori* divergence of hpAfrica1 and hpNEAfrica.; (*E*) The expansion of Nilo-Saharan speakers to the western part of Africa (~6–9 kya). The subsequent population went to Eurasian via the Levant introducing the AE2 resulting hpEurope; (*F*) The expansion of Bantu speakers, including in the western and eastern coast of Africa, which introduced hpAfrica1 to Madagascar via Mozambique channel

increasing eastwards to the Nile and Horn of Africa, which is the home of the Nilo-Saharan speaking pastoralist society. Thus, the hpNEAfrica conveniently became the marker of Nilo-Saharan language in the shape of hpNEAfrica migration wave (Moodley 2016). The hpNEAfrica population was split into two subpopulations, hspCentralNEA and hspEastNEA and the split was demonstrated roughly along the Nile valley and Sudan which straddles the Nile containing both subpopulations at high frequency. The presence of hpsCentralNEA throughout Cameroon, Angola, and Nigeria suggested the migration wave of Nilo-Saharan speakers in the Holocene humid period (~6–9 kya) and carried their own language westwards from its home in northeast Africa into the waterlogged Sahara and beyond (Nell et al. 2013). As the introduction of second *H. pylori* into Europe occurred about the same time frame compared to the hpNEAfrica expansion during Holocene periods (~6–9 kya), there is a possibility that a subsequent population of hpNEAfrica migrated to Eurasian land via the Levant and introduced the AE2, resulting from the recombination population hpEurope (Fig. 3).

Another interesting observation of this superlineage distinct between hpAfrica2 and other *H. pylori* populations is the acquisition of the *cag*PAI. The *cag*PAI was hypothesised to be acquired from the unknown source prior to the out of Africa period and alongside with the host migrated up to Pacific and crossing Bering Strait; therefore it exhibits a specific genetic diversity pattern as determined by MLST (Olbermann et al. 2010). However, the *cag*PAI has not been observed in the hpAfrica2 population. The lack of the *cag*PAI in hpAfrica2 might be due to a different wave of human migration northwards, which acquired the *cag*PAI approximately ~88–58 kya just before the out of Africa migration events (Fig. 2) (Linz et al. 2007; Moodley et al. 2012).

4.3 The hpAfrica1: The Marker of Bantu Speakers

After settling in the Nile and at the horn of Africa, subsequent human migration westwards generated another *H. pylori* population, called hpAfrica1. The distribution of hpAfrica1 was spread alongside Morocco and Algeria in the north up to South Africa in the South (Fig. 3). Based on the geographical distribution of the hpAfrica1, cluster analysis divided the hpAfrica1 into three closely related subpopulations in the west and north of Africa, called hspWAfrica (Falush et al. 2003b); central Africa, hspCAfrica (Nell et al. 2013) and southern Africa (hspSAfrica) (Falush et al. 2003b) (Table 1). This distribution alongside those countries could be associated with the expansion of Bantu speakers people within last five kya from their original homeland in Nigeria/Cameroon. The observation of the subpopulation hspCAfrica in Cameroon and Angola, but not in South Africa and Namibia, supports the probability of a migration route alongside the west coast of Africa (Fig. 2). The hspSAfrica is presumed to be evolved during Bantu speakers expansion along the east coast that brought the Nguni speakers to southern Africa. Therefore, the observation of hpSAfrica in Madagascar, which is very closely related to hpSAfrica from South Africa, suggests the migration of Bantu speakers across the Mozambique Channel during or after the migration alongside the east coast of Africa (Linz et al. 2014).

5 The Recombination During Post Colonial Expansion

Population genetics of *H. pylori* not only capture the prehistoric migrations, but also capturing recent migration events. This idea was introduced by the observation of hpAfrica1, which could be found in America. The hspWAfrica was observed in several *H. pylori* isolates from several places in America, especially in African Americans in Louisiana and Tennessee (Falush et al. 2003b). This observation suggests the recent introduction of hpAfrica1 *H. pylori* in Americas by the transatlantic slave trades in the 16th–19th centuries. In addition, the hpEurope not only colonized the Eurasian area, but was also found in Australia, Africa, and America. The fact that several European Kingdoms (e.g. Portuguese, Spanish and Great Britain) explored the world since the fifteenth century could be responsible for the export of hpEurope into several other places of the world (Falush et al. 2003b).

The high genetic recombination between more than two populations was also observed in *H. pylori* isolates from Portuguese speaking countries, including Portugal, Angola, Brazil and Cape Verde, as captured by MLST analysis (Oleastro et al. 2017). The hpEurope in Europe and Portuguese speaking countries revealed a distinct relation to each other. This distinct relation was attributed by the ancestral Africa 1 components. The observation of Ancestral Africa 1 in *H. pylori* strains from Portuguese speaking countries suggested a long history between those countries, which resulted in new recombination of *H. pylori* consisting of more than three ancestral populations. Also, the isolates from Brazil were mostly hpEurope. However, several isolates from Brazil, which were assigned as hpEurope, had a small components of EastAsian ancestral, from which hspAmerind is a subpopulation (Oleastro et al. 2017). This data suggests the replacement of hspAmerind from the native Brazilian people to the other *H. pylori* populations due to low genetic diversity compared to either hpEurope or hpAfrica1 (Dominguez-Bello et al. 2008).

Population separation was observed in *H. pylori* strains from Latin America countries, including Mexico, Nicaragua, and Colombia using MLST, Virtual Genome Fingerprint (VGF) and the *cag*PAI phylogenetic tree studies. These analyses resulted in the discovery of a new subpopulation, which is very different from the native *H. pylori* American population hspAmerind. The separation from the indigenous population could be attributed to *H. pylori*, which were isolated from the modern population of those countries instead of indigenous population. This would reflect the recent genetic admixture of pre-Columbian American groups with the European colonisers and African slaves last ~500 years due to the European expansion (Munoz-Ramirez et al. 2017). In addition, implementation of chromopainter/fineSTRUCTURE to the *H. pylori* genome isolated from Latin America discovered several new subpopulations, introduced as hspAfrica1NAmerica, hspEuropeColombia, hspAfrica1Nicaragua and hspMiscAmericas (Thorell et al. 2017). The formation of new subpopulations was attributed in the combination of recombination and drift events. The genetic drift was taken place as a result of a recent demographic bottleneck, which assumed to have a low sequence divergence. However, the pairwise sequence divergence of that new sub-population was as high as the recombinant population hpEurope. This high sequence divergence was maintained by the admixture between hpEurope and hpAfrica1 population. This high sequence divergence (which mostly attributed by the high frequency of African component) suggested that the bacteria of African origin have been particularly effective in colonizing the new continent (Thorell et al. 2017). These data suggest that the expansion of modern humans in the new environment lead to the rapid evolution of *H. pylori,* which was faster and more dynamic than the host.

6 Conclusions

In the recent years, there are remarkable findings of the human migration based on the fossil and human genetic studies, which could divide the migration waves into two major migrations, pre-60 kya and post-60 kya (reviewed by (Bae et al. 2017)). In case of *H. pylori*, the story mostly yielded to the post-60 kya "Out of Africa" era, which is regarded as the major migration event. Together, *H. pylori* has been regarded as a handy tool to trace human migrations, from the pre-historic migrations to the recent migrations. Moreover, the *H. pylori* genome with the chromopainting analysis gives us more detail in the high-resolution data of the recent migrations. The rapid and dynamic evolution compared to its host still become major features for *H. pylori,* which inform us comprehensively about the recent recombination events.

With these current data, however, it becomes obvious that several more places are needed to be discovered, such as Siberia, Mongolia, Northern part of Japan and Indonesia since those places are located at the "bridge" of the split of several populations. Interestingly, the ability of

H. pylori to colonize more than 50% of the human population became a "common sense", however, it seems not to fit to the tip of Sundaland countries such as Indonesia and Malaysia, which shows a very low prevalence of *H. pylori* (Syam et al. 2015; Miftahussurur et al. 2015b; Rahim et al. 2010). This fact leads us to the speculation that the people at the tip of Sundaland might not carry any *H. pylori* and perhaps the introduction of *H. pylori* to the Indonesian and Malay people was performed by the Chinese expansion (hspEAsia) in more recent migrations. In addition, since the "African expansion" was successfully demonstrated in America, it is also interesting to find out the "African expansion" in the different continent such as Europe and Asia.

Acknowledgements This work was supported in part by grants from the National Institutes of Health (DK62813) and the Grants-in-Aid for Scientific Research from the Ministry of Education, Culture, Sports, Science, and Technology (MEXT) of Japan (221S0002, 16H06279, 15H02657 and 16H05191) (YY). It was also supported by the Japan Society for the Promotion of Science (JSPS) Institutional Program for Core-to-Core Program; B. Africa-Asia Science Platform (YY). LAW is a doctoral students supported by the Japanese Government (The Ministry of Education, Culture, Sports, Science and Technology [MEXT]) Scholarship Program for 2015.

References

Achtman M, Azuma T, Berg DE, Ito Y, Morelli G, Pan ZJ, Suerbaum S, Thompson SA, van der Ende A, van Doorn LJ (1999) Recombination and clonal groupings within *Helicobacter pylori* from different geographical regions. Mol Microbiol 32(3):459–470

Aftab H, Miftahussurur M, Subsomwong P, Ahmed F, Khan AKA, Matsumoto T, Suzuki R, Yamaoka Y (2017) Two populations of less-virulent *Helicobacter pylori* genotypes in Bangladesh. PLoS One 12(8): e0182947. https://doi.org/10.1371/journal.pone.0182947

Allen J, O'Connell J (2014) Both half right: updating the evidence for dating first human arrivals in Sahul. Aust Archaeol 79(1):86–108. https://doi.org/10.1080/03122417.2014.11682025

Backert S, Selbach M (2008) Role of type IV secretion in *Helicobacter pylori* pathogenesis. Cell Microbiol 10 (8):1573–1581. https://doi.org/10.1111/j.1462-5822.2008.01156.x

Backert S, Tegtmeyer N, Selbach M (2010) The versatility of *Helicobacter pylori* CagA effector protein functions: the master key hypothesis. Helicobacter 15 (3):163–176. https://doi.org/10.1111/j.1523-5378.2010.00759.x

Backert S, Tegtmeyer N, Fischer W (2015) Composition, structure and function of the *Helicobacter pylori* cag pathogenicity island encoded type IV secretion system. Future Microbiol 10(6):955–965. https://doi.org/10.2217/fmb.15.32

Bae CJ, Douka K, Petraglia MD (2017) On the origin of modern humans: Asian perspectives. Science 358 (6368):eaai9067. https://doi.org/10.1126/science.aai9067

Björkholm B, Sjölund M, Falk PG, Berg OG, Engstrand L, Andersson DI (2001) Mutation frequency and biological cost of antibiotic resistance in *Helicobacter pylori*. Proc Natl Acad Sci 98(25):14607–14612

Breurec S, Guillard B, Hem S, Brisse S, Dieye FB, Huerre M, Oung C, Raymond J, Tan TS, Thiberge JM, Vong S, Monchy D, Linz B (2011a) Evolutionary history of *Helicobacter pylori* sequences reflect past human migrations in Southeast Asia. PLoS One 6(7): e22058. https://doi.org/10.1371/journal.pone.0022058

Breurec S, Guillard B, Hem S, Papadakos KS, Brisse S, Huerre M, Monchy D, Oung C, Sgouras DN, Tan TS, Thiberge JM, Vong S, Raymond J, Linz B (2011b) Expansion of European vacA and cagA alleles to East-Asian *Helicobacter pylori* strains in Cambodia. Infect Genet Evol 11(8):1899–1905. https://doi.org/10.1016/j.meegid.2011.08.007

Cavalli-Sforza LL, Menozzi P, Piazza A (1994) The history and geography of human genes. Princeton University Press, Princeton

Chomvarin C, Phusri K, Sawadpanich K, Mairiang P, Namwat W, Wongkham C, Hahnvajanawong C (2012) Prevalence of cagA EPIYA motifs in *Helicobacter pylori* among dyspeptic patients in northeast Thailand. Southeast Asian J Trop Med Public Health 43(1):105–115

Correa P, Piazuelo MB (2012) Evolutionary history of the *Helicobacter pylori* genome: implications for gastric carcinogenesis. Gut Liver 6(1):21–28. https://doi.org/10.5009/gnl.2012.6.1.21

Dominguez-Bello MG, Perez ME, Bortolini MC, Salzano FM, Pericchi LR, Zambrano-Guzman O, Linz B (2008) Amerindian *Helicobacter pylori* strains go extinct, as european strains expand their host range. PLoS One 3(10):e3307. https://doi.org/10.1371/journal.pone.0003307

Falush D, Kraft C, Taylor NS, Correa P, Fox JG, Achtman M, Suerbaum S (2001) Recombination and mutation during long-term gastric colonization by *Helicobacter pylori*: estimates of clock rates, recombination size, and minimal age. Proc Natl Acad Sci U S A 98(26):15056–15061. https://doi.org/10.1073/pnas.251396098

Falush D, Stephens M, Pritchard JK (2003a) Inference of population structure using multilocus genotype data: linked loci and correlated allele frequencies. Genetics 164(4):1567–1587

Falush D, Wirth T, Linz B, Pritchard JK, Stephens M, Kidd M, Blaser MJ, Graham DY, Vacher S, Perez-Perez GI, Yamaoka Y, Megraud F, Otto K, Reichard U, Katzowitsch E, Wang X, Achtman M, Suerbaum S (2003b) Traces of human migrations in *Helicobacter pylori* populations. Science 299 (5612):1582–1585. https://doi.org/10.1126/science.1080857

Fischer W, Windhager L, Rohrer S, Zeiller M, Karnholz A, Hoffmann R, Zimmer R, Haas R (2010) Strain-specific genes of *Helicobacter pylori*: genome evolution driven by a novel type IV secretion system and genomic island transfer. Nucleic Acids Res 38 (18):6089–6101. https://doi.org/10.1093/nar/gkq378

Furuta Y, Yahara K, Hatakeyama M, Kobayashi I (2011) Evolution of cagA oncogene of *Helicobacter pylori* through recombination. PLoS One 6(8):e23499. https://doi.org/10.1371/journal.pone.0023499

Gillespie R (2002) Dating the first Australians. Radiocarbon 44(2):455–472

Gressmann H, Linz B, Ghai R, Pleissner KP, Schlapbach R, Yamaoka Y, Kraft C, Suerbaum S, Meyer TF, Achtman M (2005) Gain and loss of multiple genes during the evolution of *Helicobacter pylori*. PLoS Genet 1(4):e43. https://doi.org/10.1371/journal.pgen.0010043

Honarmand-Jahromy S, Siavoshi F, Malekzadeh R, Sattari TN, Latifi-Navid S (2015) Multiple repeats of *Helicobacter pylori* CagA EPIYA-C phosphorylation sites predict risk of gastric ulcer in Iran. Microb Pathog 89:87–92. https://doi.org/10.1016/j.micpath.2015.09.005

Hooi JKY, Lai WY, Ng WK, Suen MMY, Underwood FE, Tanyingoh D, Malfertheiner P, Graham DY, Wong VWS, Wu JCY, Chan FKL, Sung JJY, Kaplan GG, Ng SC (2017) Global prevalence of *Helicobacter pylori* infection: systematic review and meta-analysis. Gastroenterology 153(2):420–429. https://doi.org/10.1053/j.gastro.2017.04.022

Hudjashov G, Kivisild T, Underhill PA, Endicott P, Sanchez JJ, Lin AA, Shen P, Oefner P, Renfrew C, Villems R, Forster P (2007) Revealing the prehistoric settlement of Australia by Y chromosome and mtDNA analysis. Proc Natl Acad Sci U S A 104 (21):8726–8730. https://doi.org/10.1073/pnas.0702928104

Kivi M, Tindberg Y, Sorberg M, Casswall TH, Befrits R, Hellstrom PM, Bengtsson C, Engstrand L, Granstrom M (2003) Concordance of *Helicobacter pylori* strains within families. J Clin Microbiol 41(12):5604–5608

Kocazeybek BS, Caliskan R, Erdamar Cetin S, Ergin S, Kuskucu M, Kepil N, Oyku Dinc H, Ziya Erzin Y, Saribas S, Bahar Tokman H, Kalayci F, Akgul O, Yuksel P, Karakullukcu A, Ziver T, Sirekbasan S, Caglar E, Bal K (2015) Patterns of EPIYA motifs among cagA-positive *Helicobacter pylori* strains: a case-control study in a Turkish population with Eurasian geographical features. J Med Microbiol 64 (10):1117–1123. https://doi.org/10.1099/jmm.0.000141

Lawson DJ, Hellenthal G, Myers S, Falush D (2012) Inference of population structure using dense haplotype data. PLoS Genet 8(1):e1002453. https://doi.org/10.1371/journal.pgen.1002453

Li WH, Sadler LA (1991) Low nucleotide diversity in man. Genetics 129(2):513–523

Linz B, Balloux F, Moodley Y, Manica A, Liu H, Roumagnac P, Falush D, Stamer C, Prugnolle F, van der Merwe SW, Yamaoka Y, Graham DY, Perez-Trallero E, Wadstrom T, Suerbaum S, Achtman M (2007) An African origin for the intimate association between humans and *Helicobacter pylori*. Nature 445 (7130):915–918. https://doi.org/10.1038/nature05562

Linz B, Vololonantenainab CR, Seck A, Carod JF, Dia D, Garin B, Ramanampamonjy RM, Thiberge JM, Raymond J, Breurec S (2014) Population genetic structure and isolation by distance of *Helicobacter pylori* in Senegal and Madagascar. PLoS One 9(1):e87355. https://doi.org/10.1371/journal.pone.0087355

Maiden MC, Bygraves JA, Feil E, Morelli G, Russell JE, Urwin R, Zhang Q, Zhou J, Zurth K, Caugant DA, Feavers IM, Achtman M, Spratt BG (1998) Multilocus sequence typing: a portable approach to the identification of clones within populations of pathogenic microorganisms. Proc Natl Acad Sci U S A 95 (6):3140–3145

Maixner F, Krause-Kyora B, Turaev D, Herbig A, Hoopmann MR, Hallows JL, Kusebauch U, Vigl EE, Malfertheiner P, Megraud F, O'Sullivan N, Cipollini G, Coia V, Samadelli M, Engstrand L, Linz B, Moritz RL, Grimm R, Krause J, Nebel A, Moodley Y, Rattei T, Zink A (2016) The 5300-year-old *Helicobacter pylori* genome of the Iceman. Science 351(6269):162–165. https://doi.org/10.1126/science.aad2545

Matsunari O, Miftahussurur M, Shiota S, Suzuki R, Vilaichone RK, Uchida T, Ratanachu-ek T, Tshering L, Mahachai V, Yamaoka Y (2016) Rare *Helicobacter pylori* virulence genotypes in Bhutan. Sci Rep 6:22584. https://doi.org/10.1038/srep22584

Miftahussurur M, Sharma RP, Shrestha PK, Suzuki R, Uchida T, Yamaoka Y (2015a) Molecular epidemiology of *Helicobacter pylori* infection in Nepal: specific ancestor root. PLoS One 10(7):e0134216. https://doi.org/10.1371/journal.pone.0134216

Miftahussurur M, Shiota S, Suzuki R, Matsuda M, Uchida T, Kido Y, Kawamoto F, Maimunah U, Adi P, Rezkitha Y, Nasronudin NI, Yamaoka Y (2015b) Identification of *Helicobacter pylori* infection in symptomatic patients in Surabaya, Indonesia, using five diagnostic tests. Epidemiol Infect 143(5):986–996. https://doi.org/10.1017/S095026881400154X

Miftahussurur M, Syam AF, Makmun D, Nusi IA, Zein LH, Zulkhairi AF, Uswan WB, Simanjuntak D, Uchida T, Adi P, Utari AP, Rezkitha YA, Subsomwong P, Nasronudin YY (2015c) *Helicobacter pylori* virulence genes in the five largest islands of Indonesia. Gut Pathogens 7:26. https://doi.org/10.1186/s13099-015-0072-2

Montano V, Didelot X, Foll M, Linz B, Reinhardt R, Suerbaum S, Moodley Y, Jensen JD (2015) Worldwide population structure, long-term demography, and local adaptation of *Helicobacter pylori*. Genetics 200 (3):947–963. https://doi.org/10.1534/genetics.115.176404

Moodley Y (2016) *Helicobacter pylori*: genetics, recombination, population structure, and human migrations. In: Backert S, Yamaoka Y (eds) *Helicobacter pylori* research: from bench to bedside. Springer, Tokyo, pp 3–27. https://doi.org/10.1007/978-4-431-55936-8_1

Moodley Y, Linz B, Yamaoka Y, Windsor HM, Breurec S, Wu JY, Maady A, Bernhoft S, Thiberge JM, Phuanukoonnon S, Jobb G, Siba P, Graham DY, Marshall BJ, Achtman M (2009) The peopling of the Pacific from a bacterial perspective. Science 323(5913):527–530. https://doi.org/10.1126/science.1166083

Moodley Y, Linz B, Bond RP, Nieuwoudt M, Soodyall H, Schlebusch CM, Bernhoft S, Hale J, Suerbaum S, Mugisha L, van der Merwe SW, Achtman M (2012) Age of the association between *Helicobacter pylori* and man. PLoS Pathog 8(5):e1002693. https://doi.org/10.1371/journal.ppat.1002693

Munoz-Ramirez ZY, Mendez-Tenorio A, Kato I, Bravo MM, Rizzato C, Thorell K, Torres R, Aviles-Jimenez F, Camorlinga M, Canzian F, Torres J (2017) Whole genome sequence and phylogenetic analysis show *Helicobacter pylori* strains from Latin America have followed a unique evolution pathway. Front Cell Infect Microbiol 7:50. https://doi.org/10.3389/fcimb.2017.00050

Naumann M, Sokolova O, Tegtmeyer N, Backert S (2017) *Helicobacter pylori*: a paradigm pathogen for subverting host cell signal transmission. Trends Microbiol 25(4):316–328. https://doi.org/10.1016/j.tim.2016.12.004

Nell S, Eibach D, Montano V, Maady A, Nkwescheu A, Siri J, Elamin WF, Falush D, Linz B, Achtman M, Moodley Y, Suerbaum S (2013) Recent acquisition of *Helicobacter pylori* by Baka pygmies. PLoS Genet 9 (9):e1003775. https://doi.org/10.1371/journal.pgen.1003775

Olbermann P, Josenhans C, Moodley Y, Uhr M, Stamer C, Vauterin M, Suerbaum S, Achtman M, Linz B (2010) A global overview of the genetic and functional diversity in the *Helicobacter pylori cag* pathogenicity island. PLoS Genet 6(8):e1001069. https://doi.org/10.1371/journal.pgen.1001069

Oleastro M, Rocha R, Vale FF (2017) Population genetic structure of *Helicobacter pylori* strains from Portuguese-speaking countries. Helicobacter 22(4). https://doi.org/10.1111/hel.12382

Pope KO, Terrell JE (2008) Environmental setting of human migrations in the circum-Pacific region. J Biogeogr 35(1):1–21

Posselt G, Backert S, Wessler S (2013) The functional interplay of *Helicobacter pylori* factors with gastric epithelial cells induces a multi-step process in pathogenesis. Cell Commun Signal 11:77. https://doi.org/10.1186/1478-811X-11-77

Pritchard JK, Stephens M, Donnelly P (2000) Inference of population structure using multilocus genotype data. Genetics 155(2):945–959

Rahim AA, Lee YY, Majid NA, Choo KE, Raj SM, Derakhshan MH, Graham DY (2010) *Helicobacter pylori* infection among Aborigines (the Orang Asli) in the northeastern region of Peninsular Malaysia. Am J Trop Med Hyg 83(5):1119–1122. https://doi.org/10.4269/ajtmh.2010.10-0226

Schmidt HM, Goh KL, Fock KM, Hilmi I, Dhamodaran S, Forman D, Mitchell H (2009) Distinct cagA EPIYA motifs are associated with ethnic diversity in Malaysia and Singapore. Helicobacter 14(4):256–263. https://doi.org/10.1111/j.1523-5378.2009.00684.x

Subsomwong P, Miftahussurur M, Vilaichone RK, Ratanachu-Ek T, Suzuki R, Akada J, Uchida T, Mahachai V, Yamaoka Y (2017) *Helicobacter pylori* virulence genes of minor ethnic groups in North Thailand. Gut Pathogens 9:56. https://doi.org/10.1186/s13099-017-0205-x

Suerbaum S, Smith JM, Bapumia K, Morelli G, Smith NH, Kunstmann E, Dyrek I, Achtman M (1998) Free recombination within *Helicobacter pylori*. Proc Natl Acad Sci 95(21):12619–12624

Syam AF, Miftahussurur M, Makmun D, Nusi IA, Zain LH, Zulkhairi AF, Uswan WB, Simanjuntak D, Uchida T, Adi P, Utari AP, Rezkitha YA, Subsomwong P, Nasronudin SR, Yamaoka Y (2015) Risk factors and prevalence of *Helicobacter pylori* in five largest Islands of Indonesia: a preliminary study. PLoS One 10(11):e0140186. https://doi.org/10.1371/journal.pone.0140186

Tay CY, Mitchell H, Dong Q, Goh KL, Dawes IW, Lan R (2009) Population structure of *Helicobacter pylori* among ethnic groups in Malaysia: recent acquisition of the bacterium by the Malay population. BMC Microbiol 9:126. https://doi.org/10.1186/1471-2180-9-126

Thorell K, Yahara K, Berthenet E, Lawson DJ, Mikhail J, Kato I, Mendez A, Rizzato C, Bravo MM, Suzuki R, Yamaoka Y, Torres J, Sheppard SK, Falush D (2017) Rapid evolution of distinct *Helicobacter pylori* subpopulations in the Americas. PLoS Genet 13(2): e1006546. https://doi.org/10.1371/journal.pgen.1006546

Uchida T, Nguyen LT, Takayama A, Okimoto T, Kodama M, Murakami K, Matsuhisa T, Trinh TD, Ta L, Ho DQ, Hoang HH, Kishida T, Fujioka T, Moriyama M, Yamaoka Y (2009) Analysis of virulence factors of *Helicobacter pylori* isolated from a Vietnamese population. BMC Microbiol 9:175. https://doi.org/10.1186/1471-2180-9-175

Uemura N, Okamoto S, Yamamoto S, Matsumura N, Yamaguchi S, Yamakido M, Taniyama K, Sasaki N, Schlemper RJ (2001) *Helicobacter pylori* infection and the development of gastric cancer. N Engl J Med 345 (11):784–789. https://doi.org/10.1056/NEJMoa001999

Vale FF, Vadivelu J, Oleastro M, Breurec S, Engstrand L, Perets TT, Megraud F, Lehours P (2015) Dormant phages of *Helicobacter pylori* reveal distinct populations in Europe. Sci Rep 5:14333. https://doi.org/10.1038/srep14333

Vale FF, Nunes A, Oleastro M, Gomes JP, Sampaio DA, Rocha R, Vitor JM, Engstrand L, Pascoe B, Berthenet E, Sheppard SK, Hitchings MD, Megraud F, Vadivelu J, Lehours P (2017) Genomic structure and insertion sites of *Helicobacter pylori* prophages from various geographical origins. Sci Rep 7:42471. https://doi.org/10.1038/srep42471

Wirth T, Wang X, Linz B, Novick RP, Lum JK, Blaser M, Morelli G, Falush D, Achtman M (2004) Distinguishing human ethnic groups by means of sequences from *Helicobacter pylori*: lessons from Ladakh. Proc Natl Acad Sci U S A 101 (14):4746–4751. https://doi.org/10.1073/pnas.0306629101

Yahara K, Furuta Y, Oshima K, Yoshida M, Azuma T, Hattori M, Uchiyama I, Kobayashi I (2013) Chromosome painting in silico in a bacterial species reveals fine population structure. Mol Biol Evol 30 (6):1454–1464. https://doi.org/10.1093/molbev/mst055

Yamaoka Y (2010) Mechanisms of disease: *Helicobacter pylori* virulence factors. Nat Rev Gastroenterol Hepatol 7(11):629–641. https://doi.org/10.1038/nrgastro.2010.154

Yamaoka Y, Kodama T, Kashima K, Graham DY, Sepulveda AR (1998) Variants of the 3′ region of the cagA gene in *Helicobacter pylori* isolates from patients with different *H. pylori*-associated diseases. J Clin Microbiol 36(8):2258–2263

Yamaoka Y, El-Zimaity HM, Gutierrez O, Figura N, Kim JG, Kodama T, Kashima K, Graham DY (1999) Relationship between the cagA 3′ repeat region of *Helicobacter pylori*, gastric histology, and susceptibility to low pH. Gastroenterology 117(2):342–349

Yamaoka Y, Osato MS, Sepulveda AR, Gutierrez O, Figura N, Kim JG, Kodama T, Kashima K, Graham DY (2000) Molecular epidemiology of *Helicobacter pylori*: separation of *H. pylori* from East Asian and non-Asian countries. Epidemiol Infect 124(1):91–96

Yamaoka Y, Orito E, Mizokami M, Gutierrez O, Saitou N, Kodama T, Osato MS, Kim JG, Ramirez FC, Mahachai V, Graham DY (2002) *Helicobacter pylori* in North and South America before Columbus. FEBS Lett 517(1–3):180–184

Zhang XS, Tegtmeyer N, Traube L, Jindal S, Perez-Perez-G, Sticht H, Backert S, Blaser MJ (2015) A specific A/T polymorphism in Western tyrosine phosphorylation B-motifs regulates *Helicobacter pylori* CagA epithelial cell interaction. PLoS Pathog 11:e1004621. https://doi.org/10.1371/journal.ppat.1004621

Adv Exp Med Biol - Advances in Microbiology, Infectious Diseases and Public Health (2019) 1149: 17–33
https://doi.org/10.1007/5584_2019_357

Published online: 24 April 2019

Epidemiology, Diagnosis and Risk Factors of *Helicobacter pylori* Infection

Kallirroi Kotilea, Patrick Bontems, and Eliette Touati

Abstract

Helicobacter pylori is a human-specific pathogen, which leads to gastric pathologies including gastric cancer. It is a highly unique bacterium considered as a carcinogenic agent. *H. pylori* remains a major human health problem, responsible for ~90% of the gastric cancer cases. Approximately four billion individuals have been detected for *H. pylori* infection worldwide in 2015. At the turn of the twenty-first century, the prevalence of *H. pylori* has been declining in highly industrialized countries of the Western world, whereas prevalence has plateaued at a high level in developing and newly industrialized countries. However, the infection status remains high in immigrants coming from countries with high prevalence of *H. pylori* infection. *H. pylori* can be diagnosed both by invasive and non-invasive methods. Urea breath test and stool antigens detection are among the most commonly used non-invasive ones. Although the way *H. pylori* is transmitted remains still not fully clear, the level of contamination is strongly dependent on the familial and environmental context, with a drastic impact of living conditions with poor hygiene and sanitation. However, familial socioeconomic status is the main risk factor for *H. pylori* infection among children. In addition, food and water source have a high impact on the prevalence of *H. pylori* infection worldwide. This chapter highlights the latest knowledge in the epidemiology of *H. pylori* infection, its diagnosis and critical risk factors responsible for its high prevalence in some populations and geographic areas.

Keywords

Prevalence · Invasive and non-invasive methods · Familial factors · Environment · Life habits

1 Introduction

Helicobacter pylori is a spiral-shaped and flagellated Gram-negative bacterium, which colonizes specifically the human stomach. *H. pylori* infects about 50% of the population worldwide, making it the most widespread infection in humans, especially in developing countries where its prevalence is estimated to be around 80% (Torres et al. 2000). *H. pylori* infection leads to chronic gastritis, which can evolve either into peptic ulcer

K. Kotilea and P. Bontems
Paediatric Gastroenterology Unit, Université Libre de Bruxelles, Hôpital Universitaire des Enfants Reine Fabiola, Bruxelles, Belgium
e-mail: Kallirroy.Kotilea@huderf.be; patrick.bontems@huderf.be

E. Touati (✉)
Unit of Helicobacter Pathogenesis, Institut Pasteur, Paris, France
e-mail: eliette.touati@pasteur.fr

diseases or into the development of pre-neoplastic lesions (intestinal metaplasia, dysplasia) and adenocarcinoma (Correa 1992). Until now, *H. pylori* is the only bacterium recognized as a type 1 carcinogenic agent (IARC 1994). Persistent *H. pylori* colonization and the associated chronic inflammation are critical parameters for the development of gastric malignancies. Further knowledge on the epidemiology of the infection, its pathways of transmission and risk factors could lead to public-health measures for the prevention and control of this infection. The aim of this chapter is to give an overview of the recent epidemiological data on the prevalence of *H. pylori* infection among the world population and the available diagnostic tools for its detection. The infection rate is known to be linked to socioeconomic factors. However, environmental factors including source of water and sanitary conditions as well as factors related to food can also lead to the dissemination of *H. pylori* infection. An update of these different risk factors will also be provided in this review.

2 Epidemiology of *H. pylori* Infection

A recent review with reports from 62 countries estimates that more than half the world's population is still infected with *H. pylori* (Hooi et al. 2017). This means that, based on regional prevalence estimates, there were approximately 4.4 billion individuals with *H. pylori* infection worldwide in 2015 with a wide variation in the prevalence of *H. pylori* between regions and countries. Prevalence is highest in Africa (79.1%), Latin America and the Caribbean (63.4%), and Asia (54.7%). In contrast, *H. pylori* prevalence is lowest in Northern America (37.1%) and Oceania (24.4%). At the turn to the twenty-first century, the prevalence of *H. pylori* has been declining in highly industrialized countries of the Western world, whereas prevalence has plateaued at a high level in developing and newly industrialized countries. The widening differential gap in prevalence has important implication on the future worldwide prevalence of diseases associated with *H. pylori*, including peptic ulcer and gastric cancer. These differences in *H. pylori* prevalence likely reflect the level of urbanization, sanitation, access to clean water, and varied socioeconomic status (Hooi et al. 2017). In children, a comprehensive review and meta-analysis of original pediatric studies from 2011 to 2016 performed on healthy children estimated an overall seroprevalence rate of 33% [95% confidence interval (CI) 27–38] (Zabala Torrres et al. 2017; see also Chap. 6 of this book). In the same study, a review of information available from seven cohort studies concluded that infection rates in healthy children under 5 years of age were still between 20 and 40% in high-income countries and between 30% and 50% in upper-middle income countries, indicating that the country of birth plays a role in infection prevalence. Higher rates of infection (40%), as determined by cross-sectional studies, are predominantly seen in low or low-to-middle-income areas (or in countries with severe income inequality). However, huge variations in the prevalence can be found between countries with similar living conditions. This can be observed in Europe, for example, where the prevalence of *H. pylori* infection remains high in Spain and Portugal, although the levels of sanitation and of economic development have risen in recent decades and are comparable to other European countries, where the prevalence of infection is significantly lower. Similar variations from country to country can also be seen in Asia, which cannot be fully explained by only looking at the level of development (Zamani et al. 2018). There are significant differences in the *H. pylori* prevalence even within the same country. For example, different racial groups in the United States have different *H. pylori* prevalence. It was reported that the prevalence in non-Hispanic whites ranges from 18.4% to 26.2% and that in non-whites ranges from 34.5% to 61.6% (Everhart et al. 2000; Cardenas et al. 2006).

The prevalence of *H. pylori* infection in both children and adults is, however, still decreasing in developed countries. One study from Iceland involved 205 children aged 7–17 years and found only 3.4% of infection (Asgeirsdottir et al. 2017). Furthermore, the prevalence was 2.6% among children when both parents were born in a low prevalence country compared with 17% among those with at least one parent born in a high prevalence area ($p = 0.026$). This confirms results obtained in Belgium some years ago (Mana et al. 2013). In Poland, the prevalence of *H. pylori* infection in 8661 symptomatic and untreated children from 2000 to 2013 assessed by culture was 16.1%. The highest prevalence of infection was found in the year 2000 (23.1%) and the lowest in 2010 (8.9%) (Biernat et al. 2016). However, in Latvia, no evidence of a fall in prevalence in children was found during the last 10 years and the prevalence determined by stool antigen test was 15.5% (Daugule et al. 2016).

The prevalence is also decreasing in some countries in Asia and in the Middle East. Indeed, studies from Japan have shown a considerable fall in *H. pylori* prevalence in childhood. One study from a high gastric cancer incidence area found only 85 of 1765 (4.8%) students aged 13–15 years to be infected (Kusano et al. 2017) and, in another study, the prevalence in school children aged 12–15 years was 3.1% (Nakayama et al. 2017). The same fall has been observed in Iran, where former reports of *H. pylori* infection rate indicated a global prevalence of more than 85% and recent ones estimated an overall prevalence of 54%, with a prevalence of 42% in children (Moosazadeh et al. 2016). Similar trends are seen in the Chinese city Hangzhou, where the infection rates in three age groups (3–6, 7–11, and 12–17 years) were 14.8, 20.2, and 25.8%, respectively. The overall prevalence decreased from 21.6 to 17.2% between 2007 and 2014 (Shu et al. 2017). In contrast, in Vietnam, the seroprevalence in 1094 subjects from 278 households remained stable at 51.4% in adults and 41.4% in children (Nguyen et al. 2016). Conversely, the prevalence of *H. pylori* infection remains high in newly arrived refugees attending the migrant health service in South Australia, where 922 adults and children were screened in a cross-sectional study using a monoclonal stool antigen test. *H. pylori* infection was detected in 198 of them (21.5%), almost 1.5 times that of the Australian population's estimate when both adults and children are included (Abdul Rahim et al. 2017). A systematic review involving 28 studies described the prevalence of *H. pylori* among migrants. In all but two, the prevalence was similar to or lower than in their country of origin but higher than in their country of destination. Second and later generations of migrants had a lower prevalence than the first generation (Morais et al. 2017).

As mentioned above, *H. pylori* has been identified as a Group I carcinogen by the International Agency for Research on Cancer (IARC 1994) and currently is considered as necessary, but insufficient cause of gastric adenocarcinoma (Eslick et al. 1999; Uemura et al. 2001). Approximately 89% of all gastric cancers can be attributable to *H. pylori* infection. In Africa, despite the high *H. pylori* prevalence, the reported incidence of gastric cancer was considerably lower compared with China or Japan and was postulated to be related to the predominant non-atrophic gastritis pattern in Africa; the archetypal *H. pylori Africa2* type strains largely restricted in South Africa, which lack the *cag* (cytotoxin-associated genes) pathogenicity island (PAI); and lastly intestinal parasitic infestation modulating the immune response against *H. pylori* toward a Th2 type response, which may reduce the risk of gastric cancer (Correa and Piazuelo 2011; Kodaman et al. 2014). The now defunct phenomenon known as "African enigma" was attributed to the inadequate sampling of the African population obtained through endoscopic data, limited access to health care, and a relatively short life expectancy in the population. More recent and robust data on the African gastric ulcer and cancer prevalence confirmed that it is not as low as reported previously (Graham et al. 2009).

Ongoing efforts to monitor *H. pylori* prevalence and its disease burden in a systematic manner is crucial, as it will minimize any skewed data, which can adversely affect the allocation of health care resources.

3 Diagnosis of *H. pylori* Infection

H. pylori infection can be confirmed by invasive methods, requiring gastric biopsies obtained during an endoscopy (histology, culture, PCR: polymerase chain reaction, RAP: rapid urease test) or non-invasive methods (SAT: stool antigen test, UBT: urea breath test, and serology). The invasive tests are used in clinical practice and the non-invasive ones mostly in epidemiology and to assess the outcome of an eradication treatment. It is necessary to have at least two concordant tests to confirm or deny an infection in clinical practice (Fallone et al. 2016; Malfertheiner et al. 2017; Jones et al. 2017). Indeed, false negative results of any diagnostic tests for *H. pylori* can occur since the sensitivity of any test never reaches 100% and the sensitivity is lower in case of antimicrobial use within the previous 4 weeks, of proton pump inhibitor use within the previous 2 weeks or in case of gastrointestinal bleeding. False-positives are rare, but can occur and when present may be due to the occurrence of other urease containing bacteria such as *Proteus mirabilis*, *Citrobacter freundii*, *Klebsiella pneumoniae*, *Enterobacter cloacae* and *Staphylococcus aureus* (Osaki et al. 2008). Non-invasive tests are also used in test and treat strategies, that should be restricted to adults living in high prevalence countries presenting dyspepsia without any alarm symptoms (Fallone et al. 2016; Malfertheiner et al. 2017). However, the number of patients to treat successfully to improve dyspepsia symptoms in one is around 12 (Moayyedi et al. 2006). In the context of a prudent use of antibiotics, it may appear not justified to prescribe antibiotic which will induce resistance to the drug in case of failure, be responsible of adverse events and have a high cost (although the cost-effectiveness may vary according to the cost of care in a given country).

3.1 Histology

Demonstration of the presence of *H. pylori* by histological analysis of gastric biopsies is facilitated by special stains such as Giemsa or immunohistochemical techniques (Lash and Genta 2016) using antibodies directed against surface antigens of the bacterium. This diagnostic method remains the most commonly used and allows the scoring of gastritis: updated classification of Sydney (Dixon et al. 1996) (Fig. 1), OLGA (Rugge et al. 2007) (Table 1) or OLGIM (Capelle et al. 2010) (Table 2). As the diagnostic sensitivity increases with the number of biopsies, it is advisable to take at least 2 biopsies in the antrum at the level of the large curvature, one on the small curvature and 2 at the fundus (Jones et al. 2017; Fallone et al. 2016). To properly assess atrophy, a biopsy should also be performed on the small curvature (Rugge et al. 2007).

3.2 Rapid Urease Test

The rapid urease test is based on the activity of the urease produced by live *H. pylori* (McNulty and Wise 1985). To perform the test, a gastric biopsy is placed in a medium containing urea and a colorimetric pH indicator. Following the ammonia production associated with urease activity, the pH change is indicated by the colorimetric shift of the pH indicator. The advantages of this test are its simplicity, its low cost and its ease of execution.

3.3 Culture

H. pylori culture is often difficult because this bacterium is fragile and requires microaerophilic conditions to grow. However, its culture can be facilitated by the use of a transport medium in the endoscopy room (Bontems et al. 2001; Koletzko et al. 2006; Miendje Deyi et al. 2011; Jones et al. 2017). Culture has the advantage of being able to provide information on the susceptibility of strains to antibiotics, to adapt antimicrobial therapy and to improve the rate of eradication (Miendje Deyi et al. 2011; Chan and Mackenzie 1986; Jones et al. 2017; Fallone et al. 2016). Given the possible mixed infection with susceptible and resistant strains to a given antimicrobial agents and the distribution of the bacteria in the

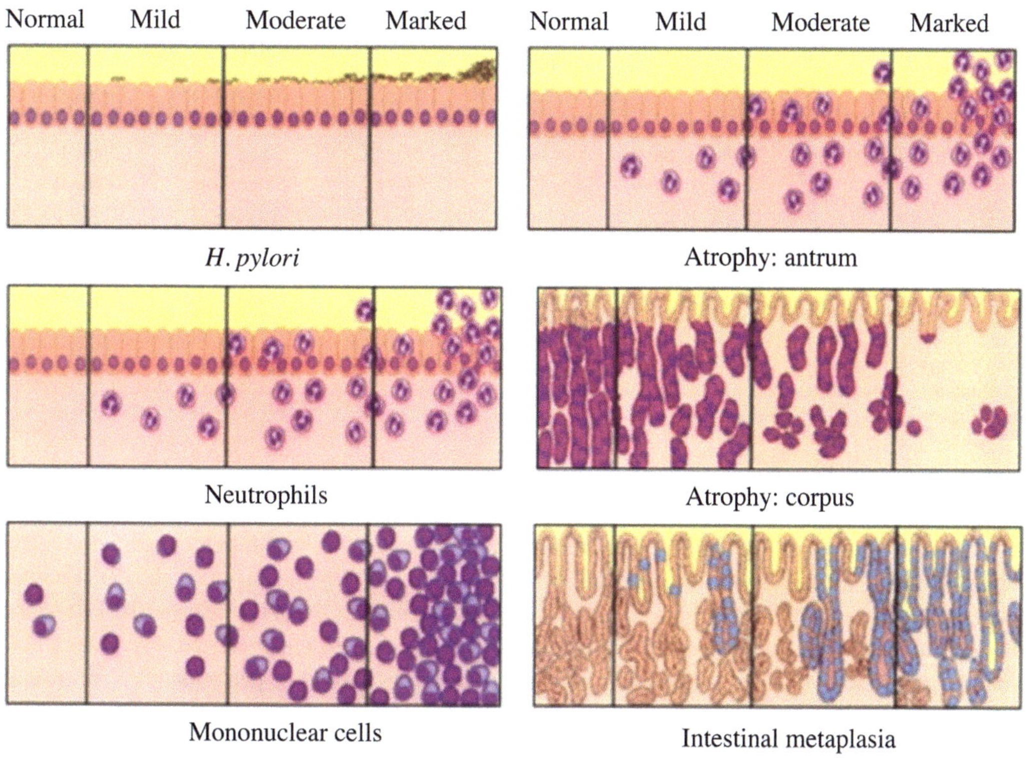

Fig. 1 The updated Sydney classification for gastric lesions. Each feature is assigned with a semi-quantitative and descriptive evaluation to 0: absence (Normal), 1 for mild, 2 for moderate and 3 for marked corresponding modification including infiltration of neutrophils and mononuclear cells, intensity of atrophy and frequence and severity of intestinal metaplasia. Taken from Dixon and co-workers (1996) with permission

Table 1 Gastritis staging according to OLGA classification

		Corpus			
	Atrophy score	No atrophy (score 0)	Mild atrophy (score 2)	Moderate atrophy (score 3)	Severe atrophy (score 4)
Antrum	No atrophy (score 0) (including incisura angularis)	Stage 0	Stage 1	Stage 1I	Stage 1I
	Mild atrophy (score 1) (including incisura angularis)	Stage I	Stage I	Stage II	Stage III
	Moderate atrophy (score 2) (including incisura angularis)	Stage II	Stage II	Stage III	Stage IV
	Severe atrophy (score 3) (including incisura angularis)	Stage III	Stage III	Stage IV	Stage IV

The Operative Link on Gastritis Assessment (OLGA) system corresponds to an histological staging for gastric inflammation, considering gastric atrophy as the histological lesion representative of disease progression. From Rugge and co-workers (2007) with permission

stomach, it is recommended to take several biopsies (at least one in the antrum and one in the fundus) (Selgrad et al. 2014; Aguilera-Correa et al. 2017; Malfertheiner et al. 2017). If the transport time to the laboratory exceeds 4 h, biopsies should be kept frozen for a maximum of 24 h. Beyond this, it is best to freeze at −70 °C or in liquid nitrogen (Miendje Deyi et al. 2011).

Table 2 Gastritis staging according to OLGIM classification

		Corpus			
	IM score	Not fat: no IM (score 0)	Mild IM (score 1)	Moderate IM (score 2)	Severe IM (score 3)
Antrum (including incisura angularis)	No IM (score 0)	Stage 0	Stage I	Stage II	Stage II
	Mild IM (score 1)	Stage I	Stage I	Stage II	Stage II
	Moderate IM (score 2)	Stage II	Stage II	Stage III	Stage IV
	Severe IM (score 3)	Stage III	Stage III	Stage IV	Stage IV

The Operative Link on Gastric Intestinal Metaplasia (OLGIM) system corresponds to an histological staging for gastric inflammation, considering intestinal metaplasia (IM) as the histological lesion representative of disease progression. From Capelle et al. (2010) with permission

3.4 Polymerase Chain Reaction (PCR)

Molecular biology techniques can replace culture for the diagnosis of *H. pylori* infection if a medical center does not have the technical capability and/or cannot send the frozen samples to a microbiology department with that expertise. These techniques also allow the detection of mutations causing resistance to certain antibiotics and the detection of infections with several strains with different susceptibility profiles to antibiotics (Miendje Deyi et al. 2011; Kalach et al. 2015).

3.5 Serology

Serology is a simple and very accessible method. A comparative study of 29 commercially available serological kits to detect *H. pylori* infection came to the conclusion that some of the available kits are excellent, with performance parameters such as sensitivity and specificity above 90% (Burucoa et al. 2013). However, local validation of these tests is still needed as their performance may vary depending on the antigenic composition of the circulating strains in a given population. The persistence, sometimes prolonged of antibodies against *H. pylori* does not allow to distinguish between an active and a cured infection. In addition, sensitivity is low in young children, although still frequently used in epidemiological studies (Westblom et al. 1992; Andersen et al. 1994; Raymond et al. 1996; Corvaglia et al. 1999; Okuda et al. 2002; Douraghi et al. 2013). However, some recent publications suggest that newer serological tests are more reliable in children (Shady et al. 2015; Kalach et al. 2017; Raj et al. 2017). Enzyme-linked immunosorbent assay (ELISA)-based methods are always preferred over rapid near-patient tests, whose performances are usually not satisfactory and with low reproducibility (Best et al. 2018).

3.6 Urea Breath Test

The marked urea breath test (UBT) consists in having the patient swallow carbon-labeled urea, low-radioactive ^{14}C or, particularly in children, the non-radioactive ^{13}C, and then assay this stable isotope in the CO_2 expired. If the patient is infected, the labeled urea is metabolized by the urease produced by *H. pylori* and the expired $^{13}CO_2$ increase is detected by a mass spectrometer (Graham et al. 1987; Vandenplas et al. 1992; Cadranel et al. 1998). This test, undeniably very sensitive, but expensive, has the advantage to detect the presence of the bacterium in the entire stomach. However, children younger than 6 years appear to have a higher rate of false positive UBT (Leal et al. 2011). The reported performance of the UBT for detection of *H. pylori* infection in a recent study, in comparison with biopsy-based histologic examination in 60 children, is low with a sensitivity of 76.2% and a specificity of 69.2% in this age group (Honar et al. 2016). Many pitfalls were underlined to explain these poor results: patient compliance, consumption of PPIs and/or antibiotics for example. Therefore, stool antigen tests with a monoclonal antibody are regarded as more convenient among young people (Honar et al. 2016; Osaki et al. 2008). It is recommended for eradication control at least 4 weeks after cessation of eradication treatment (Osaki et al. 2008; Malfertheiner et al. 2017; Jones et al. 2017).

3.7 Stool Antigen Test

Antigen testing in stool is an alternative to the respiratory test for monitoring patients after eradication treatment, epidemiology or test and treat strategies in selected populations, with the proviso that only monoclonal antibody tests with good sensitivity should be used (Makristathis et al. 1998). *H. pylori* specific antigen is tested in fresh or frozen stool samples (Guarner et al. 2010).

4 Risk Factors of *H. pylori* Infection

It is now well established that *H. pylori* infection is mostly acquired during childhood, mainly during the first 5 years of life and it is significantly influenced by geographical context and specific living conditions (Mendall et al. 1992). In developing countries, the prevalence of the infection is 30 to 50% in children and reaches 90% in adults. In contrast, in developed countries the prevalence of the infection in children is between 1–12% and reaches 30–50% in adults (Suerbaum and Michetti 2002). These differences among underdeveloped and industrialized countries are mainly due to the impact of risk factors during childhood. According to a large number of studies, the main routes of *H. pylori* transmission are person to person by oral-oral or fecal-oral routes. The level of contamination is strongly dependent of familial and environmental parameters, with a more drastic impact of living environment including poor hygiene and sanitation, which are promoting factors for *H. pylori* especially in developing countries. However, either for developing or developed countries, familial socioeconomic status is the main risk factor for *H. pylori* infection among children.

4.1 Socioeconomic Status and Environmental Conditions

4.1.1 Familial Context and Source of Transmission

The socioeconomic status is defined as occupation, family income level, parent's education level, and living conditions including crowding occupancy. Abundant evidences from many studies demonstrated that low socioeconomic status is a major risk factor in the acquisition of *H. pylori* infection (Fig. 2). This is particularly true as while a declining trend in overall prevalence over time of *H. pylori* infection among industrialized countries, poor socioeconomic conditions still remain associated with a high prevalence. There is no difference in this respect between the developed and the developing countries. For example, in the United States, the prevalence of *H. pylori* infection was approximately twice as high among black and hispanic populations compared to age-matches whites (Malaty et al. 1992). More recently, the prevalence of *H. pylori* infection has been determined in a cohort of Portuguese adolescents (EpiTeen) at the age of 13 and the incidence after a 3 years follow-up was analyzed. An inverse association was found between the prevalence of the infection and the parent's education level. The adolescents studying in private school were less likely to be infected (Bastos et al. 2013). A recent nationally representative cross-sectional study involving adults $\geq$18 years old in Turkey, where the overall prevalence of *H. pylori* infection is about 82.5%, also confirmed an inverse association of education level and *H. pylori* infection. In addition, individuals who had access to social security were at lower risk of *H. pylori* infection (Ozaydin et al. 2013). While education was the only significant factor for women, residential region, housing tenure were among risk factors for men.

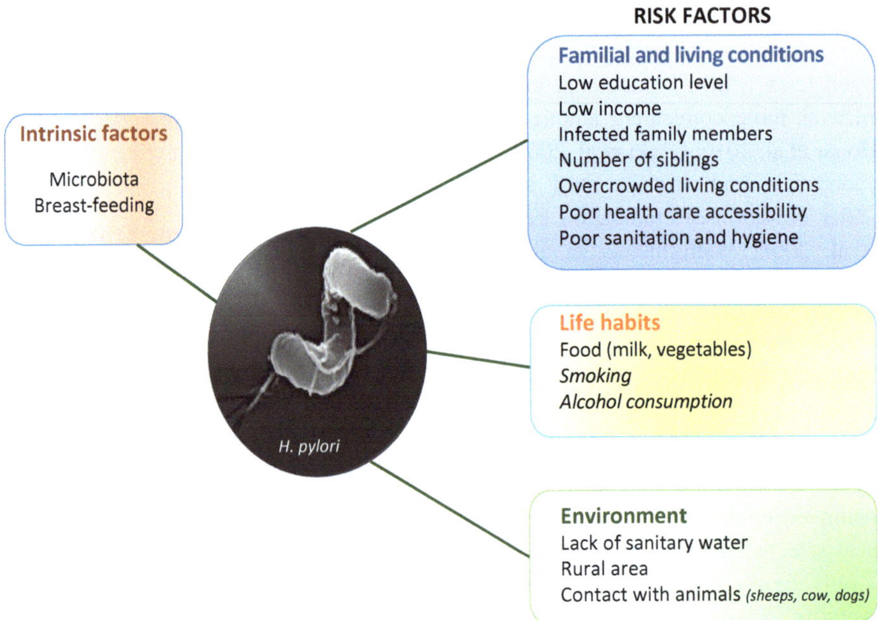

Fig. 2 Risk factors associated to *H. pylori* infection. Major risk factors associated to the transmission and spread of *H. pylori* are of multiple origins including familial and living conditions, life style habits and environmental context. Microbiota and breast-feeding correspond to individual factors suggested to modulate the transmission/acquisition of *H. pylori* infection

Although the link between several risk factors and a high prevalence of *H. pylori* infection is well established, the way of its transmission is still not fully clear. However, interpersonal and intrafamilial transmissions either oral-oral or fecal-oral appear to be the main route, as supported by a recent study which confirmed that bed sharing, and two affected parents were positively associated with the presence of *H. pylori* infection (Hasosah et al. 2015). Already in the 90s, confined-living conditions as crowded sleeping accommodation was strongly associated with the presence of infection among children (Mendall et al. 1992). In line with this, Bastos and co-workers (2013) showed that more siblings and living in a house with higher crowding index were also positive risk factors. In urban and rural Beninese populations, 406 healthy individuals including 240 and 206 subjects, respectively, were selected from 96 households. The risk of *H. pylori* infection in children was 13-fold higher when both parents were *H. pylori* positive (95% OR = 13.6, 95% Confidence Interval 3.63–51.22), compared with when only one parent was positive (95% OR = 5.3, 95% Confidence Interval 1.52–18.45) (Aguemon et al. 2005). In this study, while the spread of infection is facilitated by living conditions with high promiscuity and a high transmission from infected parents to children, a high number of siblings has been reported as another risk factor. A multilocus sequence typing (MLST) DNA analysis using the stools of parents belonging to three families showed an intrafamilial transmission in all selected families with a mother-to-child transmission in at least two families (Osaki et al. 2013). In addition, a grand-mother to child transmission has also been suggested by Urita and co-workers (2013), who tested 838 children from a small town in Japan. Furthermore, a transmission from sibling to sibling has also been proposed by a macroarray study on selected *H. pylori* coding sequences (CDS) on three families including one child with persistence of gastric symptoms after antibiotics treatment (Raymond et al. 2008). The analysis on 684 children from rural Andean area screened for *H. pylori* with the ^{13}C-urea breath test indicated that the infection is mostly transmitted among siblings, which are close in age, and most frequently from the older to youngers ones (Goodman and Correa 2000). This was also confirmed in population with low overall prevalence of *H. pylori* as reported by Sykora et al. (2009) who developed a cross-sectional population-based study on 1545 asymptomatic Czech children (aged 0–15 years), where the prevalence was 7.1%. A positive association with *H. pylori* infection was found with the number of children in a household and the lack of formal education of fathers. Also related to the social status of the family, the access to good hygiene living condition limits the prevalence of the infection, indicating that important risk factors are associated to poverty. There is no difference in this respect between the developed and the developing countries. In addition to intrafamilial transmission, spread of *H. pylori* infection can be promoted by community living conditions. Indeed, the institutionalization of children between 1 and 6 years old was significantly associated with *H. pylori* infection as reported in asymptomatic Czech children (Sykora et al. 2009). Among children living in urban area, a higher risk of *H. pylori* seropositivity was significantly found associated for those who attended day care centers or nurseries (Dore et al. 2002).

Breast-Feeding

Whether or not children were breast-fed did not have a statistically significant effect on *H. pylori* seropositivity among children living in adjacent urban and rural areas as reported by Dore and co-workers (2002) in Northern Sardinia, Italy. However, the prospective population-based study from the Czech Republic among asymptomatic children reported a higher prevalence of *H. pylori* infection among children who had never been breast-fed (Sykora et al. 2009). In contrast, no correlation among children from low socio-economic backgrounds in Lagos, Nigeria was observed either with exclusive breast-feeding or its duration, and *H. pylori* infection (Senbanjo et al. 2014). According to the history of breast-feeding (ever *vs.* never), a recent systematic review (Carreira et al. 2015), did not find a significant association between breast-feeding and *H. pylori* infection in either high- or middle-income countries, excepted children having been breast-fed for 4–6 months which showed a lower risk of *H. pylori* infection only in the middle-outcome countries. It was then proposed that breast-feeding may protect children against infection by acting as natural antibiotics. Accordingly, children whose mothers had breath milk with higher levels of anti-*H. pylori* IgA had a lower risk to be infected compared with those whose mothers had lower levels (Thomas et al. 1993).

Microbiota

Gut microbiota composition can be affected by many factors among which diet, environmental compounds, lifestyle habits, infection and disease (Rodriguez et al. 2015). A recent study reported different profiles of gastric microbiota composition when comparing between gastric cancer and chronic gastritis patients (Ferreira et al. 2018). In this study, the gastric carcinoma dysbiosis is characterized by reduced microbial diversity with a reduced *Helicobacter* abundance. A reduced microbial diversity is recognised as associated with disease states, as also reported for inflammatory diseases and cancer (Gevers et al. 2014; Ahn et al. 2013). The presence of *H. pylori* not only influence the composition of gastric microbiota, but also indirectly it modifies the intestinal microbiota, as demonstrated by Kienesberger et al. (2016) in the mouse model. As also reported by Arnold and co-workers (2011) that the presence of *H. pylori* infection induces a differential immune response, based on mouse age, that could affect the susceptibility of its host to disease and infection. Family microbiota is shared between parents and children and may play an important role in the composition of infant microbiota. Whether intestinal microbiota can affect *H. pylori* intrafamilial infection has been investigated (Osaki et al. 2018). The microbiota composition of 18 fecal specimens from five *H. pylori*-infected children and their family members were analysed in five families. The microbiota from *H. pylori*-positive children and adults showed a lower diversity than that of *H. pylori*-negative children and parents. This study indicates that the similarity in microbiota composition and its poor diversity can be a risk factor for *H. pylori* intrafamilial transmission.

4.1.2 Environmental Context

The environment may serve as a reservoir for *H. pylori* infection and directly associated risk factors may play an important role, even more important than familial socioeconomic status especially in developing countries.

Rural vs. Urban Living Conditions

The difference between urban and rural living conditions is also among factors influencing the prevalence of *H. pylori* acquisition in children from different countries. A cross-sectional study conducted among Italian children residing in different environments (urban *vs.* rural) in the north of Sardinia showed a prevalence of *H. pylori* infection of 26% among children aged 14–16 compared to 20% for those of 5–7 years-old. The prevalence was higher among children living in rural area (36%) compared to among children residing in the urban area (13%). This difference was age-independent (Dore et al. 2002). Especially in the rural areas, contact with dogs appears to be a promoting factor. While no association was seen between the prevalence of the infection and the occupation of the head of the house in rural area, the seroprevalence of *H. pylori* was directly associated with the socioeconomic status of parents in urban area. The same data are also reported on populations with different living conditions compared to European countries, as in a study on dyspeptic patients in Andkhoy (Afghanistan) where the positive prevalence for *H. pylori* has been associated with epigastric pain and rural occupancy with a positive correlation with illiteracy (Hamrah et al. 2017). In contrast, in the Beninese population, no significant differences were observed on the prevalence of the infection between rural and urban areas (Aguemon et al. 2005).

Water and Access to Sanitary and Hygiene

According to many epidemiologic studies, water could be an important source of *H. pylori* contamination (Ozaydin et al. 2013). Particularly in developing countries waterborne infection is the main infection route of *H. pylori* due to poor sanitary distribution of water among the population. As example in Peru, the analysis of drinking water samples from different locations, based on PCR assays on specific *H. pylori* genes, showed that the majority of contaminated water came from municipal water (Hulten et al. 1996). In contrast, only three of 25 municipal water samples analyzed in Sweden showed the presence of *Helicobacter spp.* DNA, while a large number of well water samples were positive in PCR assays (Hulten et al. 1996). *H. pylori* DNA was also detected in well water in Japan, whose consumers were detected positive for *H. pylori* infection (Horiuchi et al. 2001). Accordingly, a study conducted in Germany indicated a positive association for *H. pylori* infection and consumption of well water (Strebel et al. 2010), as also reported in Portugal (Amaral et al. 2017). In addition, a river water-associated *H. pylori* contamination has also been suggested in Japan (Fujimura et al. 2008).

The potential presence of live *H. pylori* infective cells in water samples is of public health concern. The analysis of 45 wastewater samples obtained from two secondary wastewater treatment plants in Valencia, Spain showed the presence of culturable *H. pylori* isolates (Moreno and Ferrus 2012). These findings support the view that fecal-contaminated water may act as a reservoir for *H. pylori* spread. It has been suggested that the ability of *H. pylori* to form biofilm may allow its survival in natural water sources and water distribution systems (Garcia et al. 2014). In addition to DNA detection, about less than 10 reports indicate that culturable form of *H. pylori* can be isolated from water samples as reviewed by Aziz and co-workers (2015). *H. pylori* cultivability in water should be limited in time with an optimum <10 h at temperatures over 20 °C (Adams et al. 2003; Azevedo et al. 2008). The acquisition of culturable phenotype in water is associated to morphological transition of the bacterium to the rod shape implicating the peptidoglycan (PG) turnover (Fernandes et al. 2017). Accordingly, *H. pylori* enters a viable, but not culturable (VBNC) state, within a few days after inoculation into water, associated to morphological changes from a spiral bacillus to coccoid form. While the exposition of mice to *H. pylori* viable strain SS1-supplemented drinking water led to infection in mice with significant gastric inflammation after 4 weeks (Boehnke et al. 2015), waterborne VBNC SS1 failed to colonize mice either through drinking water exposure or oral gavage (Boehnke et al. 2017).

4.2 Lifestyle Habits

4.2.1 Food

Several studies detected the presence of *H. pylori* DNA in dairy products and especially raw milk, that has been considered as the main source of food transmission (Momtaz et al. 2014). About 19% of raw milk and dairy product samples tested in Iran were found *H. pylori*-positive (Mousavi et al. 2014). Ovine milk and traditional cheese were the most commonly contaminated products as reported by Dore et al. (1999b), which detected the presence of *H. pylori* by PCR amplification in milk samples from sheeps, thus indicating a potential contamination of humans by milk consumption. Similar data were also reported for raw cow milk samples in Japan (Fujimura et al. 2002), as well as in Greek (Angelidis et al. 2011) and American (Dore et al. 2001) herds. Supporting these data, a recent study reports a higher prevalence of *H. pylori* positively associated with the consumption of milk (Assaad et al. 2018). The contamination of milk may be also related to lack of hygiene measure during milk processing and especially from unpasteurized milk storage in some countries. An important finding from Mousavi and co-workers (2014) was the high presence of *H. pylori* antibiotic resistant strains isolated from milk and dairy products.

Meat has also been proposed as a possible source of *H. pylori* as reservoir. Previous data identified the presence of *H. pylori* by 16SrRNA and *vacA* PCR analysis in gastric tissues from sheep, thus taking part in the human food chain. Sheeps have been proposed as a source of *H. pylori* transmission among shepherds and their family members (Dore et al. 1999a) (Papiez et al. 2003). In line with this, the prevalence of *H. pylori* infection among shepherds, who reside in Northern Sardinia and among members of their family is one of the highest in the world (98%). It may be associated with direct contact with sheeps and sheepdogs (Dore et al. 1999a, b), as also reported in Colombian Andes population (Goodman et al. 1996).

Raw vegetables may be also a source of *H. pylori* food transmission to humans. Several studies reported the detection of *H. pylori* by culture and PCR in vegetable and salad samples (Goodman et al. 1996; Yahaghi et al. 2014). Contaminated water used through washing may be likely the source of the presence of *H. pylori* in raw vegetables. Thus, the lack or deficient waste water treatment may also promote *H. pylori* food contamination and it is likely to play an important role in *H. pylori* transmission to humans. However, given the link between *H. pylori* infection and low socioeconomic status, the infection could occur predominantly among individuals with deficient nutritional status. In the Andean population in Columbia, *H. pylori* infection occurred more frequently among children of short stature for their age, with a low consumption of fruits and vegetables associated with low nutritional indicators (Goodman et al. 1997).

4.2.2 Alcohol Drinking and Smoking Habits

Studies on the association of smoking and alcohol consumption with *H. pylori* infection showed conflicting results. While Zhu and co-workers (2014) reported that the use of alcohol and tobacco had no impact on the prevalence of *H. pylori* infection, significant relationships between smoking habits (current *vs.* never) with *H. pylori* seropositivity have been reported in Japanese adults (Kikuchi et al. 1998). In Northern Ireland, smoking has been reported positively associated with the presence of *H. pylori* infection, however, no relationship with alcohol consumption has been found to be significant (Murray et al. 1997). In contrast, the study by Ogihara et al. (2000), which analyzed the impact of drinking and smoking on *H. pylori* serology in 8837 subjects working in textile companies in Japan, found a negative association between current cigarette and alcohol consumption with *H. pylori* seropositivity. The increased gastric acidity associated to smoking may be a cause of the negative association between the presence of the infection and tobacco consumption. Furthermore, more recently non-smokers and regular alcohol consumers were suggested as under less risk of *H. pylori* infection than others (Ozaydin et al. 2013).

4.3 Occupational Hazards

The occupational risk for acquiring *H. pylori* infection has been addressed in several studies among healthcare workers during contact with patients. The prevalence of *H. pylori* infection has been evaluated by immunoassay on stool samples from 249 subjects employed in a university teaching hospital according to three categories of hospital workers including personal from gastrointestinal endoscopy unit, personal and staff from other hospital units either with direct or no contact with patients. The results indicated that hospital work involving direct contact with patients constitutes a major risk factor for *H. pylori* contamination compared with hospital work without direct contact with patients (Mastromarino et al. 2005). Accordingly, nursing staff was also demonstrated as higher risk of *H. pylori* infection compared to administrative and technical staff (Triantafillidis et al. 2002). In addition, during the 5 years of following, a higher seroconversion for subjects initially *H. pylori* negative which became positive, was observed for the nursing staff category. It is interesting to notice, that also in that study, the level of education was inversely associated with the prevalence of the infection.

5 Conclusions

Even though recent research suggests that the prevalence of *H. pylori* infection trends to decrease in most of the countries, it remains high in most of developing countries as also illustrated with immigrant populations coming from countries with a high prevalence. The main risk factors remain associated with socioeconomic, familial living conditions and environmental factors. These data highlight the importance of the identification of population-specific risk factors as *H. pylori* reservoir that will allow to develop efficient preventive strategies to limit the prevalence of *H. pylori* infection among the most vulnerable populations.

References

Abdul Rahim NR, Benson J, Grocke K, Vather D, Zimmerman J, Moody T, Mwanri L (2017) Prevalence of *Helicobacter pylori* infection in newly arrived refugees attending the Migrant Health Service, South Australia. Helicobacter 22(2). https://doi.org/10.1111/hel.12360

Adams BL, Bates TC, Oliver JD (2003) *Helicobacter pylori* in a natural freshwater environment. Appl Environ Microbiol 69(12):7462–7466. https://doi.org/10.1128/AEM.69.12.7462-7466.2003

Aguemon BD, Struelens MJ, Massougbodji OEM (2005) Prevalence and risk-factors for *Helicobacter pylori* infection in urban and rural Beninese populations. Clin Microbiol Infect 11(8):611–617. https://doi.org/10.1111/j.1469-0691.2005.01189.x

Aguilera-Correa JJ, Urruzuno P, Barrio J, Martinez MJ, Agudo S, Somodevilla A, Liorca L, Alarcon T (2017) Detection of *Helicobacter pylori* and the genotypes of resistance to clarithromycin and the heterogeneous genotype to this antibiotic in biopsies obtained from symptomatic children. Diag Microbiol Infect Dis 87 (2):150–153. https://doi.org/10.1016/j.diagmicrobio.2016.03.001

Ahn J, Sinha R, Pei Z, Dominianni C, Wu J, Shi J, Goedert IJ, Hayes RB, Yang L (2013) Human gut microbiome and risk for colorectal cancer. J Natl Cancer Inst 105 (24):1907–1911. https://doi.org/10.1093/jnci/djt300

Amaral O, Fernandes I, Veiga N, Pereira C, Chaves C, Nelas P, Silva D (2017) Living conditions and *Helicobacter pylori* in adults. Biomed Res Int 2017:9082716. https://doi.org/10.1155/2017/9082716

Andersen LP, Wewer AV, Christiansen KM, Tvede M, Hansen JP, Henriksen FW, Krasilnikoff PA (1994) The humoral immune response to *Helicobacter pylori* infection in children with recurrent abdominal pain. APMIS 102(6):457–464

Angelidis AS, Tirodimos I, Bobos M, Kalamaki MS, Papageorgiou DK, Arvanitidou M (2011) Detection of *Helicobacter pylori* in raw bovine milk by fluorescence in situ hybridization (FISH). Int J Food Microbiol 151(2):252–256. https://doi.org/10.1016/j.ijfoodmicro.2011.09.007

Arnold IC, Dehzad N, Reuter S, Martin H, Becher B, Taube C, Müller A (2011) *Helicobacter pylori* infection prevents allergic asthma in mouse models through the induction of regulatory T cells. J Clin Invest 121:3088–3093. https://doi.org/10.1172/JCI45041

Asgeirsdottir GA, Kjartansdottir I, Olafsdottir AS, Hreinsson JP, Hrafnkelsson H, Johannsson E, Björnsson ES (2017) *Helicobacter pylori* infection in Icelandic children. Scand J Gastroenterol 52 (6-7):686–690. https://doi.org/10.1080/00365521.2017.1304986

Assaad S, Chaaban R, Tannous F, Costanian C (2018) Dietary habits and *Helicobacter pylori* infection: a cross sectional study at a Lebanese hospital. BMC Gastroenterol 18(1):48. https://doi.org/10.1186/s12876-018-0775-1

Azevedo NF, Almeida C, Fernandes I, Cerqueira L, Dias S, Keevil CW, Vieira JM (2008) Survival of gastric and enterohepatic helicobacter spp. in water. implications for transmission. Appl Environ Microbiol 74 (6):1805–1811. https://doi.org/10.1128/AEM.02241-07

Aziz RK, Khalifa MM, Sharaf RR (2015) Contaminated water as a source of *Helicobacter pylori* infection : a review. J Adv Res 6:539–547. https://doi.org/10.1016/j.jare.2013.07.007

Bastos J, Peleteiro B, Pinto H, Marinho A, Guimaraes JT, Ramos E, La Vecchia C, Barros H, Lunet N (2013) Prevalence, incidence and risk factors for *Helicobacter pylori* infection in a cohort of Portuguese adolescents (EpiTeen). Dig Liver Dis 45(4):290–295. https://doi.org/10.1016/j.dld.2012.11.009

Best LM, Takwoingi Y, Siddique S, Selladurai A, Gandhi A, Low B, Yaghoobi M, Gurusamy KS (2018) Non-invasive diagnostic tests for *Helicobacter pylori* infection. Cochren Database Syst Rev 3: CD012080. https://doi.org/10.1002/14651858.CD012080.pub2

Biernat MM, Iwanczak B, Binkowska A, Grabinska J, Gosciniak G (2016) The prevalence of *Helicobacter pylori* infection in symptomatic children: a 13-year observational study in the lower Silesian region. Adv Clin Exp Med 25(2):303–308. https://doi.org/10.17219/acem/44372

Boehnke KF, Eaton KA, Valdivieso M, Baker LH, Xi C (2015) Animal model reveals potential waterborne transmission of *Helicobacter pylori* infection. Helicobacter 20(5):326–333. https://doi.org/10.1111/hel.12216

Boehnke KF, Eaton KA, Fontaine C, Brewster R, Wu J, Eisenberg JNS, Valdivieso M, Baker LH, Xi C (2017) Reduced infectivity of waterborne viable but nonculturable *Helicobacter pylori* strain SS1 in mice. Helicobacter 22(4). https://doi.org/10.1111/hel.12391

Bontems P, Devaster JM, Corvaglia L, Dezsöfi A, Van Den Borre C, Goutier S, Butzler JP, Cadranel S (2001) Twelve year observation of primary and secondary antibiotic-resistant *Helicobacter pylori* strains in children. Pediatr Infect Dis J 20(11):1033–1038

Burucoa C, Delchier JC, Courillon-Mallet A, de Korwin JD, Mégraud F, Zerbib F, Raymond J, Fauchère JL (2013) Comparative evaluation of 29 commercial *Helicobacter pylori* serological kits. Helicobacter 18 (3):169–179. https://doi.org/10.1111/hel.12030

Cadranel S, Corvaglia L, Bontems P, Deprez C, Glupczynski Y, Van Riet A, Keppens E (1998) Detection of *Helicobacter pylori* infection in children with a standardized and simplified 13C-urea breath test. J Pediatr Gastroenterol Nutr 27(3):275–280

Capelle LG, De Vries AC, Haringsma J, Ter Borg F, de Vries RA, Bruno MJ, van Dekken H, Meijer J, van Grieken NC, Kuipers EJ (2010) The staging of gastritis with the OLGA system by using intestinal metaplasia as an accurate alternative for atrophic gastritis. Gastrointest Endosc 71(7):1150–1158. https://doi.org/10.1016/j.gie.2009.12.029

Cardenas VM, Mulla ZD, Ortiz M, Graham DY (2006) Iron deficiency and *Helicobacter pylori* infection in the United States. Am J Epidemiol 163(2):127–134. https://doi.org/10.1093/aje/kwj018

Carreira H, Bastos A, Peleteiro B, Lunet N (2015) Breastfeeding and *Helicobacter pylori* infection: systematic review and meta-analysis. Public Health Nutr 18 (3):500–520. https://doi.org/10.1017/S1368980014000500

Chan FT, Mackenzie AM (1986) Evaluation of primary selective media and enrichment methods for *Campylobacter* species isolation. Eur J Clin Microbiol 5 (2):162–164

Correa P (1992) Human gastric carcinogenesis: a multistep and multifactorial process. First American Cancer Society Award Lecture on Cancer Epidemiology and Prevention. Cancer Res 52(24):6735–6740

Correa P, Piazuelo MB (2011) *Helicobacter pylori* infection and gastric adenocarcinoma. US. Gastroenterol Hepatol Rev 7(1):59–64

Corvaglia L, Bontems P, Devaster JM, Heimann P, Glupczynski Y, Keppens E, Cadranel S (1999) Accuracy of serology and 13C-urea breath test for detection of *Helicobacter pylori* in children. Pediatr Infect Dis J 18(11):976–979

Daugule I, Karklina D, Rudzite D, Remberga S, Rumba-Rozenfelde I (2016) Prevalence of *Helicobacter pylori* infection among preschool children in Latvia: no significant decrease in prevalence during a ten year period. Scand J Public Health 44(4):418–422. https://doi.org/10.1177/1403494816631861

Dixon MF, Genta RM, Yardley JH, Correa P (1996) Classification and grading of gastritis. The updated Sydney System. International Workshop on the Histopathology of Gastritis, Houston 1994. Am J Surg Pathol 20 (10):1161–1181

Dore MP, Bilotta M, Vaira D, Manca A, Massarelli G, Leandro G, Atzei A, Pisanu G, Graham DY, Realdi G (1999a) High prevalence of *Helicobacter pylori* infection in shepherds. Dig Dis Sci 44(6):1161–1164

Dore MP, Sepulveda AR, Osato MS, Realdi G, Graham DY (1999b) *Helicobacter pylori* in sheep milk. Lancet 354(9173):132. https://doi.org/10.1016/S0140-6736(99)01724-9

Dore MP, Sepulveda AR, Al-Zimaity H, Yamaoka Y, Osato MS, Mototsugu K, Nieddu AM, Realdi G, Graham DY (2001) Isolation of *Helicobacter pylori* from sheep-implications for transmission to humans. Am J Gastroenterol 96(5):1396–1401. https://doi.org/10.1111/j.1572-0241.2001.03772.x

Dore MP, Malaty HM, Graham DY, Fanciulli G, Delitalia G, Realdi G (2002) Risk Factors Associated with *Helicobacter pylori* infection among children in a defined geographic area. Clin Infect Dis 35 (3):240–245. https://doi.org/10.1086/341415

Douraghi M, Nateghi Rostami M, Goudarzi H, Ghalavand Z (2013) Comparison of stool antigen immunoassay and serology for screening for *Helicobacter pylori* infection in intellectually disabled children. Microbiol Immunol 57(11):772–777. https://doi.org/10.1111/1348-0421.12099

Eslick GD, Lim LL, Byles JE, Xia HH, Talley NJ (1999) Association of *Helicobacter pylori* infection with gastric carcinoma: a meta-analysis. Am J Gastroenterol 94 (9): 2373-2379. https://doi.org/10.1111/j.1572-0241.1999.01360.x

Everhart JE, Kruszon-Moran D, Perez-Perez GI, Tralka TS, McQuillan G (2000) Seroprevalence and ethnic differences in *Helicobacter pylori* infection among adults in the United States. J Infect Dis 181:1359–1363. https://doi.org/10.1086/315384

Fallone CA, Chiba N, Van Zanten S, Fischbach L, Gisbert JP, Hunt RH, Jones NL, Render C, Leontiadis GI, Moayyedi P, Marshall JK (2016) The Toronto consensus for the treatment of *Helicobacter pylori* Infection in Adults. Gastroenterology 151(1):51–69. https://doi.org/10.1053/j.gastro.2016.04.006

Fernandes RM, Silva H, Oliveira R, Almeida C, Azevedo NF, Vieira MJ (2017) Morphological transition of *Helicobacter pylori* adapted to water. Future Microbiol 12(13):1167–1179. https://doi.org/10.2217/fmb-2016-0174

Ferreira RM, Pereira-Marques J, Pinto-Ribeiro I, Costa JL, Carneiro F, Machado JC, Figueiredo C (2018) Gastric microbial community profiling reveals a dysbiotic cancer associated microbiota. Gut 67(2):226–236. https://doi.org/10.1136/gutjnl-2017-314205

Fujimura S, Kawamura T, Kato S, Tateno H, Watanabe A (2002) Detection of *Helicobacter pylori* in cow's milk. Lett Appl Microbiol 35(6):504–507. https://doi.org/10.1046/j.1472-765X.2002.01229.x

Fujimura S, Kato S, Watanabe A (2008) Water source as a *Helicobacter pylori* transmission route: a 3-year follow-up study of Japanese children living in a unique district. J Med Microbiol 57(Pt 7):909–910. https://doi.org/10.1099/jmm.0.47683-0

Garcia A, Salas-Jara MJ, Herrera C, Gonzalez C (2014) Biofilm and *Helicobacter pylori*: from environment to human host. World J Gastroenterol 20(19):5632–5638. https://doi.org/10.3748/wjg.v20.i19.5632

Gevers D, Kugathasan S, Denson LA, Vazquez-Baeza Y, Van Treuren W, Ren B, Schwager E, Knights D, Song SJ, Yassour M, Morgan XC, Kostic AD, Luo C, Gonzalez A, McDonald D, Haberman Y, Walters T, Baker S, Rosh J, Stephens M, Heyman M, Markowitz J, Baldassano R, Griffiths A, Sylvester F, Mack D, Kim S, Crandall W, Hyams J, Huttenhower C, Knight R, Xavier RJ (2014) The treatment-naive microbiome in new onset Crohn's disease. Cell Host Microbe 15(3):382–392. https://doi.org/10.1016/j.chom.2014.02.005

Goodman KJ, Correa P (2000) Transmission of *Helicobacter pylori* among siblings. Lancet 355 (9201):358–362. https://doi.org/10.1016/S0140-6736(99)05273-3

Goodman KJ, Correa P, Tengana Aux HJ, Ramirez H, DeLany JP, Guerrero Pepinosa O, Lopez Quinones M, Collazos Parra T (1996) *Helicobacter pylori* infection in the Colombian Andes: a population-based study of transmission pathways. Am J Epidemiol 144(3):290–299

Goodman KJ, Correa P, Tengana Aux HJ, DeLany JP, Collazos T (1997) Nutritional factors and *Helicobacter pylori* infection in Colombian children. J Pediatr Gastroenterol Nutr 25(5):507–515

Graham DY, Klein PD, Evans DJ Jr, Evans DG, Alpert LC, Opekun AR, Boutton TW (1987) *Campylobacter pylori* detected noninvasively by the 13C-urea breath test. Lancet 1(8543):1174–1177. https://doi.org/10.1016/S0140-6736(87)92145-3

Graham DY, Lu H, Yamaoka Y (2009) African, Asian or Indian enigma, the East Asian *Helicobacter pylori*: facts or medical myths. J Dig Dis 10(2):77–84. https://doi.org/10.1111/j.1751-2980.2009.00368.x

Guarner J, Kalach N, Elitsur Y, Koletzko S (2010) *Helicobacter pylori* diagnostic tests in children: review of the literature from 1999 to 2009. Eur J Pediatr 169 (1):15-25. https://doi.org/10.1007/s00431-009-1033-x

Hamrah MH, Hamrah MS, Hassan Hamrah M, Kanda M, Hamrah AE, Dahi AE, Homayoun F, Hamrah MH, Fotouhi A, Sakamoto J (2017) Prevalence of *Helicobacter pylori* infection in dyspeptic patients in Andkhoy Afghanistan. Asian Pac J Cancer Prev 18 (11):3123–3127. https://doi.org/10.22034/APJCP.2017.18.11.3123

Hasosah M, Satti M, Shehzad A, Alsahafi A, Sukkar G, Alzaben A, Sunaid A, Ahmed A, AlThubiti S, Mufti A, Jacobson K (2015) Prevalence and risk factors of *Helicobacter pylori* infection in Saudi children: a three-year prospective controlled study. Helicobacter 20(1):56–63. https://doi.org/10.1111/hel.12172

Honar N, Minazadeh A, Shakibazad N, Haghighat M, Saki F, Javaherizadeh H (2016) Dagnostic accuracy of urea breath breath test for *Helicobacter pylori* infection in children with dyspepsia in comparison to histopathology. Arq Gastroenterol 53(2):108–112. https://doi.org/10.1590/S0004-28032016000200011

Hooi JKY, Lai WY, Ng WK, Suen MMY, Underwood FE, Tanyingoh D, Malfertheimer P, Graham DY, Wong VWS, Wu JCY, Chan FKL, Sung JJY, Kaplan GG,

Ng SC (2017) Global prevalence of *Helicobacter pylori* infection: systematic review and meta-analysis. Gastroenterology 153(2):420–429. https://doi.org/10.1053/j.gastro.2017.04.022

Horiuchi T, Ohkusa T, Watanabe M, Kobayashi D, Miwa H, Eishi Y (2001) *Helicobacter pylori* DNA in drinking water in Japan. Microbiol Immunol 45 (7):515–519. https://doi.org/10.1111/j.1348-0421.2001.tb02652.x

Hulten K, Hans SW, Enroth H, Klein PD, Opekun AR, Gilman RH, Evans DG, Engstrand L, Graham DY, El-Zaatari FA (1996) *Helicobacter pylori* in the drinking water in Peru. Gastroenterology 110 (4):1031–1035. https://doi.org/10.1053/gast.1996.v110.pm8612990

IARC (1994) Schistosomes, liver flukes and *Helicobacter pylori*. IARC Working Group on the Evaluation of Carcinogenic Risks to Humans. Lyon, 7–14 June 1994. IARC Monogr Eval Carcinog Risks Hum 61:1–241

Jones NL, Koletzko S, Goodman K, Bontems P, Cadranel S, Casswall T, Czinn S, Gold BD, Guarner J, Elitsur Y, Homan M, Kalach N, Kori M, Madrazo A, Mégraud F, Papadopoulo A, Rowland M, ESPGHAN NASPGHAN (2017) Joint ESPGHAN/NASPGHAN Guidelines for the Management of *Helicobacter pylori* in Children and Adolescents (Update 2016). J Pediatr Gastroenterol Nutr 64 (6):991–1003. https://doi.org/10.1097/MPG.0000000000001594

Kalach N, Gosset P, Dehecq E, Decoster A, Spyckerelle C, Papadopolos S, Dupont C, Raymond J (2015) Usefulness of gastric biopsy-based real-time polymerase chain reaction for the diagnosis of *Helicobacter pylori* infection in children. J Pediatr Gastroenterol Nutr 61 (3):307–312. https://doi.org/10.1097/MPG.0000000000000787

Kalach N, Gosset P, Dehecq E, Decoster A, Georgel AF, Spyckerelle C, Papadopoulo A, Dupont C, Raymond J (2017) A one-step immune-chromatographic *Helicobacter pylori* stool antigen test for children was quick, consistent, reliable and specific. Acta Paediatr 106(12):2025–2030. https://doi.org/10.1111/apa.13968

Kienesberger S, Cox LM, Livanos A, Zhang XS, Chung J, Perez-Perez G, Gorkiewicz G, Zechner EL, Blaser MJ (2016) Gastric *Helicobacter pylori* infection affects local and distant microbial populations and host responses. Cell Reports 14:1395–1407. https://doi.org/10.1016/j.celrep.2016.01.017

Kikuchi S, Kurosawa M, Sakiyama T (1998) *Helicobacter pylori* risk associated with sibship size and family history of gastric diseases in Japanese adults. Jpn J Cancer Res 89(11):1109-1112. https://doi.org/10.1111/j.1349-7006.1998.tb00504.x

Kodaman N, Pazos A, Schneider BG, Piazuelo MB, Mera R, Sobota RS, Sicinschi LA, Shaffer CL, Romero-Gallo J, de Sablet T, Harder RH, Bravo LE, Peek RM Jr, Wilson KT, Cover TL, Williams SM, Correa P (2014) Human and *Helicobacter pylori* coevolution shapes the risk of gastric disease. Proc Natl Acad Sci U S A 111(4):1455–1460. https://doi.org/10.1073/pnas.1318093111

Koletzko S, Richy F, Bontems P, Crone J, Kalach N, Monteiro ML, Gottrand F, Celinska-Cedro D, Roma-Giannikou E, Orderda G, Kolacek S, Urruzuno P, Martinez-Gomez MJ, Casswall T, Ashorn M, Bodanszky H, Mégraud F (2006) Prospective multicentre study on antibiotic resistance of *Helicobacter pylori* strains obtained from children living in Europe. Gut 55(12):1711–1716. https://doi.org/10.1136/gut.2006.091272

Kusano C, Gotoda T, Ishikawa MM (2017) The administrative project of *Helicobacter pylori* infection screening among junior high school students in an area of Japan with a high incidence of gastric cancer. Gastric Cancer 20(Suppl 1):16–19. https://doi.org/10.1007/s10120-017-0688-7

Lash RH, Genta RM (2016) Routine Anti-*Helicobacter* immunohistochemical staining is significantly superior to reflex staining protocols for the detection of *Helicobacter* in gastric biopsy specimens. Helicobacter 21(6):581–585. https://doi.org/10.1111/hel.12315

Leal YA, Flores LL, Fuentes-Panana EM, Cedillo-Rivera-R, Torres J (2011) 13C-urea breath test for the diagnosis of *Helicobacter pylori* infection in children: a systematic review and meta-analysis. Helicobacter 16 (4):327–337. https://doi.org/10.1111/j.1523-5378.2011.00863.x

Makristathis A, Pasching E, Schutze K, Wimmer M, Rotter ML, Hirschl AM (1998) Detection of *Helicobacter pylori* in stool specimens by PCR and antigen enzyme immunoassay. J Clin Microbiol 36(9):2772–2774

Malaty HM, Evans DG, Evans DJ Jr, Graham DY (1992) *Helicobacter pylori* in Hispanics: comparison with blacks and whites of similar age and socioeconomic class. Gastroenterology 103(3):813–816. https://doi.org/10.1016/0016-5085(92)90011-M

Malfertheiner P, Mégraud F, O'Morain CA, Gisbert JP, Kuipers EJ, Axon AT, Bazzoli F, Gasbarrini A, Atherton J, Graham DY, Hunt R, Moayyedi P, Rokkas T, Rugge M, Selgrad M, Suerbaum S, Sugano K, El-Omar EM, European Helicobacter and Microbiota Study Group and Consensus Panel (2017) Management of *Helicobacter pylori* infection-the Maastricht V/Florence consensus report. Gut 66 (1):6–30. https://doi.org/10.1136/gutjnl-2016-312288

Mana F, Vandebosch S, Miendje Deyi V, Haentjens P, Urbain D (2013) Prevalence of and risk factors for *H. pylori* infection in healthy children and young adults in Belgium anno 2010/2011. Acta Gastroenterol Belg 76(4):381–385

Mastromarino P, Conti C, Donato K, Strappini PM, Cattaruzza MS, Orsi GB (2005) Does hospital work constitute a risk factor for *Helicobacter pylori* infection? J Hosp Infect 60:261–268. https://doi.org/10.1016/j.jhin.2004.12.019

McNulty CA, Wise R (1985) Rapid diagnosis of *Campylobacter*-associated gastritis. Lancet 1 (8443):1443–1444. https://doi.org/10.1016/S0140-6736(85)91865-3

Mendall MA, Goggin PM, Molineaux N, Levy J, Toosy T, Strachan D, Northfield TC (1992) Childhood living conditions and *Helicobacter pylori* seropositivity in adult life. Lancet 339:896–897. https://doi.org/10.1016/0140-6736(92)90931-R

Miendje Deyi VY, Bontems P, Vanderpas J, De Koster E, Nitounda R, Van den Borre C, Cadranel S, Burette A (2011) Multicenter survey of routine determinations of resistance of *Helicobacter pylori* to antimicrobials over the last 20 years (1990 to 2009) in Belgium. J Clin Microbiol 49(6):2200–2209. https://doi.org/10.1128/JCM.02642-10

Moayyedi P, Soo S, Deeks J, Delaney B, Harris A, Innes M, Oakes R, Wilson S, Roalfe A, Bennett C, Forman D (2006) Eradication of *Helicobacter pylori* for non-ulcer dyspepsia. Cochrane Database Syst Rev (19, 2):Cd002096. https://doi.org/10.1002/14651858.CD002096.pub4

Momtaz H, Dabiri H, Souod N, Gholami M (2014) Study of *Helicobacter pylori* genotype status in cows, sheep, goats and human beings. BMC Gastroenterol 14(61). https://doi.org/10.1186/1471-230X-14-61

Moosazadeh M, Lankarani KB, Afshari M (2016) Meta-analysis of the prevalence of *Helicobacter pylori* infection among children and adults of Iran. Int J Prev Med 7:48. https://doi.org/10.4103/2008-7802.177893

Morais S, Costa AR, Ferro A, Lunet N, Peleteiro B (2017) Contemporary migration patterns in the prevalence of *Helicobacter pylori* infection: a systematic review. Helicobacter 22(3). https://doi.org/10.1111/hel.12372

Moreno Y, Ferrus MA (2012) Specific detection of cultivable *Helicobacter pylori* cells from wastewater treatment plants. Helicobacter 17(5):327–332. https://doi.org/10.1111/j.15235378.2012.00961.x

Mousavi S, Dehkordi FS, Rahimi E (2014) Virulence factors and antibiotic resistance of *Helicobacter pylori* isolated from raw milk and unpasteurized dairy products in Iran. J Venom Anim Toxins Incl Trop Dis 20:51. https://doi.org/10.1186/1678-9199-20-51

Murray LJ, McCrum EE, Evans AE, Bamford KB (1997) Epidemiology of *Helicobacter pylori* infection among 4742 randomly selected subjects from Northern Ireland. Int J Epidemiol 26:880–887

Nakayama Y, Lin Y, Hongo M, Hidaka H, Kikuchi S (2017) *Helicobacter pylori* infection and its related factors in junior high school students in Nagano Prefecture, Japan. Helicobacter 22(2). https://doi.org/10.1111/hel.12363

Nguyen TVH, Nguyen VB, Binh Phan TT et al (2016) Epidemiology of *Helicobacter pylori* infection in Tay Children in Vietnam. Ann Clin Lab Res 04:1–9. Available from: http://www.aclr.com.es/clinical-research/epidemiology-of-helicobacter-pylori-infection-in-tay-children-in-vietnam.pH. pylori?aid=17244

Ogihara A, Kikuchi S, Hasegawa A, Kurosawa M, Miki K, Kaneko E, Mizukoshi H (2000) Relationship between *Helicobacter pylori* infection and smoking and drinking habits. J Gastroenterol Hepatol 15(3):271–276

Okuda M, Miyashiro E, Koike M, Tanaka T, Bouoka M, Okuda S, Yoshikawa N (2002) Serodiagnosis of *Helicobacter pylori* infection is not accurate for children aged below 10. Pediatr Int 44(4):387–390. https://doi.org/10.1046/j.1442-200X.2002.01585.x

Osaki T, Mabe K, Hanawa T, Kamiya S (2008) Urease-positive bacteria in the stomach induce a false-positive reaction in a urea breath test for diagnosis of *Helicobacter pylori* infection. J Med Microbiol 57(Pt 7):814-819. https://doi.org/10.1099/jmm.0.47768-0

Osaki T, Okuda M, Ueda J, Konno M, Yonezawa H, Hojo F, Yagyu K, Lin Y, Fukuda Y, Kikuchi S, Kamiya S (2013) Multilocus sequence typing of DNA from faecal specimens for the analysis of intra-familial transmission of *Helicobacter pylori*. J Med Microbiol 62:761–765. https://doi.org/10.1099/jmm.0.053140-0

Osaki T, Zaman C, Yonezawa H, Lin Y, Okuda M, Nozaki E, Hojo F, Kurata S, Hanawa T, Kikuchi S, Kamiya S (2018) Influence of intestinal indigenous microbiota on intrafamilial infection by *Helicobacter pylori* in Japan. Front Immunol 9(287). https://doi.org/10.3389/fimmu.2018.00287

Ozaydin N, Turkyilmaz SA, Cali S (2013) Prevalence and risk factors of *Helicobacter pylori* in Turkey: a nationally-representative, cross-sectional, screening with the 13C-Urea breath test. BMC Public Health 13 (1215). https://doi.org/10.1186/1471-2458-13-1215

Papiez D, Konturek PC, Bielanski W, Plonka M, Dobrzanska M, Kaminska A, Szczyrk U, Bochenek A, Wierzchos E (2003) Prevalence of *Helicobacter pylori* infection in Polish shepherds and their families. Dig Liver Dis 35(1):10–15. https://doi.org/10.1016/S1590-8658(02)00004-X

Raj P, Thompson JF, Pan DH (2017) *Helicobacter pylori* serology testing is a useful diagnostic screening tool for symptomatic inner city children. Acta Paediatr 106 (3):470–477. https://doi.org/10.1111/apa.13724

Raymond J, Kalach N, Bergeret M, Barbet JP, Benhamou PH, Gendrel D, Dupont C (1996) Evaluation of a serological test for diagnosis of *Helicobacter pylori* infection in children. Eur J Clin Microbiol Infect Dis 15(5):415–417

Raymond J, Thiberge JM, Kalach N, Bergeret M, Dupont C, Labigne A, Dauga C (2008) Using macro-arrays to study routes of infection of *Helicobacter pylori* in three families. PLoS One 3(5):e2259. https://doi.org/10.1371/journal.pone.0002259

Rodriguez JM, Murphy K, Stanton C, Ross RP, Kober OI, Juge N, Avershina E, Rudi K, Narbad A, Janmalm MC, Marchesi JR, Collado MC (2015) The composition of the gut microbiota throughout life, with an emphasis on early life. Microb Ecol Health Dis 26 (26050). https://doi.org/10.3402/mehd.v26.26050

Rugge M, Meggio A, Pennelli G, Piscioli F, Giacomelli L, De Pretis G, Graham DY (2007) Gastritis staging in clinical practice: the OLGA staging system. Gut 56 (5):631–636. https://doi.org/10.1136/gut.2006.106666

Selgrad M, Tammer I, Langner C, Bornschein J, Meißle J, Kandulski A, Varbanova M, Wex T, Schlüter D, Malfertheimer P (2014) Different antibiotic susceptibility between antrum and corpus of the stomach, a possible reason for treatment failure of *Helicobacter pylori* infection. World J Gastroenterol 20 (43):16245–16251. https://doi.org/. https://doi.org/10.3748/wjg.v20.i43.16245

Senbanjo IO, Oshikoya KA, Njokanma OF (2014) *Helicobacter pylori* associated with breastfeeding, nutritional status and recurrent abdominal pain in healthy Nigerian children. J Infect Dev Ctries 8 (4):448–453. https://doi.org/10.3855/jidc.3196

Shady MM, Fathy HA, Ali A, Galal EM, Fathy GA, Sibaii H (2015) Comparison of serum IgG antibody test with gastric biopsy for the detection of *Helicobacter pylori* infection among Egyptian children. Open Access Maced J Med Sci 3:303–306. https://doi.org/10.3889/oamjms.2015.062

Shu X, Ping M, Yin G, Jiang M (2017) Investigation of *Helicobacter pylori* infection among symptomatic children in Hangzhou from 2007 to 2014: a retrospective study with 12,796 cases. PeerJ 5:e2937. https://doi.org/10.7717/peerj.2937

Strebel K, Rolle-Kampczyck U, Richter M, Kindler A, Richter T, Schlink U (2010) A rigorous small area modelling-study for the *Helicobacter pylori* epidemiology. Sci Total Environ 408(18):3931–3942. https://doi.org/10.1016/j.scitotenv.2010.03.045

Suerbaum S, Michetti P (2002) *Helicobacter pylori* infection. N Engl J Med 347(15):1175–1186. https://doi.org/10.1056/NEJMra020542

Sykora J, Siala K, Varvarovska J, Pazdiora P, Pomahacova R, Huml M (2009) Epidemiology of *Helicobacter pylori* infection in asymptomatic children: a prospective population-based study from the Czech Republic. Application of a monoclonal-based antigen-in-stool enzyme immunoassay. Helicobacter 14(4):286–297. https://doi.org/10.1111/j.1523-5378.2009.00689.x

Thomas JE, Austin S, Dale A, McClean P, Harding M, Coward WA, Weaver LT (1993) Protection by human milk IgA against *Helicobacter pylori* infection in infancy. Lancet 342(8863):121. https://doi.org/10.1016/0140-6736(93)91327-I

Torres J, Perez-Perez G, Goodman KJ, Atherton JC, Gold BD, Harris PR, Ia Garza AM, Guarner J, Munoz O (2000) A comprehensive review of the natural history of *Helicobacter pylori* infection in children. Arch Med Res 31(5):431–469. https://doi.org/10.1016/S0188-4409(00)00099-0

Triantafillidis JK, Gikas A, Hyphantis T, Cheracakis P, Rokkas T, Konstantellou E, Antoniou A, Mallas H, Androulakis G (2002) *Helicobacter pylori* infection in hospital workers over a 5-year period: correlation with demographic and clinical parameters. J Gastroenterol 37(12):1005–1013. https://doi.org/10.1007/s005350200170

Uemura N, Okamoto S, Yamamoto S, Matsumura N, Yamaguchi S, Yamakido M, Taniyama K, Sasaki N, Schiemper RJ (2001) *Helicobacter pylori* infection and the development of gastric cancer. N Engl J Med 345 (11):784–789. https://doi.org/10.1056/NEJMoa001999

Urita Y, Watanabe T, Kawagoe N, Takemoto I, Tanaka H, Kijima S, Kido H, Maeda T, Sugasawa Y, Miyazaki T, Honda Y, Nakanishi K, Shimada N, Nakajima H, Sugimoto M, Urita C (2013) Role of infected grandmothers in transmission of *Helicobacter pylori* to children in a Japanese rural town. J Paediatr Child Health 49(5):394–398. https://doi.org/10.1111/jpc.12191

Vandenplas Y, Blecker U, Devreker T, Keppens E, Nijs J, Cadranel S, Pipeleers-Marichal M, Goossens A, Lauwers S (1992) Contribution of the 13C-urea breath test to the detection of *Helicobacter pylori* gastritis in children. Pediatrics 90(4):608–611

Westblom TU, Madan E, Gudipati S, Midkiff BR, Czinn SJ (1992) Diagnosis of *Helicobacter pylori* infection in adult and pediatric patients by using Pyloriset, a rapid latex agglutination test. J Clin Microbiol 30(1):96–98

Yahaghi E, Khamesipour F, Mashayekhi F, Safarpoor Dehkordi F, Sakhaei MH, Masoudimanesh M, Khameneie MK (2014) *Helicobacter pylori* in vegetables and salads: genotyping and antimicrobial resistance properties. Biomed Res Int 2014:757941. https://doi.org/10.1155/2014/757941

Zabala Torres B, Lucero Y, Lagomarcino AJ, Orellana-Manzano A, George S, Torres JP, O'Ryan M (2017) Review: Prevalence and dynamics of *Helicobacter pylori* infection during childhood. Helicobacter 22(5). https://doi.org/10.1111/hel.12399

Zamani M, Ebrahimtabar F, Zamani V, Miller WH, Alizadeh-Navaei R, Shokri-Shirvani J, Derakhshan MH (2018) Systematic review with meta-analysis: the worldwide prevalence of *Helicobacter pylori* infection. Aliment Pharmacol Ther 47(7):868–876. https://doi.org/10.1111/apt.14561

Zhu Y, Zhou X, Wu J, Su J, Zhang G (2014) Risk factors and prevalence of *Helicobacter pylori* Infection in persistent high incidence area of gastric carcinoma in Yangzhong city. Gastroenterol Res Pract 2014:481365. https://doi.org/10.1155/2014/481365

Adv Exp Med Biol - Advances in Microbiology, Infectious Diseases and Public Health (2019) 1149: 35–56
https://doi.org/10.1007/5584_2019_358

Published online: 24 April 2019

Activity and Functional Importance of *Helicobacter pylori* Virulence Factors

Dionyssios Sgouras, Nicole Tegtmeyer, and Silja Wessler

Abstract

Helicobacter pylori is a very successful Gram-negative pathogen colonizing the stomach of humans worldwide. Infections with this bacterium can generate pathologies ranging from chronic gastritis and peptic ulceration to gastric cancer. The best characterized *H. pylori* virulence factors that cause direct cell damage include an effector protein encoded by the cytotoxin-associated gene A (CagA), a type IV secretion system (T4SS) encoded in the *cag*-pathogenicity island (*cag* PAI), vacuolating cytotoxin A (VacA), γ-glutamyl transpeptidase (GGT), high temperature requirement A (HtrA, a serine protease) and cholesterol glycosyl-transferase (CGT). Since these *H. pylori* factors are either surface-exposed, secreted or translocated, they can directly interact with host cell molecules and are able to hijack cellular functions. Studies on these bacterial factors have progressed substantially in recent years. Here, we review the current status in the characterization of signaling cascades by these factors *in vivo* and *in vitro*, which comprise the disruption of cell-to-cell junctions, induction of membrane rearrangements, cytoskeletal dynamics, proliferative, pro-inflammatory, as well as, pro-apoptotic and anti-apoptotic responses or immune evasion. The impact of these signal transduction modules in the pathogenesis of *H. pylori* infections is discussed.

Keywords

E-cadherin · Protease · CagA · HtrA serine protease · VacA · UreA · Adherens junction · Tight junction · Epithelial barrier · Type IV secretion T4SS

1 Introduction

The human stomach represents a highly dynamic and hostile environment for bacteria, in which the gastric pathogen *H. pylori* encounters numerous stresses, including nutrient limitations, pH fluctuations or oxidative attack (Kusters et al. 2006). Gastric colonization by *H. pylori* commonly occurs in early childhood and can persist for the entire lifetime, unless it is eradicated by antimicrobial treatment. The bacterium is a major risk factor for the development of various gastric

D. Sgouras (✉)
Laboratory of Medical Microbiology, Hellenic Pasteur Institute, Athens, Greece
e-mail: sgouras@pasteur.gr; dionsgouras@gmail.com

N. Tegtmeyer
Division of Microbiology, Department of Biology, Friedrich Alexander University Erlangen-Nuremberg, Erlangen, Germany
e-mail: nicole.tegtmeyer@fau.de

S. Wessler
Division of Microbiology, Department of Biosciences, Paris-Lodron University of Salzburg, Salzburg, Austria
e-mail: silja.wessler@sbg.ac.at

diseases and severe disorders, such as peptic ulcer disease, that can develop in about 10–15%, or gastric malignancies in 1–2% of infected individuals; occurrence of these pathologies depends on complex host-pathogen interactions and correlates to the geography of individuals (Polk and Peek 2010; Yamaoka and Graham 2014). The presence of *H. pylori* in the stomach mucosa is commonly accompanied by strong inflammatory responses, however, several immune evasion strategies by the pathogen have been described (Mejias-Luque and Gerhard 2017) presenting a prime example of a chronic bacterial infection (Ramarao et al. 2000; Pachathundikandi et al. 2016). About half of the human world population is colonized by the pathogen, associated with chronic or asymptomatic gastritis in every infected person. The pathogen has evolved multiple mechanisms to colonize and persist within the human stomach despite the harsh acidic conditions confronted in this milieu (Robinson et al. 2017). *H. pylori* is highly adapted to the stomach and grows at pH ranges between 6 and 8. Physiological, biochemical and genetic studies of *H. pylori* have identified unique properties of its metabolism, some of which are crucial for the adaptation to the gastric environment (Kusters et al. 2006). Well-known pathogenicity-associated properties of *H. pylori* comprise flagella-mediated motility, urease-driven chemotaxis and neutralization of gastric pH, counteraction of antimicrobial nitric oxide production by arginase RocF and binding of the bacteria to gastric epithelial cells using several outer-membrane proteins; the latter adhesins include BabA/B, SabA, AlpA/B, OipA, HopZ, HopQ, and others (Gobert et al. 2001; Dubois and Borén 2007; Backert et al. 2011; Roure et al. 2012; Posselt et al. 2013; Huang et al. 2015; Naumann et al. 2017).

Genetic studies have shown that *Homo sapiens* has carried *H. pylori* for more than 100,000 years and DNA sequence characteristics of the bacteria were utilized as signatures to outline multifaceted demographic events in the history of mankind, including major human migration routes (Moodley and Linz 2009). Because of this long time of co-existence with humans, it was proposed that hosting of *H. pylori* may have been advantageous for its carrier (Atherton and Blaser 2009). In our modern civilization, however, the bacterium produces a strong burden of morbidity and mortality caused by malignancies such as gastric adenocarcinoma and mucosa-associated lymphoid tissue (MALT) lymphoma (Polk and Peek 2010; Figueiredo et al. 2017). Gastric cancer represents one of the highest incident malignancies on the planet, causing over 700,000 deaths annually (Ferlay et al. 2015). The clinical consequences of *H. pylori* infection are controlled by a very complicated setup of host-pathogen interactions. The infection and development of gastric diseases is dependent on multiple parameters, including environmental factors, genetic predisposition of the host and bacterial virulence determinants. For example, the stomach microbiota, various dietary aspects, as well as important micronutrients can influence and change the equilibrium between *H. pylori*'s endeavor as a pathogen or a commensal (Amieva and El-Omar 2008; Polk and Peek 2010; Yamaoka and Graham 2014). Moreover, specific single nucleotide polymorphisms (SNPs) have been discovered in pro-inflammatory and other immune-regulatory control genes of the human genome, including tumor necrosis factor, interleukin-1β, interleukin-8, Nod-like and toll-like receptors, which can account for an increased risk of developing gastric diseases induced by *H. pylori* (Amieva and El-Omar 2008). Commonly, *H. pylori* isolates are genetically extremely variable, and this diversity also includes the presence of virulence genes, revealing different degrees of pathogenicity that affects the severity of *H. pylori* infections. Molecular mechanisms evolved in *H. pylori* to challenge host defense instruments and causing disease are under vigorous examination, by numerous research labs worldwide. Here, we review the function and activity of the major *H. pylori* virulence factors *cag* PAI carrying T4SS and CagA, VacA, HtrA, GGT and CGT.

2 Assembly and Function of the *cag* PAI-Encoded T4SS

The *cag* PAI is a genetic locus of ~40 kilobase pairs in the *H. pylori* chromosome carrying up to 32 genes that was acquired from a yet unknown ancestor by horizontal DNA transfer (Covacci and Rappuoli 2000). The *cag* PAI is present in highly virulent (type-I) *H. pylori* isolates but typically absent in less virulent (type-II) strains. Functional studies have shown that the *cag* PAI encodes a T4SS, representing a syringe-like nanostructure, spanning the inner and outer membranes of the Gram-negative bacterium. T4SS assembly involves orthologs of all 12 VirB/VirD4 subunits that were first described for the prototype *Agrobacterium tumefaciens* apparatus, and about a dozen additional Cag PAI proteins, making this system clearly unique among other T4SSs as discussed elsewhere (Backert et al. 2015; Grohmann et al. 2018). Electron microscopy has been applied to visualize the T4SS core structure which is sized approximately 41 nm in diameter, and comprises a complex of the Cag3, CagT, CagM, CagX and CagY proteins (Frick-Cheng et al. 2016). This core structure is connected with an extracellular pilus appendage in the outer membrane, which establishes host cell contact (Kwok et al. 2007; Shaffer et al. 2011). The CagL, CagI, CagY and CagA proteins have been identified as pilus-linked factors and permit binding to the host receptor integrin $\alpha_5\beta_1$, which is necessary for T4SS functionality (Kwok et al. 2007; Jimenez-Soto et al. 2009; Barden et al. 2013). The integrin $\alpha_v\beta_5$ member was also found to be exploited by *H. pylori* to induce gastrin production in a T4SS-dependent fashion (Wiedemann et al. 2012). Various translocated effector molecules and signaling effects have been reported (Fig. 1). The T4SS injects effector protein CagA (Segal et al. 1999; Stein et al. 2000; Odenbreit et al. 2000; Asahi et al. 2000; Backert et al. 2000), peptidoglycan (Viala et al. 2004), chromosomal DNA (Varga et al. 2016) and heptose-1,7-bisphosphate (HBP) into epithelial target cells, which respectively can stimulate receptor Nod1, toll-like receptor-9, TRAF-interacting protein with FHA domain (TIFA), kinase AKAP and pro-inflammatory transcription factor NF-κB in infected epithelial cells (Viala et al. 2004; Varga et al. 2016; Gall et al. 2017; Stein et al. 2017; Zimmermann et al. 2017).

Interestingly, *H. pylori* also targets the carcinoembryonic antigen-related cell adhesion molecule (CEACAM) receptors by means of adhesin HopQ for bacterial adhesion and delivery of CagA (Fig. 1) (Javaheri et al. 2016; Koniger et al. 2016). It appears that HopQ exploits the CEACAM dimer interface for binding and this interaction is required for proper T4SS function of yet unknown nature, which is required for effective stomach colonization and subsequent gastric pathogenesis (Bonsor et al. 2018; Moonens and Remaut 2017). In addition, the T4SS itself can also interact with and activate specific other host cell receptors in a CagA-independent manner, including epidermal growth factor receptor members EGFR and Her2/Neu, leading to increase cellular proliferation, anti-apoptosis and bacterial survival (Keates et al. 2001; Saha et al. 2010; Tegtmeyer et al. 2010; Sierra et al. 2018). Furthermore, the T4SS stimulates the receptor tyrosine kinase c-Met, which induces epithelial cell migration and invasion by engaging phospholipase PLCγ and mitogen-activated kinases (Fig. 1) (Churin et al. 2003; Oliveira et al. 2006).

Early studies have shown that *H. pylori* can actively inhibit its phagocytosis through professional phagocytes (Ramarao et al. 2000). These antiphagocytosis characteristics possibly play an important role in immune escape of *H. pylori* and depend on a functional *cag* PAI, since isogenic T4SS mutants abolished this feature, but does not require CagA (Ramarao et al. 2000). In addition, the pathogen was described to change the phosphorylation state of histone H3 through a CagA-independent but T4SS-dependent mechanism involving the mitotic vaccinia-related kinase 1 and Aurora B (Fehri et al. 2009). Remarkably, in epithelial cells T4SS-positive bacteria can also stimulate the NF-κB-mediated induction of AID (a DNA-editing enzyme) that leads to the

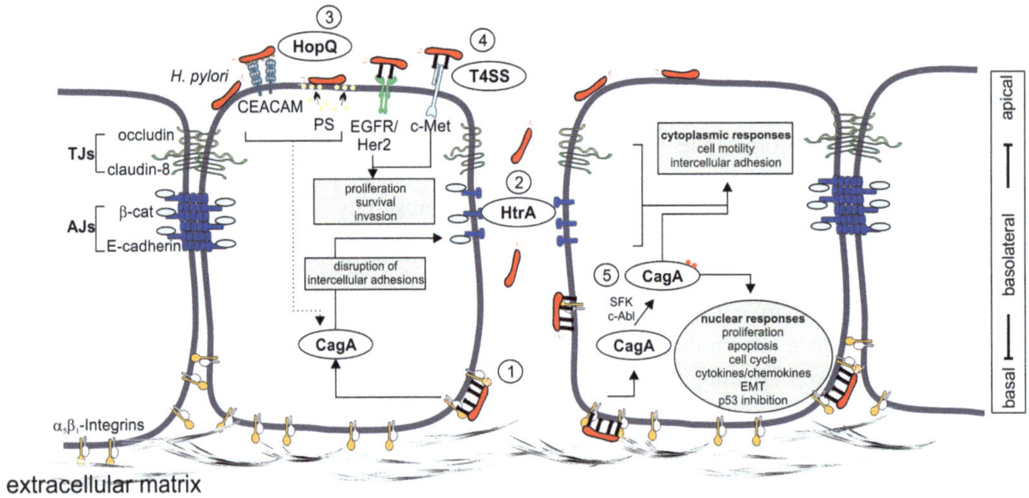

Fig. 1 T4SS-dependent effects in *H. pylori*-infected cells. Polarization of the gastric epithelium involves functional intercellular adhesion complexes, such as tight junctions (TJs) and adherens junctions (AJs). Disruption of the epithelium is facilitated by T4SS-positive *H. pylori* strains. The T4SS contact α5β1-integrins to inject the virulence factor CagA into the cytosol of infected cells (*1*), which can compete with β-catenin (β-cat) binding to the intracellular domain of the AJ cell adhesion molecule E-cadherin and contribute to the disruption of AJs. CagA translocation is enhanced by the *H. pylori*-secreted serine protease HtrA. HtrA cleaves off the extracellular domains of E-cadherin, occludin, and claudin-8 (*2*), which opens intercellular TJs and AJs. HtrA-mediated cleavage of adhesion molecules further allows binding of the T4SS to the α5β1-integrins at the basolateral domain of polarized cells. HopQ interaction with apically expressed CEACAMs is involved in efficient CagA injection (*3*). Furthermore, the T4SS can directly target receptors on the cell surface, including EGFR, Her2/Neu or c-Met, which is implicated in proliferation, cell survival and invasive growth (*4*). Cytoplasmic CagA is finally tyrosine-phosphorylated by kinases of the Src (SFK) and Abl family (*5*). Both phosphorylated and non-phosphorylated CagA induce changes in nuclear responses (e.g., proliferation, apoptosis, cell cycle arrest, synthesis of cytokines and chemokines, induction of EMT or p53 inhibition). Lastly, CagA may interfere with signaling pathways leading to cell motility, which might be facilitated by the disintegrated AJs and TJs

accumulation of mutations in p53, a well-known tumor suppressor protein (Matsumoto et al. 2007). Therefore, the activation of AID could represent a mechanism in which mutations in crucial genes could accumulate during infection and trigger gastric malignancy. Finally, the T4SS of *H. pylori* infection engages glycoprotein receptor gp130 (Lee et al. 2010), and the downstream activation of JAK2–STAT3 (Janus kinase–signal transducer and activator of transcription) signaling is linked to *H. pylori*-induced inflammation, which promotes carcinogenesis. Taken together, the T4SS located on the *cag* PAI exhibits remarkable features in its interactions with the host and is involved in causing gastric pathology. These data also demonstrate that *H. pylori* disrupts crucial cellular processes by one or more yet unidentified T4SS factors, which need to be identified in future studies.

3 CagA, a Multifunctional Master Key

CagA is an extraordinary protein of approximately 120–140 kDa, not sharing any sequence homology with other proteins known to date. It represents the most researched *H. pylori* virulence factor with over 3200 citations in PubMed (Backert and Blaser 2016). It was originally identified independently by two groups as an immunodominant protein of about 128 kDa in seropositive *H. pylori* carriers (Tummuru et al. 1993; Covacci et al. 1993). Subsequently, CagA-seropositivity in symptomatic patients was found to be associated with increased risk of gastric cancer (Blaser et al. 1995; Parsonnet et al. 1997). Its biological importance was further acknowledged when a number of research groups reported that CagA can be translocated into gastric epithelial cells, passing the membrane by means of T4SS (Covacci and Rappuoli 2000; Backert and Tegtmeyer 2017). Further work convincingly demonstrated that for the successful translocation of the CagA protein, interaction of a number of T4SS constituents with host receptor integrin α5β1 was necessary (Kwok et al. 2007; Jimenez-Soto et al. 2009; Barden et al. 2013). CagA itself can also bind to integrin α5β1 followed by its internalization into the host cell cytoplasm (Hayashi et al. 2012; Kaplan-Turkoz et al. 2012). *H. pylori* interaction with the host cell plasma membrane also includes direct binding of CagA to externalized membrane phosphatidylserine (PS), an event which is reported to be critical for CagA translocation (Fig. 1) (Murata-Kamiya et al. 2010). A partial crystal structure of N-terminal segments of the protein has been obtained (Hayashi et al. 2012; Kaplan-Turkoz et al. 2012), however, the entire C-terminal part of CagA is not yet crystallized. This part of the protein contains a number of Glu-Pro-Ile-Tyr-Ala (EPIYA)-sequence motifs which can be classified as EPIYA-A, EPIYA-B, EPIYA-C and EPIYA-D motifs, depending on their surrounding sequence (Hayashi et al. 2013). In *H. pylori* strains derived from Western countries, single EPIYA-A and EPIYA-B motifs have been reported, typically followed by one to four copies of EPIYA-C, whereas the combination of EPIYA-A and EPIYA-B with single EPIYA-D motifs has been predominantly identified in *H. pylori* isolates isolated in East-Asia (Xia et al. 2009). Strains with higher number of EPIYA-C motifs or presence of EPIYA-D have been associated with an increased risk for the development of gastric cancer (Argent et al. 2004; Jones et al. 2009; Li et al. 2017). However, the situation is not that straightforward. For instance, simultaneous infection with strains expressing diverse CagA EPIYA characteristics have been observed in adult patients (Panayotopoulou et al. 2010) and strains isolated from children do not exhibit multiple EPIYA-C motifs (Sgouras et al. 2009), suggesting that potential increments in the number of repeating EPIYA motifs in CagA occur throughout adulthood. Once intracellular, tyrosine moieties of the EPIYA motifs have been shown to be hierarchically phosphorylated by c-Src and c-Abl family host kinases (Mueller et al. 2012), thereby derailing the host cell function, effectively acting as a molecular "Trojan horse" (Covacci and Rappuoli 2000). How this deregulates downstream signaling processes was summarized in detail in other review articles (Backert et al. 2010; Senda and Hatakeyama 2016; Hatakeyama 2017; Berge and Terradot 2017; Tegtmeyer et al. 2017a). More specifically, a surprisingly high number of over 25 host cell factors have been reported to interact with CagA, in a manner that may or may not depend on EPIYA-phosphorylation, thereby suggesting that CagA can operate as a molecular master key (Backert et al. 2010). A number of key intracellular signaling pathways can be affected, relating to apoptosis and cell cycle proliferation, inflammatory response, cell motility and elongation, intercellular junction integrity or p53-inhibition (Backert et al. 2010; Hatakeyama 2017). Notable interacting targets of the "promiscuous" CagA protein have been identified in a phosphorylation-dependent manner for the SHP-2 phosphatase (Higashi et al. 2002) and in a phosphorylation-independent manner for the tight junction proteins JAM and ZO-1 (Amieva et al. 2003; Krueger et al. 2007), E-cadherin (Murata-Kamiya et al. 2007; Oliveira et al. 2009) and PAR-1 (Hayashi et al. 2012).

A more recent, holistic approach proposed that in order for *H. pylori* to control key host cell signal transduction functions, it injects the CagA protein which functions as a kinase pathway deregulator of a variety of serine/threonine and tyrosine kinases (Tegtmeyer et al. 2017a). These molecules are involved as both receptor- or non-receptor-mediated signaling elements; therefore, CagA seems to be able to manipulate a selection of fundamental cell processes such as adhesion, polarity, proliferation and motility, receptor mediated endocytosis, cytoskeletal rearrangements, apoptosis, inflammation, and cell cycle progression (Fig. 1) (Tegtmeyer et al. 2017a). CagA can accomplish such diverse strategies by activating or deactivating key kinase-dependent pathways. For instance, the Abl kinase was specifically reported to be activated by CagA (Poppe et al. 2007; Tammer et al. 2007), and so were the carboxy-terminal Src kinase (Csk) (Selbach et al. 2003; Tsutsumi et al. 2003; Selbach et al. 2009), the phosphatidylinositide 3-kinase (PI3K)/Akt pathway (Suzuki et al. 2009; Selbach et al. 2009; Wei et al. 2010; Zhang et al. 2015), the glycogen synthase kinase 3 (GSK-3) (Lee et al. 2014), the Janus kinase (JAK), a family of intracellular, non-receptor tyrosine kinases (Bauer et al. 2012), the Focal adhesion kinase (FAK) (Tegtmeyer et al. 2011), the atypical Protein Kinase C (aPKC) associated with junctional and polarity defects (Saadat et al. 2007; Zeaiter et al. 2008) and MAP kinases. On the other hand, CagA-dependent inactivation has been described for Src kinases (Selbach et al. 2003; Tsutsumi et al. 2003), the partitioning-defective Par1 kinase (Saadat et al. 2007) and the protein kinase C-related kinase 2 (PRK2) (Mishra et al. 2015). Further to GSK-3 targeting, translocated CagA has been suggested to induce epithelial mesenchymal transition (EMT) through EPIYA phosphorylation-dependent up-regulation of metalloprotease MMP-3 (Sougleri et al. 2016). In a phosphorylation-independent manner, translocated CagA has been demonstrated to promote survival of the infected epithelial cells by subverting pro-apoptotic signaling, leading to CagA-dependent p53 degradation (Tsang et al. 2010; Wei et al. 2015, 2010; Buti et al. 2011). Taken together, the CagA protein, following its endocytic translocation, can interact with a number of cellular elements, thus interfering with multiple cell functions and thereby exhibiting a versatile role in *H. pylori* pathogenesis. The elucidation of the exact molecular mechanisms and signaling of these interactions will benefit from structural studies of respective complexes involving full length CagA protein.

The application of animal models of *H. pylori* infection have further highlighted the important role that CagA may play in pathogenesis, as introduction of CagA-positive *H. pylori* into Mongolian gerbils has shown to induce gastric dysplasia and adenocarcinoma, through β-catenin activation and its nuclear accumulation, following CagA translocation (Franco et al. 2005). Further evidence on CagA tumorigenicity was provided following transgenic expression of CagA in C57BL/6J mice, under the control of the β subunit gene promoter of mouse H^+/K^+-ATPase, which resulted in abnormal proliferation of gastric epithelial and hematopoietic cells, thus contributing to the development of gastrointestinal carcinomas and leukemias/lymphomas, in a tyrosine phosphorylation-dependent manner (Ohnishi et al. 2008). Similar observations of the activation of pathways related to oncogenic potential were further supported by other transgenic model systems, including a model using *Drosophila* (Wandler and Guillemin 2012) and another with zebrafish (Neal et al. 2013).

Despite the plethora of reports describing the molecular mechanisms by which CagA can contribute to the bacterial pathogenesis, no clinical recommendations exist with regards to CagA subtyping in the management of patients (Malfertheiner et al. 2017; Chey et al. 2017), although CagA antibodies, which remain positive for a very long period of time, have been suggested to allow detection of *H. pylori* infection in gastric cancer patients when other tests are negative (Malfertheiner et al. 2017). Recent evidence provides further intriguing clues on the complex biology of CagA with regards to its clinical importance, as CagA translocation within gastric epithelial cells has been shown to be dependent on the levels of bacterial hydrogen metabolism. Clinical strains isolated from cancer patients seem to harbor significantly higher hydrogenase activity compared to those derived from patients with gastritis, thereby proposing an association between *H. pylori* hydrogenase activity and gastric carcinogenesis in humans (Wang et al. 2016). Finally, with regards to a role of CagA in pathogenicity, recent evidence suggests that variation in *cagA* gene copy numbers may serve as a novel mechanism by which *H. pylori* can modulate gastric disease development: a considerable proportion of *H. pylori* clinical strains harbor multiple *cagA* copies, which can be differentially associated with gastric disease (Jang et al. 2017). In summary, CagA will continue to intrigue by its mechanistic versatility and fascinating complexity of the evolutionary advantage it may confer to *H. pylori* pathogenesis.

4 *H. pylori* Secretes the Serine Protease HtrA to Shape the Epithelial Barrier

Depolarization of the epithelium represents a hallmark of *H. pylori*-induced gastric carcinogenesis and involves manifold complex pathogen-host interactions that have been summarized in several other review articles (Posselt et al. 2013; Wroblewski and Peek 2007; Hatakeyama 2008). The investigation of bacterial-derived proteases implicated in the disruption of the epithelial barrier function is a relatively new field of research. *H. pylori* expresses HtrA, a protein with dual function acting as a chaperone and a serine protease, which is localized in the periplasm, but is also secreted into the environment (Bumann et al. 2002; Lower et al. 2008). The extracellular localization of HtrA allows a direct interaction with host cell surface molecules. In fact, E-cadherin exposed on gastric epithelial cells was identified as the first substrate for HtrA, that has severe consequences on the epithelial integrity (Hoy et al. 2010).

E-cadherin represents an important cell adhesion molecule, which is essential for the establishment and maintenance of an intact, polarized epithelium. Alterations of E-cadherin function, either through loss-of-function mutations, epigenetically down-regulated gene expression or by protein cleavage, were identified as important steps in gastric carcinogenesis (Liu and Chu 2014; Carneiro et al. 2012). The finding of HtrA-mediated E-cadherin cleavage unravels a novel mechanism in the pathogenesis of *H. pylori* (Wessler and Backert 2017). For a long time, it was suggested that *H. pylori* initiates bacterial pathogenesis via adherence at the apical domain of the epithelium, where it translocates CagA into the cytoplasm. The observation that basolaterally exposed integrin β1 serves as a receptor for the T4SS (Kwok et al. 2007) resulted in the conclusion that *H. pylori* must open intercellular adhesion complexes, which are mainly composed of tight junctions at the transition of the apical to basolateral membrane domains and the subjacent E-cadherin-mediated adherens junctions prior to contact integrin β1 (Fig. 1). Hence, the finding that HtrA cleaves-off the ectodomain of E-cadherin uncovered an elegant mechanism by which *H. pylori* can disrupt intercellular adhesions to open the intercellular space for transmigration (Hoy et al. 2010). Consequently, HtrA-dependent E-cadherin shedding strongly enhances CagA delivery into infected host cells via integrin β1 (Tegtmeyer et al. 2017b). Additional substrates for *H. pylori* HtrA are the extracellular matrix protein fibronectin (Hoy et al. 2010) and the tight junction proteins occludin and claudin-8 (Tegtmeyer et al. 2017b). While the HtrA/E-cadherin interaction is intensively investigated (Schmidt et al. 2016a, b), HtrA-induced cleavage of fibronectin, occludin and claudin-8 needs to be examined in more detail.

H. pylori expresses HtrA ubiquitously and this protease is highly stable under extreme conditions such as high salt concentration, low pH or extreme temperature (Hoy et al. 2013). Until now, *htrA*-negative *H. pylori* isolates have not yet been described as experimental Δ*htrA* knockout mutants are lethal, underlining that the expression of HtrA is essential for bacterial survival (Tegtmeyer et al. 2016; Salama et al. 2004). These observations led to the development of potent HtrA inhibitors in the form of small molecules as well as substrate-derived peptidic inhibitors. The first described small molecule able to inhibit *Helicobacter* HtrA was developed with help of a computational homology model. This *H. pylori* HtrA inhibitor (HHI) efficiently blocked E-cadherin shedding and subsequent bacterial transmigration across a polarized epithelial monolayer (Hoy et al. 2010). Motivated by these results, a large collection of small molecule inhibitors were developed and tested on HtrA activity and *H. pylori*/epithelium interaction (Lower et al. 2011; Geppert et al. 2011; Klenner et al. 2012; Perna et al. 2014, 2015). Through the analysis of the preferred HtrA signature sites in the E-cadherin molecule, an alternative, substrate-derived peptide inhibitor was also found that selectively binds and inhibits HtrA resulting in blocked transmigration of *H. pylori* (Schmidt et al. 2016b). These studies reveal that pharmacological inhibition of *H. pylori* HtrA can represent a new option in the treatment of *H. pylori* infections.

5 *H. pylori* VacA, GGT and CGT Are Involved in Immune Suppression and Evasion

Previous work identified the *H. pylori* factors VacA, GGT and CGT, which despite a profound effect on gastric epithelial cells, seem to be able to act as immune modulators that impair the activation and proliferation of a variety of immune cells, including T cells, suggesting important roles in immune suppression and evasion (Fig. 2).

6 Vacuolating Cytotoxin A (VacA)

The vacuolating activity associated with *H. pylori* infection of epithelial cells (Leunk et al. 1988) remained controversial with relation to its relevance to pathogenesis, until a protein was purified that seemed responsible for this activity (Cover and Blaser 1992). This VacA was genetically characterized with reference to its pathological significance (Cover et al. 1994; Schmitt and Haas 1994; Telford et al. 1994; Phadnis et al. 1994). Amongst all the *Helicobacter* species known, intact VacA protein with activity associated to gastritis is only present in *H. pylori* and *H. cetorum*, the latter being isolated from marine mammals, potentially suggesting an evolutionary significance (Foegeding et al. 2016).

Early studies have indicated that VacA has a capacity to form anion-selective channels (Czajkowsky et al. 1999; Tombola et al. 1999; Iwamoto et al. 1999), so that VacA was classified as a pore-forming toxin, with vacuolating activity in cell culture assays (Vinion-Dubiel et al. 1999; McClain et al. 2003). Oligomerization into single (hexamers or heptamers) or double layered structures (12-mers or 14-mers) seems to be required for VacA activity, although VacA is believed to initially interact with the plasma membrane of host cells as a monomer, after which it oligomerizes to form a membrane channel (de Bernard et al. 1995; McClain et al. 2000). Phylogenetic analysis of *H. pylori* clinical strains has revealed the existence of several distinct groups of *vacA* alleles (Gangwer et al. 2010). Three main regions of diversity in *vacA* sequences have been recognized, namely the signal sequence region (s-region), the intermediate region (i-region) and middle region (m-region). These result in *vacA* alleles containing multiple combinations of s-, i- and m-region types, relating to variable vacuolating activity (Atherton et al. 1995; Letley and Atherton 2000; McClain et al. 2001; Letley et al. 2003; Rhead et al. 2007), and linked to a potentially higher relative risk for development of gastric cancer or peptic ulcer disease (Figueiredo et al. 2002; Cover 2016).

VacA activity on epithelial cell culture systems have revealed a multitude of effects, varying from endosomal alterations of intraluminal pH (Ricci et al. 1997; Morbiato et al. 2001) and disruption of endocytic compartment trafficking (Satin et al. 1997; Molinari et al. 1998b; Tan et al. 2011), induction of autophagy (Terebiznik et al. 2009; Yahiro et al. 2012) and enhancement of mitochondrial dysfunction, which can result either from its pore-forming ability (Willhite and Blanke 2004) or through the activation of pro-apoptotic factors (Yamasaki et al. 2006). Moreover, VacA activity has been shown to cause increased epithelial barrier alterations through augmented plasma membrane permeability to the extracellular space (Tombola et al. 2001; Debellis et al. 2001), the formation of VacA channels in the plasma membrane (Iwamoto et al. 1999; Tombola et al. 1999) and by increasing paracellular permeability (Papini et al. 1998; Pelicic et al. 1999; Amieva et al. 2003). Finally, VacA-induced effects on epithelial cells include extensive alterations in cell signaling, related to MAP kinase p38 (Nakayama et al. 2004; Hisatsune et al. 2007) and ERK1/2 activation (Nakayama et al. 2004), VEGF upregulation (Caputo et al. 2003) and β-catenin nuclear localization (Nakayama et al. 2009) with subsequent reduction in the expression of pro-survival factors (Matsumoto et al. 2011). Moreover, it has been shown to inhibit gastric acid secretion from parietal cells (Kobayashi et al. 1996; Wang et al. 2008).

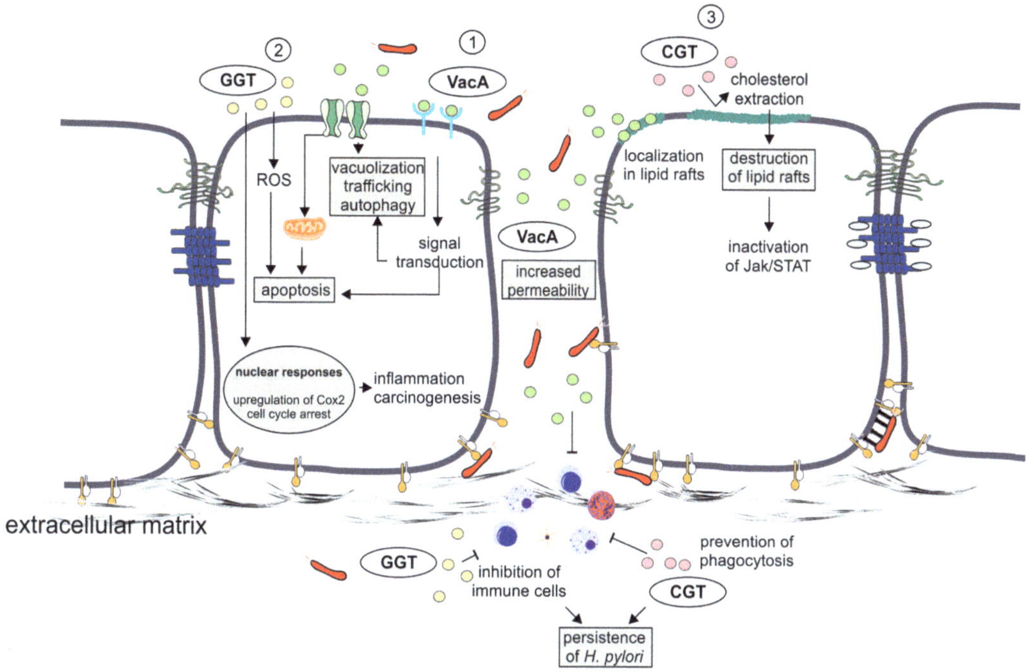

Fig. 2 An interplay of soluble *H. pylori* factors in bacterial persistence. Pleiotropic VacA is secreted by *H. pylori* and can form anion-selective channels leading to extensive vacuolization, changes in compartment trafficking, apoptosis, and autophagy. Vacualization also results in an increased permeability of the epithelial barrier through the disruption of TJs. Further, VacA-induced effects on epithelial cells lead to extensive alterations in cell signaling related to cell survival and cell death, in response to binding to cell surface receptor or localization within lipid rafts. Subsequently, VacA can inhibit the function and proliferation of T cells, B cells, eosinophils, macrophages, dendritic cells and neutrophils (*1*). Soluble GGT is responsible for the conversion of glutamine and glutathione into glutamate. This damages epithelial cells through the production of ammonia and generation of ROS, inducing a cell-cycle arrest and upregulating COX-2 in gastric epithelial cells. Similar to VacA, GGT has been described to inhibit immune cell function. Therefore, inducing *H. pylori* persistence (*2*). *H. pylori* depletes cholesterol from the cell membrane and incorporates it into the bacterial membrane where it is glycosylated by CGT. This results in a destruction of lipid rafts. It further inactivates the JAK/STAT1 signal transduction pathways in primary gastric cells. CGT has also been associated with anti-phagocytosis and T-cell inhibition (*3*)

With regards to its ability to act as an immunomodulator, VacA has been demonstrated to inhibit the function and proliferation of a variety of immune cells, such as T cells (Gebert et al. 2003; Utsch and Haas 2016), B cells (Torres et al. 2007), eosinophils (Kim et al. 2007, 2010a), macrophages (Allen et al. 2000; Zheng and Jones 2003), dendritic cells (Kim et al. 2011; Oertli et al. 2013; Djekic and Muller 2016) and neutrophils. Furthermore, the immunomodulatory activity of VacA has been demonstrated in *in vivo* experimental infection models (Oertli et al. 2013; Engler et al. 2014; Kyburz et al. 2017). Such diverse immune functions of VacA accentuate its significant role in the tempering of an immune response in order to facilitate colonization of the gastric epithelium as well as its potential immunomodulatory role on extragastric diseases (Djekic and Muller 2016).

A number of receptors have been proposed for the adhesion of VacA to host cells; however, it remains unclear whether VacA binds to a single abundant, low-affinity receptor or to multiple cell surface components (Foegeding et al. 2016). Candidates include receptor protein tyrosine phosphatase (RPTP) members α and β (Yahiro et al. 1999; Fujikawa et al. 2003; Yahiro et al. 2003, 2004), low-density lipoprotein receptor-related protein-1 (LRP1) (Yahiro et al. 2012), epidermal growth factor receptor (EGFR) (Seto et al. 1998), heparan sulphate (Utt et al. 2001), sphingomyelin (Gupta et al. 2008; Gupta et al. 2010), glycosphingolipids (Roche et al. 2007), and phospholipids (Molinari et al. 1998a). Of these, only sphingomyelin is suggested to dictate the extent to which VacA binds to the cell surface with subsequent VacA-dependent vacuolation (Foegeding et al. 2016) and sphingomyelin is thought to be the reason for VacA localization within lipid rafts (Geisse et al. 2004; Raghunathan et al. 2018).

In accordance to *in vitro* observations, animal studies have suggested that although,VacA may not be essential for gastric colonization, infection with *H. pylori* strains producing the most active forms of VacA (s1-i1) can induce more severe gastric inflammatory response and extensive metaplasia compared to strains with less active VacA of the s1-i2 or s2-i2 types (Winter et al. 2014). Whether VacA activity is related to gastric carcinogenesis due to impaired tumor surveillance, as a result of its immunomodulatory activity, or due to the augmentation of inflammatory response (Elinav et al. 2013) remains to be clarified – possibly, all three effects may attribute to the pathology.

7 Gamma-Glutamyl Transpeptidase (GGT)

The enzyme GGT catalyzes the transpeptidation and hydrolysis of the gamma-glutamyl group of glutathione and related compounds and is abundant amongst gastric *Helicobacter* species (Rossi et al. 2012). In *H. pylori*, it is synthesized as a proenzyme which is activated through autocatalysis, to form a heterodimer of two subunits of ~40 and 60 kDa, respectively (Boanca et al. 2006). Purified *H. pylori* GGT has a high hydrolyzing activity for conversion of glutamine and glutathione to glutamate with very high affinity for the substrates, indicating a central physiological role of this enzyme in glutamate biosynthesis (Shibayama et al. 2007). However, it has been shown to exhibit a pleiotropic activity both on both gastric epithelial cells and on T-cell mediated immunity. Related to its role in glutamate synthesis, a number of studies have shown that GGT is required for bacterial colonization, since knock-out mutants have exhibited a diminished (McGovern et al. 2001) or even completely abolished (Chevalier et al. 1999) ability to colonize the gastric mucosa in animal models. Furthermore, analysis of clinical strains has suggested that higher GGT activity is associated with peptic ulcer disease while lower GGT activity is more typically observed in strains causing non-ulcer dyspepsia (Gong et al. 2010). Consequently, a damaging effect of GGT on epithelial cells has been associated with the production of ammonia and generation of ROS, leading to caspase-9 and caspase-3 activation and apoptosis (Shibayama et al. 2003, 2007), ATP-depletion and necrosis (Flahou et al. 2011) as well as cell-cycle arrest at G1-S phase (Kim et al. 2010b). Moreover, it was demonstrated that GGT-induced up-regulation of EGF-related peptides and COX-2 in gastric epithelial cells could effectively contribute to the proinflammatory and procarcinogenic effect of *H. pylori* infection (Busiello et al. 2004).

In addition to the effect on epithelial cells, GGT has also been documented to modulate T-cell mediated immunity and thus contributes to immune evasion during infection. More specifically, GGT was identified as the secreted protein responsible for the G1 phase arrest of T cells through disruption of Ras MAPK-dependent signaling (Gerhard et al. 2005), independent of the VacA-dependent T cell proliferation arrest. Collectively, GGT and VacA can inhibit T cell proliferation and differentiation to Th1 and Th17 (Gerhard et al. 2005; Schmees et al. 2007; Beigier-Bompadre et al. 2011). Furthermore, GGT- and VacA-dependent effects on T-cells were suggested to be effected through dendritic cell reprogramming (Oertli et al. 2013), leading to interleukin-10 (IL-10) and IL-18 production and promotion of Treg differentiation that could further suppress Th1 and Th17 effector functions. Such activities exerted by GGT and VacA were associated with an increased protection against allergen-induced asthma, presumably by preventing airway hyper-responsiveness, bronchoalveolar eosinophilia, pulmonary inflammation and Th2 cytokine production, as was shown in mice tolerized with *H. pylori* extracts applied orally or intraperitoneally (Engler et al. 2014). Depletion of extracellular levels of glutamine by GGT could also result in the impairment of immune functions of the recruited inflammatory cells (Kabisch et al. 2016) and *H. pylori* GGT has been demonstrated to alter T lymphocyte metabolic reprogramming by depriving them from glutamine (Wustner et al. 2017).

8 Cholesterol-α-Glucosyltransferase (CGT)

H. pylori lacks the necessary components for independent sterol synthesis. During infection the bacteria migrate towards a cholesterol gradient and efficiently extract cholesterol from gastric epithelial cell membranes to incorporate glycosylated and non-glycosylated cholesterol into the bacterial membrane (Wunder et al. 2006). The enzyme cholesterol-α-glucosyltransferase (CGT) was identified to glycosylate cholesterol; it is encoded by the *hp0421* gene (Lebrun et al. 2006). The expression of CGT correlates to cholesterol depletion of host membranes, resulting in severe destruction of lipid rafts (Wunder et al. 2006). In initial studies, it was found that incorporation of non-glycosylated cholesterol could enhance phagocytosis by antigen-presenting cells (APCs) and T cell activation, which led to protection against *H. pylori* infections. In contrast, cholesteryl-glucosides abrogated the uptake of *H. pylori* by APCs. Consequently a CGT-negative *H. pylori* deletion mutant was rapidly cleared in a mouse animal model (Wunder et al. 2006), demonstrating that CGT activity can function as a new factor implicated in immune evasion and persistent infection.

The molecular mechanism of CGT-dependent immune evasion is still elusive, but it was indicated that *H. pylori* CGT can induce phagosome maturation arrest, which also involves PI3K activity (Du et al. 2016). In other cell types, such as primary gastric cells, it was proposed that decreased cholesterol levels in host cell membranes caused by *H. pylori* CGT activity not only disrupt lipid rafts, but also prevent IFNγ receptor-mediated signal transduction (Morey et al. 2017). This leads to an inactivation of JAK/STAT1 signal transduction pathways, which creates a micro-niche with lower concentrations of T-cell chemotactic attractants and anti-microbial peptides (human β-defensin 3, hBD3) (Morey et al. 2017). In summary, CGT has emerged as a novel *H. pylori* virulence factor that contributes to gastric carcinogenesis via promoting persistent infections together with T-cell inhibition.

9 Concluding Remarks

H. pylori is one of the most successful pathogens in the world, which colonizes the human gastric mucosa to induce a diverse range of gastric disorders and diseases. Since early human development, *H. pylori* coevolved with the human species through the development of a number of sophisticated strategies, leading to evasion of host surveillance and increased bacterial persistence. In particular, bacterial virulence and pathogenic factors, through their capability to specifically interfere with host cell components, contribute to a highly dynamic and complex pathomechanism. In this review we summarized the function of a number of putative bacterial virulence factors, such as T4SS, CagA, HtrA, VacA, CGT or GGT and examined the mechanisms by which they interfere with the gastric epithelial barrier and immune system. Moreover, these virulence factors seem to interact in synergy, in order to create such conditions of balance between the initial assault, the induction of tolerance and life-long bacterial persistence. These complex associations shaping coevolutionary relationships, between pathogenic *H. pylori* virulence determinants, host factors in inflammatory response genes and environmental factors warrant further careful investigation, necessary for the development of novel pharmacological compounds.

References

Allen LA, Schlesinger LS, Kang B (2000) Virulent strains of *Helicobacter pylori* demonstrate delayed phagocytosis and stimulate homotypic phagosome fusion in macrophages. J Exp Med 191(1):115–128

Amieva MR, El-Omar EM (2008) Host-bacterial interactions in *Helicobacter pylori* infection. Gastroenterology 134(1):306–323. https://doi.org/10.1053/j.gastro.2007.11.009

Amieva MR, Vogelmann R, Covacci A, Tompkins LS, Nelson WJ, Falkow S (2003) Disruption of the epithelial apical-junctional complex by *Helicobacter pylori* CagA. Science 300(5624):1430–1434. https://doi.org/10.1126/science.1081919

Argent RH, Kidd M, Owen RJ, Thomas RJ, Limb MC, Atherton JC (2004) Determinants and consequences of different levels of CagA phosphorylation for clinical isolates of *Helicobacter pylori*. Gastroenterology 127 (2):514–523

Asahi M, Azuma T, Ito S, Ito Y, Suto H, Nagai Y, Tsubokawa M, Tohyama Y, Maeda S, Omata M, Suzuki T, Sasakawa C (2000) *Helicobacter pylori* CagA protein can be tyrosine phosphorylated in gastric epithelial cells. J Exp Med 191(4):593–602

Atherton JC, Blaser MJ (2009) Coadaptation of *Helicobacter pylori* and humans: ancient history, modern implications. J Clin Invest 119(9):2475–2487. https://doi.org/10.1172/JCI38605

Atherton JC, Cao P, Peek RM Jr, Tummuru MK, Blaser MJ, Cover TL (1995) Mosaicism in vacuolating cytotoxin alleles of *Helicobacter pylori*. Association of specific vacA types with cytotoxin production and peptic ulceration. J Biol Chem 270(30):17771–17777

Backert S, Blaser MJ (2016) The role of CagA in the gastric biology of *Helicobacter pylori*. Cancer Res 76 (14):4028–4031. https://doi.org/10.1158/0008-5472.CAN-16-1680

Backert S, Tegtmeyer N (2017) Type IV secretion and signal transduction of *Helicobacter pylori* CagA through interactions with host cell receptors. Toxins (Basel) 9(4). https://doi.org/10.3390/toxins9040115

Backert S, Ziska E, Brinkmann V, Zimny-Arndt U, Fauconnier A, Jungblut PR, Naumann M, Meyer TF (2000) Translocation of the *Helicobacter pylori* CagA protein in gastric epithelial cells by a type IV secretion apparatus. Cell Microbiol 2(2):155–164

Backert S, Tegtmeyer N, Selbach M (2010) The versatility of *Helicobacter pylori* CagA effector protein functions: the master key hypothesis. Helicobacter 15 (3):163–176. https://doi.org/10.1111/j.1523-5378.2010.00759.x

Backert S, Clyne M, Tegtmeyer N (2011) Molecular mechanisms of gastric epithelial cell adhesion and injection of CagA by *Helicobacter pylori*. Cell Commun Signal CCS 9:28. https://doi.org/10.1186/1478-811X-9-28

Backert S, Tegtmeyer N, Fischer W (2015) Composition, structure and function of the *Helicobacter pylori* cag pathogenicity island encoded type IV secretion system. Future Microbiol 10(6):955–965. https://doi.org/10.2217/fmb.15.32

Barden S, Lange S, Tegtmeyer N, Conradi J, Sewald N, Backert S, Niemann HH (2013) A helical RGD motif promoting cell adhesion: crystal structures of the *Helicobacter pylori* type IV secretion system pilus protein CagL. Structure 21(11):1931–1941. https://doi.org/10.1016/j.str.2013.08.018

Bauer B, Pang E, Holland C, Kessler M, Bartfeld S, Meyer TF (2012) The *Helicobacter pylori* virulence effector CagA abrogates human beta-defensin 3 expression via inactivation of EGFR signaling. Cell Host Microbe 11 (6):576–586. https://doi.org/10.1016/j.chom.2012.04.013

Beigier-Bompadre M, Moos V, Belogolova E, Allers K, Schneider T, Churin Y, Ignatius R, Meyer TF, Aebischer T (2011) Modulation of the CD4+ T-cell response by *Helicobacter pylori* depends on known virulence factors and bacterial cholesterol and cholesterol alpha-glucoside content. J Infect Dis 204 (9):1339–1348. https://doi.org/10.1093/infdis/jir547

Berge C, Terradot L (2017) Structural insights into *Helicobacter pylori* cag protein interactions with host cell factors. Curr Top Microbiol Immunol 400:129–147. https://doi.org/10.1007/978-3-319-50520-6_6

Blaser MJ, Perez-Perez GI, Kleanthous H, Cover TL, Peek RM, Chyou PH, Stemmermann GN, Nomura A (1995) Infection with *Helicobacter pylori* strains possessing cagA is associated with an increased risk of developing adenocarcinoma of the stomach. Cancer Res 55 (10):2111–2115

Boanca G, Sand A, Barycki JJ (2006) Uncoupling the enzymatic and autoprocessing activities of *Helicobacter pylori* gamma-glutamyltranspeptidase. J Biol Chem 281(28):19029–19037. https://doi.org/10.1074/jbc.M603381200

Bonsor DA, Zhao Q, Schmidinger B, Weiss E, Wang J, Deredge D, Beadenkopf R, Dow B, Fischer W, Beckett D, Wintrode PL, Haas R, Sundberg EJ (2018) The *Helicobacter pylori* adhesin protein HopQ exploits the dimer interface of human CEACAMs to facilitate translocation of the oncoprotein CagA. EMBO J https://doi.org/10.15252/embj.201798664

Bumann D, Aksu S, Wendland M, Janek K, Zimny-Arndt-U, Sabarth N, Meyer TF, Jungblut PR (2002) Proteome analysis of secreted proteins of the gastric pathogen *Helicobacter pylori*. Infect Immun 70(7):3396–3403

Busiello I, Acquaviva R, Di Popolo A, Blanchard TG, Ricci V, Romano M, Zarrilli R (2004) *Helicobacter pylori* gamma-glutamyltranspeptidase upregulates COX-2 and EGF-related peptide expression in human gastric cells. Cell Microbiol 6(3):255–267

Buti L, Spooner E, Van der Veen AG, Rappuoli R, Covacci A, Ploegh HL (2011) *Helicobacter pylori* cytotoxin-associated gene A (CagA) subverts the apoptosis-stimulating protein of p53 (ASPP2) tumor suppressor pathway of the host. Proc Natl Acad Sci U S A 108(22):9238–9243. https://doi.org/10.1073/pnas.1106200108

Caputo R, Tuccillo C, Manzo BA, Zarrilli R, Tortora G, Blanco Cdel V, Ricci V, Ciardiello F, Romano M (2003) *Helicobacter pylori* VacA toxin up-regulates vascular endothelial growth factor expression in MKN 28 gastric cells through an epidermal growth factor receptor-, cyclooxygenase-2-dependent mechanism. Clin Cancer Res 9(6):2015–2021

Carneiro P, Fernandes MS, Figueiredo J, Caldeira J, Carvalho J, Pinheiro H, Leite M, Melo S, Oliveira P, Simoes-Correia J, Oliveira MJ, Carneiro F, Figueiredo C, Paredes J, Oliveira C, Seruca R (2012) E-cadherin dysfunction in gastric cancer--cellular consequences, clinical applications and open questions. FEBS Lett 586(18):2981–2989. https://doi.org/10.1016/j.febslet.2012.07.045

Chevalier C, Thiberge JM, Ferrero RL, Labigne A (1999) Essential role of *Helicobacter pylori* gamma-glutamyltranspeptidase for the colonization of the gastric mucosa of mice. Mol Microbiol 31(5):1359–1372

Chey WD, Leontiadis GI, Howden CW, Moss SF (2017) ACG clinical guideline: treatment of *Helicobacter pylori* infection. Am J Gastroenterol 112(2):212–239. https://doi.org/10.1038/ajg.2016.563

Churin Y, Al-Ghoul L, Kepp O, Meyer TF, Birchmeier W, Naumann M (2003) *Helicobacter pylori* CagA protein targets the c-Met receptor and enhances the motogenic response. J Cell Biol 161(2):249–255. https://doi.org/10.1083/jcb.200208039

Covacci A, Rappuoli R (2000) Tyrosine-phosphorylated bacterial proteins: Trojan horses for the host cell. J Exp Med 191(4):587–592

Covacci A, Censini S, Bugnoli M, Petracca R, Burroni D, Macchia G, Massone A, Papini E, Xiang Z, Figura N et al (1993) Molecular characterization of the 128-kDa immunodominant antigen of *Helicobacter pylori* associated with cytotoxicity and duodenal ulcer. Proc Natl Acad Sci U S A 90(12):5791–5795

Cover TL (2016) *Helicobacter pylori* diversity and gastric cancer risk. mBio 7(1):e01869–e01815. https://doi.org/10.1128/mBio.01869-15

Cover TL, Blaser MJ (1992) Purification and characterization of the vacuolating toxin from *Helicobacter pylori*. J Biol Chem 267(15):10570–10575

Cover TL, Tummuru MK, Cao P, Thompson SA, Blaser MJ (1994) Divergence of genetic sequences for the vacuolating cytotoxin among *Helicobacter pylori* strains. J Biol Chem 269(14):10566–10573

Czajkowsky DM, Iwamoto H, Cover TL, Shao Z (1999) The vacuolating toxin from *Helicobacter pylori* forms hexameric pores in lipid bilayers at low pH. Proc Natl Acad Sci U S A 96(5):2001–2006

de Bernard M, Papini E, de Filippis V, Gottardi E, Telford J, Manetti R, Fontana A, Rappuoli R, Montecucco C (1995) Low pH activates the vacuolating toxin of *Helicobacter pylori*, which becomes acid and pepsin resistant. J Biol Chem 270 (41):23937–23940

Debellis L, Papini E, Caroppo R, Montecucco C, Curci S (2001) *Helicobacter pylori* cytotoxin VacA increases alkaline secretion in gastric epithelial cells. Am J Physiol Gastrointest Liver Physiol 281(6):G1440–G1448. https://doi.org/10.1152/ajpgi.2001.281.6.G1440

Djekic A, Muller A (2016) The Immunomodulator VacA promotes immune tolerance and persistent *Helicobacter pylori* infection through its activities on T-cells and antigen-presenting cells. Toxins (Basel) 8 (6). https://doi.org/10.3390/toxins8060187

Du SY, Wang HJ, Cheng HH, Chen SD, Wang LH, Wang WC (2016) Cholesterol glucosylation by *Helicobacter pylori* delays internalization and arrests phagosome maturation in macrophages. J Microbiol Immunol Infect Wei mian yu gan ran za zhi 49(5):636–645. https://doi.org/10.1016/j.jmii.2014.05.011

Dubois A, Borén T (2007) *Helicobacter pylori* is invasive and it may be a facultative intracellular organism. Cell Microbiol 9(5):1108–1116. https://doi.org/10.1111/j.1462-5822.2007.00921.x

Elinav E, Nowarski R, Thaiss CA, Hu B, Jin C, Flavell RA (2013) Inflammation-induced cancer: crosstalk between tumours, immune cells and microorganisms. Nat Rev Cancer 13(11):759–771. https://doi.org/10.1038/nrc3611

Engler DB, Reuter S, van Wijck Y, Urban S, Kyburz A, Maxeiner J, Martin H, Yogev N, Waisman A, Gerhard M, Cover TL, Taube C, Muller A (2014) Effective treatment of allergic airway inflammation with *Helicobacter pylori* immunomodulators requires BATF3-dependent dendritic cells and IL-10. Proc Natl Acad Sci U S A 111(32):11810–11815. https://doi.org/10.1073/pnas.1410579111

Fehri LF, Rechner C, Janssen S, Mak TN, Holland C, Bartfeld S, Bruggemann H, Meyer TF (2009) *Helicobacter pylori*-induced modification of the histone H3 phosphorylation status in gastric epithelial cells reflects its impact on cell cycle regulation. Epigenetics 4(8):577–586

Ferlay J, Soerjomataram I, Dikshit R, Eser S, Mathers C, Rebelo M, Parkin DM, Forman D, Bray F (2015) Cancer incidence and mortality worldwide: sources, methods and major patterns in GLOBOCAN 2012. Int J Cancer 136(5):E359–E386. https://doi.org/10.1002/ijc.29210

Figueiredo C, Machado JC, Pharoah P, Seruca R, Sousa S, Carvalho R, Capelinha AF, Quint W, Caldas C, van Doorn LJ, Carneiro F, Sobrinho-Simoes M (2002) *Helicobacter pylori* and interleukin 1 genotyping: an opportunity to identify high-risk individuals for gastric carcinoma. J Natl Cancer Inst 94(22):1680–1687

Figueiredo C, Camargo MC, Leite M, Fuentes-Panana EM, Rabkin CS, Machado JC (2017) Pathogenesis of gastric Cancer: genetics and molecular classification. Curr Top Microbiol Immunol 400:277–304. https://doi.org/10.1007/978-3-319-50520-6_12

Flahou B, Haesebrouck F, Chiers K, Van Deun K, De Smet L, Devreese B, Vandenberghe I, Favoreel H, Smet A, Pasmans F, D'Herde K, Ducatelle R (2011) Gastric epithelial cell death caused by helicobacter suis and *Helicobacter pylori* gamma-glutamyl transpeptidase is mainly glutathione degradation-dependent. Cell Microbiol 13(12):1933–1955. https://doi.org/10.1111/j.1462-5822.2011.01682.x

Foegeding NJ, Caston RR, McClain MS, Ohi MD, Cover TL (2016) An overview of *Helicobacter pylori* VacA toxin biology. Toxins (Basel) 8(6). https://doi.org/10.3390/toxins8060173

Franco AT, Israel DA, Washington MK, Krishna U, Fox JG, Rogers AB, Neish AS, Collier-Hyams L, Perez-Perez GI, Hatakeyama M, Whitehead R, Gaus K, O'Brien DP, Romero-Gallo J, Peek RM Jr (2005) Activation of beta-catenin by carcinogenic *Helicobacter pylori*. Proc Natl Acad Sci U S A 102 (30):10646–10651. https://doi.org/10.1073/pnas.0504927102

Frick-Cheng AE, Pyburn TM, Voss BJ, McDonald WH, Ohi MD, Cover TL (2016) Molecular and structural analysis of the *Helicobacter pylori* cag type IV secretion system core complex. mBio 7(1):e02001–e02015. https://doi.org/10.1128/mBio.02001-15

Fujikawa A, Shirasaka D, Yamamoto S, Ota H, Yahiro K, Fukada M, Shintani T, Wada A, Aoyama N, Hirayama T, Fukamachi H, Noda M (2003) Mice deficient in protein tyrosine phosphatase receptor type Z are resistant to gastric ulcer induction by VacA of *Helicobacter pylori*. Nat Genet 33(3):375–381. https://doi.org/10.1038/ng1112

Gall A, Gaudet RG, Gray-Owen SD (2017) TIFA signaling in gastric epithelial cells initiates the cag type 4 secretion system-dependent innate immune response to *Helicobacter pylori*. Infection 8(4). https://doi.org/10.1128/mBio.01168-17

Gangwer KA, Shaffer CL, Suerbaum S, Lacy DB, Cover TL, Bordenstein SR (2010) Molecular evolution of the *Helicobacter pylori* vacuolating toxin gene vacA. J Bacteriol 192(23):6126–6135. https://doi.org/10.1128/jb.01081-10

Gebert B, Fischer W, Weiss E, Hoffmann R, Haas R (2003) *Helicobacter pylori* vacuolating cytotoxin inhibits T lymphocyte activation. Science 301 (5636):1099–1102. https://doi.org/10.1126/science.1086871

Geisse NA, Cover TL, Henderson RM, Edwardson JM (2004) Targeting of *Helicobacter pylori* vacuolating toxin to lipid raft membrane domains analysed by atomic force microscopy. Biochem J 381 (Pt 3):911–917. https://doi.org/10.1042/bj20031719

Geppert T, Hoy B, Wessler S, Schneider G (2011) Context-based identification of protein-protein interfaces and "hot-spot" residues. Chem Biol 18 (3):344–353. https://doi.org/10.1016/j.chembiol.2011.01.005

Gerhard M, Schmees C, Voland P, Endres N, Sander M, Reindl W, Rad R, Oelsner M, Decker T, Mempel M, Hengst L, Prinz C (2005) A secreted low-molecular-weight protein from *Helicobacter pylori* induces cell-cycle arrest of T cells. Gastroenterology 128 (5):1327–1339

Gobert AP, McGee DJ, Akhtar M, Mendz GL, Newton JC, Cheng Y, Mobley HL, Wilson KT (2001) *Helicobacter pylori* arginase inhibits nitric oxide production by eukaryotic cells: a strategy for bacterial survival. Proc Natl Acad Sci U S A 98(24):13844–13849. https://doi.org/10.1073/pnas.241443798

Gong M, Ling SS, Lui SY, Yeoh KG, Ho B (2010) *Helicobacter pylori* gamma-glutamyl transpeptidase is a pathogenic factor in the development of peptic ulcer disease. Gastroenterology 139(2):564–573. https://doi.org/10.1053/j.gastro.2010.03.050

Grohmann E, Christie PJ, Waksman G, Backert S (2018) Type IV secretion in Gram-negative and Gram-positive bacteria. Mol Microbiol 107(4):455–471. https://doi.org/10.1111/mmi.13896

Gupta VR, Patel HK, Kostolansky SS, Ballivian RA, Eichberg J, Blanke SR (2008) Sphingomyelin functions as a novel receptor for *Helicobacter pylori* VacA. PLoS Pathog 4(5):e1000073. https://doi.org/10.1371/journal.ppat.1000073

Gupta VR, Wilson BA, Blanke SR (2010) Sphingomyelin is important for the cellular entry and intracellular localization of *Helicobacter pylori* VacA. Cell Microbiol 12(10):1517–1533. https://doi.org/10.1111/j.1462-5822.2010.01487.x

Hatakeyama M (2008) Linking epithelial polarity and carcinogenesis by multitasking *Helicobacter pylori* virulence factor CagA. Oncogene 27(55):7047–7054. https://doi.org/10.1038/onc.2008.353

Hatakeyama M (2017) Structure and function of *Helicobacter pylori* CagA, the first-identified bacterial protein involved in human cancer. Proc Jpn Acad Ser B Phys Biol Sci 93(4):196–219. https://doi.org/10.2183/pjab.93.013

Hayashi T, Senda M, Morohashi H, Higashi H, Horio M, Kashiba Y, Nagase L, Sasaya D, Shimizu T, Venugopalan N, Kumeta H, Noda NN, Inagaki F, Senda T, Hatakeyama M (2012) Tertiary structure-function analysis reveals the pathogenic signaling potentiation mechanism of *Helicobacter pylori* oncogenic effector CagA. Cell Host Microbe 12(1):20–33. https://doi.org/10.1016/j.chom.2012.05.010

Hayashi T, Morohashi H, Hatakeyama M (2013) Bacterial EPIYA effectors--where do they come from? What are they? Where are they going? Cell Microbiol 15 (3):377–385. https://doi.org/10.1111/cmi.12040

Higashi H, Tsutsumi R, Muto S, Sugiyama T, Azuma T, Asaka M, Hatakeyama M (2002) SHP-2 tyrosine phosphatase as an intracellular target of *Helicobacter pylori* CagA protein. Science (New York NY) 295 (5555):683–686

Hisatsune J, Yamasaki E, Nakayama M, Shirasaka D, Kurazono H, Katagata Y, Inoue H, Han J, Sap J, Yahiro K, Moss J, Hirayama T (2007) *Helicobacter pylori* VacA enhances prostaglandin E2 production through induction of cyclooxygenase 2 expression via a p38 mitogen-activated protein kinase/activating transcription factor 2 cascade in AZ-521 cells. Infect Immun 75(9):4472–4481. https://doi.org/10.1128/iai.00500-07

Hoy B, Lower M, Weydig C, Carra G, Tegtmeyer N, Geppert T, Schroder P, Sewald N, Backert S, Schneider G, Wessler S (2010) *Helicobacter pylori* HtrA is a new secreted virulence factor that cleaves E-cadherin to disrupt intercellular adhesion. EMBO Rep 11(10):798–804. https://doi.org/10.1038/embor.2010.114

Hoy B, Brandstetter H, Wessler S (2013) The stability and activity of recombinant *Helicobacter pylori* HtrA under stress conditions. J Basic Microbiol 53 (5):402–409. https://doi.org/10.1002/jobm.201200074

Huang JY, Sweeney EG, Sigal M, Zhang HC, Remington SJ, Cantrell MA, Kuo CJ, Guillemin K, Amieva MR (2015) Chemodetection and destruction of host urea allows *Helicobacter pylori* to locate the epithelium. Cell Host Microbe 18(2):147–156. https://doi.org/10.1016/j.chom.2015.07.002

Iwamoto H, Czajkowsky DM, Cover TL, Szabo G, Shao Z (1999) VacA from *Helicobacter pylori*: a hexameric chloride channel. FEBS Lett 450(1-2):101–104

Jang S, Su H, Blum FC, Bae S, Choi YH, Kim A, Hong YA, Kim J, Kim JH, Gunawardhana N, Jeon YE, Yoo YJ, Merrell DS, Ge L, Cha JH (2017) Dynamic expansion and contraction of cagA copy number in *Helicobacter pylori* impact development of gastric disease. mBio 8(1). https://doi.org/10.1128/mBio.01779-16

Javaheri A, Kruse T, Moonens K, Mejias-Luque R, Debraekeleer A, Asche CI, Tegtmeyer N, Kalali B, Bach NC, Sieber SA, Hill DJ, Koniger V, Hauck CR, Moskalenko R, Haas R, Busch DH, Klaile E, Slevogt H, Schmidt A, Backert S, Remaut H, Singer BB, Gerhard M (2016) *Helicobacter pylori* adhesin HopQ engages in a virulence-enhancing interaction with human CEACAMs. Nat Microbiol 2:16189. https://doi.org/10.1038/nmicrobiol.2016.189

Jimenez-Soto LF, Kutter S, Sewald X, Ertl C, Weiss E, Kapp U, Rohde M, Pirch T, Jung K, Retta SF, Terradot L, Fischer W, Haas R (2009) *Helicobacter pylori* type IV secretion apparatus exploits beta1 integrin in a novel RGD-independent manner. PLoS Pathog 5(12):e1000684. https://doi.org/10.1371/journal.ppat.1000684

Jones KR, Joo YM, Jang S, Yoo YJ, Lee HS, Chung IS, Olsen CH, Whitmire JM, Merrell DS, Cha JH (2009) Polymorphism in the CagA EPIYA motif impacts development of gastric cancer. J Clin Microbiol 47 (4):959–968. https://doi.org/10.1128/JCM.02330-08

Kabisch R, Semper RP, Mejias-Luque R, Wustner S, Gerhard M (2016) *Helicobacter pylori* gamma-Glutamyltranspeptidase induces tolerogenic human dendritic cells by activation of glutamate receptors. J Immunol 196(10):4246–4252. https://doi.org/10.4049/jimmunol.1501062

Kaplan-Turkoz B, Jimenez-Soto LF, Dian C, Ertl C, Remaut H, Louche A, Tosi T, Haas R, Terradot L (2012) Structural insights into *Helicobacter pylori* oncoprotein CagA interaction with beta1 integrin. Proc Natl Acad Sci U S A 109(36):14640–14645. https://doi.org/10.1073/pnas.1206098109

Keates S, Sougioultzis S, Keates AC, Zhao D, Peek RM Jr, Shaw LM, Kelly CP (2001) cag+ *Helicobacter pylori* induce transactivation of the epidermal growth factor receptor in AGS gastric epithelial cells. J Biol Chem 276(51):48127–48134. https://doi.org/10.1074/jbc.M107630200

Kim JM, Kim JS, Lee JY, Kim YJ, Youn HJ, Kim IY, Chee YJ, Oh YK, Kim N, Jung HC, Song IS (2007) Vacuolating cytotoxin in *Helicobacter pylori* water-soluble proteins upregulates chemokine expression in human eosinophils via Ca2+ influx, mitochondrial reactive oxygen intermediates, and NF-kappaB activation. Infect Immun 75(7):3373–3381. https://doi.org/10.1128/iai.01940-06

Kim JM, Kim JS, Lee JY, Sim YS, Kim YJ, Oh YK, Yoon HJ, Kang JS, Youn J, Kim N, Jung HC, Kim S (2010a) Dual effects of *Helicobacter pylori* vacuolating cytotoxin on human eosinophil apoptosis in early and late periods of stimulation. Eur J Immunol 40 (6):1651–1662. https://doi.org/10.1002/eji.200939882

Kim KM, Lee SG, Kim JM, Kim DS, Song JY, Kang HL, Lee WK, Cho MJ, Rhee KH, Youn HS, Baik SC (2010b) *Helicobacter pylori* gamma-glutamyltranspeptidase induces cell cycle arrest at the G1-S phase transition. J Microbiol (Seoul, Korea) 48(3):372–377. https://doi.org/10.1007/s12275-010-9293-8

Kim JM, Kim JS, Yoo DY, Ko SH, Kim N, Kim H, Kim YJ (2011) Stimulation of dendritic cells with *Helicobacter pylori* vacuolating cytotoxin negatively regulates their maturation via the restoration of E2F1. Clin Exp Immunol 166(1):34–45. https://doi.org/10.1111/j.1365-2249.2011.04447.x

Klenner A, Hahnke V, Geppert T, Schneider P, Zettl H, Haller S, Rodrigues T, Reisen F, Hoy B, Schaible AM, Werz O, Wessler S, Schneider G (2012) From virtual screening to bioactive compounds by visualizing and clustering of chemical space. Mol Inform 31(1):21–26. https://doi.org/10.1002/minf.201100147

Kobayashi H, Kamiya S, Suzuki T, Kohda K, Muramatsu S, Kurumada T, Ohta U, Miyazawa M, Kimura N, Mutoh N, Shirai T, Takagi A, Harasawa S, Tani N, Miwa T (1996) The effect of *Helicobacter pylori* on gastric acid secretion by isolated parietal cells from a Guinea pig. Association with production of vacuolating toxin by *H pylori*. Scand J Gastroenterol 31(5):428–433

Koniger V, Holsten L, Harrison U, Busch B, Loell E, Zhao Q, Bonsor DA, Roth A, Kengmo-Tchoupa A, Smith SI, Mueller S, Sundberg EJ, Zimmermann W, Fischer W, Hauck CR, Haas R (2016) *Helicobacter pylori* exploits human CEACAMs via HopQ for adherence and translocation of CagA. Nat Microbiol 2:16188. https://doi.org/10.1038/nmicrobiol.2016.188

Krueger S, Hundertmark T, Kuester D, Kalinski T, Peitz U, Roessner A (2007) *Helicobacter pylori* alters the distribution of ZO-1 and p120ctn in primary human gastric epithelial cells. Pathol Res Pract 203 (6):433–444. https://doi.org/10.1016/j.prp.2007.04.003

Kusters JG, van Vliet AH, Kuipers EJ (2006) Pathogenesis of *Helicobacter pylori* infection. Clin Microbiol Rev 19(3):449–490. https://doi.org/10.1128/CMR.00054-05

Kwok T, Zabler D, Urman S, Rohde M, Hartig R, Wessler S, Misselwitz R, Berger J, Sewald N, Konig W, Backert S (2007) Helicobacter exploits integrin for type IV secretion and kinase activation. Nature 449(7164):862–866. https://doi.org/10.1038/nature06187

Kyburz A, Urban S, Altobelli A, Floess S, Huehn J, Cover TL, Muller A (2017) *Helicobacter pylori* and its secreted immunomodulator VacA protect against anaphylaxis in experimental models of food allergy. Clin Exp Allergy 47(10):1331–1341. https://doi.org/10.1111/cea.12996

Lebrun AH, Wunder C, Hildebrand J, Churin Y, Zahringer U, Lindner B, Meyer TF, Heinz E, Warnecke D (2006) Cloning of a cholesterol-alpha-glucosyltransferase from *Helicobacter pylori*. J Biol Chem 281(38):27765–27772. https://doi.org/10.1074/jbc.M603345200

Lee IO, Kim JH, Choi YJ, Pillinger MH, Kim SY, Blaser MJ, Lee YC (2010) *Helicobacter pylori* CagA phosphorylation status determines the gp130-activated SHP2/ERK and JAK/STAT signal transduction pathways in gastric epithelial cells. J Biol Chem 285 (21):16042–16050. https://doi.org/10.1074/jbc.M110.111054

Lee DG, Kim HS, Lee YS, Kim S, Cha SY, Ota I, Kim NH, Cha YH, Yang DH, Lee Y, Park GJ, Yook JI, Lee YC (2014) *Helicobacter pylori* CagA promotes snail-mediated epithelial-mesenchymal transition by reducing GSK-3 activity. Nat Commun 5:4423. https://doi.org/10.1038/ncomms5423

Letley DP, Atherton JC (2000) Natural diversity in the N terminus of the mature vacuolating cytotoxin of *Helicobacter pylori* determines cytotoxin activity. J Bacteriol 182(11):3278–3280

Letley DP, Rhead JL, Twells RJ, Dove B, Atherton JC (2003) Determinants of non-toxicity in the gastric pathogen *Helicobacter pylori*. J Biol Chem 278 (29):26734–26741. https://doi.org/10.1074/jbc.M304071200

Leunk RD, Johnson PT, David BC, Kraft WG, Morgan DR (1988) Cytotoxic activity in broth-culture filtrates of campylobacter pylori. J Med Microbiol 26 (2):93–99. https://doi.org/10.1099/00222615-26-2-93

Li Q, Liu J, Gong Y, Yuan Y (2017) Association of CagA EPIYA-D or EPIYA-C phosphorylation sites with peptic ulcer and gastric cancer risks: a meta-analysis. Medicine 96(17):e6620. https://doi.org/10.1097/md.0000000000006620

Liu X, Chu KM (2014) E-cadherin and gastric cancer: cause, consequence, and applications. Biomed Res Int 2014:637308. https://doi.org/10.1155/2014/637308

Lower M, Weydig C, Metzler D, Reuter A, Starzinski-Powitz A, Wessler S, Schneider G (2008) Prediction of extracellular proteases of the human pathogen *Helicobacter pylori* reveals proteolytic activity of the Hp1018/19 protein HtrA. PLoS One 3(10):e3510. https://doi.org/10.1371/journal.pone.0003510

Lower M, Geppert T, Schneider P, Hoy B, Wessler S, Schneider G (2011) Inhibitors of *Helicobacter pylori* protease HtrA found by 'virtual ligand' screening combat bacterial invasion of epithelia. PLoS One 6(3): e17986. https://doi.org/10.1371/journal.pone.0017986

Malfertheiner P, Megraud F, O'Morain CA, Gisbert JP, Kuipers EJ, Axon AT, Bazzoli F, Gasbarrini A, Atherton J, Graham DY, Hunt R, Moayyedi P, Rokkas T, Rugge M, Selgrad M, Suerbaum S, Sugano K, El-Omar EM, European H, Microbiota Study G, Consensus panel (2017) Management of *Helicobacter pylori* infection-the Maastricht V/Florence consensus report. Gut 66(1):6–30. https://doi.org/10.1136/gutjnl-2016-312288

Matsumoto Y, Marusawa H, Kinoshita K, Endo Y, Kou T, Morisawa T, Azuma T, Okazaki IM, Honjo T, Chiba T (2007) *Helicobacter pylori* infection triggers aberrant expression of activation-induced cytidine deaminase in gastric epithelium. Nat Med 13(4):470–476. https://doi.org/10.1038/nm1566

Matsumoto A, Isomoto H, Nakayama M, Hisatsune J, Nishi Y, Nakashima Y, Matsushima K, Kurazono H, Nakao K, Hirayama T, Kohno S (2011) *Helicobacter pylori* VacA reduces the cellular expression of STAT3 and pro-survival Bcl-2 family proteins, Bcl-2 and Bcl-XL, leading to apoptosis in gastric epithelial cells. Dig Dis Sci 56(4):999–1006. https://doi.org/10.1007/s10620-010-1420-1

McClain MS, Schraw W, Ricci V, Boquet P, Cover TL (2000) Acid activation of *Helicobacter pylori* vacuolating cytotoxin (VacA) results in toxin internalization by eukaryotic cells. Mol Microbiol 37 (2):433–442

McClain MS, Cao P, Iwamoto H, Vinion-Dubiel AD, Szabo G, Shao Z, Cover TL (2001) A 12-amino-acid segment, present in type s2 but not type s1 *Helicobacter pylori* VacA proteins, abolishes cytotoxin activity and alters membrane channel formation. J Bacteriol 183(22):6499–6508. https://doi.org/10.1128/jb.183.22.6499-6508.2001

McClain MS, Iwamoto H, Cao P, Vinion-Dubiel AD, Li Y, Szabo G, Shao Z, Cover TL (2003) Essential role of a GXXXG motif for membrane channel formation by *Helicobacter pylori* vacuolating toxin. J Biol Chem 278(14):12101–12108. https://doi.org/10.1074/jbc.M212595200

McGovern KJ, Blanchard TG, Gutierrez JA, Czinn SJ, Krakowka S, Youngman P (2001) Gamma-glutamyltransferase is a *Helicobacter pylori* virulence factor but is not essential for colonization. Infect

Immun 69(6):4168–4173. https://doi.org/10.1128/iai.69.6.4168-4173.2001

Mejias-Luque R, Gerhard M (2017) Immune evasion strategies and persistence of *Helicobacter pylori*. Curr Top Microbiol Immunol 400:53–71. https://doi.org/10.1007/978-3-319-50520-6_3

Mishra JP, Cohen D, Zamperone A, Nesic D, Muesch A, Stein M (2015) CagA of *Helicobacter pylori* interacts with and inhibits the serine-threonine kinase PRK2. Cell Microbiol 17(11):1670–1682. https://doi.org/10.1111/cmi.12464

Molinari M, Galli C, de Bernard M, Norais N, Ruysschaert JM, Rappuoli R, Montecucco C (1998a) The acid activation of *Helicobacter pylori* toxin VacA: structural and membrane binding studies. Biochem Biophys Res Commun 248(2):334–340. https://doi.org/10.1006/bbrc.1998.8808

Molinari M, Salio M, Galli C, Norais N, Rappuoli R, Lanzavecchia A, Montecucco C (1998b) Selective inhibition of Ii-dependent antigen presentation by *Helicobacter pylori* toxin VacA. J Exp Med 187 (1):135–140

Moodley Y, Linz B (2009) *Helicobacter pylori* sequences reflect past human migrations. Genome Dyn 6:62–74. https://doi.org/10.1159/000235763

Moonens K, Remaut H (2017) Evolution and structural dynamics of bacterial glycan binding adhesins. Curr Opin Struct Biol 44:48–58. https://doi.org/10.1016/j.sbi.2016.12.003

Morbiato L, Tombola F, Campello S, Del Giudice G, Rappuoli R, Zoratti M, Papini E (2001) Vacuolation induced by VacA toxin of *Helicobacter pylori* requires the intracellular accumulation of membrane permeant bases, Cl^- and water. FEBS Lett 508(3):479–483. https://doi.org/10.1016/S0014-5793(01)03133-7

Morey P, Pfannkuch L, Pang E, Boccellato F, Sigal M, Imai-Matsushima A, Dyer V, Koch M, Mollenkopf HJ, Schlaermann P, Meyer TF (2017) *Helicobacter pylori* depletes cholesterol in gastric glands to prevent interferon gamma signaling and escape the inflammatory response. Gastroenterology. https://doi.org/10.1053/j.gastro.2017.12.008

Mueller D, Tegtmeyer N, Brandt S, Yamaoka Y, De Poire E, Sgouras D, Wessler S, Torres J, Smolka A, Backert S (2012) c-Src and c-Abl kinases control hierarchic phosphorylation and function of the CagA effector protein in Western and East Asian *Helicobacter pylori* strains. J Clin Invest 122(4):1553–1566. https://doi.org/10.1172/JCI61143

Murata-Kamiya N, Kurashima Y, Teishikata Y, Yamahashi Y, Saito Y, Higashi H, Aburatani H, Akiyama T, Peek RM, Jr., Azuma T, Hatakeyama M (2007) *Helicobacter pylori* CagA interacts with E-cadherin and deregulates the beta-catenin signal that promotes intestinal transdifferentiation in gastric epithelial cells. Oncogene 26 (32):4617-4626. doi: https://doi.org/10.1038/sj.onc.1210251

Murata-Kamiya N, Kikuchi K, Hayashi T, Higashi H, Hatakeyama M (2010) *Helicobacter pylori* exploits host membrane phosphatidylserine for delivery, localization, and pathophysiological action of the CagA oncoprotein. Cell Host Microbe 7(5):399–411. https://doi.org/10.1016/j.chom.2010.04.005

Nakayama M, Kimura M, Wada A, Yahiro K, Ogushi K, Niidome T, Fujikawa A, Shirasaka D, Aoyama N, Kurazono H, Noda M, Moss J, Hirayama T (2004) *Helicobacter pylori* VacA activates the p38/activating transcription factor 2-mediated signal pathway in AZ-521 cells. J Biol Chem 279(8):7024–7028. https://doi.org/10.1074/jbc.M308898200

Nakayama M, Hisatsune J, Yamasaki E, Isomoto H, Kurazono H, Hatakeyama M, Azuma T, Yamaoka Y, Yahiro K, Moss J, Hirayama T (2009) *Helicobacter pylori* VacA-induced inhibition of GSK3 through the PI3K/Akt signaling pathway. J Biol Chem 284 (3):1612–1619. https://doi.org/10.1074/jbc.M806981200

Naumann M, Sokolova O, Tegtmeyer N, Backert S (2017) *Helicobacter pylori*: a paradigm pathogen for subverting host cell signal transmission. Trends Microbiol 25(4):316–328. https://doi.org/10.1016/j.tim.2016.12.004

Neal JT, Peterson TS, Kent ML, Guillemin K (2013) *H. pylori* virulence factor CagA increases intestinal cell proliferation by Wnt pathway activation in a transgenic zebrafish model. Dis Model Mech 6(3):802–810. https://doi.org/10.1242/dmm.011163

Odenbreit S, Puls J, Sedlmaier B, Gerland E, Fischer W, Haas R (2000) Translocation of *Helicobacter pylori* CagA into gastric epithelial cells by type IV secretion. Science 287(5457):1497–1500

Oertli M, Noben M, Engler DB, Semper RP, Reuter S, Maxeiner J, Gerhard M, Taube C, Muller A (2013) *Helicobacter pylori* gamma-glutamyl transpeptidase and vacuolating cytotoxin promote gastric persistence and immune tolerance. Proc Natl Acad Sci U S A 110 (8):3047–3052. https://doi.org/10.1073/pnas.1211248110

Ohnishi N, Yuasa H, Tanaka S, Sawa H, Miura M, Matsui A, Higashi H, Musashi M, Iwabuchi K, Suzuki M, Yamada G, Azuma T, Hatakeyama M (2008) Transgenic expression of *Helicobacter pylori* CagA induces gastrointestinal and hematopoietic neoplasms in mouse. Proc Natl Acad Sci U S A 105 (3):1003–1008. https://doi.org/10.1073/pnas.0711183105

Oliveira MJ, Costa AC, Costa AM, Henriques L, Suriano G, Atherton JC, Machado JC, Carneiro F, Seruca R, Mareel M, Leroy A, Figueiredo C (2006) *Helicobacter pylori* induces gastric epithelial cell invasion in a c-Met and type IV secretion system-dependent manner. J Biol Chem 281 (46):34888–34896. https://doi.org/10.1074/jbc.M607067200

Oliveira MJ, Costa AM, Costa AC, Ferreira RM, Sampaio P, Machado JC, Seruca R, Mareel M, Figueiredo C (2009) CagA associates with c-Met, E-cadherin, and p120-catenin in a multiproteic complex that suppresses *Helicobacter pylori*-induced cell-invasive phenotype. J Infect Dis 200(5):745–755. https://doi.org/10.1086/604727

Pachathundikandi SK, Muller A, Backert S (2016) Inflammasome activation by *Helicobacter pylori* and its implications for persistence and immunity. Curr Top Microbiol Immunol 397:117–131. https://doi.org/10.1007/978-3-319-41171-2_6

Panayotopoulou EG, Sgouras DN, Papadakos KS, Petraki K, Breurec S, Michopoulos S, Mantzaris G, Papatheodoridis G, Mentis A, Archimandritis A (2010) CagA and VacA polymorphisms are associated with distinct pathological features in *Helicobacter pylori*-infected adults with peptic ulcer and non-peptic ulcer disease. J Clin Microbiol 48 (6):2237–2239. https://doi.org/10.1128/jcm.00662-10

Papini E, Satin B, Norais N, de Bernard M, Telford JL, Rappuoli R, Montecucco C (1998) Selective increase of the permeability of polarized epithelial cell monolayers by *Helicobacter pylori* vacuolating toxin. J Clin Invest 102(4):813–820. https://doi.org/10.1172/jci2764

Parsonnet J, Friedman GD, Orentreich N, Vogelman H (1997) Risk for gastric cancer in people with CagA positive or CagA negative *Helicobacter pylori* infection. Gut 40(3):297–301

Pelicic V, Reyrat JM, Sartori L, Pagliaccia C, Rappuoli R, Telford JL, Montecucco C, Papini E (1999) *Helicobacter pylori* VacA cytotoxin associated with the bacteria increases epithelial permeability independently of its vacuolating activity. Microbiology 145 (Pt 8):2043–2050. https://doi.org/10.1099/13500872-145-8-2043

Perna AM, Reisen F, Schmidt TP, Geppert T, Pillong M, Weisel M, Hoy B, Simister PC, Feller SM, Wessler S, Schneider G (2014) Inhibiting *Helicobacter pylori* HtrA protease by addressing a computationally predicted allosteric ligand binding site. Chem Sci 5:3583–3590. https://doi.org/10.1039/c4sc01443j

Perna AM, Rodrigues T, Schmidt TP, Bohm M, Stutz K, Reker D, Pfeiffer B, Altmann KH, Backert S, Wessler S, Schneider G (2015) Fragment-based De novo design reveals a small-molecule inhibitor of *Helicobacter pylori* HtrA. Angew Chem Int Ed Engl 54(35):10244–10248. https://doi.org/10.1002/anie.201504035

Phadnis SH, Ilver D, Janzon L, Normark S, Westblom TU (1994) Pathological significance and molecular characterization of the vacuolating toxin gene of *Helicobacter pylori*. Infect Immun 62(5):1557–1565

Polk DB, Peek RM, Jr. (2010) *Helicobacter pylori*: gastric cancer and beyond. Nat Rev Cancer 10 (6):403-414. doi:https://doi.org/10.1038/nrc2857

Poppe M, Feller SM, Romer G, Wessler S (2007) Phosphorylation of *Helicobacter pylori* CagA by c-Abl leads to cell motility. Oncogene 26(24):3462–3472. https://doi.org/10.1038/sj.onc.1210139

Posselt G, Backert S, Wessler S (2013) The functional interplay of *Helicobacter pylori* factors with gastric epithelial cells induces a multi-step process in pathogenesis. Cell Commun Signal CCS 11:77. https://doi.org/10.1186/1478-811X-11-77

Raghunathan K, Foegeding NJ, Campbell AM, Cover TL, Ohi MD, Kenworthy AK (2018) Determinants of raft partitioning of the *Helicobacter pylori* pore-forming toxin VacA. Infect Immun 86(5). https://doi.org/10.1128/IAI.00872-17

Ramarao N, Gray-Owen SD, Backert S, Meyer TF (2000) *Helicobacter pylori* inhibits phagocytosis by professional phagocytes involving type IV secretion components. Mol Microbiol 37(6):1389–1404

Rhead JL, Letley DP, Mohammadi M, Hussein N, Mohagheghi MA, Eshagh Hosseini M, Atherton JC (2007) A new *Helicobacter pylori* vacuolating cytotoxin determinant, the intermediate region, is associated with gastric cancer. Gastroenterology 133 (3):926–936. https://doi.org/10.1053/j.gastro.2007.06.056

Ricci V, Sommi P, Fiocca R, Romano M, Solcia E, Ventura U (1997) *Helicobacter pylori* vacuolating toxin accumulates within the endosomal-vacuolar compartment of cultured gastric cells and potentiates the vacuolating activity of ammonia. J Pathol 183 (4):453–459. https://doi.org/10.1002/(sici)1096-9896(199712)183:4<453::aid-path950>3.0.co;2-2

Robinson K, Letley DP, Kaneko K (2017) The human stomach in health and disease: infection strategies by *Helicobacter pylori*. Curr Top Microbiol Immunol 400:1–26. https://doi.org/10.1007/978-3-319-50520-6_1

Roche N, Ilver D, Angstrom J, Barone S, Telford JL, Teneberg S (2007) Human gastric glycosphingolipids recognized by *Helicobacter pylori* vacuolating cytotoxin VacA. Microbes Infect 9(5):605–614. https://doi.org/10.1016/j.micinf.2007.01.023

Rossi M, Bolz C, Revez J, Javed S, El-Najjar N, Anderl F, Hyytiainen H, Vuorela P, Gerhard M, Hanninen ML (2012) Evidence for conserved function of gamma-glutamyltranspeptidase in helicobacter genus. PLoS One 7(2):e30543. https://doi.org/10.1371/journal.pone.0030543

Roure S, Bonis M, Chaput C, Ecobichon C, Mattox A, Barriere C, Geldmacher N, Guadagnini S, Schmitt C, Prevost MC, Labigne A, Backert S, Ferrero RL, Boneca IG (2012) Peptidoglycan maturation enzymes affect flagellar functionality in bacteria. Mol Microbiol 86(4):845–856. https://doi.org/10.1111/mmi.12019

Saadat I, Higashi H, Obuse C, Umeda M, Murata-Kamiya-N, Saito Y, Lu H, Ohnishi N, Azuma T, Suzuki A, Ohno S, Hatakeyama M (2007) *Helicobacter pylori* CagA targets PAR1/MARK kinase to disrupt epithelial cell polarity. Nature 447(7142):330–333. https://doi.org/10.1038/nature05765

Saha A, Backert S, Hammond CE, Gooz M, Smolka AJ (2010) *Helicobacter pylori* CagL activates ADAM17 to induce repression of the gastric H, K-ATPase alpha subunit. Gastroenterology 139(1):239–248. https://doi.org/10.1053/j.gastro.2010.03.036

Salama NR, Shepherd B, Falkow S (2004) Global transposon mutagenesis and essential gene analysis of *Helicobacter pylori*. J Bacteriol 186(23):7926–7935. https://doi.org/10.1128/jb.186.23.7926-7935.2004

Satin B, Norais N, Telford J, Rappuoli R, Murgia M, Montecucco C, Papini E (1997) Effect of *Helicobacter pylori* vacuolating toxin on maturation and extracellular release of procathepsin D and on epidermal growth factor degradation. J Biol Chem 272(40):25022–25028

Schmees C, Prinz C, Treptau T, Rad R, Hengst L, Voland P, Bauer S, Brenner L, Schmid RM, Gerhard M (2007) Inhibition of T-cell proliferation by *Helicobacter pylori* gamma-glutamyl transpeptidase. Gastroenterology 132(5):1820–1833. https://doi.org/10.1053/j.gastro.2007.02.031

Schmidt TP, Goetz C, Huemer M, Schneider G, Wessler S (2016a) Calcium binding protects E-cadherin from cleavage by *Helicobacter pylori* HtrA. Gut Pathogens 8:29. https://doi.org/10.1186/s13099-016-0112-6

Schmidt TP, Perna AM, Fugmann T, Bohm M, Jan H, Haller S, Gotz C, Tegtmeyer N, Hoy B, Rau TT, Neri D, Backert S, Schneider G, Wessler S (2016b) Identification of E-cadherin signature motifs functioning as cleavage sites for *Helicobacter pylori* HtrA. Sci Rep 6:23264. https://doi.org/10.1038/srep23264

Schmitt W, Haas R (1994) Genetic analysis of the *Helicobacter pylori* vacuolating cytotoxin: structural similarities with the IgA protease type of exported protein. Mol Microbiol 12(2):307–319

Segal ED, Cha J, Lo J, Falkow S, Tompkins LS (1999) Altered states: involvement of phosphorylated CagA in the induction of host cellular growth changes by *Helicobacter pylori*. Proc Natl Acad Sci U S A 96 (25):14559–14564

Selbach M, Moese S, Hurwitz R, Hauck CR, Meyer TF, Backert S (2003) The *Helicobacter pylori* CagA protein induces cortactin dephosphorylation and actin rearrangement by c-Src inactivation. EMBO J 22 (3):515–528. https://doi.org/10.1093/emboj/cdg050

Selbach M, Paul FE, Brandt S, Guye P, Daumke O, Backert S, Dehio C, Mann M (2009) Host cell interactome of tyrosine-phosphorylated bacterial proteins. Cell Host Microbe 5(4):397–403. https://doi.org/10.1016/j.chom.2009.03.004

Senda Y, Hatakeyama M (2016) CagA. In: Suzuki H, Warren R, Marshall B (eds) Helicobacter pylori. Springer, Tokyo, pp 33–47. https://doi.org/10.1007/978-4-431-55705-0_3

Seto K, Hayashi-Kuwabara Y, Yoneta T, Suda H, Tamaki H (1998) Vacuolation induced by cytotoxin from *Helicobacter pylori* is mediated by the EGF receptor in HeLa cells. FEBS Lett 431(3):347–350

Sgouras DN, Panayotopoulou EG, Papadakos K, Martinez-Gonzalez B, Roumbani A, Panayiotou J, van Vliet-Constantinidou C, Mentis AF, Roma-Giannikou E (2009) CagA and VacA polymorphisms do not correlate with severity of histopathological lesions in *Helicobacter pylori*-infected Greek children. J Clin Microbiol 47(8):2426–2434. https://doi.org/10.1128/jcm.00159-09

Shaffer CL, Gaddy JA, Loh JT, Johnson EM, Hill S, Hennig EE, McClain MS, McDonald WH, Cover TL (2011) *Helicobacter pylori* exploits a unique repertoire of type IV secretion system components for pilus assembly at the bacteria-host cell interface. PLoS Pathog 7(9):e1002237. https://doi.org/10.1371/journal.ppat.1002237

Shibayama K, Kamachi K, Nagata N, Yagi T, Nada T, Doi Y, Shibata N, Yokoyama K, Yamane K, Kato H, Iinuma Y, Arakawa Y (2003) A novel apoptosis-inducing protein from *Helicobacter pylori*. Mol Microbiol 47(2):443–451

Shibayama K, Wachino J, Arakawa Y, Saidijam M, Rutherford NG, Henderson PJ (2007) Metabolism of glutamine and glutathione via gamma-glutamyltranspeptidase and glutamate transport in *Helicobacter pylori*: possible significance in the pathophysiology of the organism. Mol Microbiol 64 (2):396–406. https://doi.org/10.1111/j.1365-2958.2007.05661.x

Sierra JC, Asim M, Verriere TG, Piazuelo MB, Suarez G, Romero-Gallo J, Delgado AG, Wroblewski LE, Barry DP (2018) Epidermal growth factor receptor inhibition downregulates *Helicobacter pylori*-induced epithelial inflammatory responses, DNA damage and gastric carcinogenesis. Gut 67(7):1247–1260. https://doi.org/10.1136/gutjnl-2016-312888

Sougleri IS, Papadakos KS, Zadik MP, Mavri-Vavagianni-M, Mentis AF, Sgouras DN (2016) *Helicobacter pylori* CagA protein induces factors involved in the epithelial to mesenchymal transition (EMT) in infected gastric epithelial cells in an EPIYA- phosphorylation-dependent manner. FEBS J 283(2):206–220. https://doi.org/10.1111/febs.13592

Stein M, Rappuoli R, Covacci A (2000) Tyrosine phosphorylation of the *Helicobacter pylori* CagA antigen after cag-driven host cell translocation. Proc Natl Acad Sci U S A 97(3):1263–1268

Stein SC, Faber E, Bats SH, Murillo T, Speidel Y, Coombs N, Josenhans C (2017) *Helicobacter pylori* modulates host cell responses by CagT4SS-dependent translocation of an intermediate metabolite of LPS inner core heptose biosynthesis. PLoS Pathog 13(7): e1006514. https://doi.org/10.1371/journal.ppat.1006514

Suzuki M, Mimuro H, Kiga K, Fukumatsu M, Ishijima N, Morikawa H, Nagai S, Koyasu S, Gilman RH, Kersulyte D, Berg DE, Sasakawa C (2009) *Helicobacter pylori* CagA phosphorylation-independent function in epithelial proliferation and inflammation. Cell Host Microbe 5(1):23–34. https://doi.org/10.1016/j.chom.2008.11.010

Tammer I, Brandt S, Hartig R, Konig W, Backert S (2007) Activation of Abl by *Helicobacter pylori*: a novel kinase for CagA and crucial mediator of host cell scattering. Gastroenterology 132(4):1309–1319. https://doi.org/10.1053/j.gastro.2007.01.050

Tan S, Noto JM, Romero-Gallo J, Peek RM, Jr., Amieva MR (2011) *Helicobacter pylori* perturbs iron trafficking in the epithelium to grow on the cell surface. PLoS Pathog 7 (5):e1002050. doi:https://doi.org/10.1371/journal.ppat.1002050
Tegtmeyer N, Hartig R, Delahay RM, Rohde M, Brandt S, Conradi J, Takahashi S, Smolka AJ, Sewald N, Backert S (2010) A small fibronectin-mimicking protein from bacteria induces cell spreading and focal adhesion formation. J Biol Chem 285(30):23515–23526. https://doi.org/10.1074/jbc.M109.096214
Tegtmeyer N, Wittelsberger R, Hartig R, Wessler S, Martinez-Quiles N, Backert S (2011) Serine phosphorylation of cortactin controls focal adhesion kinase activity and cell scattering induced by *Helicobacter pylori*. Cell Host Microbe 9(6):520–531. https://doi.org/10.1016/j.chom.2011.05.007
Tegtmeyer N, Moodley Y, Yamaoka Y, Pernitzsch SR, Schmidt V, Traverso FR, Schmidt TP, Rad R, Yeoh KG, Bow H, Torres J, Gerhard M, Schneider G, Wessler S, Backert S (2016) Characterisation of worldwide *Helicobacter pylori* strains reveals genetic conservation and essentiality of serine protease HtrA. Mol Microbiol 99(5):925–944. https://doi.org/10.1111/mmi.13276
Tegtmeyer N, Neddermann M, Asche CI, Backert S (2017a) Subversion of host kinases: a key network in cellular signaling hijacked by *Helicobacter pylori* CagA. Mol Microbiol 105(3):358–372. https://doi.org/10.1111/mmi.13707
Tegtmeyer N, Wessler S, Necchi V, Rohde M, Harrer A, Rau TT, Asche CI, Boehm M, Loessner H, Figueiredo C, Naumann M, Palmisano R, Solcia E, Ricci V, Backert S (2017b) *Helicobacter pylori* employs a unique Basolateral type IV secretion mechanism for CagA delivery. Cell Host Microbe 22 (4):552-560.e555. doi:https://doi.org/10.1016/j.chom.2017.09.005
Telford JL, Ghiara P, Dell'Orco M, Comanducci M, Burroni D, Bugnoli M, Tecce MF, Censini S, Covacci A, Xiang Z et al (1994) Gene structure of the *Helicobacter pylori* cytotoxin and evidence of its key role in gastric disease. J Exp Med 179 (5):1653–1658
Terebiznik MR, Raju D, Vazquez CL, Torbricki K, Kulkarni R, Blanke SR, Yoshimori T, Colombo MI, Jones NL (2009) Effect of *Helicobacter pylori*'s vacuolating cytotoxin on the autophagy pathway in gastric epithelial cells. Autophagy 5(3):370–379
Tombola F, Carlesso C, Szabò I, de Bernard M, Reyrat JM, Telford JL, Rappuoli R, Montecucco C, Papini E, Zoratti M (1999) *Helicobacter pylori* vacuolating toxin forms anion-selective channels in planar lipid bilayers: possible implications for the mechanism of cellular vacuolation. Biophys J 76(3):1401–1409. https://doi.org/10.1016/S0006-3495(99)77301-7
Tombola F, Morbiato L, Del Giudice G, Rappuoli R, Zoratti M, Papini E (2001) The *Helicobacter pylori* VacA toxin is a urea permease that promotes urea diffusion across epithelia. J Clin Invest 108 (6):929–937. https://doi.org/10.1172/jci13045
Torres VJ, VanCompernolle SE, Sundrud MS, Unutmaz D, Cover TL (2007) *Helicobacter pylori* vacuolating cytotoxin inhibits activation-induced proliferation of human T and B lymphocyte subsets. J Immunol 179(8):5433–5440
Tsang YH, Lamb A, Romero-Gallo J, Huang B, Ito K, Peek RM, Jr., Ito Y, Chen LF (2010) *Helicobacter pylori* CagA targets gastric tumor suppressor RUNX3 for proteasome-mediated degradation. Oncogene 29 (41):5643-5650. doi:https://doi.org/10.1038/onc.2010.304
Tsutsumi R, Higashi H, Higuchi M, Okada M, Hatakeyama M (2003) Attenuation of *Helicobacter pylori* CagA x SHP-2 signaling by interaction between CagA and C-terminal Src kinase. J Biol Chem 278 (6):3664–3670. https://doi.org/10.1074/jbc.M208155200
Tummūru MK, Cover TL, Blaser MJ (1993) Cloning and expression of a high-molecular-mass major antigen of *Helicobacter pylori*: evidence of linkage to cytotoxin production. Infect Immun 61(5):1799–1809
Utsch C, Haas R (2016) VacA's induction of VacA-containing vacuoles (VCVs) and their immunomodulatory activities on human T cells. Toxins (Basel) 8(6). https://doi.org/10.3390/toxins8060190
Utt M, Danielsson B, Wadstrom T (2001) *Helicobacter pylori* vacuolating cytotoxin binding to a putative cell surface receptor, heparan sulfate, studied by surface plasmon resonance. FEMS Immunol Med Microbiol 30(2):109–113
Varga MG, Shaffer CL, Sierra JC, Suarez G, Piazuelo MB, Whitaker ME, Romero-Gallo J, Krishna US, Delgado A, Gomez MA, Good JA, Almqvist F, Skaar EP, Correa P, Wilson KT, Hadjifrangiskou M, Peek RM (2016) Pathogenic *Helicobacter pylori* strains translocate DNA and activate TLR9 via the cancer-associated cag type IV secretion system. Oncogene 35(48):6262–6269. https://doi.org/10.1038/onc.2016.158
Viala J, Chaput C, Boneca IG, Cardona A, Girardin SE, Moran AP, Athman R, Memet S, Huerre MR, Coyle AJ, DiStefano PS, Sansonetti PJ, Labigne A, Bertin J, Philpott DJ, Ferrero RL (2004) Nod1 responds to peptidoglycan delivered by the *Helicobacter pylori* cag pathogenicity island. Nat Immunol 5(11):1166–1174. https://doi.org/10.1038/ni1131
Vinion-Dubiel AD, McClain MS, Czajkowsky DM, Iwamoto H, Ye D, Cao P, Schraw W, Szabo G, Blanke SR, Shao Z, Cover TL (1999) A dominant negative mutant of *Helicobacter pylori* vacuolating toxin (VacA) inhibits VacA-induced cell vacuolation. J Biol Chem 274(53):37736–37742
Wandler AM, Guillemin K (2012) Transgenic expression of the *Helicobacter pylori* virulence factor CagA promotes apoptosis or tumorigenesis through JNK

activation in Drosophila. PLoS Pathog 8(10): e1002939. https://doi.org/10.1371/journal.ppat.1002939

Wang F, Xia P, Wu F, Wang D, Wang W, Ward T, Liu Y, Aikhionbare F, Guo Z, Powell M, Liu B, Bi F, Shaw A, Zhu Z, Elmoselhi A, Fan D, Cover TL, Ding X, Yao X (2008) *Helicobacter pylori* VacA disrupts apical membrane-cytoskeletal interactions in gastric parietal cells. J Biol Chem 283(39):26714–26725. https://doi.org/10.1074/jbc.M800527200

Wang G, Romero-Gallo J, Benoit SL, Piazuelo MB, Dominguez RL, Morgan DR, Peek RM, Jr., Maier RJ (2016) Hydrogen metabolism in *Helicobacter pylori* plays a role in gastric carcinogenesis through facilitating CagA translocation. mBio 7 (4). doi: https://doi.org/10.1128/mBio.01022-16

Wei J, Nagy TA, Vilgelm A, Zaika E, Ogden SR, Romero-Gallo J, Piazuelo MB, Correa P, Washington MK, El-Rifai W, Peek RM, Zaika A (2010) Regulation of p53 tumor suppressor by *Helicobacter pylori* in gastric epithelial cells. Gastroenterology 139(4):1333–1343. https://doi.org/10.1053/j.gastro.2010.06.018

Wei J, Noto JM, Zaika E, Romero-Gallo J, Piazuelo MB, Schneider B, El-Rifai W, Correa P, Peek RM, Zaika AI (2015) Bacterial CagA protein induces degradation of p53 protein in a p14ARF-dependent manner. Gut 64 (7):1040–1048. https://doi.org/10.1136/gutjnl-2014-307295

Wessler S, Backert S (2017) A novel basolateral type IV secretion model for the CagA oncoprotein of *Helicobacter pylori*. Microbial cell (Graz, Austria) 5 (1):60–62. https://doi.org/10.15698/mic2018.01.611

Wiedemann T, Hofbaur S, Tegtmeyer N, Huber S, Sewald N, Wessler S, Backert S, Rieder G (2012) *Helicobacter pylori* CagL dependent induction of gastrin expression via a novel alphavbeta5-integrin-integrin linked kinase signalling complex. Gut 61 (7):986–996. https://doi.org/10.1136/gutjnl-2011-300525

Willhite DC, Blanke SR (2004) *Helicobacter pylori* vacuolating cytotoxin enters cells, localizes to the mitochondria, and induces mitochondrial membrane permeability changes correlated to toxin channel activity. Cell Microbiol 6(2):143–154

Winter JA, Letley DP, Cook KW, Rhead JL, Zaitoun AA, Ingram RJ, Amilon KR, Croxall NJ, Kaye PV, Robinson K, Atherton JC (2014) A role for the vacuolating cytotoxin, VacA, in colonization and *Helicobacter pylori*-induced metaplasia in the stomach. J Infect Dis 210(6):954–963. https://doi.org/10.1093/infdis/jiu154

Wroblewski LE, Peek RM, Jr. (2007) Orchestration of dysregulated epithelial turnover by a manipulative pathogen. Cell Host Microbe 2 (4):209-211. doi: https://doi.org/10.1016/j.chom.2007.09.011

Wunder C, Churin Y, Winau F, Warnecke D, Vieth M, Lindner B, Zahringer U, Mollenkopf HJ, Heinz E, Meyer TF (2006) Cholesterol glucosylation promotes immune evasion by *Helicobacter pylori*. Nat Med 12 (9):1030–1038. https://doi.org/10.1038/nm1480

Wustner S, Anderl F, Wanisch A, Sachs C, Steiger K, Nerlich A, Vieth M, Mejias-Luque R, Gerhard M (2017) *Helicobacter pylori* gamma-glutamyl transferase contributes to colonization and differential recruitment of T cells during persistence. Sci Rep 7(1):13636. https://doi.org/10.1038/s41598-017-14028-1

Xia Y, Yamaoka Y, Zhu Q, Matha I, Gao X (2009) A comprehensive sequence and disease correlation analyses for the C-terminal region of CagA protein of *Helicobacter pylori*. PLoS One 4(11):e7736. https://doi.org/10.1371/journal.pone.0007736

Yahiro K, Niidome T, Kimura M, Hatakeyama T, Aoyagi H, Kurazono H, Imagawa K, Wada A, Moss J, Hirayama T (1999) Activation of *Helicobacter pylori* VacA toxin by alkaline or acid conditions increases its binding to a 250-kDa receptor protein-tyrosine phosphatase beta. J Biol Chem 274 (51):36693–36699

Yahiro K, Wada A, Nakayama M, Kimura T, Ogushi K, Niidome T, Aoyagi H, Yoshino K, Yonezawa K, Moss J, Hirayama T (2003) Protein-tyrosine phosphatase alpha, RPTP alpha, is a *Helicobacter pylori* VacA receptor. J Biol Chem 278(21):19183–19189. https://doi.org/10.1074/jbc.M300117200

Yahiro K, Wada A, Yamasaki E, Nakayama M, Nishi Y, Hisatsune J, Morinaga N, Sap J, Noda M, Moss J, Hirayama T (2004) Essential domain of receptor tyrosine phosphatase beta (RPTPbeta) for interaction with *Helicobacter pylori* vacuolating cytotoxin. J Biol Chem 279(49):51013–51021. https://doi.org/10.1074/jbc.M406473200

Yahiro K, Satoh M, Nakano M, Hisatsune J, Isomoto H, Sap J, Suzuki H, Nomura F, Noda M, Moss J, Hirayama T (2012) Low-density lipoprotein receptor-related protein-1 (LRP1) mediates autophagy and apoptosis caused by *Helicobacter pylori* VacA. J Biol Chem 287(37):31104–31115. https://doi.org/10.1074/jbc.M112.387498

Yamaoka Y, Graham DY (2014) *Helicobacter pylori* virulence and cancer pathogenesis. Future Oncol 10 (8):1487–1500. https://doi.org/10.2217/fon.14.29

Yamasaki E, Wada A, Kumatori A, Nakagawa I, Funao J, Nakayama M, Hisatsune J, Kimura M, Moss J, Hirayama T (2006) *Helicobacter pylori* vacuolating cytotoxin induces activation of the proapoptotic proteins Bax and Bak, leading to cytochrome c release and cell death, independent of vacuolation. J Biol Chem 281(16):11250–11259. https://doi.org/10.1074/jbc.M509404200

Zeaiter Z, Cohen D, Musch A, Bagnoli F, Covacci A, Stein M (2008) Analysis of detergent-resistant membranes of *Helicobacter pylori* infected gastric adenocarcinoma cells reveals a role for MARK2/Par1b in CagA-mediated disruption of cellular polarity. Cell Microbiol 10(3):781–794. https://doi.org/10.1111/j.1462-5822.2007.01084.x

Zhang XS, Tegtmeyer N, Traube L, Jindal S, Perez-Perez-G, Sticht H, Backert S, Blaser MJ (2015) A specific A/T polymorphism in Western tyrosine phosphorylation B-motifs regulates *Helicobacter pylori* CagA epithelial cell interactions. PLoS Pathog 11(2):e1004621. https://doi.org/10.1371/journal.ppat.1004621

Zheng PY, Jones NL (2003) *Helicobacter pylori* strains expressing the vacuolating cytotoxin interrupt phagosome maturation in macrophages by recruiting and retaining TACO (coronin 1) protein. Cell Microbiol 5(1):25–40

Zimmermann S, Pfannkuch L, Al-Zeer MA, Bartfeld S, Koch M, Liu J, Rechner C, Soerensen M, Sokolova O, Zamyatina A, Kosma P, Maurer AP, Glowinski F, Pleissner KP, Schmid M, Brinkmann V, Karlas A, Naumann M, Rother M, Machuy N, Meyer TF (2017) ALPK1- and TIFA-dependent innate immune response triggered by the *Helicobacter pylori* type IV secretion system. Cell Rep 20(10):2384–2395. https://doi.org/10.1016/j.celrep.2017.08.039

Adv Exp Med Biol - Advances in Microbiology, Infectious Diseases and Public Health (2019) 1149: 57–75
https://doi.org/10.1007/5584_2019_359

Published online: 24 April 2019

Roles of Adhesion to Epithelial Cells in Gastric Colonization by *Helicobacter pylori*

Daniel A. Bonsor and Eric J. Sundberg

Abstract

Helicobacter pylori adherence to host epithelial cells is essential for its survival against the harsh conditions of the stomach and for successful colonization. Adherence of *H. pylori* is achieved through several related families of outer membrane proteins and proteins of a type IV secretion system (T4SS), which bridge *H. pylori* to host cells through protein-protein and other protein-ligand interactions. Local environmental conditions such as cell type, available host cell surface proteins and/or ligands, as well as responses by the host immune system force *H. pylori* to alter expression of these proteins to adapt quickly to the local environment in order to colonize and survive. Some of these host-pathogen interactions appear to function in a "catch-and-release" manner, regulated by reversible binding at varying pH and allowing *H. pylori* to detach itself from cells or debris sloughed off the gastric epithelial lining in order to return for subsequent productive interactions. Other interactions between bacterial adhesin proteins and host adhesion molecules, however, appear to function as a committed step in certain pathogenic processes, such as translocation of the CagA oncoprotein through the *H. pylori* T4SS and into host gastric epithelial cells. Understanding these adhesion interactions is critical for devising new therapeutic strategies, as they are responsible for the earliest stage of infection and its maintenance. This review will discuss the expression and regulation of several outer membrane proteins and CagL, how they engage their known host cell protein/ligand targets, and their effects on clinical outcome.

Keywords

Adhesion · Adhesin · Protein · Blood group antigen · Structure

D. A. Bonsor
Institute of Human Virology, University of Maryland School of Medicine, University of Maryland, Baltimore, MD, USA
e-mail: dbonsor@som.umaryland.edu

E. J. Sundberg (✉)
Institute of Human Virology, University of Maryland School of Medicine, University of Maryland, Baltimore, MD, USA

Department of Medicine, University of Maryland School of Medicine, University of Maryland, Baltimore, MD, USA

Department of Microbiology and Immunology, University of Maryland School of Medicine, University of Maryland, Baltimore, MD, USA
e-mail: esundberg@ihv.umaryland.edu

1 Introduction

Bacterial colonization of the stomach is fraught with danger. The constant production of hydrochloric acid and the resulting low pH, churning of the stomach, and the rapid turnover of epithelial cells (on the order of every 2–3 days) makes the stomach an inhospitable place (Belanger and Leblond 1946; Lee 1985; Schreiber et al. 2004). However, *H. pylori* has evolved mechanisms to neutralize the acid (Langenberg et al. 1984; Mobley 1996) and to move towards the epithelial cell surface through the mucosal barrier (Beier et al. 1997; Croxen et al. 2006; Keilberg and Ottemann 2016) that protects the stomach lining, where the nearly neutral pH provides a much more tolerable environment. Once encountering the gastric epithelial cell layer, *H. pylori* must anchor themselves to host cell plasma membranes in order to prevent moving back into the lumen and expulsion from the stomach. Bacterial survival depends on these mechanisms of host cell adherence.

Once *H. pylori* bacteria have adhered to host cells, they will remain within the stomach and the host will remain asymptomatic in approximately 80% of infected individuals (Blaser et al. 1995; Israel et al. 2001; Parsonnet et al. 1997). However, the remaining 20% will go on to develop gastritis, peptic ulcer disease (PUD), mucosa-associated lymphoid tissue (MALT) lymphoma and/or gastric cancer (GC) during their lifetimes (Blaser et al. 1995; Israel et al. 2001; Parsonnet et al. 1997). The more virulent strains of *H. pylori* that can translocate the CagA protein are strongly associated with these diseases (Censini et al. 1996; Parsonnet et al. 1997). CagA is an oncoprotein that interferes with cell signaling pathways through its interactions with host factors such as E-cadherin, CRK, CSK, PAR-1, SHP-2, GRB-2 and ASPP-2 (Buti et al. 2011; Lu et al. 2008; Mimuro et al. 2002; Murata-Kamiya et al. 2007; Segal et al. 1999; Selbach et al. 2009; Tegtmeyer et al. 2011; Tsutsumi et al. 2003; Zhang et al. 2015), thereby perturbing cytoskeletal organization, motility, proliferation, cell-cell contact, mitogenic gene expression and apoptosis (compare Chap. 3 of this book). CagA is encoded by the cytotoxic-associated gene pathogenicity island (*cag*PAI), a region of the *H. pylori* genome that contains ~30 genes which encode for a T4SS that delivers CagA into host cells (Censini et al. 1996; Backert et al. 2015). CagL, another member of the *cag*PAI, forms part of the T4SS injection pilus and aids in adherence of *H. pylori* to host cells and the successful translocation of CagA (Posselt et al. 2013).

To achieve adherence, the *H. pylori* genome contains over 60 outer membrane protein (OMP) genes which can be divided into five paralogous gene families (Alm et al. 2000). The largest family consists of the Hop (*H. pylori* OMP) and Hor (Hop-related) genes, which encode 33 proteins. The second family is Hof (*Helicobacter* related) with eight genes, whilst the third, Hom (*Helicobacter* outer membrane) is the smallest with four genes (Alm et al. 2000). The remaining OMPs are contained within the iron-regulated and efflux pump OMP families. Driven by gene recombination and duplication, increase in mutational rate and the exchange of DNA between different strains (Didelot et al. 2013; Kennemann et al. 2011; Morelli et al. 2010), every genome of each strain of *H. pylori* differs in the OMPs that it possesses (Alm et al. 1999; Tomb et al. 1997). In addition to the genetic differences in the OMPs between strains, each strain also regulates expression of OMPs through several different mechanisms, including phase and allelic variation, gene conversion, gene duplication, and regulation in response to pH and salt. The combination of different OMPs and their regulation allows *H. pylori* to respond to the local environment in the stomach and host immune mechanisms in order to establish and maintain colonization (Kang and Blaser 2006; Odenbreit et al. 2009).

This review summarizes our current understanding of the bacterial adhesins, from both the OMP families and the *cag*PAI, that have been demonstrated to be involved in adherence of *H. pylori* as well as their pathogenic roles in the promotion of disease. An overview of adhesins and their known ligands is shown in Fig. 1.

2 OMP Domain Organization

To date, all *H. pylori* OMPs established as *bona fide* adhesins belong to the Hop, Hor and Hom

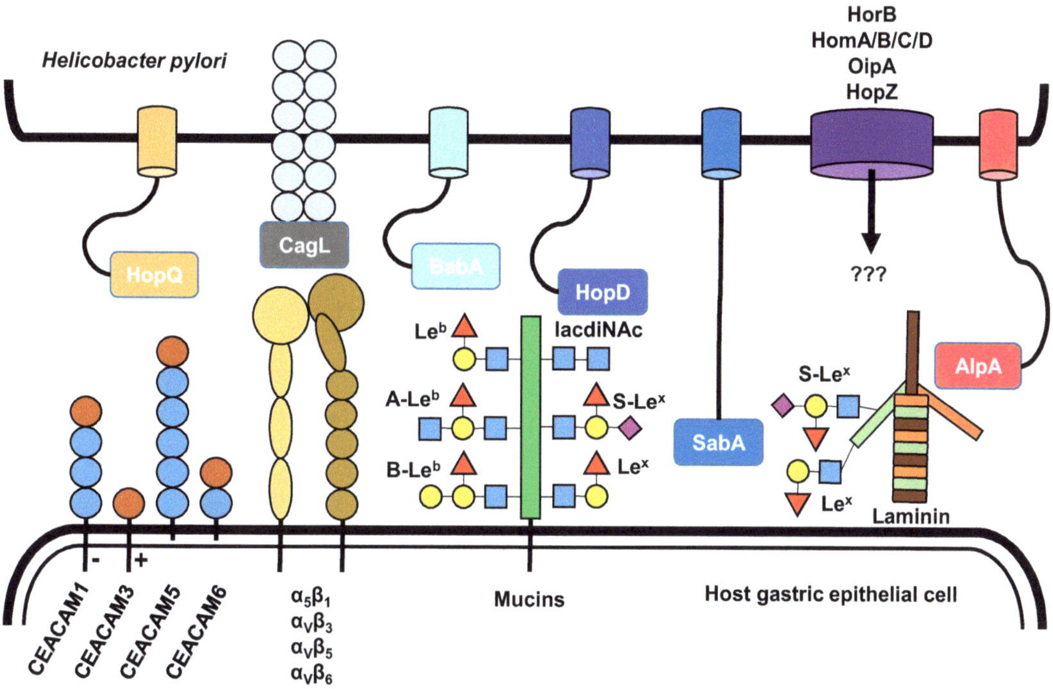

Fig. 1 Graphical representation of *H. pylori* adhesin interactions with host cell proteins. HopQ (yellow) interacts with CEACAM1, -3, -5 and -6 (IgV domain, brown, IgC2 domains, blue). CEACAM1 and CEACAM3 contain an immunoreceptor tyrosine-based inhibition (−) and activation (+) motifs, respectively. CagL (grey) is attached to the pilus tip (light grey circles) of the T4SS, where it can interact with the $\alpha 5\beta 1$, $\alpha V\beta 3$, $\alpha V\beta 5$ and $\alpha V\beta 6$ integrins. Mucins (green) are heavily glycosylated proteins and can be decorated with the blood group antigens (Le^b, A-Le^b and B-Le^B), both sialyated and asialyated Le^x, and lcadiNAc, which bind to BabA (cyan), SabA (blue) and HopD (purple), respectively. SabA also binds both sialyated and asialyated Le^x attached to Laminin. AlpA (pink) also binds Laminin. HorB, HomA/B/C/D, OipA and HopZ are outer membrane proteins of *H. pylori*. The host cell ligands of these proteins are currently unknown

families. An analysis of the genomic sequences of strains J99 and 26.695 grouped the OMPs into five separate families (Alm et al. 2000). Furthermore, comparison of the Hop and Hor families indicates that they share a common domain organization, including an N-terminal signal peptide of ~20–25 residues, a C-terminal β-barrel to anchor the protein to the outer membrane and a central domain that confers host protein specificity (Fig. 2a). The domain organization is reminiscent of autotransporters. Recently, however, several C-terminal β-barrels of the Hop proteins have been shown to be split, with a single strand found at the N-terminus, with the remaining strands found at the C-terminus (Coppens et al. 2018). Thus, the large extracellular domains are found to be inserted in the extracellular loops of the β-barrel (Fig. 2a).

3 Adhesins

Protein adhesins may recognize small host cell molecules (e.g., sugars) or larger ligands (e.g., proteins). Below, the known *H. pylori* adhesin proteins are discussed in detail, grouped according to the host cell binding partner to which they bind.

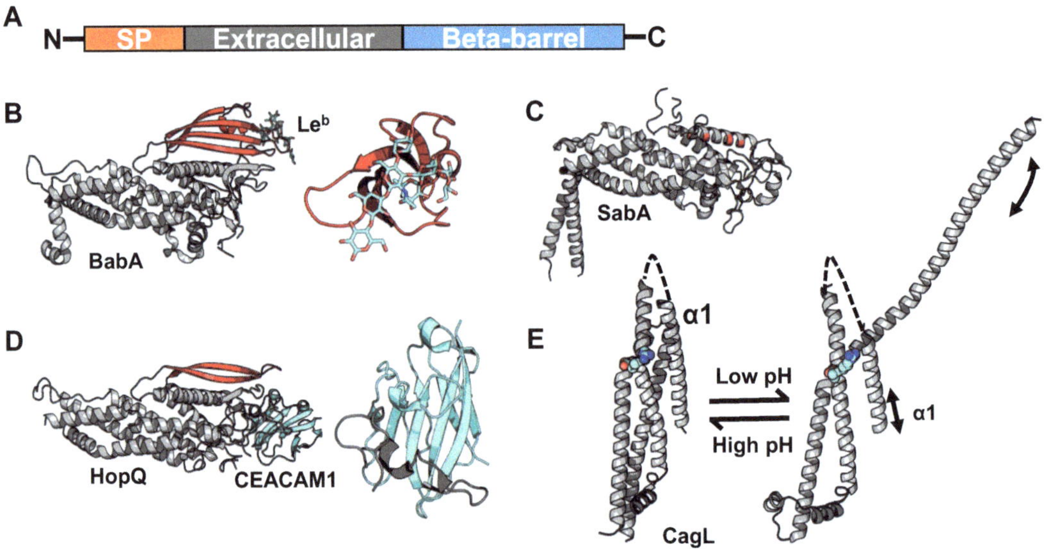

Fig. 2 Structures of *H. pylori* adhesins. (**a**) Domain organization of both the Hop and Hor proteins. *Top* – The original domain organization of Hop and Hor proteins consists of a N-terminal signal peptide (SP, orange), an extracellular domain (grey) and a C-terminal beta-barrel (blue), reminiscent of autotransporters. *Bottom* – The new domain organization of Hop and Hor proteins which show that part of the beta-barrel is formed from several residues of the N-terminus. (**b**) *Left* – Structure of BabA (PDB entry 5F9D, grey) interacting with the Lewis b blood group B heptasaccharide ligand (cyan sticks) through its insertion domain (red). *Right* – Closer inspection of the interaction. (**c**) Structure of SabA (PDB entry 4O5J, grey) with proposed s-Le^x binding site highlighted in red. (**d**) *Left* – Structure of the Type I HopQ-CEACAM1 complex (PDB entry 6AW2, grey and cyan, respectively). CEACAM1 does not contact the smaller insertion domain (red). (**e**) Conformational changes in CagL (grey) in response to pH exposes the RGD motif (cyan spheres). *Right* – At low pH, CagL exists in an extended state, with the α1 helix burying the arginine of the RGD motif (PDB entry 4X5U). *Left* – At high pH, CagL compacts and α1 moves exposing the RGD motif and allows recruitment of integrins (PDB entry 3ZCI)

3.1 Adhesins with Known Small Ligands

3.1.1 BabA/BabB/BabC

These adhesins are also known as HopS/HopT/HopU, respectively. The blood group antigen-binding adhesins (Bab) are approximately 80 kDa in size. The majority of research conducted on these proteins has been focused on BabA. BabA recognizes and can bind the mono-(H) or di-fucosylated (Lewis b, Le^b) blood group antigens from the O blood group (Boren et al. 1993; Ilver et al. 1998) and the A and B blood group antigens (A-Le^b and B-Le^b) (Aspholm-Hurtig et al. 2004), all of which are found on the surfaces of gastric epithelial cells and certain secreted mucins, including MUC1 and MUC5AC of the stomach (Linden et al. 2004; Sakamoto et al. 1989) and MUC5B in the salivary glands (Bosch et al. 2000; Veerman et al. 1997). Analysis of BabA binding to salivary proteins by matrix-assisted laser desorption ionization-mass spectrometry (MALDI-MS) identified two proteins, which could bind BabA – proline-rich glycoprotein and gp-340 (Walz et al. 2009). BabB displays no Le^b binding activity, whilst the binding specificity of BabC is unknown (Kim et al. 2015; Saberi et al. 2016). BabA exists in two allele forms, *babA2*, which encodes for the full length protein, and *babA1*, which contains a 10-bp deletion in the signal peptide that results in a frame shift (Backstrom et al. 2004; Ilver et al. 1998). The location within the chromosome is highly variable. BabA and BabB are typically found at either locus A or B, with BabC commonly found in locus C (Armitano et al. 2013;

Colbeck et al. 2006; Hennig et al. 2006). Expression of BabA is regulated by phase variation in the signal peptide through a cytosine-thymine (CT) dinucleotide repeat (Ilver et al. 1998). The CT repeats can allow slipped strand mispairing (SSM) during replication and, thus, induce a reading frame shift (Colbeck et al. 2006; Ilver et al. 1998; Solnick et al. 2004; Styer et al. 2010). Regulation can also occur in response to host mucin expression (Skoog et al. 2012) and through gene conversion of *babA* with *babB*, creating BabA/B chimeras that have varying abilities to bind Le^b (Backstrom et al. 2004; Colbeck et al. 2006; Matteo et al. 2011; Pride and Blaser 2002; Solnick et al. 2004). In models of acute *H. pylori* infections in mice, gerbils and rhesus monkeys, BabA expression was found to be lost regularly during the early stages of infection due to phase variation or gene conversion (Hansen et al. 2017; Ohno et al. 2011; Solnick et al. 2004; Styer et al. 2010). This, however, appears to be a rare event in humans (Nell et al. 2014). However, these strains were isolated from chronically infected humans, for which mutation rates are approximately ten times slower compared to strains commonly found in acute infections (Nell et al. 2014). This surge of mutations observed in acute infections allows these *H. pylori* strains to respond quickly and adapt to the host environment (Linz et al. 2014).

The affinity of the extracellular domain of BabA for the blood group antigens is rather weak. Measured affinities include: Le^b antigen hexasaccharide, $K_D = 250$ μM; Le^b antigen pentasaccharide, $K_D = 80$ μM; H antigen pentasaccharide, $K_D = 620$ μM; B-Le^b septasaccharide, $K_D = 40$ μM; and A type 1 hexasaccharide (A6–1), $K_D = 150$ μM (Hage et al. 2015; Moonens et al. 2016). These are much weaker than Le^b antigen interaction with full length BabA as measured by surface plasmon resonance (SPR) analyses and cell binding assays, which exhibits an affinity of 390 pM (Aspholm-Hurtig et al. 2004; Ilver et al. 1998; Imberty et al. 2005; Moonens et al. 2016). Cross-linking of full length BabA indicates that it oligomerizes, primarily as trimers in *H. pylori* outer membranes (Moonens et al. 2016). The structure of the extracellular domain of BabA is a 4 + 3 helical bundle fold (Moonens et al. 2016) (Fig. 2b). An 80-residue insertion domain comprised of a four-stranded sheet with a helical loop crowning the beta-strands is located between helices 4 and 5. Several structures of BabA in complex with various blood group antigens have been determined by X-ray crystallography (Hage et al. 2015; Moonens et al. 2016). In these structures, all of the glycans bind to the insertion domain, specifically to two loops (Loop 1 and Loop 2) that connect the strands and the crowning helical loop (Moonens et al. 2016). The crowning helical loop is constrained by a disulfide that, upon reduction, prevents glycan binding (Moonens et al. 2016). DL1 and DL2 differ in sequence considerably across *H. pylori* strains and, consequently, *H. pylori* isolates exhibit distinct ABO preferences and Le^b affinities (Aspholm-Hurtig et al. 2004). Most strains produce BabA generalist adhesins that promote binding to each blood group glycan; however, several strains are Le^b-only specialists. These strains are found to have a shorter Loop 1, thereby preventing binding of the N-acetylgalactosamine or galactose sugar moieties in the larger A-Le^b and B-Le^b antigens, respectively (Aspholm-Hurtig et al. 2004).

As *H. pylori* resides in the stomach, it experiences a large pH gradient. The adherence of *H. pylori* in the gastric mucosa mediated by BabA-Le^b interactions display similar affinities between pH 4.0 and 6.0 (Bugaytsova et al. 2017). Further lowering the pH results in a 1000-fold reduction in adherence. Reconditioning of the *H. pylori* to a higher pH results in the recovery of binding to gastric mucosa, demonstrating that BabA displays a reversible pH sensitivity, which has been localized to residue 199 of the crowning helical loop (Bugaytsova et al. 2017). This residue resides in yet another region of BabA that is hypervariable in length and sequence amongst *H. pylori* strains. The pH_{50} of Le^b binding (the pH value at which BabA retains 50% of its Le^b binding) was determined for tens of strains and found to vary ~2.5 pH units, from 2.3 to 4.9. Deletion of residues 199–200 of BabA in strain

17.875 resulted in an increase of the pH_{50} from 3.3 to 3.9, mimicking strains where these residues are naturally missing (Bugaytsova et al. 2017). The presence of BabA can aid in the adherence of *H. pylori*, though it is not essential as many strains exist that lack the *babA* gene (Odenbreit et al. 2009) and strains that do contain *babA* have been observed to cause PUD or GC (Gerhard et al. 1999; Yamaoka et al. 2002c). This is particularly true if the *cag*PAI is also present (Azevedo et al. 2008).

3.1.2 SabA/SabB

These adhesins are also known as HopP and HopO, respectively. The sialic acid binding (Sab) proteins are slightly smaller than BabA with a molecular weight of ~70 kDa (Alm et al. 2000). SabA recognizes and binds sialylated glycans, whilst SabB does not (Mahdavi et al. 2002). The most studied ligand is the sialyl-Lewis x (s-Le^x) sugar found attached to O-glycans. Sialyated glycans are typically found in low concentrations in normal gastric musoca (Kobayashi et al. 2009). However, SabA can bind two minor gangliosides in the stomach, Neu5Acα3-neolactohexaosylceramide and Neu5Acα3-neolactooctaosylceramide, which can promote initial infection (Benktander et al. 2018). *H. pylori* most often cause gastric inflammation by triggering IL-8 induction in host cells responding to the infection. IL-8 activates *FUT3* and *B3GNT5*, two genes involved in the biosynthesis of s-Le^x (Magalhaes et al. 2015), leading to dramatic alterations in ganglioside sialyation profiles of the gastric mucosa and an enrichment of s-Le^x (Benktander et al. 2018; Magalhaes et al. 2015). SabA has been found to interact with several host cell glycoproteins that are sialyated including MUC5B, MUC7, laminin, carbonic anhydrase VI, zinc α2-glycoprotein, parotid secretory protein and the heavy chain of secretory IgA1 (Aspholm et al. 2006; Walz et al. 2005, 2009). SabA can also bind sialylated proteins on erythrocytes, which leads to hemagglutination (Unemo et al. 2005).

Regulation of SabA expression is complex. Phase variation is observed through two SSM mechanisms: one is observed in the CT dinucleotide repeat of the signal peptide as for *babA* and the second found within a poly-thymine repeat in the promoter region (Harvey et al. 2014; Kao et al. 2012; Yamaoka et al. 2002b, 2006). This can affect transcription of *sabA* through either altering regulatory protein interactions and/or RNA polymerase. The high sequence similarities between *sabA* and *sabB* also allow gene conversion between the two (Talarico et al. 2012). SabA expression is also regulated by the external pH, through the acid-responsive ArsRS two-component signal transduction system (Goodwin et al. 2008). At low pH (pH <5.0), SabA and SabB expression are repressed, whilst at higher pH they are upregulated. The type of mucins present in the mucosa can also regulate the expression of SabA (Skoog et al. 2012). Tumor mucosa from several patients was found to consist of different mucins and glycosylation patterns. These differences were found to have an effect on SabA expression (Skoog et al. 2012). High salt concentrations can upregulate SabA expression (Loh et al. 2018); several studies have revealed a link between high salt intake and an increase GC risk in humans.

The affinity of the extracellular domain of SabA for s-Le^x is slightly tighter than BabA is for Le^b, with an affinity of 20 μM as determined by SPR (Pang et al. 2014). The same study indicated that SabA can also bind non-sialyated Lewis x (Le^x), albeit weaker with an affinity of 50 μM (Pang et al. 2014). No binding was observed between SabA and Lewis A, Le^b or Lewis Y glycans. Structurally, SabA is similar to BabA, as they share the 4 + 3 helical bundle fold (Fig. 2c). However, the insertion domain of SabA differs in sequence, is 50–70 residues shorter and some of the residues appear to be conformationally dynamic, as they are not resolved in the X-ray crystal structure. Although no high-resolution structure of a SabA-s-Le^x exists, an alanine scan of conserved residues, from a multiple sequence alignment of BabA and SabA sequences and a ligand binding site prediction program, identified a potential ligand binding pocket on the surface of SabA (Pang et al. 2014) (Fig. 2c). Two mutations (Y148A and

K152A) that had no effect on s-Le^x binding were found to weaken binding to Le^x. The Q159A mutation inhibited SabA binding to both s-Le^x and Le^x, whereas the Q162A mutation only inhibited binding to Le^x (Pang et al. 2014). This binding pocket is distinct from the insertion domain of BabA which binds blood group antigens and the carcinoembryonic antigen-related cell adhesion molecules (CEACAM) binding loop of HopQ (see below).

SabA$^+$ strains appear to be associated with GC as observed in a diverse cohort of patients (Yamaoka et al. 2006). However, another study restricted to Taiwanese patients failed to identify any significant differences with patients infected with *sabA*$^+$ and *sabA*$^-$ strains and the prevalence of gastric atrophy (Sheu et al. 2006). A similar observation is seen in a Japanese study restricted to Japanese patients (Yanai et al. 2007), suggesting that there may be geographical and environmental factors confounding the link between SabA and disease incidence and severity.

3.1.3 HopD

This GalNAcβ1-4GlcNAc glycan motif (N,N-′-diacetyllactosediamine or lacdiNAc) -binding adhesin (LabA) is a protein with a molecular weight of 77 kDa (Alm et al. 2000). LacdiNAc is only observed as an O-linked glycan on MUC5AC expressed on the superficial and foveolar epithelium of the stomach (Rossez et al. 2014). The lacdiNAc motifs comprise ~7% of human adult gastric mucin O-glycans (Kenny et al. 2012; Rossez et al. 2014). Several different strains of *H. pylori* were shown to adhere to lacdiNAc, with strain 26.695 showing the strongest adherence. *H. pylori* lysate from strain 26.695 was incubated with gastric mucins in the presence and absence of soluble lacdiNAc as a competitive binder. The supernatants were compared by SDS-PAGE analysis and revealed a prominent band in the competition experiment. Proteomic analysis identified this protein as HopD (Rossez et al. 2014). No structure has been reported for HopD, though it probably has a structure grossly similar to BabA and SabA.

3.2 Adhesins with Known Proteins

3.2.1 HopQ

HopQ is a 68 kDa protein (Alm et al. 2000) found in two allelic forms, Type I and Type II (Cao and Cover 2002). These two forms share approximately 70% sequence identity at the protein level. Both HopQ types have a significant association with GC and gastritis, with Type I HopQ also found to be associated with an increased risk of PUD (Leylabadlo et al. 2016; Ohno et al. 2009). Furthermore, HopQ Type I is found significantly more often in *cag*PAI$^+$ versus *cag*PAI$^-$ strains (Loh et al. 2008). *hopQ* was the first gene identified that is located outside of the *cag*PAI and is also essential for the translocation of the CagA oncoprotein through the T4SS (Belogolova et al. 2013; Jimenez-Soto et al. 2013). HopQ expression is regulated by salt concentrations like SabA, with higher amounts of salt leading to HopQ upregulation (Loh et al. 2018).

HopQ binds CEACAM receptors on host cell surfaces (Javaheri et al. 2016; Königer et al. 2016). Twelve CEACAMs are found in humans (Tchoupa et al. 2014) and display distinct expression patterns, with certain CEACAMs only expressed in specific tissue types (Hammarstrom 1999; Zebhauser et al. 2005). Various CEACAM members are found to possess a similar domain architecture: they are comprised of a single N-terminal IgV domain, which predominately homodimerizes, though a few can also heterodimerize; followed by a variable number of IgC2 domains and a C-terminal transmembrane helix or a glycosylphosphatidylinositol anchor (Gray-Owen and Blumberg 2006; Tchoupa et al. 2014). HopQ is found to bind the N-terminal dimerization domains of only CEACAM1, -3, -5 and -6 as determined by flow cytometry with CEACAM6 binding the weakest (Javaheri et al. 2016; Königer et al. 2016). A humanized mouse model that expresses CEACAM5 results in a significant difference in gastritis activity upon *H. pylori* infection compared to control mice (Königer et al. 2016). Isolation of *H. pylori* from these humanized mice after 6 weeks of infection display a more active T4SS as observed by the amount of CagA translocated in AGS cells *in vitro*. This is followed by a less active T4SS and lower IL-8 induction in CEACAM5 mice after 3 and 12 months of infection compared to wild-type mice (Königer et al. 2016). It is thought that to prevent clearance of the infection, *H. pylori* responds to the inflammation by lowering the activity of T4SS (i.e., a rheostat model) (Barrozo et al. 2013; Königer et al. 2016). HopQ binds one CEACAM monomer (Bonsor et al. 2018; Javaheri et al. 2016; Königer et al. 2016; Moonens et al. 2018; Tegtmeyer et al. 2019), with CEACAM1 and CEACAM3 binding affinities of ~200 and ~400 nM, respectively (Bonsor et al. 2018). CEACAM1 binding to the Type II HopQ allele is ~6-fold tighter (Moonens et al. 2018). The structure of HopQ, like those of BabA and SabA, is comprised of the common 4 + 3 helical bundle, with an insertion domain that is longer and better resolved than that of SabA, but shorter than that of BabA (Javaheri et al. 2016) (Fig. 2d). Crystal structures of the Type I HopQ in complex with CEACAM1 and CEACAM3 clearly show that CEACAMs interact with a disordered loop (in the unbound HopQ structure) of 13 residues that folds across the CEACAM dimerization interface and extends the CEACAM beta-sheet, such that Type I HopQ recognizes the monomeric form of CEACAMs and disrupts their dimerization (Bonsor et al. 2018; Moonens et al. 2018). Alanine mutagenesis of the Type I HopQ loop or the CEACAM dimerization interface failed to identify a critical residue that was important for binding (Bonsor et al. 2018), though larger substitutions in Type I HopQ such as L150N and V156N resulted in significant reduction in *H. pylori* binding to CEACAM1 expressing MKN28 cells (Moonens et al. 2018). Shortening of the loop weakened binding, whereas swapping the loop with that of BabA inactivates binding of CEACAMs and impairs translocation of CagA (Bonsor et al. 2018). The crystal structure of the Type II HopQ CEACAM1 complex is very similar to the Type I HopQ structure with two major differences. First, the disordered loop is shorter in Type II HopQ and as such does not extend the CEACAM beta-sheet. Second, the loop is more hydrophobic and results in less hydrogen bonds across the dimerization interface (Moonens et al. 2018).

The role of pH, disulfide bonds and glycans on the HopQ-CEACAM interaction have also been investigated. BabA binding to Le^b was both pH sensitive and reversible (Bugaytsova et al. 2017), whereas HopQ could still bind CEACAM1 at pH 4.0, but at lower pH values binding was neither detectable nor reversible (Bonsor et al. 2018). The CEACAM binding loop of HopQ, unlike the Le^b binding site on BabA, is not constrained by a disulfide bond, however, a disulfide exists preceding the loop and other loops proximal to the CEACAM binding site contain disulfides. Reduction of these disulfide bonds had little impact on the affinity of HopQ for CEACAM1 (Bonsor et al. 2018). CEACAMs are heavily glycosylated proteins. CEACAM1 decorated with high mannose type glycans or no glycans bind with a similar affinity to HopQ compared with the aglycosylated N-terminal domain of CEACAM1. However, CEACAM1 glycosylated with complex type glycans are found to bind eightfold tighter to HopQ, suggesting that glycans may have a role in HopQ binding through some unknown mechanism (Bonsor et al. 2018).

3.2.2 AlpA/AlpB

Also known as HopC and HopB, respectively, AlpA and AlpB share 45% sequence identity with a molecular weight of ~56–57 kDa (Alm et al. 2000). AlpA and AlpB appear to be co-expressed in all clinical strains of *H. pylori*, suggesting an essential or important function for these proteins (Odenbreit et al. 2009), in contrast with other OMPs. AlpA expression is upregulated in response to oxidative stress (Huang and Chiou 2011). Deletion of these genes results in the failed infection of guinea pigs and gerbils, the inability to adhere to human gastric tissue sections (de Jonge et al. 2004a; Senkovich et al. 2011; Sugimoto et al. 2011), and failed to stimulate secretion of IL-8 and IL-6, suggesting that these proteins are pro-inflammatory (Lu et al. 2007; Selbach et al. 2002). Stimulation of IL-8 expression may be specific to the geographic location of the strain, as deletion of AlpA and AlpB only reduced IL-8 secretion with East Asian strains (Lu et al. 2007). All strains that possess AlpA and AlpB can perturb host cell signaling pathways such as ERKs, c-Fos and cAMP-responsive element binding protein, whereas activation of Jun N-terminal Kinase, c-Jun and NF-κB signaling were specific to East Asian strains (Lu et al. 2007).

A solution of semi-purified human extracellular matrix causes aggregation of *H. pylori* (Williams et al. 2008), which is significantly reduced in a Δ*alpA/B* strain (Senkovich et al. 2011). This effect was also observed in Matrigel, a mixture of predominately collagen IV and laminin from Engelbreth-Holm-Swarm mouse sarcoma cells (Senkovich et al. 2011). Experiments with purified mouse laminin show clear binding of AlpA and AlpB by flow cytometry (Senkovich et al. 2011). Expression of AlpA and AlpB in *Escherichia coli* causes a gain of function, allowing these bacteria to adhere to mouse laminin (Senkovich et al. 2011).

3.2.3 CagL

CagL is a 25 kDa protein that forms part of the *H. pylori* T4SS, responsible for translocation of CagA into host gastric epithelial cells (Fischer et al. 2001; Kwok et al. 2007). CagL is thus strongly associated with an increased risk of GC. CagL is expressed and attached to the surfaces of the T4SS pili that form when *H. pylori* physically contacts gastric epithelial cells (Shaffer et al. 2011). CagL is essential for the translocation of CagA as deletion of CagL causes the failure of the formation of the pili (Fischer et al. 2001; Kwok et al. 2007; Shaffer et al. 2011). CagL contains an Arg-Gly-Asp (RGD) motif, a known integrin binding ligand. Various studies have demonstrated that CagL can bind to α5β1, αVβ3, αvβ5 and αvβ6 through the RGD motif (Barden and Niemann 2015; Conradi et al. 2012a, b; Kwok et al. 2007; Wiedemann et al. 2012), although three studies have shown that CagA can still be translocated in an RGD-independent manner – deletion of CagL in the P12 strain had no effect on CagA translocation nor IL-8 secretion (Jimenez-Soto et al. 2009), deletion of the RGD motif in the SU2 strain resulted in a weakened interaction between CagL and the α5β1 and αvβ5 integrins and no observed binding to αvβ6 (Bonig et al. 2016) and the $CagL^{RGA}$ mutation only weakened the interaction with αvβ5 (Wiedemann et al. 2012). CagL binds to integrins with an affinity of ~100–200 nM (Koelblen et al. 2017; Kwok et al. 2007; Wiedemann et al. 2012), independent of whether the integrin is in an extended closed or open state (Koelblen et al. 2017). This interaction is pH sensitive, with maximal binding occurring at pH 6.5 (Bonsor et al. 2015).

Several X-ray crystal structures of CagL have been determined. The first two structures revealed a four helix bundle (Barden et al. 2013; Choi et al. 2015). The RGD motif is located on a helix, which is currently the only known RGD motif not found in a loop (Fig. 2e). The following two structures identified that CagL could undergo a large conformational change, where two of the antiparallel helices become fused to form a single longer helix (Fig. 2e) and then dimerized through a domain swapped dimer mechanism (Barden et al. 2014; Bonsor et al. 2015). This conformational change is a result of low pH in both crystals and solution, though it is found not to be important for adhesion, IL-8 secretion nor CagA translocation (Bonsor et al. 2015). However, subtle conformational changes are apparent in the first helix (α1), which packs against the RGD motif (Barden et al. 2013; Bonsor et al. 2015). At low pH, α1 buries the arginine of the RGD preventing adhesion to host cells, whereas at higher pH α1 undergoes a registry shift, exposing the RGD motif and thus allows adherence, cell spreading, focal adhesion formation and heparin-binding epidermal growth factor activation (Bonsor et al. 2015; Saha et al. 2010; Tegtmeyer et al. 2010).

Several polymorphisms exist in the α1-α2 loop (residues 58–62), which may affect disease outcome in *H. pylori* infections. Worldwide, E59/I60 polymorphisms are associated significantly with GC-associated *H. pylori* isolates (Gorrell et al. 2016). Several studies indicated specific polymorphisms in local populations: in Iranian patients, the D58 polymorphism is typically observed in PUD, whereas the N58 polymorphism is associated with GC (Cherati et al.

2017); whilst in Taiwanese patients, the Y58/E59 mutations were over-represented in GC patients (Yeh et al. 2011). The role of Y58/E59 mutations is, however, debated as it has produced conflicting data. Replacement of the Y58/E59 mutation with D58/K59 resulted in a strain with a less active T4SS, lower CagA translocation and IL-8 secretion (Yeh et al. 2013). However, mutation of CagL in strain 26,695 (N58Y/E59E) inhibited CagA translocation in another study (Tegtmeyer et al. 2014). A final report investigated a larger group of polymorphisms (Y58/E59, D58/K59, D58/E59, N58/E59 and N58/K59) in both the P12 and 26.695 strains of *H. pylori* and found no significant changes in CagA translocation or IL-8 secretion (Tafreshi et al. 2015).

3.3 Adhesins with Unknown Ligands or Proteins

3.3.1 HorB

HorB or HP0127 is a 30 kDa protein. While very little is known about this protein, it is predicted to have a C-terminal β barrel domain and an N-terminal signal peptide, as well as a domain architecture similar to the Hop proteins (Snelling et al. 2007). Deletion of the gene results in a *H. pylori* strain that has a twofold reduction in adhesion, a lower production of LPS O-chains and thus the Lewis X and Y antigens attached to it. In mouse infection assays, colonization was reduced for Δ*horB* strains (Snelling et al. 2007).

3.3.2 HomA/HomB/HomC/HomD

These four proteins form the smallest OMP family and are each approximately 75 kDa in size (Oleastro et al. 2008). HomA and HomB share 90% sequence identity, and are 50% identical to HomC and HomD (Alm et al. 2000). HomC exists in three allelic forms (Kim et al. 2016). HomA and HomB occupy two defined loci within the bacterial chromosome (Alm et al. 2000). HomB can exchange positions with HomA and may have resulted from gene duplication (Alm et al. 1999, 2000; Oh et al. 2006; Tomb et al. 1997). Deletion of HomB results in *H. pylori* strains that exhibit less adherence to gastric epithelial cells and cause lower IL-8 secretion (Oleastro et al. 2008). One strain tested had two copies of HomB. Sequential deletion of the genes led to a further decrease in adherence (Oleastro et al. 2008). All four proteins are predicted to be 24-stranded β barrels (Servetas et al. 2018). The sequence variation between HomA and HomB is predicted to occur within the extracellular loops. Similar variance is observed in HomC (Servetas et al. 2018).

HomB is strongly associated with PUD (Oleastro et al. 2008). It also correlates with the presence of the *cagA*, *babA*, *hopQ* and *oipA* genes (Oleastro et al. 2008). Correlation is also observed with one of the *homC* alleles and the presence of *babA* at locus A (see BabA), suggesting that HomC may play a similar role in disease as BabA (Kim et al. 2016).

3.3.3 OipA

Outer inflammatory protein A, or HopH, is relatively small compared to the other abovementioned OMP adhesins, with a molecular weight of ~34 kDa (Alm et al. 2000). No structural data exists for this protein, but like the other Hop proteins is predicted to have a C-terminal β barrel and an N-terminal signal peptide (Alm et al. 2000). Therefore, the extracellular domain would be small, consisting of approximately 75 residues. *OipA* regulation is achieved through phase variation in the CT nucleotide repeat of the signal peptide and is found either in an *oipA*-off or *oipA*-on state (Saunders et al. 1998; Yamaoka et al. 2000). OipA is believed to stimulate IL-8 secretion from host cells and cause inflammation, hence its name, though the evidence for this is conflicting. Several studies showed that *oipA* mutants did not alter IL-8 secretion *in vitro* or inflammation in gerbils (Dossumbekova et al. 2006; Franco et al. 2008). However, other studies clearly show that OipA does in fact cause IL-8 secretion and promotes inflammation (Yamaoka et al. 2000, 2002a). Furthermore, it has been shown that IL-8 secretion by OipA is regulated through PI3K/Akt activation and inactivation of FoxO1/3a (Tabassam et al. 2012). This conflict is thought to be compounded by the fact that OipA is strongly associated with the *cag*PAI, which stimulates IL-8 secretion. Indeed, two studies have shown that greater than 95% of *cag*PAI$^+$ strains contain an *oipA*-on allele, whilst no *cag*PAI$^-$ strain has been found with an *oipA*-on allele (Ando et al. 2002; Farzi et al. 2018; Odenbreit et al. 2009). Mutation of the *H. pylori* J68 strain (*cag*PAI$^-$/ *oipA*-off) to produce an *oipA*-on strain, results in a bacterium with increased adherence, but did not alter IL-8 secretions of host cells (Horridge et al. 2017). Mutation of the 26.695 strain (*cag*PAI$^+$/*oipA*-on) to produce an *oipA*-off strain failed to stimulate IL-8 secretion, demonstrating that OipA is essential, but insufficient for IL-8 secretion (Horridge et al. 2017). The 26.695 *cag*PAI$^+$/*oipA*-off mutant strain was also found incapable of CagA translocation (Horridge et al. 2017). This is the second adhesin gene that does not reside with the *cag*PAI locus (HopQ was the first, see above), which has been shown to be essential for a functional T4SS. As the *oipA*-on allele correlates so strongly with *cag*PAI$^+$ strains, OipA is found to be associated with PUD, GC and MALT lymphoma (Dabiri et al. 2009). Purified OipA is found to trigger apoptosis of gastric epithelial cell lines through increasing Bax and cleaved Caspase 3 concentrations and lowering of Bcl-2 (Teymournejad et al. 2017).

3.3.4 HopZ

HopZ is a ~74 kDa protein (Alm et al. 2000; Peck et al. 1999). Like *hopQ*, its closest homologue (Oleastro and Menard 2013), *hopZ* exists as two alleles, which can undergo recombination (Kennemann et al. 2011; Peck et al. 1999). HopZ expression is regulated through SSM within the CT dinucleotide repeats in the signal peptide. The switch from off-to-on is influenced by colonization density and during infection (Kennemann et al. 2011, 2012; Peck et al. 1999; Yamaoka et al. 2002b). Deletion of *hopZ* results in lower adherence of *H. pylori* to gastric epithelial cell lines (Yamaoka et al. 2002b), while mutation of the gene does not affect colonization of guinea pig stomachs (de Jonge et al. 2004a). However, in germ-free transgenic mice, *hopZ* deletion reduces *H. pylori* survival in their stomachs (Giannakis et al. 2009). HopZ does not appear to be associated with gastric disease, except for MALT lymphoma where lower expression levels of HopZ are found (Chiarini et al. 2009; de Jonge et al. 2004b; Kennemann et al. 2012). Indeed, the *hopZ*-off state is found to be associated with MALT lymphoma (Lehours et al. 2004).

4 Conclusions

H. pylori has successfully adapted to life in the human stomach. The ever changing local environment (churning of the stomach, acidic environment, constant shedding of epithelial cells) in the stomach has placed substantial selective pressure on *H. pylori* to escape the acidic lumen on the stomach and swim towards the gastric epithelial lining, where it can use a repertoire of over 60 outer membrane proteins to achieve adherence to host cells. This large selection of adhesion molecules allows *H. pylori* to rapidly change the proteins that it presents on its cell surface to adhere the host proteins and/or ligands that are present in the local environment. Furthermore, altering its outer membrane proteins aids in lessening of the immune response by the host and, thus, its elimination. However, some of the proteins are associated with a more severe clinical outcome. This review has discussed several important adhesion-host protein/ligand interactions which are important in *H. pylori* colonization, survival and disease. These data could provide druggable or vaccine targets for the eradication of *H. pylori*.

References

Alm RA, Ling LS, Moir DT, King BL, Brown ED, Doig PC, Smith DR, Noonan B, Guild BC, de Jonge BL, Carmel G, Tummino PJ, Caruso A, Uria-Nickelsen M, Mills DM, Ives C, Gibson R, Merberg D, Mills SD, Jiang Q, Taylor DE, Vovis GF, Trust TJ (1999) Genomic-sequence comparison of two unrelated isolates of the human gastric pathogen *Helicobacter pylori*. Nature 397:176–180. https://doi.org/10.1038/16495

Alm RA, Bina J, Andrews BM, Doig P, Hancock RE, Trust TJ (2000) Comparative genomics of *Helicobacter pylori*: analysis of the outer membrane protein families. Infect Immun 68:4155–4168. https://doi.org/10.1128/IAI.68.7.4155-4168.2000

Ando T, Peek RM, Pride D, Levine SM, Takata T, Lee YC, Kusugami K, van der Ende A, Kuipers EJ, Kusters JG, Blaser MJ (2002) Polymorphisms of *Helicobacter pylori* HP0638 reflect geographic origin and correlate with cagA status. J Clin Microbiol 40:239–246. https://doi.org/10.1128/JCM.40.1.239-246.2002

Armitano RI, Matteo MJ, Goldman C, Wonaga A, Viola LA, De Palma GZ, Catalano M (2013) *Helicobacter pylori* heterogeneity in patients with gastritis and peptic ulcer disease. Infect Genet Evol 16:377–385. https://doi.org/10.1016/j.meegid.2013.02.024

Aspholm M, Olfat FO, Norden J, Sonden B, Lundberg C, Sjostrom R, Altraja S, Odenbreit S, Haas R, Wadstrom T, Engstrand L, Semino-Mora C, Liu H, Dubois A, Teneberg S, Arnqvist A, Boren T (2006) SabA is the *H. pylori* hemagglutinin and is polymorphic in binding to sialylated glycans. PLoS Pathog 2: e110. https://doi.org/10.1371/journal.ppat.0020110

Aspholm-Hurtig M, Dailide G, Lahmann M, Kalia A, Ilver D, Roche N, Vikstrom S, Sjostrom R, Linden S, Backstrom A, Lundberg C, Arnqvist A, Mahdavi J, Nilsson UJ, Velapatino B, Gilman RH, Gerhard M, Alarcon T, Lopez-Brea M, Nakazawa T, Fox JG, Correa P, Dominguez-Bello MG, Perez-Perez GI, Blaser MJ, Normark S, Carlstedt I, Oscarson S, Teneberg S, Berg DE, Boren T (2004) Functional adaptation of BabA, the *H. pylori* ABO blood group antigen binding adhesin. Science 305:519–522. https://doi.org/10.1126/science.1098801

Azevedo M, Eriksson S, Mendes N, Serpa J, Figueiredo C, Resende LP, Ruvoen-Clouet N, Haas R, Boren T, Le Pendu J, David L (2008) Infection by *Helicobacter pylori* expressing the BabA adhesin is influenced by the secretor phenotype. J Pathol 215:308–316. https://doi.org/10.1002/path.2363

Backert S, Tegtmeyer N, Fischer W (2015) Composition, structure and function of the *Helicobacter pylori* cag pathogenicity island encoded type IV secretion system. Future Microbiol 10(6):955–965. https://doi.org/10.2217/fmb.15.32

Backstrom A, Lundberg C, Kersulyte D, Berg DE, Boren T, Arnqvist A (2004) Metastability of *Helicobacter pylori bab* adhesin genes and dynamics in Lewis b antigen binding. Proc Natl Acad Sci U S A 101:16923–16928. https://doi.org/10.1073/pnas.0404817101

Barden S, Niemann HH (2015) Adhesion of several cell lines to *Helicobacter pylori* CagL is mediated by integrin alphaVbeta6 via an RGDLXXL motif. J Mol Biol 427:1304–1315. https://doi.org/10.1016/j.jmb.2015.01.006

Barden S, Lange S, Tegtmeyer N, Conradi J, Sewald N, Backert S, Niemann HH (2013) A helical RGD motif promoting cell adhesion: crystal structures of the *Helicobacter pylori* type IV secretion system pilus protein CagL. Structure 21:1931–1941. https://doi.org/10.1016/j.str.2013.08.018

Barden S, Schomburg B, Conradi J, Backert S, Sewald N, Niemann HH (2014) Structure of a three-dimensional domain-swapped dimer of the *Helicobacter pylori* type IV secretion system pilus protein CagL. Acta Crystallogr D Biol Crystallogr 70:1391–1400. https://doi.org/10.1107/S1399004714003150

Barrozo RM, Cooke CL, Hansen LM, Lam AM, Gaddy JA, Johnson EM, Cariaga TA, Suarez G, Peek RM Jr, Cover TL, Solnick JV (2013) Functional plasticity in the type IV secretion system of *Helicobacter pylori*. PLoS Pathog 9:e1003189. https://doi.org/10.1371/journal.ppat.1003189

Beier D, Spohn G, Rappuoli R, Scarlato V (1997) Identification and characterization of an operon of *Helicobacter pylori* that is involved in motility and stress adaptation. J Bacteriol 179:4676–4683. https://doi.org/10.1128/jb.179.15.4676-4683.1997

Belanger LF, Leblond CP (1946) A method for locating radioactive elements in tissues by covering histological sections with a photographic emulsion. Endocrinology 39:8–13. https://doi.org/10.1210/endo-39-1-8

Belogolova E, Bauer B, Pompaiah M, Asakura H, Brinkman V, Ertl C, Bartfeld S, Nechitaylo TY, Haas R, Machuy N, Salama N, Churin Y, Meyer TF (2013) *Helicobacter pylori* outer membrane protein HopQ identified as a novel T4SS-associated virulence factor. Cell Microbiol 15:1896–1912. https://doi.org/10.1111/cmi.12158

Benktander J, Barone A, Johansson MM, Teneberg S (2018) *Helicobacter pylori* SabA binding gangliosides of human stomach. Virulence 9:738–751. https://doi.org/10.1080/21505594.2018.1440171

Blaser MJ, Perez-Perez GI, Kleanthous H, Cover TL, Peek RM, Chyou PH, Stemmermann GN, Nomura A (1995) Infection with *Helicobacter pylori* strains possessing *cagA* is associated with an increased risk of developing adenocarcinoma of the stomach. Cancer Res 55:2111–2115

Bonig T, Olbermann P, Bats SH, Fischer W, Josenhans C (2016) Systematic site-directed mutagenesis of the *Helicobacter pylori* CagL protein of the Cag type IV secretion system identifies novel functional domains. Sci Rep 6:38101. https://doi.org/10.1038/srep38101

Bonsor DA, Pham KT, Beadenkopf R, Diederichs K, Haas R, Beckett D, Fischer W, Sundberg EJ (2015) Integrin engagement by the helical RGD motif of the *Helicobacter pylori* CagL protein is regulated by pH-induced displacement of a neighboring helix. J Biol Chem 290:12929–12940. https://doi.org/10.1074/jbc.M115.641829

Bonsor DA, Zhao Q, Schmidinger B, Weiss E, Wang J, Deredge D, Beadenkopf R, Dow B, Fischer W, Beckett D, Wintrode PL, Haas R, Sundberg EJ (2018) The *Helicobacter pylori* adhesin protein HopQ exploits the dimer interface of human CEACAMs to facilitate translocation of the oncoprotein CagA. EMBO J. https://doi.org/10.15252/embj.201798664

Boren T, Falk P, Roth KA, Larson G, Normark S (1993) Attachment of *Helicobacter pylori* to human gastric epithelium mediated by blood group antigens. Science 262:1892–1895. https://doi.org/10.1126/science.8018146

Bosch JA, de Geus EJ, Ligtenberg TJ, Nazmi K, Veerman EC, Hoogstraten J, Amerongen AV (2000) Salivary MUC5B-mediated adherence (ex vivo) of *Helicobacter pylori* during acute stress. Psychosom Med 62:40–49

Bugaytsova JA, Bjornham O, Chernov YA, Gideonsson P, Henriksson S, Mendez M, Sjostrom R, Mahdavi J, Shevtsova A, Ilver D, Moonens K, Quintana-Hayashi MP, Moskalenko R, Aisenbrey C, Bylund G, Schmidt A, Aberg A, Brannstrom K, Koniger V, Vikstrom S, Rakhimova L, Hofer A, Ogren J, Liu H, Goldman MD, Whitmire JM, Aden J, Younson J, Kelly CG, Gilman RH, Chowdhury A, Mukhopadhyay AK, Nair GB, Papadakos KS, Martinez-Gonzalez B, Sgouras DN, Engstrand L, Unemo M, Danielsson D, Suerbaum S, Oscarson S, Morozova-Roche LA, Olofsson A, Grobner G, Holgersson J, Esberg A, Stromberg N, Landstrom M, Eldridge AM, Chromy BA, Hansen LM, Solnick JV, Linden SK, Haas R, Dubois A, Merrell DS, Schedin S, Remaut H, Arnqvist A, Berg DE, Boren T (2017) *Helicobacter pylori* adapts to chronic infection and gastric disease via pH-responsive BabA-mediated adherence. Cell Host Microbe 21:376–389. https://doi.org/10.1016/j.chom.2017.02.013

Buti L, Spooner E, Van der Veen AG, Rappuoli R, Covacci A, Ploegh HL (2011) *Helicobacter pylori* cytotoxin-associated gene A (CagA) subverts the apoptosis-stimulating protein of p53 (ASPP2) tumor suppressor pathway of the host. Proc Natl Acad Sci U S A 108:9238–9243. https://doi.org/10.1073/pnas.1106200108

Cao P, Cover TL (2002) Two different families of *hopQ* alleles in *Helicobacter pylori*. J Clin Microbiol 40:4504–4511. https://doi.org/10.1128/JCM.40.12.4504-4511.2002

Censini S, Lange C, Xiang Z, Crabtree JE, Ghiara P, Borodovsky M, Rappuoli R, Covacci A (1996) *cag*, a pathogenicity island of *Helicobacter pylori*, encodes type I-specific and disease-associated virulence factors. Proc Natl Acad Sci U S A 93:14648–14653. https://doi.org/10.1073/pnas.93.25.14648

Cherati MR, Shokri-Shirvani J, Karkhah A, Rajabnia R, Nouri HR (2017) Helicobacter pylori cagL amino acid polymorphism D58E59 pave the way toward peptic ulcer disease while N58E59 is associated with gastric cancer in north of Iran. Microb Pathog 107:413–418. https://doi.org/10.1016/j.micpath.2017.04.025

Chiarini A, Cala C, Bonura C, Gullo A, Giuliana G, Peralta S, D'Arpa F, Giammanco A (2009) Prevalence of virulence-associated genotypes of *Helicobacter pylori* and correlation with severity of gastric pathology in patients from western Sicily, Italy. Eur J Clin Microbiol Infect Dis 28:437–446. https://doi.org/10.1007/s10096-008-0644-x

Choi JM, Choi YH, Sudhanva MS, Devakumar S, Lee KH, Cha JH, Lee SH (2015) Crystal structure of CagL from *Helicobacter pylori* K74 strain. Biochem Biophys Res Commun 460:964–970. https://doi.org/10.1016/j.bbrc.2015.03.135

Colbeck JC, Hansen LM, Fong JM, Solnick JV (2006) Genotypic profile of the outer membrane proteins BabA and BabB in clinical isolates of *Helicobacter pylori*. Infect Immun 74:4375–4378. https://doi.org/10.1128/IAI.00485-06

Conradi J, Huber S, Gaus K, Mertink F, Royo Gracia S, Strijowski U, Backert S, Sewald N (2012a) Cyclic RGD peptides interfere with binding of the *Helicobacter pylori* protein CagL to integrins alphaVbeta3 and alpha5beta1. Amino Acids 43:219–232. https://doi.org/10.1007/s00726-011-1066-0

Conradi J, Tegtmeyer N, Woźna M, Wissbrock M, Michalek C et al (2012b) An RGD helper sequence in CagL of *Helicobacter pylori* assists in interactions with integrins and injection of CagA. Front Cell Infect Microbiol 2(70). https://doi.org/10.3389/fcimb.2012.00070

Coppens F, Castaldo G, Debraekeleer A, Subedi S, Moonens K, Lo A, Remaut H (2018) Hop-family *Helicobacter* outer membrane adhesins form a novel class of type 5-like secretion proteins with an interrupted beta-barrel domain. Mol Microbiol. https://doi.org/10.1111/mmi.14075

Croxen MA, Sisson G, Melano R, Hoffman PS (2006) The *Helicobacter pylori* chemotaxis receptor TlpB (HP0103) is required for pH taxis and for colonization of the gastric mucosa. J Bacteriol 188:2656–2665. https://doi.org/10.1128/JB.188.7.2656-2665.2006

Dabiri H, Maleknejad P, Yamaoka Y, Feizabadi MM, Jafari F, Rezadehbashi M, Nakhjavani FA, Mirsalehian A, Zali MR (2009) Distribution of *Helicobacter pylori cagA*, *cagE*, *oipA* and *vacA* in different major ethnic groups in Tehran, Iran. J Gastroenterol Hepatol 24:1380–1386. https://doi.org/10.1111/j.1440-1746.2009.05876.x

de Jonge R, Durrani Z, Rijpkema SG, Kuipers EJ, van Vliet AH, Kusters JG (2004a) Role of the *Helicobacter pylori* outer-membrane proteins AlpA and AlpB in colonization of the guinea pig stomach. J Med Microbiol 53:375–379. https://doi.org/10.1099/jmm.0.45551-0

de Jonge R, Pot RG, Loffeld RJ, van Vliet AH, Kuipers EJ, Kusters JG (2004b) The functional status of the *Helicobacter pylori* sabB adhesin gene as a putative marker for disease outcome. Helicobacter 9:158–164. https://doi.org/10.1111/j.1083-4389.2004.00213.x

Didelot X, Nell S, Yang I, Woltemate S, van der Merwe S, Suerbaum S (2013) Genomic evolution and transmission of *Helicobacter pylori* in two South African families. Proc Natl Acad Sci U S A

110:13880–13885. https://doi.org/10.1073/pnas.1304681110
Dossumbekova A, Prinz C, Mages J, Lang R, Kusters JG, Van Vliet AH, Reindl W, Backert S, Saur D, Schmid RM, Rad R (2006) *Helicobacter pylori* HopH (OipA) and bacterial pathogenicity: genetic and functional genomic analysis of *hopH* gene polymorphisms. J Infect Dis 194:1346–1355. https://doi.org/10.1086/508426
Farzi N, Yadegar A, Aghdaei HA, Yamaoka Y, Zali MR (2018) Genetic diversity and functional analysis of *oipA* gene in association with other virulence factors among *Helicobacter pylori* isolates from Iranian patients with different gastric diseases. Infect Genet Evol 60:26–34. https://doi.org/10.1016/j.meegid.2018.02.017
Fischer W, Puls J, Buhrdorf R, Gebert B, Odenbreit S, Haas R (2001) Systematic mutagenesis of the *Helicobacter pylori cag* pathogenicity island: essential genes for CagA translocation in host cells and induction of interleukin-8. Mol Microbiol 42:1337–1348. https://doi.org/10.1046/j.1365-2958.2001.02714.x
Franco AT, Johnston E, Krishna U, Yamaoka Y, Israel DA, Nagy TA, Wroblewski LE, Piazuelo MB, Correa P, Peek RM Jr (2008) Regulation of gastric carcinogenesis by *Helicobacter pylori* virulence factors. Cancer Res 68:379–387. https://doi.org/10.1158/0008-5472.CAN-07-0824
Gerhard M, Lehn N, Neumayer N, Boren T, Rad R, Schepp W, Miehlke S, Classen M, Prinz C (1999) Clinical relevance of the *Helicobacter pylori* gene for blood-group antigen-binding adhesin. Proc Natl Acad Sci U S A 96:12778–12783. https://doi.org/10.1073/pnas.96.22.12778
Giannakis M, Backhed HK, Chen SL, Faith JJ, Wu M, Guruge JL, Engstrand L, Gordon JI (2009) Response of gastric epithelial progenitors to *Helicobacter pylori* isolates obtained from Swedish patients with chronic atrophic gastritis. J Biol Chem 284:30383–30394. https://doi.org/10.1074/jbc.M109.052738
Goodwin AC, Weinberger DM, Ford CB, Nelson JC, Snider JD, Hall JD, Paules CI, Peek RM Jr, Forsyth MH (2008) Expression of the *Helicobacter pylori* adhesin SabA is controlled via phase variation and the ArsRS signal transduction system. Microbiology 154:2231–2240. https://doi.org/10.1099/mic.0.2007/016055-0
Gorrell RJ, Zwickel N, Reynolds J, Bulach D, Kwok T (2016) *Helicobacter pylori* CagL Hypervariable motif: a global analysis of geographical diversity and association with gastric cancer. J Infect Dis 213:1927–1931. https://doi.org/10.1093/infdis/jiw060
Gray-Owen SD, Blumberg RS (2006) CEACAM1: contact-dependent control of immunity. Nat Rev Immunol 6:433–446. https://doi.org/10.1038/nri1864
Hage N, Howard T, Phillips C, Brassington C, Overman R, Debreczeni J, Gellert P, Stolnik S, Winkler GS, Falcone FH (2015) Structural basis of Lewis^{b} antigen binding by the *Helicobacter pylori* adhesin BabA. Sci Adv 1:e1500315. https://doi.org/10.1126/sciadv.1500315
Hammarstrom S (1999) The carcinoembryonic antigen (CEA) family: structures, suggested functions and expression in normal and malignant tissues. Semin Cancer Biol 9:67–81. https://doi.org/10.1006/scbi.1998.0119
Hansen LM, Gideonsson P, Canfield DR, Boren T, Solnick JV (2017) Dynamic expression of the BabA Adhesin and its BabB Paralog during *Helicobacter pylori* infection in rhesus macaques. Infect Immun 85. https://doi.org/10.1128/IAI.00094-17
Harvey VC, Acio CR, Bredehoft AK, Zhu L, Hallinger DR, Quinlivan-Repasi V, Harvey SE, Forsyth MH (2014) Repetitive sequence variations in the promoter region of the adhesin-encoding gene sabA of helicobacter pylori affect transcription. J Bacteriol 196:3421–3429. https://doi.org/10.1128/JB.01956-14
Hennig EE, Allen JM, Cover TL (2006) Multiple chromosomal loci for the *babA* gene in *Helicobacter pylori*. Infect Immun 74:3046–3051. https://doi.org/10.1128/IAI.74.5.3046-3051.2006
Horridge DN, Begley AA, Kim J, Aravindan N, Fan K, Forsyth MH (2017) Outer inflammatory protein a (OipA) of *Helicobacter pylori* is regulated by host cell contact and mediates CagA translocation and interleukin-8 response only in the presence of a functional cag pathogenicity island type IV secretion system. Pathog Dis 75. https://doi.org/10.1093/femspd/ftx113
Huang CH, Chiou SH (2011) Proteomic analysis of upregulated proteins in *Helicobacter pylori* under oxidative stress induced by hydrogen peroxide. Kaohsiung J Med Sci 27:544–553. https://doi.org/10.1016/j.kjms.2011.06.019
Ilver D, Arnqvist A, Ogren J, Frick IM, Kersulyte D, Incecik ET, Berg DE, Covacci A, Engstrand L, Boren T (1998) *Helicobacter pylori* adhesin binding fucosylated histo-blood group antigens revealed by retagging. Science 279:373–377. https://doi.org/10.1126/science.279.5349.373
Imberty A, Mitchell EP, Wimmerova M (2005) Structural basis of high-affinity glycan recognition by bacterial and fungal lectins. Curr Opin Struct Biol 15:525–534. https://doi.org/10.1016/j.sbi.2005.08.003
Israel DA, Salama N, Arnold CN, Moss SF, Ando T, Wirth HP, Tham KT, Camorlinga M, Blaser MJ, Falkow S, Peek RM Jr (2001) *Helicobacter pylori* strain-specific differences in genetic content, identified by microarray, influence host inflammatory responses. J Clin Invest 107:611–620. https://doi.org/10.1172/JCI11450
Javaheri A, Kruse T, Moonens K, Mejias-Luque R, Debraekeleer A, Asche CI, Tegtmeyer N, Kalali B, Bach NC, Sieber SA, Hill DJ, Koniger V, Hauck CR, Moskalenko R, Haas R, Busch DH, Klaile E, Slevogt H, Schmidt A, Backert S, Remaut H, Singer BB, Gerhard M (2016) Helicobacter pylori adhesin HopQ engages in a virulence-enhancing interaction

with human CEACAMs. Nat Microbiol 2:16189. https://doi.org/10.1038/nmicrobiol.2016.189

Jimenez-Soto LF, Kutter S, Sewald X, Ertl C, Weiss E, Kapp U, Rohde M, Pirch T, Jung K, Retta SF, Terradot L, Fischer W, Haas R (2009) *Helicobacter pylori* type IV secretion apparatus exploits beta1 integrin in a novel RGD-independent manner. PLoS Pathog 5:e1000684. https://doi.org/10.1371/journal.ppat.1000684

Jimenez-Soto LF, Clausen S, Sprenger A, Ertl C, Haas R (2013) Dynamics of the Cag-type IV secretion system of *Helicobacter pylori* as studied by bacterial co-infections. Cell Microbiol 15:1924–1937. https://doi.org/10.1111/cmi.12166

Kang J, Blaser MJ (2006) Bacterial populations as perfect gases: genomic integrity and diversification tensions in *Helicobacter pylori*. Nat Rev Microbiol 4:826–836. https://doi.org/10.1038/nrmicro1528

Kao CY, Sheu BS, Sheu SM, Yang HB, Chang WL, Cheng HC, Wu JJ (2012) Higher motility enhances bacterial density and inflammatory response in dyspeptic patients infected with *Helicobacter pylori*. Helicobacter 17:411–416. https://doi.org/10.1111/j.1523-5378.2012.00974.x

Keilberg D, Ottemann KM (2016) How *Helicobacter pylori* senses, targets and interacts with the gastric epithelium. Environ Microbiol 18:791–806. https://doi.org/10.1111/1462-2920.13222

Kennemann L, Didelot X, Aebischer T, Kuhn S, Drescher B, Droege M, Reinhardt R, Correa P, Meyer TF, Josenhans C, Falush D, Suerbaum S (2011) *Helicobacter pylori* genome evolution during human infection. Proc Natl Acad Sci U S A 108:5033–5038. https://doi.org/10.1073/pnas.1018444108

Kennemann L, Brenneke B, Andres S, Engstrand L, Meyer TF, Aebischer T, Josenhans C, Suerbaum S (2012) In vivo sequence variation in HopZ, a phase-variable outer membrane protein of *Helicobacter pylori*. Infect Immun 80:4364–4373. https://doi.org/10.1128/IAI.00977-12

Kenny DT, Skoog EC, Linden SK, Struwe WB, Rudd PM, Karlsson NG (2012) Presence of terminal N-acetylgalactosaminebeta1-4N-acetylglucosamine residues on O-linked oligosaccharides from gastric MUC5AC: involvement in *Helicobacter pylori* colonization? Glycobiology 22:1077–1085. https://doi.org/10.1093/glycob/cws076

Kim A, Servetas SL, Kang J, Kim J, Jang S, Cha HJ, Lee WJ, Kim J, Romero-Gallo J, Peek RM Jr, Merrell DS, Cha JH (2015) *Helicobacter pylori bab* Paralog distribution and association with *cagA*, *vacA*, and *homA/B* genotypes in American and South Korean clinical isolates. PLoS One 10:e0137078. https://doi.org/10.1371/journal.pone.0137078

Kim A, Servetas SL, Kang J, Kim J, Jang S, Choi YH, Su H, Jeon YE, Hong YA, Yoo YJ, Merrell DS, Cha JH (2016) *Helicobacter pylori* outer membrane protein, HomC, shows geographic dependent polymorphism that is influenced by the Bab family. J Microbiol 54:846–852. https://doi.org/10.1007/s12275-016-6434-8

Kobayashi M, Lee H, Nakayama J, Fukuda M (2009) Roles of gastric mucin-type O-glycans in the pathogenesis of *Helicobacter pylori* infection. Glycobiology 19:453–461. https://doi.org/10.1093/glycob/cwp004

Koelblen T, Berge C, Cherrier MV, Brillet K, Jimenez-Soto L, Ballut L, Takagi J, Montserret R, Rousselle P, Fischer W, Haas R, Fronzes R, Terradot L (2017) Molecular dissection of protein-protein interactions between integrin alpha5beta1 and the *Helicobacter pylori* Cag type IV secretion system. FEBS J 284:4143–4157. https://doi.org/10.1111/febs.14299

Königer V, Holsten L, Harrison U, Busch B, Loell E, Zhao Q, Bonsor DA, Roth A, Kengmo-Tchoupa A, Smith SI, Mueller S, Sundberg EJ, Zimmermann W, Fischer W, Hauck CR, Haas R (2016) Helicobacter pylori exploits human CEACAMs via HopQ for adherence and translocation of CagA. Nat Microbiol 2:16188. https://doi.org/10.1038/nmicrobiol.2016.188

Kwok T, Zabler D, Urman S, Rohde M, Hartig R, Wessler S, Misselwitz R, Berger J, Sewald N, Konig W, Backert S (2007) *Helicobacter* exploits integrin for type IV secretion and kinase activation. Nature 449:862–866. https://doi.org/10.1038/nature06187

Langenberg ML, Tytgat GN, Schipper MEI (1984) *Campylobacter*-like organisms in the stomach of patients and healthy individuals. Lancet 1:1348–1349. https://doi.org/10.1016/S0140-6736(84)91836-1

Lee ER (1985) Dynamic histology of the antral epithelium in the mouse stomach: III. Ultrastructure and renewal of pit cells. Am J Anat 172:225–240. https://doi.org/10.1002/aja.1001720305

Lehours P, Menard A, Dupouy S, Bergey B, Richy F, Zerbib F, Ruskone-Fourmestraux A, Delchier JC, Megraud F (2004) Evaluation of the association of nine *Helicobacter pylori* virulence factors with strains involved in low-grade gastric mucosa-associated lymphoid tissue lymphoma. Infect Immun 72:880–888. https://doi.org/10.1128/IAI.72.2.880-888.2004

Leylabadlo HE, Yekani M, Ghotaslou R (2016) *Helicobacter pylori hopQ* alleles (type I and II) in gastric cancer. Biomed Rep 4:601–604. https://doi.org/10.3892/br.2016.634

Linden S, Mahdavi J, Hedenbro J, Boren T, Carlstedt I (2004) Effects of pH on *Helicobacter pylori* binding to human gastric mucins: identification of binding to non-MUC5AC mucins. Biochem J 384:263–270. https://doi.org/10.1042/BJ20040402

Linz B, Windsor HM, McGraw JJ, Hansen LM, Gajewski JP, Tomsho LP, Hake CM, Solnick JV, Schuster SC, Marshall BJ (2014) A mutation burst during the acute phase of *Helicobacter pylori* infection in humans and rhesus macaques. Nat Commun 5:4165. https://doi.org/10.1038/ncomms5165

Loh JT, Torres VJ, Algood HM, McClain MS, Cover TL (2008) *Helicobacter pylori* HopQ outer membrane protein attenuates bacterial adherence to gastric epithelial cells. FEMS Microbiol Lett 289:53–58. https://doi.org/10.1111/j.1574-6968.2008.01368.x

Loh JT, Beckett AC, Scholz MB, Cover TL (2018) High-salt conditions alter transcription of *Helicobacter pylori* genes encoding outer membrane proteins. Infect Immun 86. https://doi.org/10.1128/IAI.00626-17

Lu H, Wu JY, Beswick EJ, Ohno T, Odenbreit S, Haas R, Reyes VE, Kita M, Graham DY, Yamaoka Y (2007) Functional and intracellular signaling differences associated with the *Helicobacter pylori* AlpAB adhesin from Western and East Asian strains. J Biol Chem 282:6242–6254. https://doi.org/10.1074/jbc.M611178200

Lu HS, Saito Y, Umeda M, Murata-Kamiya N, Zhang HM, Higashi H, Hatakeyama M (2008) Structural and functional diversity in the PAR1b/MARK2-binding region of *Helicobacter pylori* CagA. Cancer Sci 99:2004–2011. https://doi.org/10.1111/j.1349-7006.2008.00950.x

Magalhaes A, Marcos-Pinto R, Nairn AV, Dela Rosa M, Ferreira RM, Junqueira-Neto S, Freitas D, Gomes J, Oliveira P, Santos MR, Marcos NT, Xiaogang W, Figueiredo C, Oliveira C, Dinis-Ribeiro M, Carneiro F, Moremen KW, David L, Reis CA (2015) *Helicobacter pylori* chronic infection and mucosal inflammation switches the human gastric glycosylation pathways. Biochim Biophys Acta 1852:1928–1939. https://doi.org/10.1016/j.bbadis.2015.07.001

Mahdavi J, Sonden B, Hurtig M, Olfat FO, Forsberg L, Roche N, Angstrom J, Larsson T, Teneberg S, Karlsson KA, Altraja S, Wadstrom T, Kersulyte D, Berg DE, Dubois A, Petersson C, Magnusson KE, Norberg T, Lindh F, Lundskog BB, Arnqvist A, Hammarstrom L, Boren T (2002) *Helicobacter pylori* SabA adhesin in persistent infection and chronic inflammation. Science 297:573–578. https://doi.org/10.1126/science.1069076

Matteo MJ, Armitano RI, Romeo M, Wonaga A, Olmos M, Catalano M (2011) *Helicobacter pylori bab* genes during chronic colonization. Int J Mol Epidemiol Genet 2:286–291

Mimuro H, Suzuki T, Tanaka J, Asahi M, Haas R, Sasakawa C (2002) Grb2 is a key mediator of *helicobacter pylori* CagA protein activities. Mol Cell 10:745–755. https://doi.org/10.1016/S1097-2765(02)00681-0

Mobley HL (1996) The role of *Helicobacter pylori* urease in the pathogenesis of gastritis and peptic ulceration. Aliment Pharmacol Ther 10(Suppl 1):57–64. https://doi.org/10.1046/j.1365-2036.1996.22164006.x

Moonens K, Gideonsson P, Subedi S, Bugaytsova J, Romao E, Mendez M, Norden J, Fallah M, Rakhimova L, Shevtsova A, Lahmann M, Castaldo G, Brannstrom K, Coppens F, Lo AW, Ny T, Solnick JV, Vandenbussche G, Oscarson S, Hammarstrom L, Arnqvist A, Berg DE, Muyldermans S, Boren T, Remaut H (2016) Structural insights into polymorphic ABO glycan binding by *Helicobacter pylori*. Cell Host Microbe 19:55–66. https://doi.org/10.1016/j.chom.2015.12.004

Moonens K, Hamway Y, Neddermann M, Reschke M, Tegtmeyer N, Kruse T, Kammerer R, Mejias-Luque R, Singer BB, Backert S, Gerhard M, Remaut H (2018) *Helicobacter pylori* adhesin HopQ disrupts trans dimerization in human CEACAMs. EMBO J. https://doi.org/10.15252/embj.201798665

Morelli G, Didelot X, Kusecek B, Schwarz S, Bahlawane C, Falush D, Suerbaum S, Achtman M (2010) Microevolution of *Helicobacter pylori* during prolonged infection of single hosts and within families. PLoS Genet 6:e1001036. https://doi.org/10.1371/journal.pgen.1001036

Murata-Kamiya N, Kurashima Y, Teishikata Y, Yamahashi Y, Saito Y, Higashi H, Aburatani H, Akiyama T, Peek RM Jr, Azuma T, Hatakeyama M (2007) *Helicobacter pylori* CagA interacts with E-cadherin and deregulates the beta-catenin signal that promotes intestinal transdifferentiation in gastric epithelial cells. Oncogene 26:4617–4626. https://doi.org/10.1038/sj.onc.1210251

Nell S, Kennemann L, Schwarz S, Josenhans C, Suerbaum S (2014) Dynamics of Lewis^{b} binding and sequence variation of the *babA* adhesin gene during chronic *Helicobacter pylori* infection in humans. MBio 5. https://doi.org/10.1128/mBio.02281-14

Odenbreit S, Swoboda K, Barwig I, Ruhl S, Boren T, Koletzko S, Haas R (2009) Outer membrane protein expression profile in *Helicobacter pylori* clinical isolates. Infect Immun 77:3782–3790. https://doi.org/10.1128/IAI.00364-09

Oh JD, Kling-Backhed H, Giannakis M, Xu J, Fulton RS, Fulton LA, Cordum HS, Wang C, Elliott G, Edwards J, Mardis ER, Engstrand LG, Gordon JI (2006) The complete genome sequence of a chronic atrophic gastritis *Helicobacter pylori* strain: evolution during disease progression. Proc Natl Acad Sci U S A 103:9999–10004. https://doi.org/10.1073/pnas.0603784103

Ohno T, Sugimoto M, Nagashima A, Ogiwara H, Vilaichone RK, Mahachai V, Graham DY, Yamaoka Y (2009) Relationship between *Helicobacter pylori hopQ* genotype and clinical outcome in Asian and Western populations. J Gastroenterol Hepatol 24:462–468. https://doi.org/10.1111/j.1440-1746.2008.05762.x

Ohno T, Vallstrom A, Rugge M, Ota H, Graham DY, Arnqvist A, Yamaoka Y (2011) Effects of blood group antigen-binding adhesin expression during *Helicobacter pylori* infection of Mongolian gerbils. J Infect Dis 203:726–735. https://doi.org/10.1093/infdis/jiq090

Oleastro M, Menard A (2013) The role of *Helicobacter pylori* outer membrane proteins in adherence and pathogenesis. Biology (Basel) 2:1110–1134. https://doi.org/10.3390/biology2031110

Oleastro M, Cordeiro R, Ferrand J, Nunes B, Lehours P, Carvalho-Oliveira I, Mendes AI, Penque D,

Monteiro L, Megraud F, Menard A (2008) Evaluation of the clinical significance of *homB*, a novel candidate marker of *Helicobacter pylori* strains associated with peptic ulcer disease. J Infect Dis 198:1379–1387. https://doi.org/10.1086/592166

Pang SS, Nguyen ST, Perry AJ, Day CJ, Panjikar S, Tiralongo J, Whisstock JC, Kwok T (2014) The three-dimensional structure of the extracellular adhesion domain of the sialic acid-binding adhesin SabA from *Helicobacter pylori*. J Biol Chem 289:6332–6340. https://doi.org/10.1074/jbc.M113.513135

Parsonnet J, Friedman GD, Orentreich N, Vogelman H (1997) Risk for gastric cancer in people with CagA positive or CagA negative *Helicobacter pylori* infection. Gut 40:297–301. https://doi.org/10.1136/gut.40.3.297

Peck B, Ortkamp M, Diehl KD, Hundt E, Knapp B (1999) Conservation, localization and expression of HopZ, a protein involved in adhesion of *Helicobacter pylori*. Nucleic Acids Res 27:3325–3333. https://doi.org/10.1093/nar/27.16.3325

Posselt G, Backert S, Wessler S (2013) The functional interplay of *Helicobacter pylori* factors with gastric epithelial cells induces a multi-step process in pathogenesis. Cell Commun Signal 11:77. https://doi.org/10.1186/1478-811X-11-77

Pride DT, Blaser MJ (2002) Concerted evolution between duplicated genetic elements in *Helicobacter pylori*. J Mol Biol 316:629–642. https://doi.org/10.1006/jmbi.2001.5311

Rossez Y, Gosset P, Boneca IG, Magalhaes A, Ecobichon C, Reis CA, Cieniewski-Bernard C, Joncquel Chevalier Curt M, Leonard R, Maes E, Sperandio B, Slomianny C, Sansonetti PJ, Michalski JC, Robbe-Masselot C (2014) The lacdiNAc-specific adhesin LabA mediates adhesion of *Helicobacter pylori* to human gastric mucosa. J Infect Dis 210:1286–1295. https://doi.org/10.1093/infdis/jiu239

Saberi S, Schmidt A, Eybpoosh S, Esmaili M, Talebkhan Y, Mohajerani N, Oghalaie A, Eshagh Hosseini M, Mohagheghi MA, Bugaytova J, Boren T, Mohammadi M (2016) *Helicobacter pylori* strains from duodenal ulcer patients exhibit mixed *babA/B* genotypes with low levels of BabA adhesin and Lewis^b binding. Dig Dis Sci 61:2868–2877. https://doi.org/10.1007/s10620-016-4217-z

Saha A, Backert S, Hammond CE, Gooz M, Smolka AJ (2010) *Helicobacter pylori* CagL activates ADAM17 to induce repression of the gastric H, K-ATPase alpha subunit. Gastroenterology 139:239–248. https://doi.org/10.1053/j.gastro.2010.03.036

Sakamoto S, Watanabe T, Tokumaru T, Takagi H, Nakazato H, Lloyd KO (1989) Expression of Lewis^a, Lewis^b, Lewis^x, Lewis^y, siayl-Lewis^a, and sialyl-Lewis^x blood group antigens in human gastric carcinoma and in normal gastric tissue. Cancer Res 49:745–752

Saunders NJ, Peden JF, Hood DW, Moxon ER (1998) Simple sequence repeats in the *Helicobacter pylori* genome. Mol Microbiol 27:1091–1098. https://doi.org/10.1046/j.1365-2958.1998.00768.x

Schreiber S, Konradt M, Groll C, Scheid P, Hanauer G, Werling HO, Josenhans C, Suerbaum S (2004) The spatial orientation of *Helicobacter pylori* in the gastric mucus. Proc Natl Acad Sci U S A 101:5024–5029. https://doi.org/10.1073/pnas.0308386101

Segal ED, Cha J, Lo J, Falkow S, Tompkins LS (1999) Altered states: involvement of phosphorylated CagA in the induction of host cellular growth changes by *Helicobacter pylori*. Proc Natl Acad Sci U S A 96:14559–14564. https://doi.org/10.1073/pnas.96.25.14559

Selbach M, Moese S, Meyer TF, Backert S (2002) Functional analysis of the *Helicobacter pylori cag* pathogenicity island reveals both VirD4-CagA-dependent and VirD4-CagA-independent mechanisms. Infect Immun 70:665–671. https://doi.org/10.1128/IAI.70.2.665-671.2002

Selbach M, Paul FE, Brandt S, Guye P, Daumke O, Backert S, Dehio C, Mann M (2009) Host cell interactome of tyrosine-phosphorylated bacterial proteins. Cell Host Microbe 5:397–403. https://doi.org/10.1016/j.chom.2009.03.004

Senkovich OA, Yin J, Ekshyyan V, Conant C, Traylor J, Adegboyega P, McGee DJ, Rhoads RE, Slepenkov S, Testerman TL (2011) *Helicobacter pylori* AlpA and AlpB bind host laminin and influence gastric inflammation in gerbils. Infect Immun 79:3106–3116. https://doi.org/10.1128/IAI.01275-10

Servetas SL, Kim A, Su H, Cha JH, Merrell DS (2018) Comparative analysis of the Hom family of outer membrane proteins in isolates from two geographically distinct regions: the United States and South Korea. Helicobacter 23:e12461. https://doi.org/10.1111/hel.12461

Shaffer CL, Gaddy JA, Loh JT, Johnson EM, Hill S, Hennig EE, McClain MS, McDonald WH, Cover TL (2011) *Helicobacter pylori* exploits a unique repertoire of type IV secretion system components for pilus assembly at the bacteria-host cell interface. PLoS Pathog 7:e1002237. https://doi.org/10.1371/journal.ppat.1002237

Sheu BS, Odenbreit S, Hung KH, Liu CP, Sheu SM, Yang HB, Wu JJ (2006) Interaction between host gastric Sialyl-Lewis X and *H. pylori* SabA enhances *H. pylori* density in patients lacking gastric Lewis B antigen. Am J Gastroenterol 101:36–44. https://doi.org/10.1111/j.1572-0241.2006.00358.x

Skoog EC, Sjoling A, Navabi N, Holgersson J, Lundin SB, Linden SK (2012) Human gastric mucins differently regulate *Helicobacter pylori* proliferation, gene expression and interactions with host cells. PLoS One 7:e36378. https://doi.org/10.1371/journal.pone.0036378

Snelling WJ, Moran AP, Ryan KA, Scully P, McGourty K, Cooney JC, Annuk H, O'Toole PW (2007) HorB (HP0127) is a gastric epithelial cell adhesin.

Helicobacter 12:200–209. https://doi.org/10.1111/j.1523-5378.2007.00499.x

Solnick JV, Hansen LM, Salama NR, Boonjakuakul JK, Syvanen M (2004) Modification of *Helicobacter pylori* outer membrane protein expression during experimental infection of rhesus macaques. Proc Natl Acad Sci U S A 101:2106–2111. https://doi.org/10.1073/pnas.0308573100

Styer CM, Hansen LM, Cooke CL, Gundersen AM, Choi SS, Berg DE, Benghezal M, Marshall BJ, Peek RM Jr, Boren T, Solnick JV (2010) Expression of the BabA adhesin during experimental infection with *Helicobacter pylori*. Infect Immun 78:1593–1600. https://doi.org/10.1128/IAI.01297-09

Sugimoto M, Ohno T, Graham DY, Yamaoka Y (2011) *Helicobacter pylori* outer membrane proteins on gastric mucosal interleukin 6 and 11 expression in Mongolian gerbils. J Gastroenterol Hepatol 26:1677–1684. https://doi.org/10.1111/j.1440-1746.2011.06817.x

Tabassam FH, Graham DY, Yamaoka Y (2012) *Helicobacter pylori*-associated regulation of forkhead transcription factors FoxO1/3a in human gastric cells. Helicobacter 17:193–202. https://doi.org/10.1111/j.1523-5378.2012.00939.x

Tafreshi M, Zwickel N, Gorrell RJ, Kwok T (2015) Preservation of *Helicobacter pylori* CagA translocation and host cell proinflammatory responses in the face of CagL Hypervariability at amino acid residues 58/59. PLoS One 10:e0133531. https://doi.org/10.1371/journal.pone.0133531

Talarico S, Whitefield SE, Fero J, Haas R, Salama NR (2012) Regulation of *Helicobacter pylori* adherence by gene conversion. Mol Microbiol 84:1050–1061. https://doi.org/10.1111/j.1365-2958.2012.08073.x

Tchoupa AK, Schuhmacher T, Hauck CR (2014) Signaling by epithelial members of the CEACAM family – mucosal docking sites for pathogenic bacteria. Cell Commun Signal 12:27. https://doi.org/10.1186/1478-811X-12-27

Tegtmeyer N, Hartig R, Delahay RM, Rohde M, Brandt S, Conradi J, Takahashi S, Smolka AJ, Sewald N, Backert S (2010) A small fibronectin-mimicking protein from bacteria induces cell spreading and focal adhesion formation. J Biol Chem 285:23515–23526. https://doi.org/10.1074/jbc.M109.096214

Tegtmeyer N, Wessler S, Backert S (2011) Role of the *cag*-pathogenicity island encoded type IV secretion system in *Helicobacter pylori* pathogenesis. FEBS J 278:1190–1202. https://doi.org/10.1111/j.1742-4658.2011.08035.x

Tegtmeyer N, Lind J, Schmid B, Backert S (2014) *Helicobacter pylori* CagL Y58/E59 mutation turns-off type IV secretion-dependent delivery of CagA into host cells. PLoS One 9:e97782. https://doi.org/10.1371/journal.pone.0097782

Tegtmeyer N, Harrer A, Schmitt V, Singer BB, Backert S (2019) Expression of CEACAM1 or CEACAM5 in AZ-521 cells restores the type IV secretion deficiency for translocation of CagA by *Helicobacter pylori*. Cell Microbiol 21:e12965. https://doi.org/10.1111/cmi.12965

Teymournejad O, Mobarez AM, Hassan ZM, Talebi Bezmin Abadi A (2017) Binding of the *Helicobacter pylori* OipA causes apoptosis of host cells via modulation of Bax/Bcl-2 levels. Sci Rep 7:8036. https://doi.org/10.1038/s41598-017-08176-7

Tomb JF, White O, Kerlavage AR, Clayton RA, Sutton GG, Fleischmann RD, Ketchum KA, Klenk HP, Gill S, Dougherty BA, Nelson K, Quackenbush J, Zhou L, Kirkness EF, Peterson S, Loftus B, Richardson D, Dodson R, Khalak HG, Glodek A, McKenney K, Fitzegerald LM, Lee N, Adams MD, Hickey EK, Berg DE, Gocayne JD, Utterback TR, Peterson JD, Kelley JM, Cotton MD, Weidman JM, Fujii C, Bowman C, Watthey L, Wallin E, Hayes WS, Borodovsky M, Karp PD, Smith HO, Fraser CM, Venter JC (1997) The complete genome sequence of the gastric pathogen *Helicobacter pylori*. Nature 388:539–547. https://doi.org/10.1038/41483

Tsutsumi R, Higashi H, Higuchi M, Okada M, Hatakeyama M (2003) Attenuation of *Helicobacter pylori* CagA x SHP-2 signaling by interaction between CagA and C-terminal Src kinase. J Biol Chem 278:3664–3670. https://doi.org/10.1074/jbc.M208155200

Unemo M, Aspholm-Hurtig M, Ilver D, Bergstrom J, Boren T, Danielsson D, Teneberg S (2005) The sialic acid binding SabA adhesin of *Helicobacter pylori* is essential for nonopsonic activation of human neutrophils. J Biol Chem 280:15390–15397. https://doi.org/10.1074/jbc.M412725200

Veerman EC, Bolscher JG, Appelmelk BJ, Bloemena E, van den Berg TK, Nieuw Amerongen AV (1997) A monoclonal antibody directed against high M (r) salivary mucins recognizes the SO3-3Gal beta 1-3GlcNAc moiety of sulfo-Lewis[a]: a histochemical survey of human and rat tissue. Glycobiology 7:37–43. https://doi.org/10.1093/glycob/7.1.37

Walz A, Odenbreit S, Mahdavi J, Boren T, Ruhl S (2005) Identification and characterization of binding properties of *Helicobacter pylori* by glycoconjugate arrays. Glycobiology 15:700–708. https://doi.org/10.1093/glycob/cwi049

Walz A, Odenbreit S, Stuhler K, Wattenberg A, Meyer HE, Mahdavi J, Boren T, Ruhl S (2009) Identification of glycoprotein receptors within the human salivary proteome for the lectin-like BabA and SabA adhesins of *Helicobacter pylori* by fluorescence-based 2-D bacterial overlay. Proteomics 9:1582–1592. https://doi.org/10.1002/pmic.200700808

Wiedemann T, Hofbaur S, Tegtmeyer N, Huber S, Sewald N, Wessler S, Backert S, Rieder G (2012) *Helicobacter pylori* CagL dependent induction of gastrin expression via a novel alphavbeta5-integrin-integrin linked kinase signalling complex. Gut 61:986–996. https://doi.org/10.1136/gutjnl-2011-300525

Williams JC, McInnis KA, Testerman TL (2008) Adherence of *Helicobacter pylori* to abiotic surfaces is influenced by serum. Appl Environ Microbiol 74:1255–1258. https://doi.org/10.1128/AEM.01958-07

Yamaoka Y, Kwon DH, Graham DY (2000) A M (r) 34,000 proinflammatory outer membrane protein (oipA) of *Helicobacter pylori*. Proc Natl Acad Sci U S A 97:7533–7538. https://doi.org/10.1073/pnas.130079797

Yamaoka Y, Kikuchi S, el-Zimaity HM, Gutierrez O, Osato MS, Graham DY (2002a) Importance of *Helicobacter pylori oipA* in clinical presentation, gastric inflammation, and mucosal interleukin 8 production. Gastroenterology 123:414–424. https://doi.org/10.1053/gast.2002.34781

Yamaoka Y, Kita M, Kodama T, Imamura S, Ohno T, Sawai N, Ishimaru A, Imanishi J, Graham DY (2002b) *Helicobacter pylori* infection in mice: role of outer membrane proteins in colonization and inflammation. Gastroenterology 123:1992–2004. https://doi.org/10.1053/gast.2002.37074

Yamaoka Y, Souchek J, Odenbreit S, Haas R, Arnqvist A, Boren T, Kodama T, Osato MS, Gutierrez O, Kim JG, Graham DY (2002c) Discrimination between cases of duodenal ulcer and gastritis on the basis of putative virulence factors of *Helicobacter pylori*. J Clin Microbiol 40:2244–2246. https://doi.org/10.1128/JCM.40.6.2244-2246.2002

Yamaoka Y, Ojo O, Fujimoto S, Odenbreit S, Haas R, Gutierrez O, El-Zimaity HM, Reddy R, Arnqvist A, Graham DY (2006) *Helicobacter pylori* outer membrane proteins and gastroduodenal disease. Gut 55:775–781. https://doi.org/10.1136/gut.2005.083014

Yanai A, Maeda S, Hikiba Y, Shibata W, Ohmae T, Hirata Y, Ogura K, Yoshida H, Omata M (2007) Clinical relevance of *Helicobacter pylori sabA* genotype in Japanese clinical isolates. J Gastroenterol Hepatol 22:2228–2232. https://doi.org/10.1111/j.1440-1746.2007.04831.x

Yeh YC, Chang WL, Yang HB, Cheng HC, Wu JJ, Sheu BS (2011) H. pylori cagL amino acid sequence polymorphism Y58E59 induces a corpus shift of gastric integrin alpha5beta1 related with gastric carcinogenesis. Mol Carcinog 50:751–759. https://doi.org/10.1002/mc.20753

Yeh YC, Cheng HC, Yang HB, Chang WL, Sheu BS (2013) H. pylori CagL-Y58/E59 prime higher integrin alpha5beta1 in adverse pH condition to enhance hypochlorhydria vicious cycle for gastric carcinogenesis. PLoS One 8:e72735. https://doi.org/10.1371/journal.pone.0072735

Zebhauser R, Kammerer R, Eisenried A, McLellan A, Moore T, Zimmermann W (2005) Identification of a novel group of evolutionarily conserved members within the rapidly diverging murine Cea family. Genomics 86:566–580. https://doi.org/10.1016/j.ygeno.2005.07.008

Zhang XS, Tegtmeyer N, Traube L, Jindal S, Perez-Perez-G, Sticht H, Backert S, Blaser MJ (2015) A specific A/T polymorphism in Western tyrosine phosphorylation B-motifs regulates *Helicobacter pylori* CagA epithelial cell interactions. PLoS Pathog 11(2):e1004621. https://doi.org/10.1371/journal.ppat.1004621

Adv Exp Med Biol - Advances in Microbiology, Infectious Diseases and Public Health (2019) 1149: 77–106
https://doi.org/10.1007/5584_2019_360

Published online: 3 May 2019

Immune Cell Signaling by *Helicobacter pylori*: Impact on Gastric Pathology

Nicole Blaser, Steffen Backert,
and Suneesh Kumar Pachathundikandi

Abstract

Helicobacter pylori represents a highly successful colonizer of the human stomach. Infections with this Gram-negative bacterium can persist lifelong, and although in the majority of cases colonization is asymptomatic, it can trigger pathologies ranging from chronic gastritis and peptic ulceration to gastric cancer. The interaction of the bacteria with the human host modulates immune responses in different ways to enable bacterial survival and persistence. *H. pylori* uses various pathogenicity-associated factors such as VacA, NapA, CGT, GGT, lipopolysaccharide, peptidoglycan, heptose 1,7-bisphosphate, ADP-heptose, cholesterol glucosides, urease and a type IV secretion system for controlling immune signaling and cellular functions. It appears that *H. pylori* manipulates multiple extracellular immune receptors such as integrin-β_2 (CD18), EGFR, CD74, CD300E, DC-SIGN, MINCLE, TRPM2, T-cell and Toll-like receptors as well as a number of intracellular receptors including NLRP3, NOD1, NOD2, TIFA and ALPK1. Consequently, downstream signaling pathways are hijacked, inducing tolerogenic dendritic cells, inhibiting effector T cell responses and changing the gastrointestinal microbiota. Here, we discuss in detail the interplay of bacterial factors with multiple immunoregulatory cells and summarize the main immune evasion and persistence strategies employed by *H. pylori*.

Keywords

T4SS · TLR · PAMP · PRR · Inflammasome

N. Blaser, S. Backert, and S. K. Pachathundikandi (✉)
Department of Biology, Institute for Microbiology, Friedrich Alexander University Erlangen-Nuremberg, Erlangen, Germany
e-mail: pachathundikandi.suneesh@fau.de

1 Introduction

Hallmarks of microbial infections are inflammation and subsequent changes in the affected tissue. Generally, most infections are cleared by the host immune system through innate and adaptive responses. Microbial invasion can be detected by a plethora of factors belonging to the innate immune machinery. The well-studied pattern recognition receptors (PRRs) detect pathogen associated molecular patterns (PAMPs) and this interaction leads to various arms of signal transduction to produce a timely response by the immune system to control the infection (Takeuchi and Akira 2010). In addition, inherent signals of anti-inflammation are needed to subsequently down-regulate the immune activity and avoid unnecessary damage to host tissue. Many pathogens have developed strategies to evade host immune responses to variable degrees. For instance, the untimely induction of both pro- and anti-inflammatory responses will jeopardize the control of infection and avoid a return to homeostasis. *H. pylori* is an example of a pathogen that effectively manipulates the host's immune response upon infection. It colonizes the human gastric mucosa and is associated with gastritis, peptic ulcer and gastric cancer. When present, *H. pylori* is effectively detected by the host innate immune system, which in response produces pro- and anti-inflammatory cytokines and other inflammatory mediators (White et al. 2015). These responses ought to be sufficient to result in adaptive immunity against this pathogen, but they cannot effectively clear the infection, which allows *H. pylori* to colonize an individual from childhood (when most primary infections occur) to last an entire life span.

In 10–20% of cases, infection with *H. pylori* is associated with the development of peptic ulcers and about 1–2% develop gastric cancer or gastric mucosa-associated lymphoid tissue (MALT) lymphoma (Wroblewski et al. 2010; Bauer and Meyer 2011). A major risk factor for these *H. pylori*-associated diseases is local chronic gastric inflammation in response to colonization (Dunn et al. 1997; White et al. 2015; Gobert and Wilson 2016). The immune response during *H. pylori* infection is characterized by the infiltration of several types of immune cells, for instance anti-*H. pylori* T-cells were found in the gastric mucosa of infected individuals (D'Elios et al. 1997). The prolonged pathogenesis by this bacterium is based on its adaptation to and survival in the harsh conditions of the stomach, to which *H. pylori* responds by differential regulation of its gene expression during colonization (Wang and Maier 2015; Gieseler et al. 2005). Analyzing the bacterial gene expression of *H. pylori* derived from mouse stomachs or infected cultured murine cells showed an up-regulation of the cytotoxin-associated gene A (*cagA*) encoding the CagA effector protein and of *vacA* encoding the vacuolating toxin VacA (Singh et al. 2012). Both gene products are well-characterized virulence factors of *H. pylori* (Posselt et al. 2013). CagA, which was amongst the first virulence factors to be discovered for this pathogen, can hamper the maturation of dendritic cells (DCs), as demonstrated using both cultured human cell lines and a mouse model, suggesting it employs an immune regulatory effect (Tanaka et al. 2010; Käbisch et al. 2014). Effects on cellular vacuolation, apoptosis or immune cell inhibition are described for VacA, while the *cag* pathogenicity island (PAI), encoding a type IV secretion system (T4SS), is crucial for delivery of CagA across the bacterial membrane into the host cells (Backert et al. 2011; Bridge and Merrell 2013; Naumann et al. 2017). It appears that the inflammatory response during *H. pylori* infection is mainly controlled by the *cag*PAI, in line with the observation that *cag*PAI-positive strains (which have been designated as type-I) are more virulent compared to *cag*PAI-negative (type-II) isolates (Tegtmeyer et al. 2011). Later studies showed that only strains positive for CagA and VacA are able to drive immune cell tolerance during infection, presumably to promote chronic persistence of the pathogen (Oertli et al. 2013; Käbisch et al. 2014). Studies investigating the impact of *H. pylori* have shown that VacA can directly interact with T-cells, B-cells, monocytes and macrophages, which trigger both immune stimulatory and suppressive activities (Boncristiano et al. 2003; Gebert et al. 2003; Singh et al. 2012).

In addition, *H. pylori* is able to weaken the gastric epithelial barrier function and can induce epithelial apoptosis (Backert et al. 2017, 2018). Disruption of the epithelial integrity leads to increased amounts of bacterial virulence factors in the lamina propria, where they can interact with immune cells (Mai et al. 1991, 1992). Disrupted epithelial barrier function and epithelial apoptosis can be induced by VacA, but also by the secreted enzyme γ-glutamyl transpeptidase (GGT), which also contributes to the virulence of *H. pylori* (Cover et al. 2003; Shibayama et al. 2003). Besides its importance during colonization as demonstrated by *in vivo* models, GGT was further shown to promote cell cycle arrest, increase production of reactive oxygen species and induce the secretion of inflammatory cytokines leading to apoptosis and necrosis of gastric epithelial cells (Chevalier et al. 1999; McGovern et al. 2001; Oertli et al. 2013). The enzyme GGT, which is highly conserved between *H. pylori* strains, catalyzes the hydrolysis of glutamine to glutamate and ammonia and further the hydrolysis and transpeptidation of various γ-glutamyl compounds (Shibayama et al. 2007; Song et al. 2011). Apart from their direct functional effect, virulence factors can also modulate the cellular signaling and homeostasis. For example, cholesterol, a common constituent of the host cellular membranes, is extracted and converted to glucosides by cholesterol-α-glucosyltransferase (CGT) of *H. pylori*, which prevents phagocytosis and subsequent antigen presentation. Furthermore, cholesterol depletion dampened specific cellular signaling like interferon gamma (IFNγ) by depleting its receptors from lipid rafts (Wunder et al. 2006; Lai et al. 2008, 2011; Morey et al. 2018). Antigen presentation was tightly regulated at different levels by different mechanisms like interference on phagosome maturation, downregulated expression of antigen presenting MHC (major histocompatibility complex) molecules, co-stimulatory factors, and T-cell differentiation to regulatory phenotypes (Molinari et al. 1998; Ramarao et al. 2000; Allen 2007; Beswick et al. 2007; Wang et al. 2010). The interaction of *H. pylori* proteins like urease subunit B (UreB), heat shock protein 60 (Hsp60) and neutrophil activating protein A (NapA) with Toll-like-receptor 2 (TLR2) resulted in varied responses (Amedei et al. 2006; Zhao et al. 2007; Koch et al. 2015). Moreover, common PAMPs like lipopolysaccharide (LPS), peptidoglycan, DNA and RNA of *H. pylori* were shown to induce various pro- and anti-inflammatory signals through their respective PRRs (Ishihara et al. 2004; Viala et al. 2004; Allison et al. 2009; Rad et al. 2007; Nagashima et al. 2015). *H. pylori* infection also regulates the inflammasome for the secretion of cytokines interleukin 1β (IL-1β) and IL-18, which ultimately favors bacterial survival and persistence (Kim et al. 2013; Koch et al. 2015; Ng et al. 2016; Pachathundikandi and Backert 2018). The induction of pro- and anti-inflammatory states in this infection skewed for both reduction in bacterial colonization and immune-pathologies, but never resulted in resolution of inflammation without prophylactic methods (Garhart et al. 2002; Matsumoto et al. 2005; Kao et al. 2010; Quiding-Järbrink et al. 2010; Cook et al. 2014). Here we review the overall interplay of these various *H. pylori* factors with the host's immune system.

2 The Role of *H. pylori* GGT on Immune Tolerance

H. pylori GGT induces immune tolerance through altering DC processes, and GGT enzymatic activity is needed for this immune regulation, as was first demonstrated in infected mice (Oertli et al. 2013). Subsequently, by infecting human DCs with *H. pylori* Käbisch and co-workers (2016) showed that GGT promotes the progression of naïve T-cells to regulatory T-cells. As a result of the enzyme's activity, levels of glutamate increase in the stomach and these promote the activation of glutamate receptors that are expressed on DCs; this induces immune tolerance during *H. pylori* infection (Shibayama et al. 2007; Käbisch et al. 2016). Recently, Wüstner et al. (2015) have shown that activated T-cells are highly sensitive to glutamine concentrations in the extracellular space. In addition, previous studies have identified an inhibitory effect of insufficient glutamine levels on T-cell proliferation (Yaqoob and Calder 1997). Moreover, the expression of transcription factors that stimulate T-cell receptor signaling and influence their differentiation and expansion is decreased in the presence of active GGT, suggesting a hampering effect of *H. pylori* GGT on the activation and proliferation of these immune cells (Man et al. 2013; Yao et al. 2013).

Käbisch and co-workers (2016) showed, by infecting human DCs with *H. pylori* wild-type and an isogenic Δ*ggt* deletion mutant, respectively, that *H. pylori* can suppress the secretion of IL-6. Moreover, using glutamate receptor inhibitors it was shown that glutamate might have a regulatory impact on IL-6 secretion, influencing downstream T-cell responses to the pathogen. Conversely, insufficient glutamine levels resulting from inactive GGT might affect the production of IL-2 or IFNγ, indicating that by modulating its GGT activity *H. pylori* can actively influence the secretion of specific cytokines (Carr et al. 2010). This ability is not restricted to *H. pylori*; during infection with *Helicobacter suis,* supplementation with glutamate showed protection against bacterial-induced pathologies and suppression of inflammatory cell infiltration, underlying the importance of GGT in immune regulation (De Bruyne et al. 2016).

3 Interference of VacA with T-Cell Receptor/IL-2 and Nuclear Factor of Activated T-Cells (NFAT) Signaling

Although both VacA and GGT act on T-cells and are involved in inducing tolerance, they accomplish this via entirely different pathways (Fig. 1). Various studies have shown that VacA hampers the proliferation of T-cells, for which multiple mechanisms have been proposed (Boncristiano et al. 2003; Gebert et al. 2003; Sundrud et al. 2004). Differential binding and uptake of VacA in epithelial and immune cells was described more than 10 years ago (Gauthier et al. 2005). In immunoprecipitation studies, the CD18 receptor of human T-cells was identified to be targeted by VacA and seemed to act as a receptor or co-receptor for the toxin (Sewald et al. 2008). Indeed, in human primary T-cells, CD18 expression is essential for the uptake of VacA and subsequent cellular effects. However, a determinative difference between the VacA internalization exists between human and murine T-cells, indicating that specific pathways are responsible for the VacA-dependent effects in the human host (Sewald et al. 2008).

In general, T-cell receptor (TCR) activation by antigen-presenting cells, together with co-stimulation, is the first step for a proper T-cell response (Zheng et al. 2008; Smith-Garvin et al. 2009; Brownlie and Zamoyska 2013). As summarized by others, TCR engagement causes several phosphorylation events leading to the recruitment of effector proteins (Roifman and Grunebaum 2013). This results in actin cytoskeleton rearrangements, activation of Ras GTPase, and calcium mobilization (Nika et al. 2010; Roifman and Grunebaum 2013). Subsequently, various transcription factors become activated, such as nuclear factor of activated T-cells (NFAT) and nuclear factor kappa-light-chain-enhancer of activated B cells (NF-κB), which initiate the transcription and secretion of pro-inflammatory or pleiotropic cytokines, *e.g.*, IFNγ, IL-2, or of proliferative genes, respectively (Nika et al. 2010; Roifman and Grunebaum 2013). IL-2 represents a key mediator in the activation and proliferation of T-cells, and binds to receptor IL-2R, which not only has a crucial role in T-cell activation, but also promotes the progression of self-tolerance and regulates the functionality of natural killer (NK) cells (Nika et al. 2010; Roifman and Grunebaum 2013).

The effect of VacA on T-cell processing was investigated by the use of Jurkat cells, a transformed human T-cell line. Genes associated with apoptosis, signal transduction, NF-κB-dependent signaling or inflammation, *e.g.* IL-8 and IL-2R, were found to be upregulated (Takeshima et al. 2009). The stimulatory effect of VacA on the local immune response of Jurkat cells was confirmed using isogenic *H. pylori* ΔvacA mutants (Takeshima et al. 2009). Although it is well known that IL-2 stimulates T-cell proliferation in Jurkat cells, inhibition of proliferation was found in human $CD4^+$ T-cells, but these different responses are neither dependent on IL-2 expression nor on NFAT activation (Sundrud et al. 2004). In addition, a role of the N-terminal hydrophobic domain of VacA in mediating signaling to human T-cells and Jurkat cells was demonstrated (Sundrud et al. 2004). Making use of the known blocking effect of NPPBs (non-specific chloride channel inhibitors) on VacA activity, it was shown that VacA might hamper T-cell activation by a channel-independent mechanism in Jurkat cells (Boncristiano et al. 2003). In this process, two regions in the VacA protein, named i1 and i2, seem to be crucial for cell type specificity; moreover, VacA was shown to inhibit the activation of NFAT in T-cells by preventing the nuclear translocation (Boncristiano et al. 2003; Gebert et al. 2003; González-Rivera et al. 2012). Further, VacA can bind to receptors being expressed on

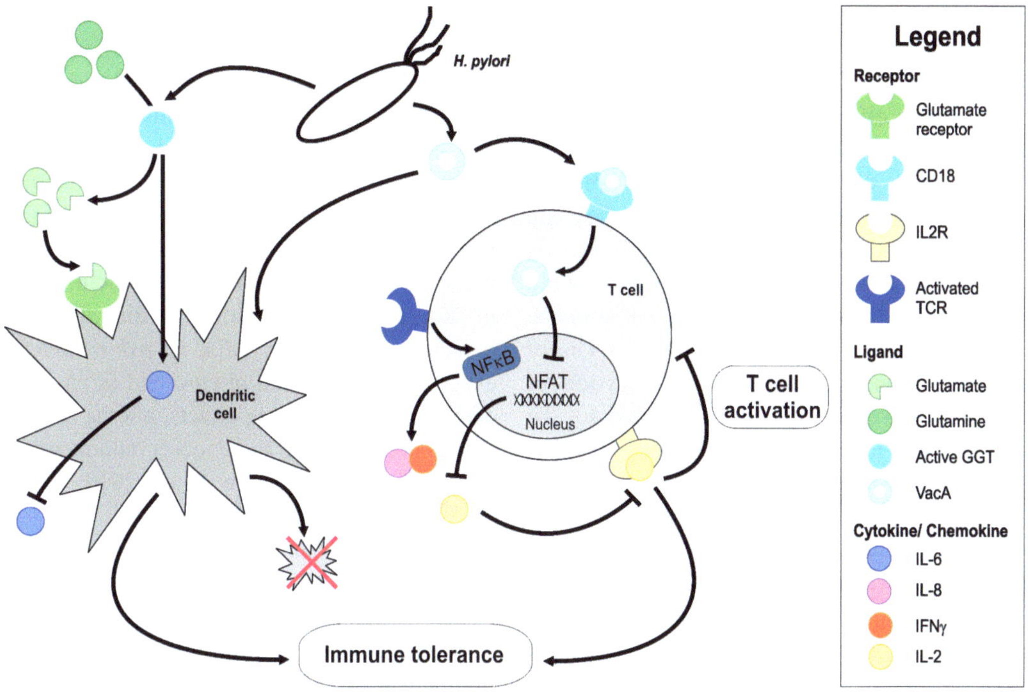

Fig. 1 *H. pylori* GGT- and VacA-mediated responses in immune cells of the host. GGT was found to activate the glutamate receptor on DCs and T-cells by increasing the glutamate concentration in the microenvironment through conversion of glutamine to glutamate. This interaction suppressed the secretion of pro-inflammatory cytokine IL-6 and induced tolerance in DCs. Moreover, GGT-primed tolerant DCs co-cultured with CD4$^+$ T-cells differentiated into Treg cells. It was also reported that GGT can manipulate T-cell proliferation through inducing cell cycle arrest. VacA binds to the CD18 receptor on T-cells for cellular entry and this inhibits the NFAT signaling and subsequent IL-2 production, which ultimately suppresses proliferation. Altogether, tolerated DCs, suppression of T-cells expansion and production of Tregs during interaction with *H. pylori* factors pave the way for immune tolerance and persistent *H. pylori* infection

T-cells beyond β-integrin, which results in the inhibition of IL-2 secretion and decreased NFAT activation (Sewald et al. 2008, 2011). Nevertheless, it was found that extracellular calcium ions are essential for the TCR-dependent NF-κB signaling. VacA is able to regulate the calcium ion influx, forcing NF-κB activation and thus elevating the pro-inflammatory response in human eosinophils (Kim et al. 2007a). Thus, *H. pylori* may manipulate the calcium balance during infection, which might be responsible for the VacA-dependent effect on NF-κB in T-cells (Liu et al. 2016). The role of VacA in NFAT- and IL-2R-dependent signaling is summarized in Fig. 1.

Interestingly, CagA acts as an antagonist of VacA with respect to NFAT activation (Yokoyama et al. 2005). Microarray-based analysis of CagA-transfected human gastric epithelial cells demonstrated that CagA activates NFAT signaling through induction of nuclear translocation of cytoplasmic NFAT (Yokoyama et al. 2005). For this, the EPIYA (Glu-Pro-Ile-Tyr-Ala)-containing region of CagA protein is essential, but CagA phosphorylation (which typically takes place during infection) is not required (Yokoyama et al. 2005).

4 Activities of VacA on Dendritic Cells and Macrophages

The apoptotic response of gastric epithelial cells and eosinophils upon exposure to VacA, as well as its immune regulatory effect on T-cells, have been well characterized (Calore et al. 2010; Kim et al. 2010; Käbisch et al. 2016). However, VacA is also known to affect the maturation of DCs (Kim et al. 2011; Oertli et al. 2013). The manipulating effect of *H. pylori* VacA on DCs and the role in immune cell tolerance is schematically shown in Fig. 1. Immature DCs are located in peripheral tissue where they can be activated to undergo maturation by various antigens, including microbial peptides (Zanotti et al. 2002). Mature DCs can initiate an immune response by activating other immune cells. One of the earlier reports showed that expression of *cag*PAI or *vacA* genes was not required for the activation and maturation of DCs during *H. pylori* infection (Kranzer et al. 2005). However, in murine bone-marrow and splenic DCs derived from infected animals, the STAT3 (signal transducer and activator of transcription-3) pathway is activated (Kao et al. 2010; Oertli et al. 2012; Rizzuti et al. 2015). In this model, stimulation by *H. pylori* and its secreted virulence factors result in increased levels of the pro- and anti-inflammatory cytokines IL-1β and IL-10, respectively, comparable in strength to induction by *Escherichia coli* LPS (Kao et al. 2010; Oertli et al. 2012; Rizzuti et al. 2015). This suggests that these cytokines affect STAT3 regulation during *H. pylori* infection. Moreover, chemical or genetic inhibition of STAT3 led to an up-regulated DC maturation, indicating that STAT3 inhibits DC activation (Melillo et al. 2010). Since IL-10 promotes STAT3 activation, increasing the amount of secreted IL-10 would hamper the activation of DCs (Braun et al. 2013). These authors have shown that in addition to IL-10, IL-6 can activate the STAT3 pathway. Furthermore, it was shown that IL-6 mediated STAT3 activation leads to a transient pro-inflammatory response, while IL-10 based effects might continuously act anti-inflammatory (Braun et al. 2013). It can be concluded that a cytokine imbalance results from various *H. pylori* virulence factors that might alter activation or maturation of DCs.

Contrasting findings have been reported regarding the apoptotic effect of *H. pylori* virulence factors on DCs. In monocyte-derived DCs exposed to *H. pylori*, the induction of apoptosis was not detected (Galgani et al. 2004). However, analyzing the direct effect of VacA on human DCs, Kim et al. (2015a) identified that VacA can induce endoplasmic reticulum (ER) stress which can lead to apoptosis. Moreover, ER stress seems to occur earlier than the induction of apoptosis, so that ER stress might be the critical inducer for the regulation of apoptotic processes in DCs (Kim et al. 2015b). Regarding the development of tolerogenic DCs, *H. pylori* infection experiments performed *in vitro* and *in vivo* have demonstrated promoting effects, but the host cell mechanisms behind these observations remain currently unclear (Calore et al. 2010; Necchi et al. 2009).

Apart from affecting DC maturation, *H. pylori* virulence factors including VacA are further known to act on monocytes and macrophages during infection. As Allen (2007) already summarized, a variable ability of human monocytes or macrophages to kill *H. pylori* was experimentally shown. Compared to monocytes, macrophages exhibit a reduced capacity to kill *H. pylori* (Allen 2007; Borlace et al. 2008) for as yet unknown reasons. To investigate a possible protective effect by CagA or VacA against killing, primary human monocytes and macrophages were infected with strains of *H. pylori* differing in their CagA expression and VacA activity; however, after 48 h of infection no correlation was found between the number of viable bacteria and thus the cell's ability to kill *H. pylori*, and CagA expression or VacA activity (Borlace et al. 2008). The *cag*PAI plays no role in survival of the bacteria inside phagocytic cells (Odenbreit et al. 2001) and the *vacA* status of the bacteria is non-determinative for the phagosome fusion in human monocytes (Rittig et al. 2003). In contrast, other studies have indicated that VacA is crucial for intracellular survival and phagosome maturation (Petersen et al. 2001; Terebiznik et al. 2006). A link between urease activity and VacA for the survival in macrophages was indicated by Schwartz and Allen (2006). Infection of the monocytic cell line THP-1 by *H. pylori* for 8 h revealed an elevated amount of oncostatin M (belonging to the group of IL-6 cytokines) in the supernatant, indicative of a pro-inflammatory response (Zeaiter et al. 2011). However, using isogenic deletion mutants it was shown that this pro-inflammatory response is independent of VacA, CagA or T4SS (Zeaiter et al. 2011). Thus, other virulence factors of *H. pylori* can be postulated to affect the pro- and anti-inflammatory cytokine secretion in monocytes and macrophages, but their nature remains to be investigated in future studies.

5 GGT Manipulates T Cell Proliferation and Cell Cycle Progression

Various studies described an *H. pylori*-associated inhibitory effect on cell growth leading to cell cycle arrest (Lew et al. 1991; Wagner et al. 1997; Chiou et al. 2003). Both CagA and VacA inhibit T-cell activation, while proliferation of these cells can also be inhibited by incubation with bacterial supernatants or purified VacA (Gebert et al. 2003; Sundrud et al. 2004). Moreover, as discussed above, infection with *H. pylori* leads to inhibition of T-cell proliferation by initiating apoptotic pathways (Wang et al. 2001). However, studies by Gebert and co-workers (2003) have shown that deletion of *vacA* did not hamper the effect on T-cell proliferation. In contrast, Sewald et al. (2008) have determined that VacA binds only to active human T-cells, leading to internalization of the virulence factor, whereas it does not bind to non-activated T-cells (Sewald et al. 2008). Moreover, this functional uniqueness of human T-cells compared to murine humanized T-cells indicates an important specificity of *H. pylori* VacA to host immune cells (Sewald et al. 2008). Since it was shown that during infection with *H. pylori* the proliferation of T-cells is inhibited by inducing cell cycle arrest, virulence factors other than CagA or VacA may be responsible for this phenotype (Gerhard et al. 2005).

Apart from its well-characterized importance for colonization *in vivo*, GGT has been shown to play a role in *H. pylori*-mediated apoptosis by mitochondrial pathways in epithelial cells (Kim et al. 2007b). In addition, it was demonstrated that *H. pylori* GGT is linked to blockage of T-cell proliferation and function, leading to immune evasion (Wüstner et al. 2015). It had previously been described that GGT is involved in proliferation of T-cells and in initiation of G1 arrest (Schmees et al. 2007). Using *H. pylori* isogenic Δggt deletion mutants, it was shown that the gene is crucial to suppress the proliferation of antigen-stimulated primary human T-cells and this effect could be replicated with recombinant *H. pylori* GGT, but not by mammalian GGT (Schmees et al. 2007). By incubating AGS gastric epithelial cells with recombinant *H. pylori* protein, GGT inhibition of cell cycle progression at G1 to S phase transition was demonstrated (Wüstner et al. 2015). These authors suggested that the arrest might depend on the growth characteristics of the target cells. Additionally, *H. pylori* seems to alter the expression of cell-cycle specific mediators, which would result in a dysregulated cell cycle progression during *H. pylori* infection (Wüstner et al. 2015). However, the exact mechanisms behind these processes require further investigation.

6 Function of the Neutrophil Activating Protein NapA

Recently, a correlation between the *H. pylori* virulence factor NapA and *H. pylori*-associated peptic ulcer disease was determined, although no correlation between other *H. pylori* associated diseases such as gastritis could be detected, neither for NapA nor for the virulence factors CagA, VacA, UreA, or UreB (Oktem-Okullu et al. 2015). It was already known that NapA is crucial for activation of neutrophils, which seems to promote damage of stomach tissue (Dundon et al. 2001). As recently reviewed elsewhere, NapA interacts with TLRs during infection (Amedei et al. 2006; Pachathundikandi et al. 2015). This TLR signaling induces neutrophil trans-endothelial migration and is involved in complex host cell processes, and in the case of *H. pylori* this results in persistent colonization and chronicity of infection (Brisslert et al. 2005). Using purified recombinant *H. pylori* NapA or bacteria secreting the protein, these authors were able to show that treatment induces transendothelial transmigration of neutrophils in a CagA- and VacA-independent manner (Brisslert et al. 2005). The continuous recruitment of neutrophils to the site of infection had already been shown by others (Go 1997; Del Giudice et al. 2001). It is likely, however, that other chemotactic effector molecules, in addition to NapA, are involved in the migration of neutrophils (Satin et al. 2000; Brisslert et al. 2005). An additional role for NapA was shown in the T-cell dependent immune responses (Amedei et al. 2006). The effects of NapA on transmigration of neutrophils and the role in reactive oxygen species release are summarized in Fig. 2.

Previously, it was shown that iron can be stored in the cavity of NapA, although crystal structure analyses failed to detect iron (Tonello et al. 1999; Zanotti et al. 2002). The presence of high amounts of basic residues typically present in chemokines or cytokines might be crucial for the neutrophil activating function of NapA (Collaborative Computational Project, Number 4 1994; Yang et al. 1994; Zanotti et al. 2002). NapA and urease both seem to be important for the recruitment of neutrophils to the site of infection and may be involved in the response to oxidative stress (Wang et al. 2006). In a further study, expression of NapA and the outer membrane protein (Omp) 18, which is expressed in a limited number of *H. pylori* strains, correlated

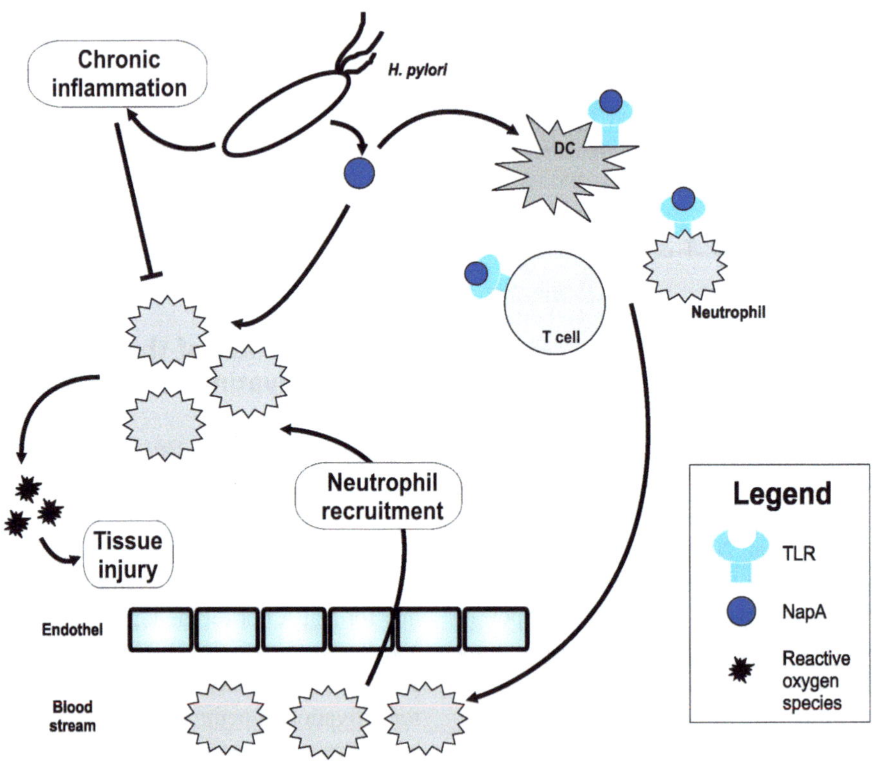

Fig. 2 Neutrophil immunity control during *H. pylori* infection. NapA of *H. pylori* is named for its role in the recruitment and activation of neutrophils during infection. NapA-induced activation of neutrophils leads to the production of reactive oxygen species and related tissue injury, which supports chronic inflammation. In addition, NapA can bind to TLR2 on neutrophils and monocytes to induce pro-inflammatory cytokine production and furthermore induces a pro-inflammatory Th1 phenotype. Therefore, NapA-mediated above effects ultimately result in epithelial barrier disruption and increased inflammatory reactions

with IFNγ-mediated immune response (Shan et al. 2015). During infection with a Δ*omp18* deletion strain, NapA expression was upregulated when the cells were co-incubated with IFNγ, suggesting NapA expression might be modulated by Omp18 in an IFNγ-dependent manner (Shan et al. 2015). However, the immune regulatory mechanisms behind the Omp18-dependent effect on NapA expression and neutrophil activation remained as yet unclear.

7 Role of Cholesterol in *H. pylori* Interactions with Immune Cells

Cholesterol is an important component of mammalian cellular membranes and has physiological roles in fat metabolism. In contrast, the membranes of many prokaryotes do not contain cholesterol, and most bacteria do not possess the genes required for cholesterol synthesis. *H. pylori* has evolved mechanisms to extract cholesterol from the host's cellular membranes and converts into glucosides by using a specific enzyme, CGT (Wunder et al. 2006). It was a remarkable finding that a common component of host membranes is utilized by a pathogen, to employ for virulence. Cholesterol acquisition appears to be essential for *H. pylori* survival in the human host and for prevention of effective host immune attacks (Wunder et al. 2006). *H. pylori* cells exhibit a high affinity for cholesterol and the chemotactic bacteria follow a cholesterol gradient, even responding to 20 times lower cholesterol concentration than are normally present in serum. It has been shown that the bacteria extract cholesterol from cell membranes when co-cultured with epithelial cell lines and convert to α-glucosides such as cholesteryl-α-glucoside, cholesterylacyl-α-glucoside or cholesteryl-phosphatidyl-α-glucoside by means of the CGT glucosyltransferase (encoded by gene HP0421, also known as *cgt* or *capJ*). This enzyme is necessary for prevention of phagocytosis by macrophages. However, when the bacteria were artificially loaded with high levels of cholesterol, this actually increased phagocytosis rates, indicating that only a high enough ratio of metabolized α-glucosides per cholesterol (unconverted) prevents engulfment. In addition, cholesterol glucosylation was also found to reduce the T-cell responses against *H. pylori* (Wunder et al. 2006). A recent study supported the role of cholesterol acquisition and its modification in *H. pylori* virulence. A murine macrophage cell line was used to show that phagocytosis of wild-type bacteria was delayed and phagosome maturation in infected cells was inhibited, but not when the cells were infected with a knock-out Δ*cgt/capJ* mutant. The interference in phagocytosis and phagosome trafficking was dependent on lipid raft formation and phosphoinositide-3-kinase (PI3K) signaling (Du et al. 2016).

It was further reported that CagA co-fractionates with redistributed MARK2/Par1b in the detergent-resistant membrane fraction of infected AGS cells (Zeaiter et al. 2008). In addition to CagA, VacA was reported to be co-localizing in the lipid raft regions during infection and cholesterol depletion disrupted this localization (Nakayama et al. 2006; Lai et al. 2008). The methyl-β-cyclodextrin mediated cholesterol depletion disrupted CagA translocation and phosphorylation in infected cells. CagA-induced cellular elongation and IL-8 secretion were also severely affected in this process. This shows that the capacity to bind and extract cholesterol acts as a point of delivery for virulence factors in the lipid raft regions of infected cells. However, the overall adherence of the bacteria to epithelial cells is not affected by cholesterol depletion; this can be explained by the presence of other cholesterol-independent adherence factors of *H. pylori* (Lai et al. 2008).

A study involving various N-terminal and C-terminal truncation mutants of CagA revealed that the EPIYA containing C-terminal domain may directly interact with cholesterol to induce IL-8 secretion (Lai et al. 2011). Mutation of the *capJ* gene prevented CagA phosphorylation, c-Src and FAK (focal adhesion kinase-1) dephosphorylation and the subsequent elongation of *H. pylori* infected AGS cells. By complementation with CapJ or supply of exogenous CGs (cholesterol glucosides), the CagA function in the *capJ* mutant was restored. Infection with wild-type *H. pylori* showed that lipid raft components were recruited to the site of attachment and exogenously supplied fluorescence tagged cholesterol/CGs co-localized in this area (Wang et al. 2012).

The cholesterol acquisition and its subsequent conversion were thought to be independent processes, but the discovery of the flotillin-like protein (HP0248) in detergent resistant membrane fractions of *H. pylori* changed that view. HP0248 was found to be important for cholesterol sequestration, as infection with a Δ*HP0248* mutant severely affected cholesterol accumulation in the bacterial membrane (Hutton et al. 2017). This mutation also affected CagA translocation, cell scattering and IL-8 secretion, which indicates the protein has an important role in virulence, and this was confirmed using a mice model of infection (Hutton et al. 2017). Moreover, it was shown that the T4SS pilus protein CagL interacts with the host cell receptor $\alpha_5\beta_1$ integrin during CagA delivery (Kwok et al. 2007; Conradi et al. 2012; Barden et al. 2013). This $\alpha_5\beta_1$ integrin co-localizes with cholesterol-rich micro-domains within the membrane, which supports the concept of CagA delivery in a cholesterol-dependent manner. In addition, $\alpha_5\beta_1$ integrin in cholesterol-rich micro-domains is required for the delivery of peptidoglycan to NOD1 (nucleotide binding oligomerization domain containing 1) recognition and NF-κB activation. The depletion of cholesterol from cell surfaces with methyl-β-cyclodextrin resulted in reduced NF-κB activation and IL-8 secretion in *H. pylori* infected epithelial cells (Hutton et al. 2010).

H. pylori growth in the presence of cholesterol substantially increased its resistance to antimicrobials, although phosphorylation of lipid-A played a major role in this process (McGee et al. 2011). A recent study discovered another important aspect of cholesterol glucosylation in the immune response against *H. pylori*. Wild-type but not Δ*cgt/capJ* bacteria blocked IFNγ signaling through decreased phosphorylation of Janus kinase (JAK) and STAT1 in infected primary gastric cells and gastric epithelial cell lines (Morey et al. 2018). The disruption of IFNγ signaling was due to the destruction of lipid rafts (mediated by cholesterol depletion), while cholesterol coating of infected cells regained the signaling activation. It was found that *H. pylori* infection disrupted the distribution of receptors IFNGR1 and IFNGR2 in lipid rafts, which provides an explanation for these observations. Lipid raft disruption also inhibited the responses to IFNβ, IL-6 and IL-22 and subsequent signaling (Morey et al. 2018). Cytokine-induced hBD3 (human-β-defensin 3) expression was also downregulated during *H. pylori* infection. Infection of cells with wild-type or Δ*cgt/capJ H. pylori* induced almost identical changes in gene expression. However, cholesterol depletion by wild-type bacteria suppressed the immune responses in infected cells, while non-infected cells (thereby not suffering from cholesterol depletion) in the vicinity may get inflamed by increased cell signaling, induced by bacterial factors that were released in the microenvironment (Morey et al. 2018). The above data show that cholesterol acquisition from host cell membrane helps *H. pylori* to deliver virulence factors, interferes with phagocytosis and also contributes in the manipulation of important host cell signaling mechanisms for the benefit of survival and persistence.

8 *H. pylori* Manipulates Antigen Presentation and Bacterial Recognition

Infecting bacteria are normally subject to phagocytosis, but many reports have shown that *H. pylori* interferes with this defensive process. In one of the earlier reports, Ramarao and co-workers (2000) showed that *H. pylori* inhibits the phagocytic function of neutrophils and monocytes, an inhibition that was dependent on proteins VirB7 and VirB11 of the T4SS. It was also documented that phagocytosis of *H. pylori* type-I strains by macrophages got delayed, whereas type-II strains were easily phagocytosed (Allen 2007). In addition, *H. pylori* prevented phagosome maturation and instead resulted in formation of a hybrid phagosome-endosome-lysosome with no or strongly reduced degradation (Borlace et al. 2011). This is a clear indication for reduced antigen epitope production and MHCII presentation and provides an example of a successful bacterial adaptive response at the most apt time to avoid infection clearance. *H. pylori* VacA can inhibit CD74 (Ii)-dependent MHCII antigen presentation, but not the independent pathway of recycling MHCII presentation (Molinari et al. 1998). Moreover, CD74 is up-regulated on the surface of gastric cells during *H. pylori* infection and was reported to act as a receptor for *H. pylori* urease. CD74 is well known for its role in antigen presentation as it directs MHCII molecules to the endosomes, where it is partly digested to relieve the cleft of the MHCII molecule for peptide loading. It is plausible that CD74 binding to *H. pylori* urease interferes with MHCII localization and antigen loading for activation of adaptive immunity (Beswick et al. 2005; Beswick and Reyes 2009).

At least in mice, the important role of TLRs in the development of adaptive immunity against *H. pylori* infection is evident from several reports (reviewed by Pachathundikandi et al. 2015, summarized in Fig. 3). One of the major findings was the TLR-mediated MyD88 (myeloid differentiation-88) signaling on antigen presentation and co-stimulation in infected DCs for development of an adaptive immune response against *H. pylori* infection (Rad et al. 2007). In contrast, it was also found that at the initial phase of infection *H. pylori* can reside and replicate in macrophages, DCs or epithelial cells inside double-membrane auto-phagosomes, which later fuse with lysosomes and are almost fully degraded at 48 h of infection (Wang et al. 2009, 2010). *H. pylori* infection reduced the cell surface expression of MHCII, CD80 and CD86 instead these molecules were co-localized with the *H. pylori* containing vacuoles, which supports the interference on antigen loading and trafficking to the cell surface. Moreover, formalin-fixed bacteria exhibited the same effect. However, in infected DCs, MHCI expression was enhanced (Wang et al. 2010). These authors found that TLR2 and TLR4 are necessary for the interference on MHCII trafficking and subsequent antigen presentation, which suggests an important role of *H. pylori* LPS in this mechanism (Wang et al. 2010).

Recently, it was also reported that miR-30b was upregulated in patients with chronic *H. pylori* infection and was targeting autophagy proteins ATG12 (autophagy related protein 12) and BECN1 (Beclin 1). The control of autophagy in *H. pylori* infected cells resulted in increased intracellular survival and bacterial replication (Tang et al. 2012). This shows a different-level control on autophagy and intracellular growth of *H. pylori,* which ultimately influences the antigen presentation. Furthermore, CD300E was identified as a new factor involved in *H. pylori* mediated interference on antigen presentation (Pagliari et al. 2017). Differentiation of monocytes to macrophages normally reduces the expression of CD300E, however, *H. pylori* infection increased the expression of this protein by down-regulating miR-4270, a miRNA that targets *CD300E* mRNA. This increased expression and activation of CD300E by *H. pylori* alone and agonistic antibody treatment drastically reduced the expression of MHCII molecules, thereby reducing antigen presentation and that prevented activation of T-cells (Pagliari et al. 2017). The above studies clearly affirm the interference on antigen presentation in *H. pylori* infection.

The interference on antigen presentation could be responsible for both the reduced recognition of antigens by T-cells and the ineffective adaptive immune response against *H. pylori*. DC maturation is an important step in the transition from antigen-capturing to antigen presentation, which is necessary for T-cell priming and activation. VacA suppresses DC maturation by downregulating the surface expression of MHCII, CD40, CD80 and CD86, apart from reducing migratory power of DCs (Molinari et al. 1998). The increased expression of PD1-L1 (Programmed cell death protein ligand-1)/B7-H1 in gastric epithelial cells was reported in *H. pylori* infection, which is independent of the virulence factors CagA, VacA or urease, but is enhanced by IFNγ (Das et al. 2006). In addition, gastric epithelial cells are expressing MHCII and co-stimulatory molecules such as CD80 and CD86, and this antigen presenting capacity helps to activate naïve T-cells at the epithelial contact site. This presumably leads to clonal differentiation to enable well-developed adaptive responses against bacterial infection. However, *H. pylori* induced expression of PD1-L1 and the binding of this to PD1 on T-cells led to the suppression of antigen-specific T-cells, while it supported the maintenance of $FoxP3^+$ Treg cells (Beswick et al. 2007; Zhang et al. 2016). Induced expression of PD1-L1 in gastric epithelial cells along with other factors appears to maintain the Treg pool reported in gastric mucosa of *H. pylori* infected individuals (Cook et al. 2014; Hussain et al. 2016).

9 TLR Signaling in Immune Cells Induced by *H. pylori*

TLRs constitute a group of host cell surface and subcellular transmembrane proteins, which detect intruding microbes or microbial products outside the cells through their varied presence on the cell surface (TLR1, TLR2, TLR4, TLR5, and TLR10) or in intracellular compartments (TLR3, TLR7, TLR8, and TLR9) (Beutler 2009). These type I transmembrane glycosylated protein receptors are germ line encoded proteins composed of an ectodomain containing leucine-rich repeats, a transmembrane region, and an intracytoplasmic Toll/IL-1 receptor (TIR) domain. TLRs sense the presence of various PAMP or MAMP (microbe associated molecular patterns) and can thus detect bacteria, viruses, or fungi, which leads to the induction of downstream signals. TLR signaling can be divided into MyD88 dependent and TRIF (TIR domain containing adaptor inducing interferon-β) dependent signaling, both of which ultimately leads to activation of NF-κB, AP-1 and interferon regulatory factors (IRFs) for the production of chemokines, cytokines and type I interferons that activate the host immune system to control an infection (Pachathundikandi et al. 2011, 2015).

Early studies reported the difference on *H. pylori* LPS activation of NF-κB through TLR4 in epithelial cells and monocytes (Ishihara et al. 2004). In uninfected individuals, TLR4 expression was noticed only in lamina propria mononuclear cells, but *H. pylori* infection induced the expression of TLR4 and MD2 in gastric epithelial cells as well. Gastric epithelial cell lines expressing TLR4 and MD2 did not activate NF-κB when treated with *H. pylori* LPS, whereas THP1 cells responded with a robust activation (Ishihara et al. 2004). TLR4 and CD14 were involved in the activation of NF-κB in infected THP1 monocytes but not in the MKN45 gastric epithelial cell line. Mouse macrophages with a point mutation in their TLR4 gene showed decreased activation of NF-κB and TNFα (tumor necrosis factor-α) secretion upon *H. pylori* infection compared to wild type macrophages. Moreover, incubating monocytes with *H. pylori* culture supernatant resulted in NF-κB activation, but bacterial contact or a functional *cag*PAI was necessary for activation of epithelial cells (Maeda et al. 2001). Our studies showed that infection of THP1 cells with *H. pylori* increased the expression of TLR2 and TLR5 as well as secretion of IL-8 and TNFα in a *cag*PAI dependent manner (Pachathundikandi et al. 2011). In contrast, IL-8 production in mouse macrophages was mediated through TLR4 in response to *H. pylori* LPS, whereas intact bacterial cells of *H. pylori, Helicobacter hepaticus* or *Helicobacter felis* resulted in a response mediated through TLR2. These bacterial infections also induced IL-6 expression in a TLR2 dependent manner (Mandell et al. 2004).

Apart from LPS, at least three *H. pylori* proteins have been identified that can activate TLR2. *H. pylori* Hsp60 (also called GroEL) was shown to induce IL-8 secretion mediated by MAPK signaling in a human monocytic cell line (Zhao et al. 2007). NapA induced the production of IL-12 and IL-23 in primary neutrophils and monocytes and induced a Th1 response in T-cell clones prepared from healthy donors, while it shifted the Th2 response to Th1 in cells obtained from allergic donors (Amedei et al. 2006). Moreover, T-cell clones prepared from gastric mucosa of infected patients were found to be of the cytotoxic Th1 phenotype and produced TNFα (Amedei et al. 2006). Further, *H. pylori* UreB could activate TLR2, resulting in an increased

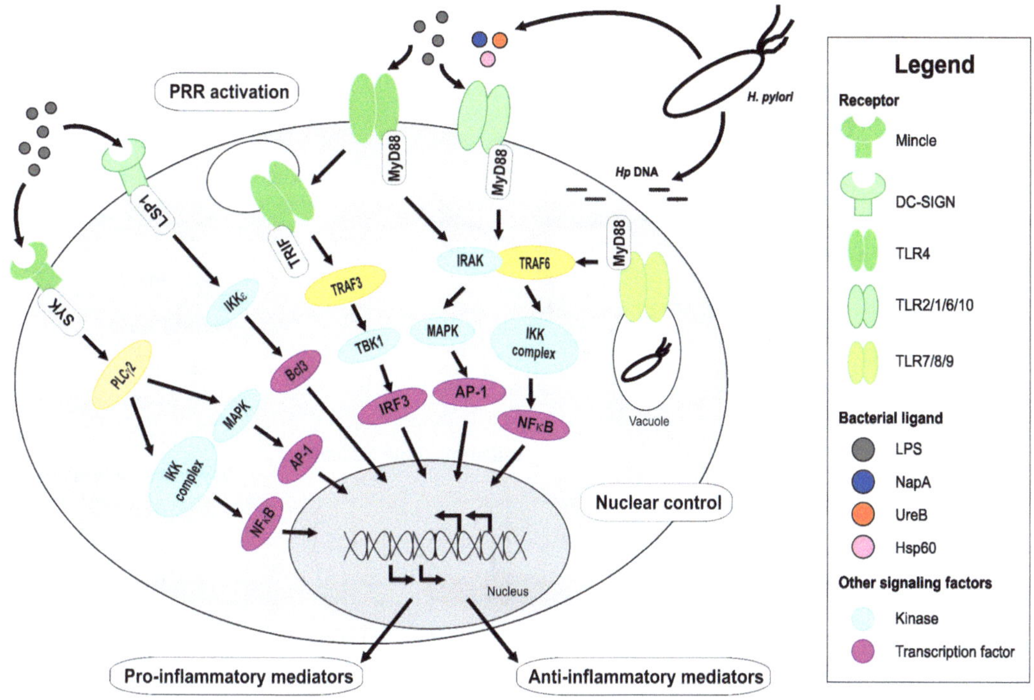

Fig. 3 The pro- and anti-inflammatory signaling through different PRRs during *H. pylori* infection. *H. pylori* is reported to interact with various PRRs such as TLRs and DC-SIGN during infection with host cells. This ultimately leads to the induction of various signaling pathways to activate or modify functions of transcription factors such as NF-κB, AP-1 and IRFs. These multiple induction and modification mechanisms resulted in the production of pro- and anti-inflammation states in *H. pylori* infection. Thus, different signaling mediated variation in immune responses may ultimately determine the outcome of associated diseases

expression of NLRP3 (NOD-like receptor pyrin domain-containing-3) and inflammasome assembly, while Δ*ureB* mutant bacteria inhibited caspase-1 activation in murine and human DCs (Koch et al. 2015). Thus roles for HSP60, NapA and UreB in TLR2 activation have all been demonstrated (summarized in Figs. 3 and 4).

H. pylori infection of mouse DCs induced the expression of IL-12 and IL-10 through TLR4/MyD88 signaling, whereas secretion of IFNα was increased substantially in *myd88* deficient cells. Moreover, IL-6 and IL-1β expression was decreased in *tlr2* deficient cells infected with *H. pylori* (Obonyo et al. 2007). We found that HEK293 cells stably expressing TLR2 (HEK293-TLR2) differentially expressed *IL-1β* but not *IL-6* during infection with *H. pylori*, however, *TNFα* expression was induced in both HEK293-TLR2 and HEK293-TLR10 cells after infection (Pachathundikandi and Backert 2016). *H. pylori* LPS activated HEK293 cells jointly overexpressing TLR2 and TLR10 to induce NF-κB signaling for *IL-8* and *TNFα* expression (Nagashima et al. 2015). We have also shown highly induced expression of TLR10 in THP1 monocytes infected with *H. pylori* (Pachathundikandi and Backert 2016). It was reported that induction of pro-IL-1β expression in DCs by *H. pylori* depends on TLR2 and NOD1, while infected *tlr2* deficient cells suppressed pro-IL-1β expression more than *nod1* deficient cells (Kim et al. 2013). Moreover, this work demonstrated a cumulative effect in *tlr2-nod1* double deficient cells, which suggests that these two receptors have redundant roles in pro-IL-1β expression during *H. pylori* infection (Kim et al. 2013).

Mice deficient in *myd88* produced decreased gastric inflammation in response to *H. pylori* infection, but the gastric colonization levels

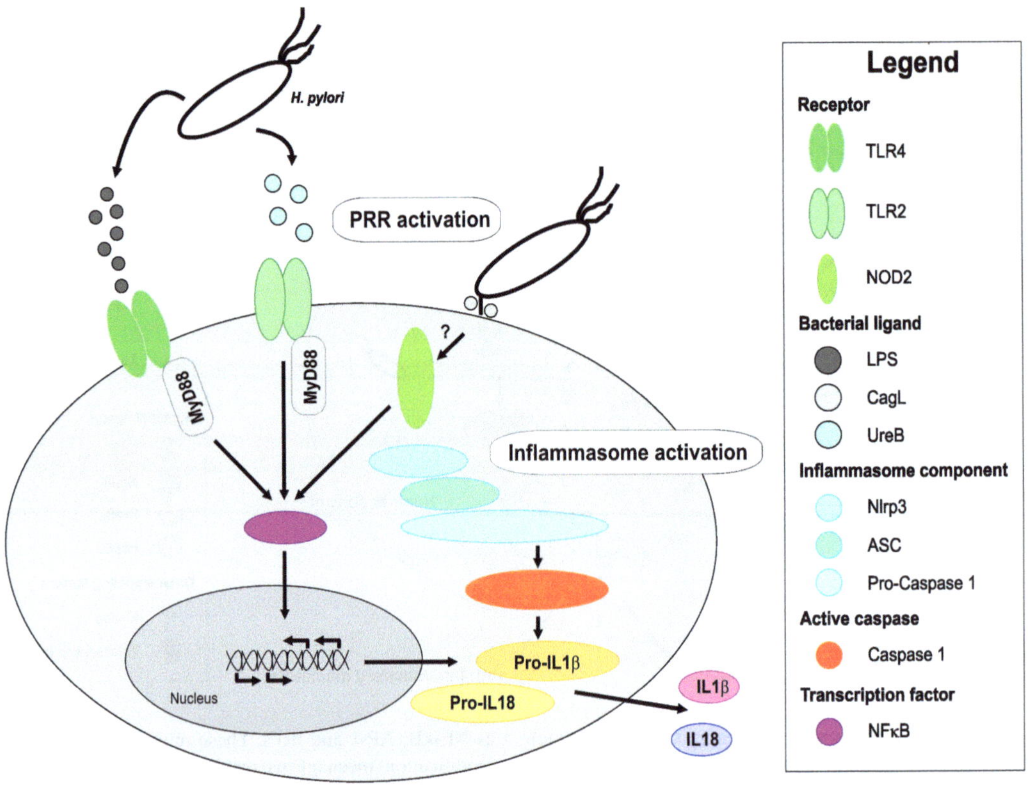

Fig. 4 TLR2- and NOD2-mediated inflammasome activation during *H. pylori* infection. It was found that *H. pylori* UreB activated TLR2 signaling for NLRP3 expression and assembly of the inflammasome in mice. However, pro-IL1β expression was induced by LPS through TLR4 signaling. The inflammasome induced activation of caspase-1 resulting in the production of active IL-1β and IL-18 and exerted various effects described in the text

were increased (Rad et al. 2007). The secretion of pro-inflammatory IL-6 and IL-12p40 cytokines by murine DCs infected with *H. pylori* lysates was largely dependent on TLR2 activation, whereas infection with live bacteria induced anti-inflammatory IL-10 secretion in a TLR2-dependent manner. *H. pylori* DNA or RNA was able to induce strong cytokine secretion in wildtype and *tlr2/4* deficient DC cells, an effect that was completely abrogated in *tlr2/4/9* deficient cells. However, *tlr9* deficient cells showed equivalent amounts of cytokines secretion when compared to wild type cells upon exposure to bacterial DNA or RNA (Rad et al. 2007). *H. pylori* RNA detection by DCs was dependent on TLR8 alone or possibly in combination with TLR7. In addition, exposure of DCs to *H. pylori* RNA resulted in type-I IFN production that was independent of TRIF or MyD88 but dependent on RIG1 activation (Rad et al. 2009).

The production of Treg cells ($CD25^{+}Foxp3^{+}$) in response to *H. pylori* presence is also proposed to be mediated through TLR involvement (Rad et al. 2006). TLR2-mediated activation of B-cells in mice infected with *H. felis* prevented immunopathology and pre-neoplastic changes through the production of a Treg cell population (Sayi et al. 2011). Instead, the adoptive transfer of *H. felis*-specific effector T-cells aggravated the immunopathology to produce pre-neoplastic changes and reduced colonization in infected *tcrβ/rag1* knockout, immune-deficient mice. However, co-transfer of effector T-cells and Treg cells to these infected mice exhibited alleviation of symptoms in *tcrβ* deficient mice, but not in *rag1* deficient mice, which suggests B-cells are involved in this process. *H. felis* induced the activation of TLR2 in B-cells which led to an increased expression of co-stimulatory molecules CD40, CD80, CD86 and secretion of IL-10, IL-6, TNFα, and antibodies (Sayi et al. 2011). When murine DCs deficient in *tlr2* were infected with *H. pylori*, the cells produced more IFNγ and less IL-17 and IL-10 compared to wild type DCs. In addition, *tlr2* deficient mice expressed more IFNγ *in vivo* and less FoxP3, IL-10 and IL-17A in their infected gastric mucosa compared to infected wild type animals. These observations may explain the increased gastritis and lower colonization of bacteria in infected *tlr2* deficient mice. TLR2 mediated signaling during *H. pylori* infection produced tolerogenic DCs that dampened the immuno-pathological Th1 response and allowed higher colonization levels to be reached (Sun et al. 2013).

H. pylori infection increased the IRAK-M expression in *tlr2* deficient DCs. Infection of these DCs induced a more pronounced pro-inflammatory response through higher expression of MHCII, TNFα and MIP2 and reduced IL-10 expression, although Treg and Th17 responses were comparable to those seen in wild-type mice (Shiu et al. 2013). TLR9 expression in the gastric tissue was increased after *H. pylori* infection in mice and this elevated expression was mostly observed in macrophages, DCs and $CD3^+$ cells. The *tlr9* deficient mice showed increased myeloperoxidase (MPO), TNFα and IFNγ expression during initial phase of *H. pylori* infection, but colonization levels were similar in wild type and deficient mice. The treatment of exogenous recombinant IFNα reduced the pro-inflammatory changes in the infected *tlr9* deficient mice (Otani et al. 2012).

The reports summarized here demonstrate that TLR activation plays a crucial role in innate and adaptive immune responses against *H. pylori*. The above data show that immune cell signaling in *H. pylori* infection is mainly carried out through TLR2, TLR4 and TLR9 receptors, with minor roles for TLR7, TLR8 and TLR10 (summarized in Figs. 3 and 4). These receptors are expressed at epithelial and immune cells alike, which are the two major cell types *H. pylori* interacts in the gastric mucosa. The most striking feature of this interaction is the dual role of TLR2 on activating both pro-inflammatory and anti-inflammatory responses, which may partly explain the varied disease outcome of infected individuals.

10 Pro- and Anti-inflammatory Signaling by *H. pylori*

The interaction between bacteria and their host leads to the activation of various pro- and anti-inflammatory signal transduction pathways from various PRRs, resulting in the activation of number of transcription factors in the host cells (Backert and Naumann 2010; White et al. 2015). Gastric epithelial cells, interacting with *H. pylori*, respond with various signals to initiate an inflammatory process in an attempt to control the infection. Several studies reported the involvement of different pathways in this process (summarized in Fig. 3). *H. pylori* T4SS has been shown to function as the conduit for entry of different factors such as effector protein CagA, heptose 1,7-bisphosphate (HBP), ADP-heptose, peptidoglycan as well as bacterial DNA, which ultimately activate these pathways (Backert et al. 2000; Viala et al. 2004; Varga et al. 2016; Gall et al. 2017; Pfannkuch et al. 2018). Some earlier studies ruled out the involvement of CagA in the NF-κB pathway, although it was reported that CagA can potentiate the NF-κB response through a protein interaction cascade of Grb2 → Ras→Raf → Mek → Erk for IL-8 secretion (Brandt et al. 2005). The transgenic CagA expression in mice, induced the PAR1 mediated IκB (Inhibitor kappa B) sequestering to lower the NF-κB threshold for activation (Suzuki et al. 2015). It was experimentally shown that outer membrane vesicles (OMVs) from *H. pylori* could deliver peptidoglycan to the cytosol of exposed cells, resulting in NOD1 mediated NF-κB activation and IL-8 and CXCL2 production (Kaparakis et al. 2010). Moreover, *H. pylori* induced NOD1-dependent IFNβ secretion in turn activated the expression of IFN-stimulated gene factor 3 (ISGF3) and CXCL10; which protected the mice from infection (Kaparakis et al. 2010).

Activation of NOD1 by *H. pylori* resulted in responses that were augmented by IFNγ to produce various chemokines such as IL-8, CXCL10, CCL2, CCL3, CCL4, and CCL5. *H. pylori* infection further activated phosphorylation of IFNγ signaling factor STAT1 and expression of IRF1 via a NOD1-dependent mechanism. Furthermore, infection activated the production of Th1 cells and elevated the secretion of high amounts of IFNγ, which amplified the NOD1 response in a feedback manner. This synergistic action of NOD1 and IFNγ exacerbated the immune responses in the gastric mucosa during *H. pylori* infection; an observation in line with the finding that gastric cancer patients display upregulated expression of NOD1, IRF1 and IL-8 (Allison et al. 2013). In contrast, NOD1 activation by *H. pylori* suppressed the transcription factor CDX2 (caudal homeobox 2) in both normal and cancerous gastric epithelial cells; this factor is involved in intestinal metaplasia (Asano et al. 2016). Bile duct epithelial cells responded to *H. pylori* through NOD1 and MyD88 pathways to produce activation of NF-κB and IL-8 secretion, which was inhibited by pre-treatment with antibodies against $\alpha_5\beta_1$ integrin (Boonyanugomol et al. 2013). Trans-epithelial neutrophil migration was reported to be dependent on NOD1 mediated IL-8 secretion in *H. pylori* infection (Kim et al. 2015a). Finally, a recent study reported that $\alpha_5\beta_1$ integrin- and Src-mediated JNK/ERK signaling for NF-κB and AP-1 activation following *H. pylori* infection was independent of NOD1 and CagA, but required active CagL (Gorrell et al. 2013).

HBP is a bacterial metabolic intermediate of LPS biosynthesis. It was proposed that HBP can enter the cytosol of infected epithelial cells via the T4SS, and activates a novel signaling cascade involving alpha-kinase 1 (ALPK1) and the phosphorylation-dependent oligomerization of the TNF-α receptor-associated factor (TRAF)-interacting protein with forkhead-associated domain (TIFA) for NF-κB activation. TIFA deficiency or HBP mutants of *H. pylori* almost completely abrogated NF-κB mediated IL-8 production in infected epithelial cells. The major difference between *H. pylori* activated ALPK1-mediated TIFA signaling platform from that observed with other bacteria is the presence of TRAF2 instead of TRAF6 for the NF-κB activation (Stein et al. 2017; Gall et al. 2017; Zimmermann et al. 2017). However, very recent studies indicate that the translocated metabolite activating NF-κB may not be HBP, but ADP-heptose, produced downstream by the same LPS biosynthesis pathway (Zhou et al. 2018; Pfannkuch et al. 2018).

H. pylori infection can also activate EGFR (epidermal growth factor receptor) signaling for COX2 (cyclooxygenase 2) expression and PGE2 (prostaglandin E2) secretion in a *cag*PAI-dependent manner. *Egfr* deficiency reduced this COX2 induction and PGE2 expression in infected murine cells. Likewise, *H. pylori*-induced COX2 expression increased the survival of epithelial cell lines; this would suggest the existence of an EGFR mediated pro-cancerous signaling axis (Sierra et al. 2013). In addition, EGFR signaling is also involved in the pro-inflammatory reaction and epithelial DNA damage in *H. pylori* infection (Sierra et al. 2018). Furthermore, non-phosphorylated CagA activated two other transcription factors such as SRF (serum response factor) and ELK1 (ETS domain containing protein) (Hirata et al. 2002). ELK1 and SRF activation induced the intestinal cell specific marker Villin expression in *H. pylori* infected gastric epithelial cells (Rieder et al. 2005). Apart from that, *H. pylori* infection of gastric cancer cell lines increased the expression of SIAH2 (seven in absentia homologue 2) through the ETS2 (E26 oncogene homolog 2) and TWIST1 (twist related protein 1) transcription factors. The stable expression of SIAH2 increased the invasiveness and migration capacity of gastric cancer cells (Das et al. 2016).

H. pylori can also induce host anti-inflammatory responses, whose role is to dampen the ability to clear an infection, which is somehow essential for general inflammation control by the pathogen. The binding of *H. pylori* LPS to

DC-SIGN, a c-type lectin receptor (CLR), resulted in the signaling which blocked Th1 cell development and decreased production of IL-12 and IL-6 (Bergman et al. 2004; Gringhuis et al. 2009). DC-SIGN signaling can revert the TLR mediated pro-inflammatory cytokine expression to favor an anti-inflammatory cytokine IL-10 production through Raf1, however Raf1 deficiency did not suppress IL-10 expression in *H. pylori* infection (Gringhuis et al. 2007, 2009). When human DCs (derived from monocytes) were infected with *H. pylori,* it produced IL-10 in a DC-SIGN, TLR2 and TLR4 dependent manner, and interestingly all these receptors were reported to bind bacterial LPS (Chang et al. 2012). It was observed that, p38 signaling mediated NF-κB activation led to histone modification for upregulated expression of IL-10 in *H. pylori* infection (Chang et al. 2012). However, it was observed that IL-10 and CD40 expression reduced in *H. pylori* infected DCs obtained from gastric cancer patients, while infected T-cells exhibited decreased IL-17 expression (Chang et al. 2012). Recently, it was found that *H. pylori* can induce Th2 response through DC-SIGN mediated activation of an atypical NF-κB pathway. DC-SIGN interaction with fucose-containing moieties from *H. pylori* and *Schistosoma mansoni* induced the phosphorylation of LSP1 (lymphocyte specific protein 1) for IKKε activation to effect nuclear translocation of Bcl3 (B-cell lymphoma encoded protein 3) and association with p50 NF-κB subunit. This pathway resulted in the downregulation of pro-inflammatory cytokine production and upregulation of anti-inflammatory cytokine IL-10 and chemokines for specific recruitment of Th2 cells (Gringhuis et al. 2014).

Anti-inflammatory responses through CLR during *H. pylori* infection are not limited to DC-SIGN signaling, as MINCLE (macrophage inducible C-type lectin) is reported to induce IL-10 production in infected human macrophages; it was found that LPS can function as the ligand for this induction (Devi et al. 2015). The induced expression of small RNA miR-223–3p as well as secreted IL-10 in infected human immune cells were found to significantly downregulate the expression of inflammasome forming NLRP3, and this can interfere with the production of pro-inflammatory active IL-1β (Pachathundikandi and Backert 2018). The IL-25 (IL-17E) signaling induced Th2 response significantly reduced the inflammation in mice during infection with *H. pylori,* while IL-23 signaling increased this inflammation (Horvath Jr et al. 2012, 2013). Moreover, it was reported that *H. pylori* could induce HO-1 (heme oxygensase 1) expression through phospho-CagA dependent p38 signaling and nulcear factor E2-related factor 2 (NRF-2) activation. The HO-1 expression was found to be upregulated in tissue from infected patients as well as from infected mice (Gobert et al. 2014). Furthermore, deficiency of *hmox1* exacerbated the inflammation through increased M1 macrophage, Th1 and Th17 responses, which resulted in reduced *H. pylori* colonization (Gobert et al. 2014). Similarly, deficiency in *trpm*2 (transient receptor potential cation channel subfamily M member 2), a calcium channel protein, increased gastric inflammation and reduced the bacterial colonization in infected mice (Beceiro et al. 2017). TRPM2 deficient macrophages had altered calcium levels and produced more inflammatory mediators like IL-12, IL-6, IL-1β and TNFα. The oxidative stress was also increased, which attributed to the high expression of NADPH oxidase and iNOS due to enhanced MAPK signaling in TRPM2 deficiency. In addition, TRMP2 deficiency polarized the macrophages to a more pro-inflammatory M1 phenotype (Beceiro et al. 2017). Transgenic mice lacking Nogin, the BMP (bone morphogenetic protein) signaling inhibitor, showed exacerbated expression of pro-inflammatory cytokines during *H. pylori* infection, which suggests for an anti-inflammatory role of BMP in this infection (Takabayashi et al. 2014). *H. pylori* arginase (encoded by *roc*F) was found to suppress the expression of NF-κB family transcription factors and their corresponding cytokine genes. Infection with Δ*rocF* mutants induced more IL-8 in epithelial cells than wild type bacteria, which confirms a role for immunosuppression during infection (Kim et al. 2012). The gastric tissue TGFβ and its receptor levels were both decreased in patients infected with *H. pylori,* which is known to have anti-inflammatory activities through suppression of T-cells, B-cells, macrophages and NK cells (Shih et al. 2005; Jo et al. 2010). Furthermore, the model proposed by Li and co-workers (2015) incorporated decreased TGFβ and receptors in acute *H. pylori* infection, while chronic infection increased their signaling activity. In conclusion, the above described pro- and anti-inflammatory signaling mechanisms and their timely or untimely induction would by and large determine the outcome of *H. pylori* infection.

11 Inflammasome Activation Through TLR2 and NOD2 Signal Transduction

The formation of the inflammasome is an important aspect of the innate immune response, constituting multiprotein scaffolds for the recruitment of zymogen pro-Caspase 1 to become activated through proximity induced auto-proteolysis and subsequent production of active IL-1β and IL-18. Proteins belonging to the family of Nod-like receptors (NLRs) are the major contributors in this process. In particular, NLR pyrin domain containing 1 (NLRP1), NLRP3 and NLR card domain containing 4 (NLRC4) are the three major and best studied inflammasome forming proteins. The inflammasome can also form by activity of non-NLR factors; AIM2 (absent in melanoma 2) is the major non-NLR inflammasome forming factor for active IL-1β and IL-18 production. Other NLR proteins have also been implicated for the formation of the inflammasome, however, the mechanistic details of these are yet to be discovered (Backert 2016).

As could be expected, *H. pylori* infection activates the formation of the inflammasome both in mouse and human cells (summarized in Fig. 4). The production of active IL-1β and IL-18 have multiple roles in the fine tuning of the host's immune responses against the bacteria. Apart from its known role as a pyrogen, IL-1β can recruit and activate different leukocytes and result in production of multiple mediators of inflammation. IL-18 is involved in the modulation of T-cell responses, which depends on the presence of other mediators of inflammation. For instance, IL-18 can turn the fate of T-helper cells into either Th1 or Th2 phenotype and counter the pro-inflammatory response induced by IL-1β (Dinarello 2009). The activation of an inflammasome requires the optimal expression of crucial components such as NLRs, pro-Caspase 1, IL-1β and IL-18. Generally, TLR signaling activation of NF-κB induces the expression of these proteins. In case of *H. pylori* infection, TLR2 signaling was reported to be the rate-limiting step in the activation of NLRP3 inflammasome (Koch et al. 2015). A study using a transposon mutant library of *H. pylori* revealed that urease deficient bacteria were no longer able of such activation in DCs (Koch et al. 2015). Infection of DCs with specific Δ*ure*A and Δ*ure*B mutants pinpointed the role of UreB in this activation (Koch et al. 2015). However, pro-IL-1β expression was induced through *H. pylori* LPS mediated activation of TLR4, as particular LPS mutants failed to do so. This shows the need for concerted action of TLRs in the activation of the inflammasome in *H. pylori* infection. The *tlr2* and *nlrp3* deficient mice showed an increase in IFNγ producing $CD4^+$ cells and reduced colonization of *H. pylori* (Koch et al. 2015). Mice deficient in either *casp1, il18,* or *il18r* genes also revealed lower *H. pylori* colonization rates. This shows that TLR2 mediated NLRP3 inflammasome activation in mice benefits *H. pylori* colonization and persistence. Wild-type mice infected with Δ*ure*B mutants had more IFNγ producing $CD4^+$ cells compared to animals infected with wild type bacteria; likewise, *tlr2* or *nlrp3* deficient mice produced lower levels of these cells when colonized with either type of bacteria (Koch et al. 2015). Moreover, it was reported that animals with allergic asthma had reduced pathologies when neonatal mice were infected with wild-type *H. pylori,* but absence of this effect in case of the Δ*ureB* mutant suggests that UreB-induced immunity can protect the host from allergic asthma (Koch et al. 2015). This observed protection was abrogated by treatment of blocking antibodies raised against IL-18.

Moreover, adoptive transfer of CD25$^+$ Treg cells from mice infected with wild type *H. pylori* alleviated the asthma pathologies (Koch et al. 2015).

Neither VacA nor CagA were essential for inflammasome activated IL-1β secretion, but a bacterial mutant lacking T4SS pilus protein CagL resulted in decreased activation, which hints to a role of intact T4SS in this process (Kim et al. 2013). In this study, expression of *il1β* and *nlrp3* was found to depend on TLR2 and NOD2. DCs from either *tlr2* or *nod2* deficient mice significantly reduced the activation of the inflammasome, while in double deficient mice this effect was augmented (Kim et al. 2013). DCs deficient of *casp1* expressed similar amounts of pro-IL-1β as wild type mice upon infection with the bacteria, but were no longer able to secrete active IL-1β (Kim et al. 2013). These murine infection experiments showed that *H. pylori* colonization was increased in *il1, il1r,* and *casp1* deficient mice compared to wild type mice, but *nlrp3* deficient mice resulted in normal colonization, which contrasts with the above-mentioned study (Kim et al. 2013). In conclusion, TLR2, along with NOD2, is involved in the activation of the NLRP3 inflammasome during *H. pylori* infection and this process requires bacterial factors such as UreB, LPS and CagL.

12 Resolution of Inflammation by *H. pylori*

Inflammation typically undergoes three phases: acute inflammation, onset of resolution, and finally resolution to regain homeostasis. The acute inflammatory stage is marked by infiltration of neutrophils and monocytes at the inflamed site, producing classically activated M1 macrophages and resulting in the production of more pro-inflammatory mediators. This results in long-term adaptive immunity against the invader and works as a memory system to prevent future attack. At the onset of resolution, the second stage is characterized by reduced secretion of pro-inflammatory cytokines and chemokines. Neutrophils start to produce microparticles and the acute-phase production of lipid pro-inflammatory mediators (e.g. prostaglandins) switches to production of pro-resolution lipid mediators such as lipoxins, resolvins, maresins and protectins. In addition, more anti-inflammatory cytokines such as IL-10 and TGFβ are produced during this stage and thereby induces more M2 macrophages, these cells are necessary for resolution to gradually restore homeostasis with assistance of other factors. (Ortega-Gómez et al. 2013; Sugimoto et al. 2016).

H. pylori infection is a chronic condition that in humans typically starts during childhood and lasts for a life time if not eradicated by antibiotic therapy. However, the majority of infected individuals do not develop associated complications like peptic ulcer, gastric cancer or MALT lymphoma, despite the fact that gastritis and local inflammation of the gastric mucosa probably appears in most infected individuals (Wroblewski et al. 2010; Bauer and Meyer 2011). The colonization of gastric mucosa by *H. pylori* and its interaction with the epithelium produces a strong chemokine response and attracts large amounts of neutrophils and other immune cells to the site (Dunn et al. 1997; White et al. 2015; Gobert and Wilson 2016). This inflammation continues unless the bacteria are eradicated by therapy, which suggests that resolution processes are limited in this infection. It is known that long-term infection with *H. pylori* produces a robust Th1 immune response and that can control the infection to a certain degree (White et al. 2015). The presence of the pro-resolution factors, such as Treg cells and IL-10, was demonstrated, but this was not sufficient for total clearance and resolution, although these factors are able to reduce the immunopathologies associated with *H. pylori* (Kao et al. 2010; Cook et al. 2014; Hussain et al. 2016). The reduced immune-pathology mediated by IL-10 could be attributed to its pro-resolution effort, but lack of complete *H. pylori* clearance prevents recovery to homeostasis. In addition, human *H. pylori* infection results in a mixed M1/M2 macrophage response, whereas in the mouse infection model a clear M1 polarization is observed (Quiding-Järbrink et al. 2010; Gobert et al. 2014; Beceiro et al. 2017). This partial pro-resolution in humans may be due to mixed immune effector cell populations, which hinder bacterial clearance and resolution of the infection.

There are a limited number of studies available that describe resolution of *H. pylori* infection, which are either based on observations with vaccinated mice or describe eradication with antibiotic therapy. One early murine study showed that prophylactic immunization reduced colonization levels with several logs 2 weeks after dosage, and after 52 weeks their gastric tissue resembled that of uninfected mice (Garhart et al. 2002). Unimmunized infected mice also reduced colonization levels eventually, but these animals developed gastritis later on. This shows that prophylactic immunization against *H. pylori* doesn't prevent infection but helps to reduce colonization levels and at longer time, either clear the infection or at least resolve the inflammation; however, this effect cannot be excluded to mice specific (Garhart et al. 2002).

Involvement of IL-10 has been studied by infecting *il10*$^{-/-}$ deficient mice. The immune-regulatory role of IL-10 was found to be important in the control of immune responses against *H. pylori*. The IL-10 deficiency enhanced the immune responses and inflammation with a highly significant reduction in bacterial colonization during the early phase, while it reduced gastritis and colonization levels at prolonged infection (Matsumoto et al. 2005). These observations suggest that a reduction in IL-10 producing Treg cells might increase the chance of bacterial eradication and assist resolution in *H. pylori* infection (Matsumoto et al. 2005). It was also shown that infecting *irf1* deficient mice did not result in gastritis or atrophy, despite colonization with very high numbers of bacteria. These mice also failed to produce Th1 and Th2 responses, which correlates with the reduced immunopathology (Sommer et al. 2001). The above data indicate multiple levels of involvement in *H. pylori* induced gastric inflammation and associated pathologies. If we were able to modulate both colonization and inflammation by targeting crucial checkpoints of *H. pylori* infection, it might be possible to clear the infection without the need of antimicrobial drugs.

13 Conclusions

H. pylori currently colonizes about 50% of the world population, although the proportion of infected individuals is decreasing over time. The bacteria are responsible for a significant global health burden, including peptic ulceration and gastric malignancies. More than 30 years of research on the bacteria-host cell interactions taking place during infection have provided amazing insights into the biology of *H. pylori*, with considerable progress being made in the past few years. In this chapter, we have reviewed the interactions of an array of bacterial factors with a wide selection of host signaling modules. Upon the initial intimate contact between bacteria and gastric epithelial cells, specific bacterial factors interfere with selected host receptors and other factors to manipulate the downstream cell signaling. We have highlighted our current understanding how bacterial factors such as VacA, GGT, UreB, NapA, Hsp60, LPS, peptidoglycan and ADP-heptose can hijack host cell signaling modules and downstream signal transduction pathways. It can be presumed that there exists very complex crosstalk between bacterial ligands and their corresponding host cell receptors to influence cellular responses, which ensures chronic colonization and gastric disease development. It remains a highly interesting challenge to further unravel the important actions of *H. pylori* virulence factors, such as translocation of ADP-heptose. In addition, single-nucleotide polymorphisms (SNPs) and other genetic host differences have been reported in multiple cell receptors and immune-regulatory factors including in IL-1β, IL-1 receptor, TLRs, interleukins or TNFα, which also control the clinical outcome of infections by *H. pylori* (see also Chapter "The role of host genetic polymorphisms in *Helicobacter pylori* mediated disease outcome" of this book). In addition to their ongoing uncovering, it can be expected that in the near future additional genetic polymorphisms in

H. pylori populations may also be discovered, now that novel high-throughput sequencing methods have become widely available. This will definitely promote the ongoing assessment and treatment schemes for *H. pylori* infection. Therefore, it appears that *H. pylori* interactions with the host immune system will continue to be an attractive and gratifying subject for future researchers.

Acknowledgments This work was supported by grant A04 in CRC-1181 of the German Science Foundation (DFG) to S.B.

Conflict of Interest The authors declare no conflict of interest.

References

Allen LA (2007) Phagocytosis and persistence of *Helicobacter pylori*. Cell Microbiol 9(4):817–828. https://doi.org/10.1111/j.1462-5822.2007.00906.x

Allison CC, Kufer TA, Kremmer E, Kaparakis M, Ferrero RL (2009) *Helicobacter pylori* induces MAPK phosphorylation and AP-1 activation via a NOD1-dependent mechanism. J Immunol 183 (12):8099–8109. https://doi.org/10.4049/jimmunol.0900664

Allison CC, Ferrand J, McLeod L, Hassan M, Kaparakis-Liaskos M, Grubman A, Bhathal PS, Dev A, Sievert W, Jenkins BJ, Ferrero RL (2013) Nucleotide oligomerization domain 1 enhances IFN-γ signaling in gastric epithelial cells during *Helicobacter pylori* infection and exacerbates disease severity. J Immunol 190(7):3706–3715. https://doi.org/10.4049/jimmunol.1200591

Amedei A, Cappon A, Codolo G, Cabrelle A, Polenghi A, Benagiano M, Tasca E, Azzurri A, D'Elios MM, Del Prete G, de Bernard M (2006) The neutrophil-activating protein of *Helicobacter pylori* promotes Th1 immune responses. J Clin Invest 116 (4):1092–1101. https://doi.org/10.1172/JCI27177

Asano N, Imatani A, Watanabe T, Fushiya J, Kondo Y, Jin X, Ara N, Uno K, Iijima K, Koike T, Strober W, Shimosegawa T (2016) Cdx2 expression and intestinal metaplasia induced by *Helicobacter pylori* infection of gastric cells is regulated by NOD1-mediated innate immune responses. Cancer Res 76(5):1135–1145. https://doi.org/10.1158/0008-5472.CAN-15-2272

Backert S (2016) Inflammasome signaling and bacterial infections, 1st edn. Springer, Heidelberg, pp 1–22. https://doi.org/10.1007/978-3-319-41171-2

Backert S, Naumann M (2010) What a disorder: proinflammatory signaling pathways induced by helicobacter pylori. Trends Microbiol 18 (11):479–486. https://doi.org/10.1016/j.tim.2010.08.003

Backert S, Ziska E, Brinkmann V, Zimny-Arndt U, Fauconnier A, Jungblut PR, Naumann M, Meyer TF (2000) Translocation of the *Helicobacter pylori* CagA protein in gastric epithelial cells by a type IV secretion apparatus. Cell Microbiol 2(2):155–164. https://doi.org/10.1046/j.1462-5822.2000.00043.x

Backert S, Clyne M, Tegtmeyer N (2011) Molecular mechanisms of gastric epithelial cell adhesion and injection of CagA by *Helicobacter pylori*. Cell Commun Signal 9:28. https://doi.org/10.1186/1478-811X-9-28

Backert S, Schmidt TP, Harrer A, Wessler S (2017) Exploiting the gastric epithelial barrier: *Helicobacter pylori*'s attack on tight and adherens junctions. Curr Top Microbiol Immunol 400:195–226. https://doi.org/10.1007/978-3-319-50520-6_9

Backert S, Bernegger S, Skórko-Glonek J, Wessler S (2018) Extracellular HtrA serine proteases: an emerging new strategy in bacterial pathogenesis. Cell Microbiol 20(6):e12845. https://doi.org/10.1111/cmi.12845

Barden S, Lange S, Tegtmeyer N, Conradi J, Sewald N, Backert S, Niemann HH (2013) A helical RGD motif promoting cell adhesion: crystal structures of the *Helicobacter pylori* type IV secretion system pilus protein CagL. Structure 21(11):1931–1941. https://doi.org/10.1016/j.str.2013.08.018

Bauer B, Meyer TF (2011) The human gastric pathogen *Helicobacter pylori* and its association with gastric cancer and ulcer disease. Ulcers 2011:1–23. https://doi.org/10.1155/2011/340157

Beceiro S, Radin JN, Chatuvedi R, Piazuelo MB, Horvarth DJ, Cortado H, Gu Y, Dixon B, Gu C, Lange I, Koomoa DL, Wilson KT, Algood HM, Partida-Sánchez S (2017) TRPM2 ion channels regulate macrophage polarization and gastric inflammation during *Helicobacter pylori* infection. Mucosal Immunol 10 (2):493–507. https://doi.org/10.1038/mi.2016.60

Bergman MP, Engering A, Smits HH, van Vliet SJ, van Bodegraven AA, Wirth HP, Kapsenberg ML, Vandenbroucke-Grauls CM, van Kooyk Y, Appelmelk BJ (2004) *Helicobacter pylori* modulates the T helper cell 1/T helper cell 2 balance through phase-variable interaction between lipopolysaccharide and DC-SIGN. J Exp Med 200(8):979–990. https://doi.org/10.1084/jem.20041061

Beswick EJ, Reyes VE (2009) CD74 in antigen presentation, inflammation, and cancers of the gastrointestinal tract. World J Gastroenterol 15(23):2855–2861. https://doi.org/10.3748/wjg.15.2855

Beswick EJ, Das S, Pinchuk IV, Adegboyega P, Suarez G, Yamaoka Y, Reyes VE (2005) *Helicobacter pylori*-induced IL-8 production by gastric epithelial cells up-regulates CD74 expression. J Immunol 175 (1):171–176. https://doi.org/10.4049/jimmunol.175.1.171

Beswick EJ, Pinchuk IV, Das S, Powell DW, Reyes VE (2007) Expression of the programmed death ligand 1, B7-H1, on gastric epithelial cells after *Helicobacter pylori* exposure promotes development of CD4+ CD25 + FoxP3+ regulatory T cells. Infect Immun 75 (9):4334–4341. https://doi.org/10.1128/IAI.00553-07

Beutler BA (2009) TLRs and innate immunity. Blood 113 (7):1399–1407. https://doi.org/10.1182/blood-2008-07-019307

Boncristiano M, Paccani SR, Barone S, Ulivieri C, Patrussi L, Ilver D, Amedei A, D'Elios MM, Telford JL, Baldari CT (2003) The *Helicobacter pylori* vacuolating toxin inhibits T cell activation by two independent mechanisms. J Exp Med 198 (12):1887–1897. https://doi.org/10.1084/jem.20030621

Boonyanugomol W, Chomvarin C, Hahnvajanawong C, Sripa B, Kaparakis-Liaskos M, Ferrero RL (2013) *Helicobacter pylori cag* pathogenicity island (*cag*PAI) involved in bacterial internalization and IL-8 induced responses via NOD1- and MyD88-dependent mechanisms in human biliary epithelial cells. PLoS One 8:e77358. https://doi.org/10.1371/journal.pone.0077358

Borlace GN, Butler RN, Brooks DA (2008) Monocyte and macrophage killing of *Helicobacter pylori*: relationship to bacterial virulence factors. Helicobacter 13:380–387. https://doi.org/10.1111/j.1523-5378.2008.00625.x

Borlace GN, Jones HF, Keep SJ, Butler RN, Brooks DA (2011) *Helicobacter pylori* phagosome maturation in primary human macrophages. Gut Pathog 3(1):3. https://doi.org/10.1186/1757-4749-3-3

Brandt S, Kwok T, Hartig R, König W, Backert S (2005) NF-kappaB activation and potentiation of proinflammatory responses by the *Helicobacter pylori* CagA protein. Proc Natl Acad Sci U S A 102 (26):9300–9305. https://doi.org/10.1073/pnas.0409873102

Braun DA, Fribourg M, Sealfon SC (2013) Cytokine response is determined by duration of receptor and signal transducers and activators of transcription 3 (STAT3) activation. J Biol Chem 288:2986–2993. https://doi.org/10.1074/jbc.M112.386573

Bridge DR, Merrell DS (2013) Polymorphism in the *Helicobacter pylori* CagA and VacA toxins and disease. Gut Microbes 4:101–117. https://doi.org/10.4161/gmic.23797

Brisslert M, Enarsson K, Lundin S, Karlsson A, Kusters JG, Svennerholm AM, Backert S, Quiding-Järbrink M (2005) *Helicobacter pylori* induce neutrophil transendothelial migration: role of the bacterial HP-NAP. FEMS Microbiol Lett 249(1):95–103. https://doi.org/10.1016/j.femsle.2005.06.008

Brownlie RJ, Zamoyska R (2013) T cell receptor signaling networks: branched, diversified and bounded. Nat Rev Immunol 13:257–269. https://doi.org/10.1038/nri3403

Calore F, Genisset C, Casellato A, Rossato M, Codolo G, Esposti MD, Scorrano L, de Bernard M (2010) Endosome-mitochondria juxtaposition during apoptosis induced by *H pylori* VacA. Cell Death Differ 17:1707–1716. https://doi.org/10.1038/cdd.2010.42

Carr EL, Kelman A, Wu GS, Gopaul R, Senkevitch E, Aghvanyan A, Turay AM, Frauwirth KA (2010) Glutamine uptake and metabolism are coordinately regulated by ERK/MAPK during T lymphocyte activation. J Immunol 185:1037–1044. https://doi.org/10.4049/jimmunol.0903586

Chang LL, Wang SW, Wu IC, Yu FJ, Su YC, Chen YP, Wu DC, Kuo CH, Hung CH (2012) Impaired dendritic cell maturation and IL-10 production following *H. pylori* stimulation in gastric cancer patients. Appl Microbiol Biotechnol 96(1):211–220. https://doi.org/10.1007/s00253-012-4034-z

Chevalier C, Thiberge JM, Ferrero RL, Labigne A (1999) Essential role of *Helicobacter pylori* gamma-glutamyltranspeptidase for the colonization of the gastric mucosa of mice. Mol Microbiol 31:1359–1372. https://doi.org/10.1046/j.1365-2958.1999.01271.x

Chiou CC, Chan CC, Kuo YP, Chan EC (2003) *Helicobacter pylori* inhibits activity of *cdc*2 kinase and delays G2/M to G1 progression in gastric adenocarcinoma cell line. Scand J Gastroenterol 38 (2):147–152. https://doi.org/10.1080/00365520310000627

Collaborative Computational Project, Number 4 (1994) The CCP4 suite: programs for protein crystallography. Acta Crystallogr Sect D Biol Crystallogr 50:760–763. https://doi.org/10.1107/S0907444994003112

Conradi J, Tegtmeyer N, Woźna M, Wissbrock M, Michalek C, Gagell C, Cover TL, Frank R, Sewald N, Backert S (2012) An RGD helper sequence in CagL of *Helicobacter pylori* assists in interactions with integrins and injection of CagA. Front Cell Infect Microbiol 2:70. https://doi.org/10.3389/fcimb.2012.00070

Cook KW, Letley DP, Ingram RJ, Staples E, Skjoldmose H, Atherton JC, Robinson K (2014) CCL20/CCR6-mediated migration of regulatory T cells to the *Helicobacter pylori*-infected human gastric mucosa. Gut 63:1550–1559. https://doi.org/10.1136/gutjnl-2013-306253

Cover TL, Krishna US, Israel DA, Peek RM (2003) Induction of gastric epithelial cell apoptosis by *Helicobacter pylori* vacuolating cytotoxin. Cancer Res 63 (5):951–957

D'Elios MM, Manghetti M, De Carli M, Costa F, Baldari CT, Burroni D, Telford JL, Romagnani S, Del Prete G (1997) T helper 1 effector cells specific for *Helicobacter pylori* in the gastric antrum of patients with peptic ulcer disease. J Immunol 158:962–967

Das S, Suarez G, Beswick EJ, Sierra JC, Graham DY, Reyes VE (2006) Expression of B7-H1 on gastric epithelial cells: its potential role in regulating T cells during *Helicobacter pylori* infection. J Immunol 176:3000–3009. https://doi.org/10.4049/jimmunol.176.5.3000

Das L, Kokate SB, Rath S, Rout N, Singh SP, Crowe SE, Mukhopadhyay AK, Bhattacharyya A (2016) ETS2 and Twist1 promote invasiveness of *Helicobacter pylori*-infected gastric cancer cells by inducing Siah2. Biochem J 473(11):1629–1640. https://doi.org/10.1042/BCJ20160187

De Bruyne E, Ducatelle R, Foss D, Sanchez M, Joosten M, Zhang G, Smet A, Pasmans F, Haesebrouck F, Flahou B (2016) Oral glutathione supplementation drastically reduces *Helicobacter*-induced gastric pathologies. Sci Rep 6:20169. https://doi.org/10.1038/srep20169

Del Giudice G, Covacci A, Telford JL, Montecucco C, Rappuoli R (2001) The design of vaccines against *Helicobacter pylori* and their development. Annu Rev Immunol 19:523–563. https://doi.org/10.1146/annurev.immunol.19.1.523

Devi S, Rajakumara E, Ahmed N (2015) Induction of Mincle by *Helicobacter pylori* and consequent anti-inflammatory signaling denote a bacterial survival strategy. Sci Rep 5:15049. https://doi.org/10.1038/srep15049

Dinarello CA (2009) Immunological and inflammatory functions of the interleukin-1 family. Annu Rev Immunol 27:519–550. https://doi.org/10.1146/annurev.immunol.021908.132612

Du SY, Wang HJ, Cheng HH, Chen SD, Wang LH, Wang WC (2016) Cholesterol glucosylation by *Helicobacter pylori* delays internalization and arrests phagosome maturation in macrophages. J Microbiol Immunol Infect 49:636–645. https://doi.org/10.1016/j.jmii.2014.05.011

Dundon WG, de Bernard M, Montecucco C (2001) Virulence factors of *Helicobacter pylori*. Int J Med Microbiol 290(8):647–658. https://doi.org/10.1016/S1438-4221(01)80002-3

Dunn BE, Cohen H, Blaser MJ (1997) Helicobacter pylori. Clin Microbiol Rev 10(4):720–741

Galgani M, Busiello I, Censini S, Zappacosta S, Racioppi L, Zarrilli R (2004) *Helicobacter pylori* induces apoptosis of human monocytes but not monocyte-derived dendritic cells: role of the cag pathogenicity island. Infect Immun 72:4480–4485. https://doi.org/10.1128/IAI.72.8.4480-4485.2004

Gall A, Gaudet RG, Gray-Owen SD, Salama NR (2017) TIFA signaling in gastric epithelial cells initiates the *cag* type 4 secretion system-dependent innate immune response to *Helicobacter pylori* infection. MBio 8(4). pii: e01168–17. https://doi.org/10.1128/mBio.01168-17

Garhart CA, Redline RW, Nedrud JG, Czinn SJ (2002) Clearance of *Helicobacter pylori* infection and resolution of postimmunization gastritis in a kinetic study of prophylactically immunized mice. Infect Immun 70:3529–3538. https://doi.org/10.1128/IAI.70.7.3529-3538.2002

Gauthier NC, Monzo P, Kaddai V, Doye A, Ricci V, Boquet P (2005) *Helicobacter pylori* VacA cytotoxin: a probe for a clathrin-independent and Cdc42-dependent pinocytic pathway routed to late endosomes. Mol Biol Cell 16:4852–4866. https://doi.org/10.1091/mbc.E05-05-0398

Gebert B, Fischer W, Weiss E, Hoffmann R, Haas R (2003) *Helicobacter pylori* vacuolating cytotoxin inhibits T lymphocyte activation. Science 301:1099–1102. https://doi.org/10.1126/science.1086871

Gerhard M, Schmees C, Voland P, Endres N, Sander M, Reindl W, Rad R, Oelsner M, Decker T, Mempel M, Hengst L, Prinz C (2005) A secreted low-molecular-weight protein from *Helicobacter pylori* induces cell-cycle arrest of T cells. Gastroenterology 128(5):1327–1339. https://doi.org/10.1053/J.GASTRO.2005.03.018

Gieseler S, König B, König W, Backert S (2005) Strain-specific expression profiles of virulence genes in *Helicobacter pylori* during infection of gastric epithelial cells and granulocytes. Microbes Infect 7(3):437–447. https://doi.org/10.1016/j.micinf.2004.11.018

Go MF (1997) What are the host factors that place an individual at risk for *Helicobacter pylori*-associated disease. Gastroenterology 113:S15–S20. https://doi.org/10.1016/S0016-5085(97)80005-4

Gobert AP, Wilson KT (2016) The immune battle against *Helicobacter pylori* infection: NO offense. Trends Microbiol 24(5):366–376. https://doi.org/10.1016/j.tim.2016.02.005

Gobert AP, Verriere T, Asim M, Barry DP, Piazuelo MB, de Sablet T, Delgado AG, Bravo LE, Correa P, Peek RM Jr, Chaturvedi R, Wilson KT (2014) Heme oxygenase-1 dysregulates macrophage polarization and the immune response to *Helicobacter pylori*. J Immunol 193(6):3013–3022. https://doi.org/10.4049/jimmunol.1401075

González-Rivera C, Algood HMS, Radin JN, McClain MS, Cover TL (2012) The intermediate region of *Helicobacter pylori* VacA is a determinant of toxin potency in a Jurkat T cell assay. Infect Immun 80:2578–2588. https://doi.org/10.1128/IAI.00052-12

Gorrell RJ, Guan J, Xin Y, Tafreshi MA, Hutton ML, McGuckin MA, Ferrero RL, Kwok T (2013) A novel NOD1- and CagA-independent pathway of interleukin-8 induction mediated by the *Helicobacter pylori* type IV secretion system. Cell Microbiol 15(4):554–570. https://doi.org/10.1111/cmi.12055

Gringhuis SI, den Dunnen J, Litjens M, van Het Hof B, van Kooyk Y, Geijtenbeek TB (2007) C-type lectin DC-SIGN modulates toll-like receptor signaling via Raf-1 kinase-dependent acetylation of transcription factor NF-kappaB. Immunity 26(5):605–616. https://doi.org/10.1016/j.immuni.2007.03.012

Gringhuis SI, den Dunnen J, Litjens M, van der Vlist M, Geijtenbeek TB (2009) Carbohydrate-specific signaling through the DC-SIGN signalosome tailors immunity to *Mycobacterium tuberculosis*, HIV-1 and *Helicobacter pylori*. Nat Immunol 10(10):1081–1088. https://doi.org/10.1038/ni.1778

Gringhuis SI, Kaptein TM, Wevers BA, Mesman AW, Geijtenbeek TB (2014) Fucose-specific DC-SIGN signalling directs T helper cell type-2 responses via IKKε- and CYLD-dependent Bcl3 activation. Nat Commun 5:3898. https://doi.org/10.1038/ncomms4898

Hirata Y, Maeda S, Mitsuno Y, Tateishi K, Yanai A, Akanuma M, Yoshida H, Kawabe T, Shiratori Y, Omata M (2002) *Helicobacter pylori* CagA protein activates serum response element-driven transcription independently of tyrosine phosphorylation. Gastroenterology 123(6):1962–1971. https://doi.org/10.1053/gast.2002.37044

Horvath DJ Jr, Washington MK, Cope VA, Algood HM (2012) IL-23 contributes to control of chronic *Helicobacter pylori* infection and the development of T helper responses in a mouse model. Front Immunol 3:56. https://doi.org/10.3389/fimmu.2012.00056

Horvath DJ Jr, Radin JN, Cho SH, Washington MK, Algood HM (2013) The interleukin-17 receptor B subunit is essential for the Th2 response to *Helicobacter pylori*, but not for control of bacterial burden. PLoS One 8(3):e60363. https://doi.org/10.1371/journal.pone.0060363

Hussain K, Letley DP, Greenaway AB, Kenefeck R, Winter JA, Tomlinson W, Rhead J, Staples E, Kaneko K, Atherton JC, Robinson K (2016) *Helicobacter pylori*-mediated protection from allergy is associated with IL-10-secreting peripheral blood regulatory T cells. Front Immunol 7:71. https://doi.org/10.3389/fimmu.2016.00071

Hutton ML, Kaparakis-Liaskos M, Turner L, Cardona A, Kwok T, Ferrero RL (2010) *Helicobacter pylori* exploits cholesterol-rich microdomains for induction of NF-kappaB-dependent responses and peptidoglycan delivery in epithelial cells. Infect Immun 78 (11):4523–4531. https://doi.org/10.1128/IAI.00439-10

Hutton ML, D'Costa K, Rossiter AE, Wang L, Turner L, Steer DL, Masters SL, Croker BA, Kaparakis-Liaskos-M, Ferrero RL (2017) A *Helicobacter pylori* homolog of eukaryotic flotillin is involved in cholesterol accumulation, epithelial cell responses and host colonization. Front Cell Infect Microbiol 6(7):219. https://doi.org/10.3389/fcimb.2017.00219

Ishihara S, Rumi MA, Kadowaki Y, Ortega-Cava CF, Yuki T, Yoshino N, Miyaoka Y, Kazumori H, Ishimura N, Amano Y, Kinoshita Y (2004) Essential role of MD-2 in TLR4-dependent signaling during *Helicobacter pylori*-associated gastritis. J Immunol 173:1406–1416. https://doi.org/10.4049/jimmunol.173.2.1406

Jo Y, Han SU, Kim YJ, Kim JH, Kim ST, Kim SJ, Hahm KB (2010) Suppressed gastric mucosal TGF-beta1 increases susceptibility to *H. pylori*-induced gastric inflammation and ulceration: a stupid host defense response. Gut Liver 4(1):43–53. https://doi.org/10.5009/gnl.2010.4.1.43

Käbisch R, Mejías-Luque R, Prinz C, Gerhard M (2014) *Helicobacter pylori* cytotoxin-associated gene a impairs human dendritic cell maturation and function through IL-10-mediated activation of STAT3. J Immunol 192:316–323. https://doi.org/10.4049/jimmunol.1302476

Käbisch R, Semper RP, Wüstner S, Gerhard M, Mejías-Luque R (2016) *Helicobacter pylori* γ-glutamyltranspeptidase induces tolerogenic human dendritic cells by activation of glutamate receptors. J Immunol 196:4246–4252. https://doi.org/10.4049/jimmunol.1501062

Kao JY, Zhang M, Miller MJ, Mills JC, Wang B, Liu M, Eaton KA, Zou W, Berndt BE, Cole TS, Takeuchi T, Owyang SY, Luther J (2010) *Helicobacter pylori* immune escape is mediated by dendritic cell–induced Treg skewing and Th17 suppression in mice. Gastroenterology 138:1046–1054. https://doi.org/10.1053/J.GASTRO.2009.11.043

Kaparakis M, Turnbull L, Carneiro L, Firth S, Coleman HA, Parkington HC, Le Bourhis L, Karrar A, Viala J, Mak J, Hutton ML, Davies JK, Crack PJ, Hertzog PJ, Philpott DJ, Girardin SE, Whitchurch CB, Ferrero RL (2010) Bacterial membrane vesicles deliver peptidoglycan to NOD1 in epithelial cells. Cell Microbiol 12:372–385. https://doi.org/10.1111/j.1462-5822.2009.01404.x

Kim JM, Kim JS, Lee JY, Kim YJ, Youn HJ, Kim IY, Chee YJ, Oh YK, Kim N, Jung HC, Song IS (2007a) Vacuolating cytotoxin in *Helicobacter pylori* water-soluble proteins upregulates chemokine expression in human eosinophils via Ca2+ influx, mitochondrial reactive oxygen intermediates, and NF-kappaB activation. Infect Immun 75:3373–3381. https://doi.org/10.1128/IAI.01940-06

Kim KM, Lee SG, Park MG, Song JY, Kang HL, Lee WK, Cho MJ, Rhee KH, Youn HS, Baik SC (2007b) γ-Glutamyltranspeptidase of *Helicobacter pylori* induces mitochondria-mediated apoptosis in AGS cells. Biochem Biophys Res Commun 355:562–567. https://doi.org/10.1016/J.BBRC.2007.02.021

Kim JM, Kim JS, Lee JY, Sim YS, Kim YJ, Oh YK, Yoon HJ, Kang JS, Youn J, Kim N, Jung HC, Kim S (2010) Dual effects of *Helicobacter pylori* vacuolating cytotoxin on human eosinophil apoptosis in early and late periods of stimulation. Eur J Immunol 40:1651–1662. https://doi.org/10.1002/eji.200939882

Kim JM, Kim JS, Yoo DY, Ko SH, Kim N, Kim H, Kim YJ (2011) Stimulation of dendritic cells with *Helicobacter pylori* vacuolating cytotoxin negatively regulates their maturation via the restoration of E2F1. Clin Exp Immunol 166:34–45. https://doi.org/10.1111/j.1365-2249.2011.04447.x

Kim SH, Sierra RA, McGee DJ, Zabaleta J (2012) Transcriptional profiling of gastric epithelial cells infected with wild type or arginase-deficient *Helicobacter pylori*. BMC Microbiol 12:175. https://doi.org/10.1186/1471-2180-12-175

Kim DJ, Park JH, Franchi L, Backert S, Núñez G (2013) The cag pathogenicity island and interaction between TLR2/NOD2 and NLRP3 regulate IL-1β production in *Helicobacter pylori* infected dendritic cells. Eur J Immunol 43(10):2650–2658. https://doi.org/10.1002/eji.201243281

Kim BJ, Kim JY, Hwang ES, Kim JG (2015a) Nucleotide binding oligomerization domain 1 is an essential signal transducer in human epithelial cells infected with *Helicobacter pylori* that induces the transepithelial migration of neutrophils. Gut Liver 9:358–369. https://doi.org/10.5009/gnl13218

Kim JM, Kim JS, Kim N, Ko SH, Jeon JI, Kim YJ (2015b) *Helicobacter pylori* vacuolating cytotoxin induces apoptosis via activation of endoplasmic reticulum stress in dendritic cells. J Gastroenterol Hepatol 30:99–108. https://doi.org/10.1111/jgh.12663

Koch KN, Hartung ML, Urban S, Kyburz A, Bahlmann AS, Lind J, Backert S, Taube C, Müller A (2015) *Helicobacter* urease-induced activation of the TLR2/NLRP3/IL-18 axis protects against asthma. J Clin Invest 125:3297–3302. https://doi.org/10.1172/JCI79337

Kranzer K, Söllner L, Aigner M, Lehn N, Deml L, Rehli M, Schneider-Brachert W (2005) Impact of *Helicobacter pylori* virulence factors and compounds on activation and maturation of human dendritic cells. Infect Immun 73:4180–4189. https://doi.org/10.1128/IAI.73.7.4180-4189.2005

Kwok T, Zabler D, Urman S, Rohde M, Hartig R, Wessler S, Misselwitz R, Berger J, Sewald N, König W, Backert S (2007) *Helicobacter* exploits integrin for type IV secretion and kinase activation. Nature 449:862–866. https://doi.org/10.1038/nature06187

Lai CH, Chang YC, Du SY, Wang HJ, Kuo CH, Fang SH, Fu HW, Lin HH, Chiang AS, Wang WC (2008) Cholesterol depletion reduces *Helicobacter pylori* CagA translocation and CagA-induced responses in AGS cells. Infect Immun 76:3293–3303. https://doi.org/10.1128/IAI.00365-08

Lai CH, Wang HJ, Chang YC, Hsieh WC, Lin HJ, Tang CH, Sheu JJ, Lin CJ, Yang MS, Tseng SF, Wang WC (2011) *Helicobacter pylori* CagA-mediated IL-8 induction in gastric epithelial cells is cholesterol-dependent and requires the C-terminal tyrosine phosphorylation-containing domain. FEMS Microbiol Lett 323(2):155–163. https://doi.org/10.1111/j.1574-6968.2011.02372.x

Lew DJ, Dulić V, Reed SI (1991) Isolation of three novel human cyclins by rescue of G1 cyclin (cln) function in yeast. Cell 66:1197–1206. https://doi.org/10.1016/0092-8674(91)90042-W

Li N, Xie C, Lu NH (2015) Transforming growth factor-β: an important mediator in *Helicobacter pylori*-associated pathogenesis. Front Cell Infect Microbiol 5:77. https://doi.org/10.3389/fcimb.2015.00077

Liu X, Berry CT, Ruthel G, Madara JJ, MacGillivray K, Gray CM, Madge LA, McCorkell KA, Breiting DP, Hershberg U, May MJ, Freedman BD (2016) T cell receptor-induced nuclear factor κB (NF-κB) signaling and transcriptional activation are regulated by STIM1- and Orai1-mediated calcium entry. J Biol Chem 291 (16):8440–8452. https://doi.org/10.1074/jbc.M115.713008

Maeda S, Akanuma M, Mitsuno Y, Hirata Y, Ogura K, Yoshida H, Shiratori Y, Omata M (2001) Distinct mechanism of *Helicobacter pylori*-mediated NF-kappa B activation between gastric cancer cells and monocytic cells. J Biol Chem 276:44856–44864. https://doi.org/10.1074/jbc.M105381200

Mai UE, Perez-Perez GI, Wahl LM, Wahl SM, Blaser MJ, Smith PD (1991) Soluble surface proteins from *Helicobacter pylori* activate monocytes/macrophages by lipopolysaccharide-independent mechanism. J Clin Invest 87:894–900. https://doi.org/10.1172/JCI115095

Mai UE, Perez-Perez GI, Allen JB, Wahl SM, Blaser MJ, Smith PD (1992) Surface proteins from *Helicobacter pylori* exhibit chemotactic activity for human leukocytes and are present in gastric mucosa. J Exp Med 175:517–525. https://doi.org/10.1084/jem.175.2.517

Man K, Miasari M, Shi W, Xin A, Henstridge DC, Preston S, Pellegrini M, Belz GT, Smyth GK, Febbraio MA, Nutt SL, Kallies A (2013) The transcription factor IRF4 is essential for TCR affinity–mediated metabolic programming and clonal expansion of T cells. Nat Immunol 14:1155–1165. https://doi.org/10.1038/ni.2710

Mandell L, Moran AP, Cocchiarella A, Houghton J, Taylor N, Fox JG, Wang TC, Kurt-Jones EA (2004) Intact gram-negative *Helicobacter pylori*, *Helicobacter felis*, and *Helicobacter hepaticus* bacteria activate innate immunity via toll-like receptor 2 but not toll-like receptor 4. Infect Immun 72(11):6446–6454. https://doi.org/10.1128/IAI.72.11.6446-6454.2004

Matsumoto Y, Blanchard TG, Drakes ML, Basu M, Redline RW, Levine AD, Czinn SJ (2005) Eradication of *Helicobacter pylori* and resolution of gastritis in the gastric mucosa of IL-10-deficient mice. Helicobacter 10:407–415. https://doi.org/10.1111/j.1523-5378.2005.00349.x

McGee DJ, George AE, Trainor EA, Horton KE, Hildebrandt E, Testerman TL (2011) Cholesterol enhances *Helicobacter pylori* resistance to antibiotics and LL-37. Antimicrob Agents Chemother 55:2897–2904. https://doi.org/10.1128/AAC.00016-11

McGovern KJ, Blanchard TG, Gutierrez JA, Czinn SJ, Krakowka S, Youngman P (2001) Gamma-Glutamyltransferase is a *Helicobacter pylori* virulence factor but is not essential for colonization. Infect Immun 69:4168–4173. https://doi.org/10.1128/IAI.69.6.4168-4173.2001

Melillo JA, Song L, Bhagat G, Blazquez AB, Plumlee CR, Lee C, Berin C, Reizis B, Schindler C (2010) Dendritic cell (DC)-specific targeting reveals Stat3 as a negative regulator of DC function. J Immunol 184:2638–2645. https://doi.org/10.4049/jimmunol.0902960

Molinari M, Salio M, Galli C, Norais N, Rappuoli R, Lanzavecchia A, Montecucco C (1998) Selective inhibition of ii-dependent antigen presentation by *Helicobacter pylori* toxin VacA. J Exp Med 187:135–140. https://doi.org/10.1084/jem.187.1.135

Morey P, Pfannkuch L, Pang E, Boccellato F, Sigal M, Imai-Matsushima A, Dyer V, Koch M, Mollenkopf HJ, Schlaermann P, Meyer TF (2018) *Helicobacter pylori* depletes cholesterol in gastric glands to prevent interferon gamma signaling and escape the inflammatory iesponse. Gastroenterology 154(5):1391–1404.e9. https://doi.org/10.1053/j.gastro.2017.12.008

Nagashima H, Iwatani S, Cruz M, Jiménez Abreu JA, Uchida T, Mahachai V, Vilaichone RK, Graham DY, Yamaoka Y (2015) Toll-like receptor 10 in *Helicobacter pylori* infection. J Infect Dis 212 (10):1666–1676. https://doi.org/10.1093/infdis/jiv270

Nakayama M, Hisatsune J, Yamasaki E, Nishi Y, Wada A, Kurazono H, Sap J, Yahiro K, Moss J, Hirayama T (2006) Clustering of *Helicobacter pylori* VacA in lipid rafts, mediated by its receptor, receptor-like protein tyrosine phosphatase beta, is required for intoxication in AZ-521 cells. Infect Immun 74(12):6571–6580. https://doi.org/10.1128/IAI.00356-06

Naumann M, Sokolova O, Tegtmeyer N, Backert S (2017) *Helicobacter pylori*: a paradigm pathogen for subverting host cell signal transmission. Trends Microbiol 25:316–328. https://doi.org/10.1016/J.TIM.2016.12.004

Necchi V, Manca R, Ricci V, Solcia E (2009) Evidence for transepithelial dendritic cells in human *H. pylori* active gastritis. Helicobacter 14:208–222. https://doi.org/10.1111/j.1523-5378.2009.00679.x

Ng GZ, Menheniott TR, Every AL, Stent A, Judd LM, Chionh YT, Dhar P, Komen JC, Giraud AS, Wang TC, McGuckin MA, Sutton P (2016) *The MUC1 mucin protects against Helicobacter pylori* pathogenesis in mice by regulation of the NLRP3 inflammasome. Gut 65(7):1087–1099. https://doi.org/10.1136/gutjnl-2014-307175

Nika K, Soldani C, Salek M, Paster W, Gray A, Etzensperger R, Fugger L, Polzella P, Cerundolo V, Dushek O, Höfer T, Viola A, Acuto O (2010) Constitutively active Lck kinase in T cells drives antigen receptor signal transduction. Immunity 32 (6):766–777. https://doi.org/10.1016/j.immuni.2010.05.011

Obonyo M, Sabet M, Cole SP, Ebmeyer J, Uematsu S, Akira S, Guiney DG (2007) Deficiencies of myeloid differentiation factor 88, toll-like receptor 2 (TLR2), or TLR4 produce specific defects in macrophage cytokine secretion induced by *Helicobacter pylori*. Infect Immun 75:2408–2414. https://doi.org/10.1128/IAI.01794-06

Odenbreit S, Gebert B, Puls J, Fischer W, Haas R (2001) Interaction of *Helicobacter pylori* with professional phagocytes: role of the cag pathogenicity island and translocation, phosphorylation and processing of CagA. Cell Microbiol 3:21–31. https://doi.org/10.1046/j.1462-5822.2001.00088.x

Oertli M, Sundquist M, Hitzler I, Engler DB, Arnold IC, Reuter S, Maxeiner J, Hansson M, Taube C, Quiding-Järbrink M, Müller A (2012) DC-derived IL-18 drives Treg differentiation, murine *Helicobacter pylori*-specific immune tolerance, and asthma protection. J Clin Invest 122:1082–1096. https://doi.org/10.1172/JCI61029

Oertli M, Noben M, Engler DB, Semper RP, Reuter S, Maxeiner J, Gerhard M, Taube C, Müller A (2013) *Helicobacter pylor*i γ-glutamyl transpeptidase and vacuolating cytotoxin promote gastric persistence and immune tolerance. Proc Natl Acad Sci U S A 110:3047–3052. https://doi.org/10.1073/pnas.1211248110

Oktem-Okullu S, Tiftikci A, Saruc M, Cicek B, Vardareli E, Tozun N, Kocagoz T, Sezerman U, Yavuz AS, Sayi-Yazgan A (2015) Multiplex-PCR-based screening and computational modeling of virulence factors and T-cell mediated immunity in *Helicobacter pylori* infections for accurate clinical diagnosis. PLoS One 10:e0136212. https://doi.org/10.1371/journal.pone.0136212

Ortega-Gómez A, Perretti M, Soehnlein O (2013) Resolution of inflammation: an integrated view. EMBO Mol Med 5(5):661–674. https://doi.org/10.1002/emmm.201202382

Otani K, Tanigawa T, Watanabe T, Nadatani Y, Sogawa M, Yamagami H, Shiba M, Watanabe K, Tominaga K, Fujiwara Y, Arakawa T (2012) Toll-like receptor 9 signaling has anti-inflammatory effects on the early phase of *Helicobacter pylori*-induced gastritis. Biochem Biophys Res Commun 426:342–349. https://doi.org/10.1016/j.bbrc.2012.08.080

Pachathundikandi SK, Backert S (2016) Differential expression of interleukin-1β during *Helicobacter pylori* infection of toll-like receptor 2 (TLR2)- and TLR10-expressing HEK293 cell lines. J Infect Dis 214:166–167. https://doi.org/10.1093/infdis/jiw154

Pachathundikandi SK, Backert S (2018) *Helicobacter pylori* controls NLRP3 expression by regulating hsa-miR-223-3p and IL-10 in cultured and primary human immune cells. Innate Immun 24(1):11–23. https://doi.org/10.1177/1753425917738043

Pachathundikandi SK, Brandt S, Madassery J, Backert S (2011) Induction of TLR-2 and TLR-5 expression by *Helicobacter pylori* switches *cag*PAI-dependent signaling leading to the secretion of IL-8 and TNF-α. PLoS One 6:e19614. https://doi.org/10.1371/journal.pone.0019614. Erratum: PLoS One (2015) 10(12): e0144365

Pachathundikandi SK, Lind J, Tegtmeyer N, El-Omar EM, Backert S (2015) Interplay of the gastric pathogen *Helicobacter pylori* with toll-like receptors. Biomed Res Int 2015:192420. https://doi.org/10.1155/2015/192420

Pagliari M, Munari F, Toffoletto M, Lonardi S, Chemello F, Codolo G, Millino C, Della Bella C, Pacchioni B, Vermi W, Fassan M, de Bernard M, Cagnin S (2017) *Helicobacter pylori* affects the antigen presentation activity of macrophages modulating the expression of the immune receptor CD300E through miR-4270. Front Immunol 8:1288. https://doi.org/10.3389/fimmu.2017.01288

Petersen AM, Sorensen K, Blom J, Krogfelt KA (2001) Reduced intracellular survival of *Helicobacter pylori vacA* mutants in comparison with their wild-types indicates the role of VacA in pathogenesis. FEMS Immunol Med Microbiol 30:103–108. https://doi.org/10.1111/j.1574-695X.2001.tb01556.x

Pfannkuch L, Hurwitz R, Traulsen J, Kosma P, Schmid M, Meyer TF (2018) ADP heptose, a novel pathogen-associated molecular pattern associated with *Helicobacter pylori* type 4 secretion. bioRxiv. https://doi.org/10.1101/405951

Posselt G, Backert S, Wessler S (2013) The functional interplay of *Helicobacter pylori* factors with gastric epithelial cells induces a multi-step process in pathogenesis. Cell Commun Signal 11:77. https://doi.org/10.1186/1478-811X-11-77

Quiding-Järbrink M, Raghavan S, Sundquist M (2010) Enhanced M1 macrophage polarization in human *Helicobacter pylori*-associated atrophic gastritis and in vaccinated mice. PLoS One 5(11):e15018. https://doi.org/10.1371/journal.pone.0015018

Rad R, Brenner L, Bauer S, Schwendy S, Layland L, da Costa CP, Reindl W, Dossumbekova A, Friedrich M, Saur D, Wagner H, Schmid RM, Prinz C (2006) CD25+/Foxp3+ T cells regulate gastric inflammation and *Helicobacter pylori* colonization *in vivo*. Gastroenterology 131:525–537. https://doi.org/10.1053/j.gastro.2006.05.001

Rad R, Brenner L, Krug A, Voland P, Mages J, Lang R, Schwendy S, Reindl W, Dossumbekova A, Ballhorn W, Wagner H, Schmid RM, Bauer S, Prinz C (2007) Toll-like receptor-dependent activation of antigen-presenting cells affects adaptive immunity to *Helicobacter pylori*. Gastroenterology 133:150–163. https://doi.org/10.1053/j.gastro.2007.04.071

Rad R, Ballhorn W, Voland P, Eisenächer K, Mages J, Rad L, Ferstl R, Lang R, Wagner H, Schmid RM, Bauer S, Prinz C, Kirschning CJ, Krug A (2009) Extracellular and intracellular pattern recognition receptors cooperate in the recognition of *Helicobacter pylori*. Gastroenterology 136:2247–2257. https://doi.org/10.1053/j.gastro.2009.02.066

Ramarao N, Gray-Owen SD, Backert S, Meyer TF (2000) *Helicobacter pylori* inhibits phagocytosis by professional phagocytes involving type IV secretion components. Mol Microbiol 37(6):1389–1404. https://doi.org/10.1046/j.1365-2958.2000.02089.x

Rieder G, Tessier AJ, Qiao XT, Madison B, Gumucio DL, Merchant JL (2005) *Helicobacter*-induced intestinal metaplasia in the stomach correlates with Elk-1 and serum response factor induction of villin. J Biol Chem 280(6):4906–4912. https://doi.org/10.1074/jbc.M413399200

Rittig MG, Shaw B, Letley DP, Thomas RJ, Argent RH, Atherton JC (2003) *Helicobacter pylori*-induced homotypic phagosome fusion in human monocytes is independent of the bacterial *vacA* and *cag* status. Cell Microbiol 5:887–899. https://doi.org/10.1046/j.1462-5822.2003.00328.x

Rizzuti D, Ang M, Sokollik C, Wu T, Abdullah M, Greenfield L, Fattouh R, Reardon C, Tang M, Diao J, Schindler C, Cattral M, Jones NL (2015) *Helicobacter pylori* inhibits dendritic cell maturation via interleukin-10-mediated activation of the signal transducer and activator of transcription 3 pathway. J Innate Immun 7:199–211. https://doi.org/10.1159/000368232

Roifman CM, Grunebaum E (2013) 35 – Primary T-cell immunodeficiencies. Clin Immunol:437–453. https://doi.org/10.1016/B978-0-7234-3691-1.00052-0

Satin B, Del Giudice G, Della Bianca V, Dusi S, Laudanna C, Tonello F, Kelleher D, Rappuoli R, Montecucco C, Rossi F (2000) The neutrophil-activating protein (HP-NAP) of *Helicobacter pylori* is a protective antigen and a major virulence factor. J Exp Med 191:1467–1476. https://doi.org/10.1084/jem.191.9.1467

Sayi A, Kohler E, Toller IM, Flavell RA, Müller W, Roers A, Müller A (2011) TLR-2-activated B cells suppress *Helicobacter*-induced preneoplastic gastric immunopathology by inducing T regulatory-1 cells. J Immunol 186:878–890. https://doi.org/10.4049/jimmunol.1002269

Schmees C, Prinz C, Treptau T, Rad R, Hengst L, Voland P, Bauer S, Brenner L, Schmid RM, Gerhard M (2007) Inhibition of T-cell proliferation by *Helicobacter pylori* γ-glutamyl transpeptidase. Gastroenterology 132:1820–1833. https://doi.org/10.1053/J.GASTRO.2007.02.031

Schwartz JT, Allen LA (2006) Role of urease in megasome formation and *Helicobacter pylori* survival in macrophages. J Leukoc Biol 79:1214–1225. https://doi.org/10.1189/jlb.0106030

Sewald X, Gebert-Vogl B, Prassl S, Barwig I, Weiss E, Fabbri M, Osicka R, Schiemann M, Busch DH, Semmrich M, Holzmann B, Sebo P, Haas R (2008) Integrin subunit CD18 is the T-lymphocyte receptor for the *Helicobacter pylori* vacuolating cytotoxin. Cell Host Microbe 3:20–29. https://doi.org/10.1016/J.CHOM.2007.11.003

Sewald X, Jiménez-Soto L, Haas R (2011) PKC-dependent endocytosis of the *Helicobacter pylori* vacuolating cytotoxin in primary T lymphocytes. Cell Microbiol 13:482–496. https://doi.org/10.1111/j.1462-5822.2010.01551.x

Shan Y, Lu X, Han Y, Li X, Wang X, Shao C, Wang L, Liu Z, Tang W, Sun Y, Jia J (2015) *Helicobacter pylori* outer membrane protein 18 (Hp1125) is involved in persistent colonization by evading interferon-γ signaling. Biomed Res Int 2015:571280. https://doi.org/10.1155/2015/571280

Shibayama K, Kamachi K, Nagata N, Yagi T, Nada T, Doi Y, Shibata N, Yokoyama K, Yamane K, Kato H, Iinuma Y, Arakawa Y (2003) A novel apoptosis-inducing protein from *Helicobacter pylori*. Mol Microbiol 47:443–451. https://doi.org/10.1046/j.1365-2958.2003.03305.x

Shibayama K, Wachino J, Arakawa Y, Saidijam M, Rutherford NG, Henderson PJF (2007) Metabolism of glutamine and glutathione via γ-glutamyltranspeptidase and glutamate transport in *Helicobacter pylori*: possible significance in the pathophysiology of the organism. Mol Microbiol 64:396–406. https://doi.org/10.1111/j.1365-2958.2007.05661.x

Shih SC, Tseng KW, Lin SC, Kao CR, Chou SY, Wang HY, Chang WH, Chu CH, Wang TE, Chien CL (2005) Expression patterns of transforming growth factor-beta and its receptors in gastric mucosa of patients with refractory gastric ulcer. World J Gastroenterol 11 (1):136–141. https://doi.org/10.3748/wjg.v11.i1.136

Shiu J, Czinn SJ, Kobayashi KS, Sun Y, Blanchard TG (2013) IRAK-M expression limits dendritic cell activation and proinflammatory cytokine production in response to *Helicobacter pylori*. PLoS One 8(6): e66914. https://doi.org/10.1371/journal.pone.0066914

Sierra JC, Hobbs S, Chaturvedi R, Yan F, Wilson KT, Peek RM Jr, Polk DB (2013) Induction of COX-2 expression by *Helicobacter pylori* is mediated by activation of epidermal growth factor receptor in gastric epithelial cells. Am J Physiol Gastrointest Liver Physiol 305(2):G196–G203. https://doi.org/10.1152/ajpgi.00495.2012

Sierra JC, Asim M, Verriere TG, Piazuelo MB, Suarez G, Romero-Gallo J, Delgado AG, Wroblewski LE, Barry DP, Peek RM Jr, Gobert AP, Wilson KT (2018) Epidermal growth factor receptor inhibition downregulates *Helicobacter pylori*-induced epithelial inflammatory responses, DNA damage and gastric carcinogenesis. Gut 67(7):1247–1260. https://doi.org/10.1136/gutjnl-2016-312888

Singh A, Hodgson N, Yan M, Joo J, Gu L, Sang H, Gregory-Bryson E, Wood WG, Ni Y, Smith K, Jackson SH, Coleman WG (2012) Screening *Helicobacter pylori* genes induced during infection of mouse stomachs. World J Gastroenterol 18:4323–4334. https://doi.org/10.3748/wjg.v18.i32.4323

Smith-Garvin JE, Koretzky GA, Jordan MS (2009) T cell activation. Annu Rev Immunol 27:591–619. https://doi.org/10.1146/annurev.immunol.021908.132706

Sommer F, Faller G, Röllinghoff M, Kirchner T, Mak TW, Lohoff M (2001) Lack of gastritis and of an adaptive immune response in interferon regulatory factor-1-deficient mice infected with *Helicobacter pylori*. Eur J Immunol 31:396–402. https://doi.org/10.1002/1521-4141(200102)31:2<396::AID-IMMU396>3.0.CO;2-Y

Song JY, Choi YJ, Kim JM, Kim YR, Jo JS, Park JS, Park HJ, Song YG, Lee KH, Kang HL, Baik SC, Youn HS, Cho MJ, Rhee KH, Lee WK (2011) Purification and characterization of *Helicobacter pylori* γ-glutamyltranspeptidase. J Bacteriol Virol 41:255–265. https://doi.org/10.4167/jbv.2011.41.4.255

Stein SC, Faber E, Bats SH, Murillo T, Speidel Y, Coombs N, Josenhans C (2017) *Helicobacter pylori* modulates host cell responses by CagT4SS-dependent translocation of an intermediate metabolite of LPS inner core heptose biosynthesis. PLoS Pathog 13(7): e1006514. https://doi.org/10.1371/journal.ppat.1006514

Sugimoto MA, Sousa LP, Pinho V, Perretti M, Teixeira MM (2016) Resolution of inflammation: what controls its onset? Front Immunol 7:160. https://doi.org/10.3389/fimmu.2016.00160

Sun X, Zhang M, El-Zataari M, Owyang SY, Eaton KA, Liu M, Chang YM, Zou W, Kao JY (2013) TLR2 mediates *Helicobacter pylori*-induced tolerogenic immune response in mice. PLoS One 8:e74595. https://doi.org/10.1371/journal.pone.0074595

Sundrud MS, Torres VJ, Unutmaz D, Cover TL (2004) Inhibition of primary human T cell proliferation by *Helicobacter pylori* vacuolating toxin (VacA) is independent of VacA effects on IL-2 secretion. Proc Natl Acad Sci U S A 101:7727–7732. https://doi.org/10.1073/pnas.0401528101

Suzuki N, Murata-Kamiya N, Yanagiya K, Suda W, Hattori M, Kanda H, Bingo A, Fujii Y, Maeda S, Koike K, Hatakeyama M (2015) Mutual reinforcement of inflammation and carcinogenesis by the *Helicobacter pylori* CagA oncoprotein. Sci Rep 5:10024. https://doi.org/10.1038/srep10024

Takabayashi H, Shinohara M, Mao M, Phaosawasdi P, El-Zaatari M, Zhang M, Ji T, Eaton KA, Dang D, Kao J, Todisco A (2014) Anti-inflammatory activity of bone morphogenetic protein signaling pathways in stomachs of mice. Gastroenterology 147(2):396–406.e7. https://doi.org/10.1053/j.gastro.2014.04.015

Takeshima E, Tomimori K, Takamatsu R, Ishikawa C, Kinjo F, Hirayama T, Fujita J, Mori N (2009) *Helicobacter pylori* VacA activates NF-κB in T cells via the classical but not alternative pathway. Helicobacter 14:271–279. https://doi.org/10.1111/j.1523-5378.2009.00683.x

Takeuchi O, Akira S (2010) Pattern recognition receptors and inflammation. Cell 140(6):805–820. https://doi.org/10.1016/j.cell.2010.01.022

Tanaka H, Yoshida M, Nishiumi S, Ohnishi N, Kobayashi K, Yamamoto K, Fujita T, Hatakeyama M, Azuma T (2010) The CagA protein of *Helicobacter pylori* suppresses the functions of dendritic cell in mice. Arch Biochem Biophys 498:35–42. https://doi.org/10.1016/J.ABB.2010.03.021

Tang B, Li N, Gu J, Zhuang Y, Li Q, Wang HG, Fang Y, Yu B, Zhang JY, Xie QH, Chen L, Jiang XJ, Xiao B, Zou QM, Mao XH (2012) Compromised autophagy by MIR30B benefits the intracellular survival of *Helicobacter pylori*. Autophagy 8:1045–1057. https://doi.org/10.4161/auto.20159

Tegtmeyer N, Wessler S, Backert S (2011) Role of the cag-pathogenicity island encoded type IV secretion system in *Helicobacter pylori* pathogenesis. FEBS J 278:1190–1202. https://doi.org/10.1111/j.1742-4658.2011.08035.x

Terebiznik MR, Vazquez CL, Torbicki K, Banks D, Wang T, Hong W, Blanke SR, Colombo MI, Jones NL (2006) *Helicobacter pylori* VacA toxin promotes bacterial intracellular survival in gastric epithelial cells. Infect Immun 74:6599–6614. https://doi.org/10.1128/IAI.01085-06

Tonello F, Dundon WG, Satin B, Molinari M, Tognon G, Grandi G, Del Giudice G, Rappuoli R, Montecucco C

(1999) The *Helicobacter pylori* neutrophil-activating protein is an iron-binding protein with dodecameric structure. Mol Microbiol 34:238–246. https://doi.org/10.1046/j.1365-2958.1999.01584.x

Varga MG, Shaffer CL, Sierra JC, Suarez G, Piazuelo MB, Whitaker ME, Romero-Gallo J, Krishna US, Delgado A, Gomez MA, Good JA, Almqvist F, Skaar EP, Correa P, Wilson KT, Hadjifrangiskou M, Peek RM (2016) Pathogenic *Helicobacter pylori* strains translocate DNA and activate TLR9 via the cancer-associated *cag* type IV secretion system. Oncogene 35:6262–6269. https://doi.org/10.1038/onc.2016.158

Viala J, Chaput C, Boneca IG, Cardona A, Girardin SE, Moran AP, Athman R, Mémet S, Huerre MR, Coyle AJ, DiStefano PS, Sansonetti PJ, Labigne A, Bertin J, Philpott DJ, Ferrero RL (2004) Nod1 responds to peptidoglycan delivered by the *Helicobacter pylori cag* pathogenicity island. Nat Immunol 5:1166–1174. https://doi.org/10.1038/ni1131

Wagner S, Beil W, Westermann J, Logan RP, Bock CT, Trautwein C, Bleck JS, Manns MP (1997) Regulation of gastric epithelial cell growth by *Helicobacter pylori*: evidence for a major role of apoptosis. Gastroenterology 113:1836–1847

Wang G, Maier RJ (2015) A novel DNA-binding protein plays an important role in *Helicobacter pylori* stress tolerance and survival in the host. J Bacteriol 197:973–982. https://doi.org/10.1128/JB.02489-14

Wang J, Brooks EG, Bamford KB, Denning TL, Pappo J, Ernst PB (2001) Negative selection of T cells by *Helicobacter pylori* as a model for bacterial strain selection by immune evasion. J Immunol 167:926–934. https://doi.org/10.4049/JIMMUNOL.167.2.926

Wang G, Hong Y, Olczak A, Maier SE, Maier RJ (2006) Dual roles of *Helicobacter pylori* NapA in inducing and combating oxidative stress. Infect Immun 74:6839–6846. https://doi.org/10.1128/IAI.00991-06

Wang YH, Wu JJ, Lei HY (2009) When *Helicobacter pylori* invades and replicates in the cells. Autophagy 5(4):540–542. https://doi.org/10.4161/auto.5.4.8167

Wang YH, Gorvel JP, Chu YT, Wu JJ, Lei HY (2010) *Helicobacter pylori* impairs murine dendritic cell responses to infection. PLoS One 5:e10844. https://doi.org/10.1371/journal.pone.0010844

Wang HJ, Cheng WC, Cheng HH, Lai CH, Wang WC (2012) *Helicobacter pylori* cholesteryl glucosides interfere with host membrane phase and affect type IV secretion system function during infection in AGS cells. Mol Microbiol 83:67–84. https://doi.org/10.1111/j.1365-2958.2011.07910.x

White JR, Winter JA, Robinson K (2015) Differential inflammatory response to *Helicobacter pylori* infection: etiology and clinical outcomes. J Inflamm Res 8:137–147. https://doi.org/10.2147/JIR.S64888

Wroblewski LE, Peek RM, Wilson KT, Wilson KT (2010) *Helicobacter pylori* and gastric cancer: factors that modulate disease risk. Clin Microbiol Rev 23:713–739. https://doi.org/10.1128/CMR.00011-10

Wunder C, Churin Y, Winau F, Warnecke D, Vieth M, Lindner B, Zähringer U, Mollenkopf HJ, Heinz E, Meyer TF (2006) Cholesterol glucosylation promotes immune evasion by *Helicobacter pylori*. Nat Med 12:1030–1038. https://doi.org/10.1038/nm1480

Wüstner S, Mejías-Luque R, Koch MF, Rath E, Vieth M, Sieber SA, Haller D, Gerhard M (2015) *Helicobacter pylori* γ-glutamyltranspeptidase impairs T-lymphocyte function by compromising metabolic adaption through inhibition of cMyc and IRF4 expression. Cell Microbiol 17:51–61. https://doi.org/10.1111/cmi.12335

Yang Y, Mayo KH, Daly TJ, Barry JK, La Rosa GJ (1994) Subunit association and structural analysis of platelet basic protein and related proteins investigated by 1H NMR spectroscopy and circular dichroism. J Biol Chem 269:20110–20118

Yao S, Buzo BF, Pham D, Jiang L, Taparowsky EJ, Kaplan MH, Sun J (2013) Interferon regulatory factor 4 sustains CD8(+) T cell expansion and effector differentiation. Immunity 39:833–845. https://doi.org/10.1016/j.immuni.2013.10.007

Yaqoob P, Calder PC (1997) Glutamine requirement of proliferating T lymphocytes. Nutrition 13:646–651. https://doi.org/10.1016/S0899-9007(97)83008-0

Yokoyama K, Higashi H, Ishikawa S, Fujii Y, Kondo S, Kato H, Azuma T, Wada A, Hirayama T, Aburatani H, Hatakeyama M (2005) Functional antagonism between *Helicobacter pylori* CagA and vacuolating toxin VacA in control of the NFAT signaling pathway in gastric epithelial cells. Proc Natl Acad Sci U S A 102:9661–9666. https://doi.org/10.1073/pnas.0502529102

Zanotti G, Papinutto E, Dundon W, Battistutta R, Seveso M, Giudice G, Rappuoli R, Montecucco C (2002) Structure of the neutrophil-activating protein from *Helicobacter pylori*. J Mol Biol 323:125–130. https://doi.org/10.1016/S0022-2836(02)00879-3

Zeaiter Z, Cohen D, Müsch A, Bagnoli F, Covacci A, Stein M (2008) Analysis of detergent-resistant membranes of *Helicobacter pylori* infected gastric adenocarcinoma cells reveals a role for MARK2/Par1b in CagA-mediated disruption of cellular polarity. Cell Microbiol 10:781–794. https://doi.org/10.1111/j.1462-5822.2007.01084.x

Zeaiter Z, Diaz H, Stein M, Huynh HQ (2011) *Helicobacter pylori* induces expression and secretion of oncostatin M in macrophages *in vitro*. Dig Dis Sci 56:689–697. https://doi.org/10.1007/s10620-010-1341-z

Zhang B, Chikuma S, Hori S, Fagarasan S, Honjo T (2016) Nonoverlapping roles of PD-1 and FoxP3 in maintaining immune tolerance in a novel autoimmune pancreatitis mouse model. Proc Natl Acad Sci U S A 113:8490–8495. https://doi.org/10.1073/pnas.1608873113

Zhao Y, Yokota K, Ayada K, Yamamoto Y, Okada T, Shen L, Oguma K (2007) *Helicobacter pylori* heat-shock protein 60 induces interleukin-8 via a toll-like receptor (TLR)2 and mitogen-activated protein (MAP) kinase pathway in human monocytes. J Med Microbiol 56:154–164. https://doi.org/10.1099/jmm.0.46882-0

Zheng Y, Zha Y, Gajewski TF (2008) Molecular regulation of T-cell anergy. EMBO Rep 9:50–55. https://doi.org/10.1038/sj.embor.7401138

Zhou P, She Y, Dong N, Li P, He H, Borio A, Wu Q, Lu S, Ding X, Cao Y, Xu Y, Gao W, Dong M, Ding J, Wang DC, Zamyatina A, Shao F (2018) Alpha-kinase 1 is a cytosolic innate immune receptor for bacterial ADP-heptose. Nature 561:122–126. https://doi.org/10.1038/s41586-018-0433-3

Zimmermann S, Pfannkuch L, Al-Zeer MA, Bartfeld S, Koch M, Liu J, Rechner C, Soerensen M, Sokolova O, Zamyatina A, Kosma P, Mäurer AP, Glowinski F, Pleissner KP, Schmid M, Brinkmann V, Karlas A, Naumann M, Rother M, Machuy N, Meyer TF (2017) ALPK1- and TIFA-dependent innate immune response triggered by the *Helicobacter pylori* type IV secretion system. Cell Rep 20(10):2384–2395. https://doi.org/10.1016/j.celrep.2017.08.039

Adv Exp Med Biol - Advances in Microbiology, Infectious Diseases and Public Health (2019) 1149: 107–120
https://doi.org/10.1007/5584_2019_361

Published online: 30 April 2019

Helicobacter pylori Infection in Children and Adolescents

Masumi Okuda, Yingsong Lin, and Shogo Kikuchi

Abstract

About one-third of all children worldwide is infected with *H. pylori* and its prevalence is low in developed and high in developing countries. *H. pylori* is mainly acquired during childhood and transmission of the bacterium commonly proceeds from person to person, especially among family members. The most frequent transmission route is from the mother to children. Various gastrointestinal and extra-gastrointestinal diseases are reported to be associated with *H. pylori* in children and adolescents, but the strongest recommendation for testing and treating is introduced only with children and adolescents having peptic ulcer disease. Iron deficiency anemia and chronic immune thrombocytopenic purpura are also considered for testing and treating, but the effectiveness is somewhat controversial. Invasive diagnosis is recommended, whereas none of the available diagnostic tests have 100% accuracy for reliable diagnosis, and therefore at least two or more tests should be performed. Urea breath test is the most reliable among the non-invasive tests. Because the number of antibiotics-resistant *H. pylori* strains is increasing, it is desirable to conduct a drug susceptibility test before treatment and to select the corresponding regime. *H. pylori* has been proven to be a major cause of gastric cancer and 'screen-and-treat' strategies are recommended in communities at high risk of gastric cancer. However, the application to children and adolescents is controversial. An effective vaccine is desirable, but not yet available. Screen-and-treat for adolescents has started in a few areas in Japan, where conditions are well established. New prevention strategies for gastric cancer are awaited worldwide.

Keywords

Child · Epidemiology · Manifestation · Treatment · Screen

M. Okuda (✉)
Department of Pediatrics, Hyogo College of Medicine, Nishinomiya, Hyogo, Japan

Department of General Medicine and Community Health Science, Hyogo College of Medicine, Sasayama, Hyogo, Japan
e-mail: okuda@naxnet.or.jp

Y. Lin and S. Kikuchi
Department of Public Health, Aichi Medical University School of Medicine, Nagakute, Aichi, Japan

1 Introduction

Helicobacter pylori infection is acquired mainly during childhood and usually persists unless treated. Most of infected children and adolescents are asymptomatic, but it can cause peptic ulcer disease, iron deficiency anemia (IDA), and chronic immune thrombocytopenic purpura (cITP). The features of infection in children and adolescents differ from that in adults in terms of epidemiology, clinical manifestations, related diseases and treatment. In addition, it also depends on the area in which they live. In this article, we discuss the epidemiology, associated diseases, diagnosis and treatment of *H. pylori* infections in children and adolescents including prevention of gastric cancer.

2 Epidemiology and Transmission in Children and Adolescents

2.1 Prevalence of *H. pylori* Infection in Children and Adolescents

In recent years, the prevalence of *H. pylori* infection in children and adolescents is decreasing in Western countries as well as in various developing countries (Table 1). However, in some areas, the infection rate is over 60% even in children. It was recently suggested that about one-third of all children worldwide have been infected with *H. pylori* (Zabala Torrres et al. 2017).

2.2 Acquisition and Transmission

According to current knowledge, *H. pylori* infection mostly appears during infancy. In a study of 248 Gambian children, the positive rate of urea breath test (UBT) was 19% at 3 months and reached to 84% at 30 months of age (Thomas et al. 1999). From recent studies in developing countries, *H. pylori* infection commonly occurs in the first year of life (Jafar et al. 2013; Kienesberger et al. 2018). The frequency of colonization among 180 children in a high prevalence area of *H. pylori* infection in Germany using a stool antigen test was 8.9%, 36.4% and 31.9% in 1, 2, and 4 year-old children, respectively (Rothenbacher et al. 2000). In a prospective follow-up study using fecal antigen test in Japan, four infants were shown to be infected with *H. pylori* and the acquisition age was 4 months (1 infant), 8 months (2 infants), and 12 months (1 infant), respectively (Okuda et al. 2007). In another prospective follow-up study held in Japan, 51 children born from *H. pylori*-positive mothers were followed using fecal antigen test every 4–6 months from birth to 5 years of age. Five children became infected with *H. pylori* and the acquisition age was 1 year 2 months, 1 year 3 months, 1 year 6 months, 1 year 8 months, and 4 years 4 months, respectively (Konno et al. 2005). These studies suggest that the acquisition of *H. pylori* mainly occurs until the second year of life. In developed countries, fewer acquisition in adolescents and adults has been reported (Rowland et al. 2018). In a further study held in Japan, 430 children with negative fecal antigen test were followed and no children became infected after a year (Okuda et al. 2015). But in the high prevalence area, the overall prevalence of *H. pylori* infection in 844 investigated asymptomatic children was 31.6% and it was increased with age (19.9, 37.0 and 51.5%, in age groups 0–5, 6–10, and 11–15, respectively) and 296 of 569 negative results were followed up for 3 years and 62 children acquired infection (median age of acquisition was 6.0 years) (Oleastro et al. 2011).

The transmission route of the bacterium is commonly from person to person, especially among family members (Osaki et al. 2015) and the most frequent path is from the mother to children (Konno et al. 2005; Yokota et al. 2015). Transmission occurs primarily via the oral-to-oral and possibly via gastro-oral routes mediated by refluxed gastric juice or vomitus, and fecal-oral transmission may also occur. It should be also noted that spontaneous eradication (loss of infection) has been observed. For example, it was reported in 2.9% of patients during the year in a high prevalence area (Zhou et al. 2018), but extremely low incidence was seen in developed countries (Okuda et al. 2015; Rowland et al. 2018).

3 Clinical Manifestations

Various diseases including gastrointestinal and extra-gastrointestinal ones are reported to be associated with *H. pylori* in children and

Table 1 Prevalence of *H. pylori* infection in children and adolescence

Country (Area)	Studied period	Population (n)	Age (years)	Method	Positivity (%)	References
Germany	2006	1905	14	UBT	6.5	Bauer et al. (2011)
Lisbon, Portugal	2002–2003	844	0~15	FA	31.6	Oleastro et al. (2011)
Portugal	2003–2004	1312	13	SA	66.2	Bastos et al. (2013)
Iran	2007	458	0.3~15	FA	64.2	Jafar et al. (2013)
Turkey	2007–2013	500	3~16	UBT	49	Çınar et al. (2015)
Netherland	Unkown	4467	4~8	SA	10	den Hollander et al. (2015)
Hyogo, Japan	2010–2011	1524	0~11	FA	1.8	Okuda et al. (2015)
Nagano, Japan	2007–2013	3263	16, 17	UA	4.2	Akamatsu et al. (2015)
China	2009–2011	3491	0~18	FA	6.8	Ding et al. (2015)
Taiwan	2010–2012	715	0~15	SA	6	Wu et al. (2015)
USA	1999–2000	1806	3~13	SA	7.1	Krueger et al. (2015)
Latvia	2009–2010	101	1~6	FA	15.5	Daugule et al. (2016)
Bhutan	2015–2016	327	4~19	SA	66	Wangda et al. (2017)
Akita, Japan	2015	1765	13~15	UA and then UBT	4.8	Kusano et al. (2017)
Nagano, Japan	2011–2013	454	12~15	SA and then UBT	3.1	Nakayama et al. (2017)
Shanghai, China	2014	867	7~12	UBT	24.1	Zhou et al. (2018)
Sudan	2012	431	6~18	SA	21.8	Abbas et al. (2018)

Abbreviations used: *FA* Fecal antigen test, *UA* Urine antibody test, *UBT* Urea breath test, *SA* Serum antibody test

adolescents and many trials are performed worldwide. In our previous nation survey study for treatment of *H. pylori* in Japan (Okuda et al. 2017), 332 patients aged 2–18 years were analyzed (Table 2). *H. pylori*-associated gastritis (HpAG), IDA, and duodenal ulcer (DU) were most frequent, constituting 75.8% of the cases, while gastric ulcer (GU) was seen only in 7.2% of the cases. The cases were divided into three groups according to age: 1–6 years old, 7–12 years old and 13–18 years old. cITP, HpAG, and IDA were the most frequent diseases among 1–6, 7–12, and 13–18-year-old children, respectively (Okuda et al. 2017).

3.1 Gastrointestinal Manifestations

3.1.1 Gastritis and Peptic Ulcer

The most relevant gastrointestinal manifestation associated with *H. pylori* in children is DU. This circumstance can be explained by the course of *H. pylori* infection (Kato and Asaka 2010). First, *H. pylori* colonizes the antrum and can cause antral gastritis. In this stage, gastric acid secretion is accelerated and the acid flows into the duodenum and DU is likely to occur. And persistent infection causes spread of gastritis and progress to pan-gastritis including nodular gastritis has been often observed in adolescents. Among the patients with *H. pylori* infection, corpus-predominant gastritis is associated with severe gastric atrophy and intestinal metaplasia, which exhibit a significantly higher risk for gastric cancer, especially the intestinal type (Kato and Asaka 2010). While the diffuse-type gastric cancer arose from pan-gastritis, gastric cancer usually does not occur from antrum predominant gastritis. In another study, the prevalence of *H. pylori* in non-nodular gastritis, nodular gastritis, DU, and GU was 28.8%, 98.5%, 83.0%, and 44.2%, respectively (Kato et al. 2004). Interestingly, *H. pylori* infection was more associated with DU compared to GU in children and adolescents.

Table 2 Treatment of *H. pylori*-associated diseases in Japan[a]

	Total	(%)	1–6 years	(%)	7–12 years	(%)	13–18 years	(%)
HpAG	89	26.8	3	9.1	47	38.2	39	22.2
IDA	83	25.0	5	15.2	15	12.2	63	35.8
DU	82	24.7	7	21.2	37	30.1	38	21.6
GU	24	7.2	3	9.1	9	7.3	12	6.8
GU and DU	6	1.8	0	0	1	0.8	5	2.8
ITP	29	8.7	12	36.4	10	8.1	7	4.0
MALToma	2	0.6	0	0	0	0	2	1.1
Gastric cancer	1	0.3	0	0	0	0	1	0.6
Others	16	4.8	3	9.1	4	3.3	9	5.1
Total	332		33		123		176	

Abbreviations used: *HpAG* H. pylori associated gastritis, *IDA* iron deficiency anemia, *DU* duodenal ulcer, *GU* gastric ulcer, *ITP* idiopathic thrombocytopenic purpura

[a]Data are from Okuda et al. (2017)

And in the ESPGHAN/NASPGHAN guidelines 2016, a strong recommendation for testing and treating of *H. pylori* has been given to be performed in children with gastric or duodenal peptic ulcer diseases (PUD) (Jones et al. 2017).

3.1.2 Atrophic Gastritis

Among the *H. pylori*-positive patients, persistent infection can cause a corpus-predominant gastritis and is associated with gastric atrophy and intestinal metaplasia in adults (Kato and Asaka 2010). However, the actual condition in children is not well understood. Eighty-two Mexican children (mean age 8.3 ± 4.8 years) with chronic gastritis (36 *H. pylori*-positive, 46 negative) were examined and atrophy was diagnosed in 7 (9%), and intestinal metaplasia in 5 (6%) of 82 cases by histology. Atrophy was more frequent in *H. pylori*-positive ($P < .0001$), whereas intestinal metaplasia showed no significant correlation with the *H. pylori* status (Villarreal-Calderon et al. 2014). For a study of *H. pylori*-infected 131 children aged 1–16 years in Japan, atrophy was evaluated according to the Updated Sydney system (Dixon et al. 1996). The prevalence of grade 2 or 3 atrophy in the antrum was 10.7% and in the corpus 4.3% (Kato et al. 2006). Among a group of high school students (aged 16–17 years) in Japan, screening for *H. pylori* was performed. Sixty students were positive for *H. pylori* infection and atrophic gastritis was found in 36 students (60%). The endoscopic degree of atrophic gastritis according to the Kimura-Takemoto classification (Kimura and Takemoto 1969) was closed type (C-1: 9, C-2: 23, C-3: 4) in all cases (Akamatsu et al. 2015).

3.1.3 Dyspepsia and Other Abdominal Complaints

The relation between functional dyspepsia or recurrent abdominal pain (RAP) symptoms and *H. pylori* is currently controversial. In the ESPGHAN/NASPGHAN guidelines 2016, treatment to eliminate *H. pylori* infection is not expected to improve symptoms in children, except in cases of PUD. Diagnosis of *H. pylori* infection is not recommended in patients other than PUD (Jones et al. 2017). A population-based case-control study to evaluate the association between *H. pylori* infection and RAP was held in Iran. Among 1558 children aged 6–13 years, 145 children with RAP (according to the criteria by Apley and Naish (1958)) were compared with 145 age-matched *H. pylori* negative controls recruited from the same area. There was no significant difference between the RAP symptoms in *H. pylori*-positive children versus *H. pylori*-negative ones (Alimohammadi et al. 2017). On the other hand, a study for evaluating the effect of *H. pylori* eradication on dyspepsia symptom scores in Turkish children with functional dyspepsia was reported. The symptom scores were lower in *H. pylori*-eradicated group than *H. pylori*-uneradicated group (Ünlüsoy Aksu et al. 2018).

3.1.4 Gastric Cancer

The association between gastric cancer and *H. pylori* infection remains unclear in children. Also, we found no reports in Pubmed discussing an association between *H. pylori* and gastric cancer in children. In our survey and review, 80 cases of gastric cancer patients under 16 years old were identified. Of the 80 patients, only three children underwent testing for *H. pylori* infection, and two were *H. pylori*-positive (Okuda et al. 2017; Okuda et al. 2019). One of the positives was a 12 years old girl and the symptoms at the first visit were epigastric pain and tarry stool. Her father and grandfather also had a history of gastric cancer, suggesting a genetically associated disorder in the family. Histology revealed moderately to poorly differentiated adenocarcinoma and signet-ring cell carcinoma (Okuda et al. 2019). Another positive case was a 15 years old girl, who also had a family history of gastric cancer (Okuda et al. 2017). A *H. pylori*-negative case was a 11 years old girl with adenocarcinoma at the esophagogastric junction (Sasaki et al. 1999), which is commonly less associated with *H. pylori* infection.

The vital statistics data allowed us to examine the trends in gastric cancer deaths in children and adolescents (Cancer Registry and Statistics 2019), which showed the rapid decrease in gastric cancer deaths in patients under 20 years of age (Fig. 1) and may reflect the decreasing prevalence of *H. pylori* infection in the Japanese population (Wang et al. 2017).

3.2 Extra-Gastrointestinal Manifestations

3.2.1 Iron Deficiency Anemia (IDA)

Recently, five meta-analyses including both pediatric and adult patients have shown an association between *H. pylori* infection and IDA (Pacifico et al. 2014; Muhsen and Cohen 2008; Qu et al. 2010; Huang et al. 2010; Yuan et al. 2010). And according to the systematic review and meta-analysis to examine the prevalence of depleted iron stores among persons infected with *H. pylori* compared to uninfected ones; the impact of *H. pylori* treatment plus iron therapy on ferritin and hemoglobin levels compared to iron therapy was assessed. Current evidence indicates increased likelihood of depleted iron stores in relation to *H. pylori* infection. *H. pylori* eradication therapy, added to iron therapy, might be beneficial in increasing ferritin and hemoglobin levels (Muhsen and Cohen 2008). The Maastricht V/Florence Consensus Report, based on the results of these studies, recommends to search and treat *H. pylori* infection in unexplained IDA cases (Malfertheiner et al. 2017). But in the ESPGHAN/NASPGHAN guidelines 2016, it is recommended against diagnostic testing for *H. pylori* infection as part of the initial investigation in children with IDA (Jones et al. 2017). However, in regions with high prevalence of *H. pylori* infection, the recommendation may not be applicable. Following the consensus, infection should only be searched for refractory anemia, when other causes have been ruled out. The term ‘refractory’ is, unfortunately, not clearly defined. In our practice, the term is used in case of failure of or recurrence after intravenous supplementation (Kotilea et al. 2018). As mentioned before, since many IDA patients associated with *H. pylori* show severe anemia in Japan, they need to be treated for *H. pylori* for dissolving the symptoms of IDA.

3.2.2 Chronic Immune Thrombocytopenic Purpura (cITP)

cITP is an autoimmune disease characterized by autoantibody-mediated platelet destruction lasting for more than 6 months (Blanchette and Price 2003). In adults, cITP is one of the extragastric disorders, for which diagnosis and treatment of *H. pylori* infection are indicated (Malfertheiner et al. 2017), but the effect of anti-bacterial treatment for pediatric cITP is not fully clarified. Because of spontaneous complete remission, defined as an increase in the number of PLT to 100×10^9/L or higher for at least 3 months off PLT-enhancing therapy, that occurred in 36% (102/282) of cases (Blanchette and Price 2003), it is difficult to evaluate the effect of treatment of *H. pylori*. In addition, as the studies on the efficacy

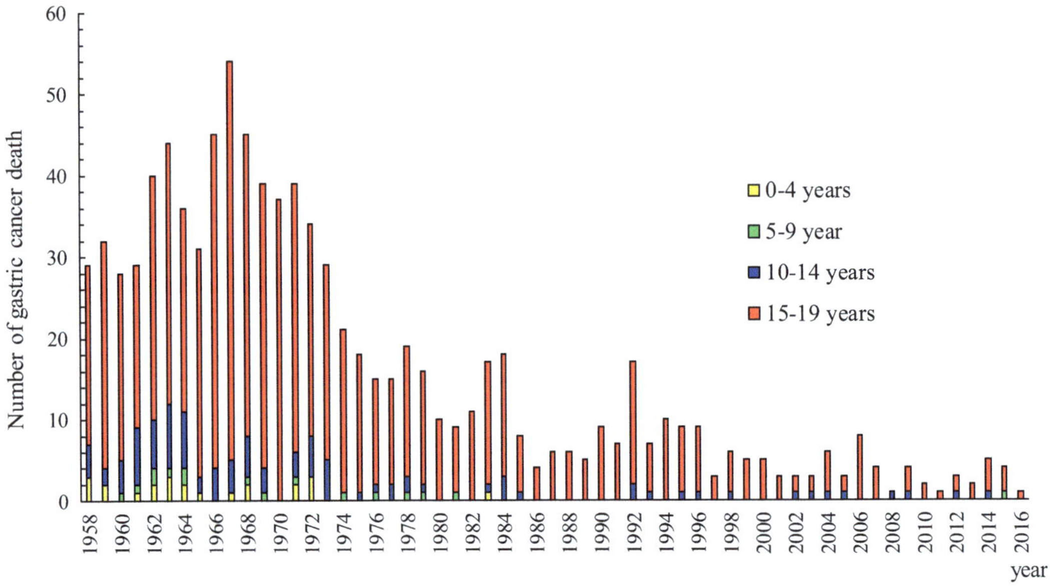

Fig. 1 Statistics of gastric cancer-related deaths in patients aged <20 years in Japan from 1958–2016. (These data are available in Japanese at https://ganjoho.jp/reg_stat/statistics/dl/index.html (Cancer Registry and Statistics, cited 18 January 2019))

of *H. pylori* treatment due to childhood cITP included small sample sizes and the spontaneous recovery of cITP occurred in about one third in childhood (Pacifico et al. 2014), it is difficult to judge a significant difference. A prospective, controlled, multicenter study was performed in Italy. In 244 cases of cITP younger than 18 years of age, 50 (20%) had *H. pylori* infection and 37 received treatment of *H. pylori* and completed follow-up. Eradication was successful in 33/37 patients (89%). PLT recovery was demonstrated in 13/33 (39%) patients after eradication, whereas spontaneous remission was observed in 17/166 (10%) *H. pylori*-negative patients ($P < 0.005$). They concluded that it may be appropriate to diagnose *H. pylori* infection and eventually eradicate it in children with cITP (Russo et al. 2011). A multicenter randomized controlled trial (RCT) was conducted in Thailand, and 16 of 55 cITP children were shown to be infected with *H. pylori* using UBT. Patients with *H. pylori* infection were randomized into two groups: *H. pylori*-treated and non-treated groups. At 6 months, PLT recovery was demonstrated in one patient of the treatment group as well as one in the control group, and they concluded no beneficial effect of *H. pylori* eradication on PLT recovery in children, in which cITP was identified (Treepongkaruna et al. 2009). Further evidences are needed for better management of cITP children with *H. pylori* infection.

3.2.3 Allergic Disorders

The inverse association of allergic disorders and *H. pylori* infection has been discussed with various results. Several studies have demonstrated inverse relationships between asthma and *H. pylori* presence (Reibman et al. 2008; Chen and Blaser 2008) and others have reported no association (Tsang et al. 2000; Bodner et al. 2000). Recent studies in children also showed the various conclusions as follows; *H. pylori* seropositivity protects against childhood asthma and inversely correlates to its clinical and functional severity (Fouda et al. 2018) or prevalence of *H. pylori* infection did not differ significantly between children with or without allergy (Daugule et al. 2017). For the association with eosinophilic esophagitis (EoE), *H. pylori* was not inversely associated with EoE, neither in children nor in adults (Molina-Infante et al. 2018). Thus, more studies with larger cohorts are required to investigate the potential correlation between *H. pylori* infection and allergies in children.

4 Diagnosis

H. pylori infection can be diagnosed both by invasive or non-invasive tests in children similar to diagnosis in adults. Bacterial culture is 100% specific, but none of the other available diagnostic tests have 100% accuracy. For reliable diagnosis, at least two or more tests should be performed. In the ESPGHAN/NASPGHAN guidelines 2016, it is recommended that the diagnosis of *H. pylori* infection should be based on either (a) histopathology (*H. pylori*–positive gastritis) plus at least 1 other positive biopsy-based test or (b) positive culture (Jones et al. 2017). For the diagnosis at upper gastrointestinal endoscopy, at least 6 gastric biopsies need to be obtained. Two biopsies should be obtained from the antrum and 2 biopsies from the corpus for the histopathological evaluation applying the updated Sydney system classification, at least 1 biopsy from the antrum and 1 from the corpus for culture (if available) and at least 1 biopsy for any additional diagnostic tests from the antrum (rapid-urease, or molecular-based assays) (Jones et al. 2017). Biopsy-based culture should be selected for the *H. pylori* antimicrobial susceptibility testing because antibiotics resistant *H. pylori* strains are prevalent. In the case of cITP, non-invasive testing should be selected. Among the non-invasive tests, UBT has the best sensitivity in all age groups and excellent specificity (Mégraud et al. 2005) and UBT appears to be an excellent test for children and adolescents. Monoclonal stool antigen test (SAT) is also an alternative (Malfertheiner et al. 2017). Before testing, waiting at least 2 weeks after stopping proton pump inhibitor (PPI) and 4 weeks after stopping antibiotics treatment (Jones et al. 2017) is needed. Antibody-based tests using serum, urine or saliva should not be used to detect the current *H pylori* infection and diagnosis by single antibody test should not be recommended.

The judgement of *H. pylori* treatment success is assessed at least 4 weeks after completion of therapy using UBT or SAT. Antibody-based tests should not be used in post-treatment evaluation in children or adolescents.

5 Treatment

Indications for treatment of *H. pylori* infection in children and adolescents should be taken into consideration by regional infection rate and pathogenicity of the bacterium. In addition, variation of treatment regimes must be considered, for example in cases where antibiotics resistances have been noted (Jones et al. 2017; Mabe et al. 2018).

5.1 Treatment Regimes

In children and adolescents, when antibiotics resistant *H. pylori* strains are prevalent, it is desirable to conduct a drug susceptibility test before treatment. In the ESPGHAN/NASPGHAN guidelines for management of *H. pylori* in children and adolescents (Update 2016), recommended options are listed (Table 3) and treatment period is 14 days (Jones et al. 2017). And fixed-dose (standard dose) drugs, according to the bodyweight of the child, need to be used for eradication regimes of *H. pylori* infection in children (Jones et al. 2017; Kotilea et al. 2018) as shown in Table 4 and high dosing regimes for amoxicillin (AMX) in Table 5. The PPI-based triple therapy for treatment of Japanese children and adolescents is shown in Table 6. The duration of treatment is 7 days and dosage of antibiotics is less in the Japanese regime compared to the regime in ESPGHAN/NASPGHAN guidelines. In the Japanese national health insurance system, the treatment period is only allowed for 7 days. In the multicenter, randomized trial held between 2012 and 2015, PPI-based triple therapy containing metronidazole (MTZ) [Lansoprazole (30 mg) - AMX (750 mg) – MTZ (250 mg) twice a day] for 7 days was good performance without susceptibility test, but containing

Table 3 Recommended options for first-line therapy for *H. pylori* infection[a]

H. pylori antimicrobial susceptibility	Suggested treatment
Known	
Susceptible to CLR and to MTZ	PPI- AMX – CLR 14 days with standard dose[b] or sequential therapy[b] for 10 days[c]
Resistant to CLR, susceptible to MTZ	PPI- AMX- MTZ 14 days or bismuth-based[b]
Resistant to MTZ, susceptible to CLR	PPI- AMX-CLR 14 days or bismuth-based[b]
Resistant to CLR and to MTZ	PPI- AMX-MTZ 14 days with high dose for amoxicillin (Table 5) or bismuth-based[b]
Unkown	
	High-dose PPI- AMX-MTZ 14 days or bismuth-based[b,d]

Abbreviations used: *AMX* amoxicillin, *CLR* clarithromycin, *MTZ* metronidazole, *PPI* proton pump inhibitor
[a]Data are from Jones et al. (2017)
[b]In the case of penicillin allergy: if the strain is susceptible to CLR and MTZ, use standard dose triple therapy with MTZ in place of AMX; if the strain is resistant to CLR, then use bismuth-based therapy with tetracycline instead of AMX if older than 8 years
[c]Sequential therapy: PPI with AMX for 5 days followed by PPI with CLR and MTZ for 5 days with standard dose
[d]Or concomitant therapy (PPI- AMX-MTZ-CLR) for 14 days

Table 4 Standard dosing regimen[a]

Drug	Weight groups (kg)	Morning dose (mg)	Evening dose (mg)
Esomeprazole	15–24	20	10
	25–34	20	20
	>35	40	20
AMX	15–24	500	500
	25–34	750	750
	>35	1000	1000
CLR	15–24	250	250
	25–34	500	250
	>35	500	500
MTZ	15–24	250	250
	25–34	500	250
	>35	500	500

Abbreviations used: *AMX* amoxicillin, *CLR* clarithromycin, *MTZ* metronidazole
[a]Data are from Jones et al. (2017) and Kotilea et al. (2018)

Table 5 High dosing regimens for amoxicillin

Bodyweight range, kg	Morning dose, mg	Evening dose, mg
15–24	750	750
25–34	1000	1000
>35	1500	1500

clarithromycin (CLR) [Lansoprazole (30 mg) – AMX (750 mg) – CLR (200 mg) twice a day] regimen was low eradication rate (Table 7) (Mabe et al. 2018). On the other hand, the eradication rate of potassium-competitive acid blocker (P-CAB) (20 mg) – AMX (750 mg) – CLR (200 mg) twice a day for 7 days was 85.7% as shown in Table 5 (Kusano et al. 2018). The 7 days treatment regime, particularly using MTZ is satisfactory for the treatment of *H. pylori* in Japanese

Table 6 Standard regiment (PPI based triple therapy) for treatment of *H. pylori* for children and adolescents in Japan[a]

		Dose	Maximal Adult dose	Duration
First line or CLR sensitive	AMX	25 mg/kg twice a day	750 mg twice a day	7 days
	CLR	5~10 mg/kg twice a day	200 (400) mg twice a day	
	PPI[b]			
Second line or CLR resistance	AMX	25 mg/kg twice a day	750 mg twice a day	7 days
	MTZ	5~10 mg/kg twice a day	250 mg twice a day	
	PPI[b]			

Abbreviations used: *AMX* amoxicillin, *CLR* clarithromycin, *MTZ* metronidazole
[a]In the case of penicillin allergy, 7 days triple therapy of CLR, MTZ and PPI is selected
[b]PPI Lansoprazole (0.75 mg/kg (max 30 mg) twice a day) or Omeprazole (0.5 mg/kg (max 20 mg) twice a day) or Rabeprazole (0.25 mg/kg (max 10 mg) twice a day)

Table 7 Eradication rates of 7 days treatment and compliance in Japanese adolescents (junior or senior high school students) unkown of susceptibility[a]

	ITT	PP
Mabe et al. (2018)		
PPI-AC	26/43 (60.5%)	26/41 (63.4%)
PPI-AM	57/58 (98.3%)	55/55 (100%)
Kusano et al. (2018)		
P-CAB-AC	96/118 (81.3%)	96/112 (85.7%)

Abbreviations used: *ITT* intention to treat, *PP* per protocole
[a]PPI-AC: Lansoprazole (30 mg) – AMX (750 mg) – CLR (200 mg) twice a day
PPI-AM: Lansoprazole (30 mg) – AMX (750 mg) – MTZ (250 mg) twice a day
P-CAB-AC: potassium-competitive acid blocker (20 mg) – AMX (750 mg) – CLR (200 mg) twice a day

children and adolescents. In consideration of drug administration period and dosage, *H. pylori* strains in Japanese children and adolescents may be more susceptible to eradication treatment compared with the strains in children and adolescents living in Europe and North America.

5.2 Drug Resistance

According to the increase of *H. pylori* antimicrobial resistances worldwide, the successful treatment rate by particularly CLR-containing regimes has been decreasing. In a study of the northern Portuguese pediatric center held from 2013 to 2017, the resistance rates for MTZ and CLR were 3.3% and 23.3%, respectively. There was no resistance observed for AMX and levofloxacin (LVFX) (Silva et al. 2018). From 2012–2014, 545 *H. pylori* strains isolated from children in Hangzhou (China), the total resistance rates of *H. pylori* to CLR, MTZ, and LEV were 20.6%, 68.8%, and 9.0%, respectively (Shu et al. 2017). No resistance to AMX, gentamicin, and furazolidone was detected (Shu et al. 2017). High resistant rates against CLR in children from Japan were reported as 36.1% (Kato and Fujimura 2010) and 43.4% (Okuda et al. 2017), respectively.

5.3 Reinfection After Successful Eradication

In developing countries, the reinfection rate after eradication in children is high. In a trial of population-based "screen and treat" in the low-income country Bolivia with a high *H. pylori* prevalence, the reinfection rate 1 year after successful treatment was 20% in children younger than 10 years, compared to 8% in older children and adolescents (Sivapalasingam et al. 2014). In Germany, on the other hand, reinfection rate in children (aged 1.8–18 years) was only 2.3% per year (Feydt-Schmidt et al. 2002).

6 Screen-and-Treat for Preventing Future Gastric Cancer

H. pylori has been proven to be a major cause of gastric cancer development (Kikuchi et al. 1995; Uemura et al. 2001; Malfertheiner et al. 2017).

Strategies for preventing gastric cancer are proposed to be different for adolescents and elderly people because the incidence of gastric cancer occurrence after *H. pylori* treatment differs between the two age groups (Asaka 2013, IARC *Helicobacter pylori* Working Group (2014). Indeed, 'a screen-and-treat' approach is recommended for younger people that includes universal *H. pylori* testing and immediate bacterial eradication in those with a positive result (Asaka 2013). In the 'Management of *Helicobacter pylori* infection—the Maastricht V/Florence Consensus Report', 'screen-and-treat' strategies are recommended in communities at high risk of gastric cancer (Malfertheiner et al. 2017). They are considered to be cost-effective as to the expected level of adverse events and compliance (Malfertheiner et al. 2017). But the application to children and adolescents is controversial. The 'ESPGHAN/NASPGHAN Guidelines for the Management of *Helicobacter pylori* in Children and Adolescents' declare "We recommend against a "test and treat" strategy for *H. pylori* infection in children" (Jones et al. 2017). Because current evidence indicates that *H. pylori* infection does not cause any symptoms excepted for PUD cases, performing a noninvasive test to detect infection and treating if the test is positive are not warranted (Jones et al. 2017).

Gastric cancer is one of the common causes of cancer-related death in Japan, and a 'screen-and-treat' strategy for *H. pylori* infection for the prevention of gastric cancer is anticipated (Asaka et al. 2014). Several studies have suggested that in the elderly or in patients with gastric atrophy, treatment for *H. pylori* infection only slightly reduced the incidence of gastric cancer (Fukase et al. 2008; Take et al. 2015). Nonetheless, younger patients or those with mild baseline gastric mucosal atrophy were shown to be significant factors for the prophylactic effect of eradication therapy against gastric cancer (Take et al. 2007). Therefore, in a gastric cancer prevalent area such as Japan, China and Taiwan, it is necessary to eradicate *H. pylori* in youth for preventing gastric cancer. In addition, several local governments are currently performing the 'screen and treat' for junior or senior high school students in Japan (Akamatsu et al. 2015; Kusano et al. 2017; Kakiuchi et al. 2019). The screening was performed in two steps. At the first step, antibody test (mainly urine) is performed and the second step was performed using UBT or fecal antigen test, when the first test was positive. Fortunately, *H. pylori* treatment of Japanese children and adolescents is not difficult (low dose and short duration) and few side effects were reported (Okuda et al. 2017). And because the prevalence of *H. pylori* in Japanese children and adolescents is very low (about 2~5%) (Akamatsu et al. 2015; Okuda et al. 2015; Kusano et al. 2017) it may allow performing the 'screen and treat' of *H. pylori* infection. In Japan, the cumulative gastric cancer risk based on 2014 data was 11% in male and 5% in female (Cancer Registry and Statistics 2019). Infected Japanese children have been progressing gastritis and some of them develop surely gastric cancer. So, we started to protect adolescents from gastric cancer using the 'screen and treat' strategy. But this strategy must be performed carefully and it is important to know that it is not always possible in countries other than Japan. Prevention of gastric cancer worldwide using vaccine is awaited.

7 Conclusions

It is suggested that one-third of children worldwide are infected with *H. pylori*. Depending on the country or region, the infection time, transmission route, the pathogenicity and the drug susceptibility are different. The management of *H. pylori* infection in children and adolescents should be considered individually using evidence obtained in each country or region.

References

Abbas M, Sharif FA, Osman SM, Osman AM, El Sanousi SM, Magzoub M, Ibrahim ME (2018) Prevalence and associated symptoms of *Helicobacter pylori* infection among schoolchildren in Kassala state, east of Sudan. Interdiscip Perspect Infect Dis 15:4325752. https://doi.org/10.1155/2018/4325752

Akamatsu T, Okamura T, Iwaya Y, Suga T (2015) Screening to identify and eradicate *Helicobacter pylori* infection in teenagers in Japan. Gastroenterol Clin N Am 44:667–676

Alimohammadi H, Fouladi N, Salehzadeh F, Alipour SA, Javadi MS (2017) Childhood recurrent abdominal pain and *Helicobacter pylori* infection, Islamic Republic of Iran. East Mediterr Health J 22:860–864

Apley J, Naish N (1958) Recurrent abdominal pains: a field survey of 1,000 school children. Arch Dis Child 33:165–170

Asaka M (2013) A new approach for elimination of gastric cancer deaths in Japan. Int J Cancer 132:1272–1276

Asaka M, Kato M, Sakamoto N (2014) Roadmap to eliminate gastric cancer with *Helicobacter pylori* eradication and consecutive surveillance in Japan. J Gastroenterol 49:1–8

Bastos J, Peleteiro B, Pinto H, Marinho A, Guimarães JT, Ramos E, La Vecchia C, Barros H, Lunet N (2013) Prevalence, incidence and risk factors for *Helicobacter pylori* infection in a cohort of Portuguese adolescents (EpiTeen). Dig Liver Dis 45:290–295. https://doi.org/10.1016/j.dld.2012.11.009

Bauer S, Krumbiegel P, Richter M, Richter T, Röder S, Rolle-Kampczyk U, Herbarth O (2011) Influence of sociodemographic factors on *Helicobacter pylori* prevalence variability among schoolchildren in Leipzig, Germany. A long-term follow-up study. Cent Eur J Public Health 19:42–45

Blanchette VS, Price V (2003) Childhood chronic immune thrombocytopenic purpura: unresolved issues. J Pediatr Hematol Oncol 25(Suppl 1):S28–S33

Bodner C, Anderson WJ, Reid TS, Godden DJ (2000) Childhood exposure to infection and risk of adult onset wheeze and atopy. Thorax 55:383–387

Cancer Registry and Statistics [Cited 18 Jan 2019.] Available from: https://ganjoho.jp/reg_stat/statistics/dl/index.html (in Japanese)

Chen Y, Blaser MJ (2008) *Helicobacter pylori* colonization is inversely associated with childhood asthma. J Infect Dis 198:553–560. https://doi.org/10.1086/590158

Çınar A, Sadıç M, Atılgan Hİ, Baskın A, Koca G, Demirel K, Korkmaz M (2015) Prevalence of *Helicobacter Pylori* infection in school and pre-school aged children with C-14 urea breath test and the association with familial and environmental factors. Mol Imaging Radionucl Ther 24:66–70

Daugule I, Karklina D, Rudzite D, Remberga S, Rumba-Rozenfelde I (2016) Prevalence of *Helicobacter pylori* infection among preschool children in Latvia: no significant decrease in prevalence during a ten year period. Scand J Public Health 44:418–422

Daugule I, Karklina D, Remberga S, Rumba-Rozenfelde I (2017) *Helicobacter pylori* infection and risk factors in relation to allergy in children. Pediatr Gastroenterol Hepatol Nutr 20:216–221. https://doi.org/10.5223/pghn.2017.20.4.216

den Hollander WJ, Holster IL, van Gilst B, van Vuuren AJ, Jaddoe VW, Hofman A, Perez-Perez GI, Kuipers EJ, Moll HA, Blaser MJ (2015) Intergenerational reduction in *Helicobacter pylori* prevalence is similar between different ethnic groups living in a Western city. Gut 64:1200–1208

Ding Z, Zhao S, Gong S, Li Z, Mao M, Xu X, Zhou L (2015) Prevalence and risk factors of *Helicobacter pylori* infection in asymptomatic Chinese children: a prospective, cross-sectional, population-based study. Aliment Pharmacol Ther 42:1019–1026

Dixon MF, Genta RM, Yardley JH, Correa P (1996) Classification and grading of gastritis. The updated Sydney system. International workshop on the histopathology of gastritis, Houston 1994. Am J Surg Pathol 20:1161–1181

Feydt-Schmidt A, Kindermann A, Konstantopoulos N, Demmelmair H, Ballauff A, Findeisen A, Koletzko S (2002 Oct) Reinfection rate in children after successful *Helicobacter pylori* eradication. Eur J Gastroenterol Hepatol 14(10):1119–1123

Fouda EM, Kamel TB, Nabih ES, Abdelazem AA (2018) *Helicobacter pylori* seropositivity protects against childhood asthma and inversely correlates to its clinical and functional severity. Allergol Immunopathol (Madr) 46:76–81. https://doi.org/10.1016/j.aller.2017.03.004

Fukase K, Kato M, Kikuchi S, Inoue K, Uemura N, Okamoto S, Terao S, Amagai K, Hayashi S, Asaka M, Japan Gast Study Group (2008) Effect of eradication of *Helicobacter pylori* on incidence of metachronous gastric carcinoma after endoscopic resection of early gastric cancer: an open-label, randomised controlled trial. Lancet 372:392–397

Huang X, Qu X, Yan W, Huang Y, Cai M, Hu B, Wu L, Lin H, Chen Z, Zhu C, Lu L, Sun X, Rong L, Jiang Y, Sun D, Zhong L, Xiong P (2010) Iron deficiency anaemia can be improved after eradication of *Helicobacter pylori*. Postgrad Med J 86:272–278

IARC *Helicobacter pylori* Working Group (2014). *Helicobacter pylori* eradication as a strategy for preventing gastric Cancer. Lyon, France: International Agency for Research on Cancer (IARC Working Group Reports, No. 8). Available from: http://www.iarc.fr/en/publications/pdfsonline/wrk/wrk8/index.php

Jafar S, Jalil A, Soheila N, Sirous S (2013) Prevalence of *helicobacter pylori* infection in children, a population-based cross-sectional study in West Iran. Iran J Pediatr 23:13–18

Jones NL, Koletzko S, Goodman K, Bontems P, Cadranel S, Casswall T, Czinn S, Gold BD, Guarner J, Elitsur Y, Homan M, Kalach N, Kori M, Madrazo A, Megraud F, Papadopoulou A, Rowland M, ESPGHAN NASPGHAN (2017) Joint ESPGHAN/NASPGHAN guidelines for the Management of *Helicobacter pylori* in children and adolescents (update 2016). J Pediatr Gastroenterol Nutr 64:991–1003. https://doi.org/10.1097/MPG.0000000000001594

Kakiuchi T, Matsuo M, Endo H, Nakayama A, Sato K, Takamori A, Sasaki K, Takasaki M, Hara M, Sakata Y, Okuda M, Kikuchi S, Eguchi Y, Takahashi H, Anzai K, Fujimoto KA (2019) A *Helicobacter pylori* screening and treatment program to eliminate gastric cancer among junior high school students in Saga Prefecture: a preliminary report. J Gastroenterol. https://doi.org/10.1007/s00535-019-01559-9. [Epub ahead of print]

Kato M, Asaka M (2010) Recent knowledge of the relationship between *Helicobacter pylori* and gastric cancer and recent progress of gastroendoscopic diagnosis and treatment for gastric cancer. Jpn J Clin Oncol 40:828–837. https://doi.org/10.1093/jjco/hyq119

Kato S, Fujimura S (2010) Primary antimicrobial resistance of *Helicobacter pylori* in children during the past 9 years. Pediatr Int 52:187–190

Kato S, Nishino Y, Ozawa K, Konno M, Maisawa S, Toyoda S, Tajiri H, Ida S, Fujisawa T, Iinuma K (2004) The prevalence of *Helicobacter pylori* in Japanese children with gastritis or peptic ulcer disease. J Gastroenterol 39:734–738

Kato S, Nakajima S, Nishino Y, Ozawa K, Minoura T, Konno M, Maisawa S, Toyoda S, Yoshimura N, Vaid A, Genta RM (2006) Association between gastric atrophy and *Helicobacter pylori* infection in Japanese children: a retrospective multicenter study. Dig Dis Sci 51:99–104

Kienesberger S, Perez-Perez GI, Olivares AZ, Bardhan P, Sarker SA, Hasan KZ, Sack RB, Blaser MJ (2018) When is *Helicobacter pylori* acquired in populations in developing countries? A birth-cohort study in Bangladeshi children. Gut Microbes 9:252–263. https://doi.org/10.1080/19490976.2017.1421887

Kikuchi S, Wada O, Nakajima T, Nishi T, Kobayashi O, Konishi T, Inaba Y (1995) Serum anti-*Helicobacter pylori* antibody and gastric carcinoma among young adults research group on prevention of gastric carcinoma among young adults. Cancer 75:2789–2793

Kimura K, Takemoto T (1969) An endoscopic recognition of the atrophic border and its significance in chronic gastritis. Endoscopy 1:87–97

Konno M, Fujii N, Yokota S, Sato K, Takahashi M, Sato K, Mino E, Sugiyama T (2005) Five-year follow-up study of mother-to-child transmission of *Helicobacter pylori* infection detected by a random amplified polymorphic DNA fingerprinting method. J Clin Microbiol 43:2246–2250

Kotilea K, Kalach N, Homan M, Bontems P (2018) *Helicobacter pylori* infection in pediatric patients: update on diagnosis and eradication strategies. Paediatr Drugs 20:337–351. https://doi.org/10.1007/s40272-018-0296-y

Krueger WS, Hilborn ED, Converse RR, Wade TJ (2015) Environmental risk factors associated with *Helicobacter pylori* seroprevalence in the United States: a cross-sectional analysis of NHANES data. Epidemiol Infect 143:2520–2531. https://doi.org/10.1017/S0950268814003938. Epub 2015 Jan 16

Kusano C, Gotoda T, Ishikawa H, Moriyama M (2017) The administrative project of *Helicobacter pylori* infection screening among junior high school students in an area of Japan with a high incidence of gastric cancer. Gastric Cancer 20(Suppl 1):16–19

Kusano C, Gotoda T, Suzuki S, Ikehara H, Moriyama M (2018) Safety of first-line triple therapy with a potassium-competitive acid blocker for *Helicobacter pylori* eradication in children. J Gastroenterol 53:718–724. https://doi.org/10.1007/s00535-017-1406-2

Mabe K, Okuda M, Kikuchi S, Amagai K, Yoshimura R, Kato M, Sakamoto N, Asaka M, Japan Gast Study Group (2018) Randomized controlled trial: PPI-based triple therapy containing metronidazole versus clarithromycin as first-line treatment for *Helicobacter pylori* in adolescents and young adults in Japan. J Infect Chemother 24:538–543

Malfertheiner P, Megraud F, O'Morain CA, Gisbert JP, Kuipers EJ, Axon AT, Bazzoli F, Gasbarrini A, Atherton J, Graham DY, Hunt R, Moayyedi P, Rokkas T, Rugge M, Selgrad M, Suerbaum S, Sugano K, El-Omar EM, Helicobacter E, Microbiota Study Group and Consensus panel (2017) Management of *Helicobacter pylori* infection-the Maastricht V/Florence consensus report. Gut 66:6–30. https://doi.org/10.1136/gutjnl-2016-312288

Mégraud F, European Paediatric Task Force on *Helicobacter pylori* (2005) Comparison of non-invasive tests to detect *Helicobacter pylori* infection in children and adolescents: results of a multicenter European study. J Pediatr 146:198–203

Molina-Infante J, Gutierrez-Junquera C, Savarino E, Penagini R, Modolell I, Bartolo O, Prieto-García A, Mauro A, Alcedo J, Perelló A, Guarner-Argente C, Alcaide N, Vegas AM, Barros-García P, Murzi-Pulgar-M, Perona M, Gisbert JP, Lucendo AJ, Upper GI Tract Study Group from the Spanish Gastroenterological Association (AEG) (2018) *Helicobacter pylori* infection does not protect against eosinophilic esophagitis: results from a large multicenter case-control study. Am J Gastroenterol 113:972–979. https://doi.org/10.1038/s41395-018-0035-6. Epub 2018 Mar 15

Muhsen K, Cohen D (2008) *Helicobacter pylori* infection and iron stores: a systematic review and meta-analysis. Helicobacter 13:323–340

Nakayama Y, Lin Y, Hongo M, Hidaka H, Kikuchi S (2017) *Helicobacter pylori* infection and its related factors in junior high school students in Nagano Prefecture, Japan. Helicobacter 22. https://doi.org/10.1111/hel.12363

Okuda M, Miyashiro E, Booka M, Tsuji T, Nakazawa T (2007) *Helicobacter pylori* colonization in the first 3 years of life in Japanese children. Helicobacter 12:324–327

Okuda M, Osaki T, Lin Y, Yonezawa H, Maekawa K, Kamiya S, Fukuda Y, Kikuchi S (2015) Low prevalence and incidence of *Helicobacter pylori* infection in children: a population-based study in Japan. Helicobacter 20:133–138

Okuda M, Kikuchi S, Mabe K, Osaki T, Kamiya S, Fukuda Y, Kato M (2017) Nation-wide survey of

Helicobacter pylori treatment for children and adolescents in Japan. Pediatr Int 59:57–61

Okuda M, Nomura K, Kato M, Lin Y, Mabe K, Miyamoto R, Okumura A, Kikuchi S (2019) Gastric cancer in children and adolescents in Japan. Pediatr Int 61:80–86. https://doi.org/10.1111/ped.13720

Oleastro M, Pelerito A, Nogueira P, Benoliel J, Santos A, Cabral J, Lopes AI, Ramalho PM, Monteiro L (2011) Prevalence and incidence of *Helicobacter pylori* infection in a healthy pediatric population in the Lisbon area. Helicobacter 16:363–372

Osaki T, Konno M, Yonezawa H, Hojo F, Zaman C, Takahashi M, Fujiwara S, Kamiya S (2015) Analysis of intra-familial transmission of *Helicobacter pylori* in Japanese families. J Med Microbiol 64:67–73. https://doi.org/10.1099/jmm.0.080507-0

Pacifico L, Osborn JF, Tromba V, Romaggioli S, Bascetta S, Chiesa C (2014) *Helicobacter pylori* infection and extragastric disorders in children: a critical update. World J Gastroenterol 20:1379–1401

Qu X-H, Huang X-L, Xiong P, Zhu CY, Huang YL, Lu LG, Sun X, Rong L, Zhong L, Sun DY, Lin H, Cai MC, Chen ZW, Hu B, Wu LM, Jiang YB, Yan WL (2010) Does *Helicobacter pylori* infection play a role in iron deficiency anemia? A meta-analysis. World J Gastroenterol 16:886–896

Reibman J, Marmor M, Filner J, Fernandez-Beros ME, Rogers L, Perez-Perez GI, Blaser MJ (2008) Asthma is inversely associated with *Helicobacter pylori* status in an urban population. PLoS One 3:e4060. https://doi.org/10.1371/journal.pone.0004060

Rothenbacher D, Inceoglu J, Bode G, Brenner H (2000) Acquisition of *Helicobacter pylori* infection in a high-risk population occurs within the first 2 years of life. J Pediatr 136:744–748

Rowland M, Clyne M, Daly L, O'Connor H, Bourke B, Bury G, O'Dowd T, Connolly L, Ryan J, Shovlin S, Dolan B, Drumm B (2018) Long-term follow-up of the incidence of *Helicobacter pylori*. Clin Microbiol Infect 24:980–984. https://doi.org/10.1016/j.cmi.2017.10.020

Russo G, Miraglia V, Branciforte F, Matarese SM, Zecca M, Bisogno G, Parodi E, Amendola G, Giordano P, Jankovic M, Corti A, Nardi M, Farruggia P, Battisti L, Baronci C, Palazzi G, Tucci F, Ceppi S, Nobili B, Ramenghi U, De Mattia D, Notarangelo L, AIEOP-ITP Study Group (2011) Effect of eradication of *Helicobacter pylori* in children with chronic immune thrombocytopenia: a prospective, controlled, multicenter study. Pediatr Blood Cancer 56:273–278. https://doi.org/10.1002/pbc.22770

Sasaki H, Sasano H, Ohi R, Imaizumi M, Shineha R, Nakamura M, Shibuya D, Hayashi Y (1999) Adenocarcinoma at the esophageal gastric junction arising in an 11-year-old girl. Pathol Int 49:1109–1113

Shu X, Ping M, Yin G, Jiang M (2017) Investigation of *Helicobacter pylori* infection among symptomatic children in Hangzhou from 2007 to 2014: a retrospective study with 12,796 cases. PeerJ 5:e2937. https://doi.org/10.7717/peerj.2937

Silva GM, Silva HM, Nascimento J, Gonçalves JP, Pereira F, Lima R (2018) *Helicobacter pylori* antimicrobial resistance in a pediatric population. Helicobacter 23:e12528. https://doi.org/10.1111/hel.12528

Sivapalasingam S, Rajasingham A, Macy JT, Friedman CR, Hoekstra RM, Ayers T, Gold B, Quick RE (2014) Recurrence of *Helicobacter pylori* infection in Bolivian children and adults after a population-based "screen and treat" strategy. Helicobacter 19:343–348. https://doi.org/10.1111/hel.12137

Take S, Mizuno M, Ishiki K, Nagahara Y, Yoshida T, Yokota K, Oguma K (2007) Baseline gastric mucosal atrophy is a risk factor associated with the development of gastric cancer after *Helicobacter pylori* eradication therapy in patients with peptic ulcer diseases. J Gastroenterol 42(Suppl 17):21–27

Take S, Mizuno M, Ishiki K, Hamada F, Yoshida T, Yokota K, Okada H, Yamamoto K (2015) Seventeen-year effects of eradicating *Helicobacter pylori* on the prevention of gastric cancer in patients with peptic ulcer; a prospective cohort study. J Gastroenterol 50:638–644

Thomas JE, Dale A, Harding M, Coward WA, Cole TJ, Weaver LT (1999) *Helicobacter pylori* colonization in early life. Pediatr Res 45:218–223

Treepongkaruna S, Sirachainan N, Kanjanapongkul S, Winaichatsak A, Sirithorn S, Sumritsopak R, Chuansumrit A (2009) Absence of platelet recovery following *Helicobacter pylori* eradication in childhood chronic idiopathic thrombocytopenic purpura: a multicenter randomized controlled trial. Pediatr Blood Cancer 53:72–77. https://doi.org/10.1002/pbc.21991

Tsang KW, Lam WK, Chan KN, Hu W, Wu A, Kwok E, Zheng L, Wong BC, Lam SK (2000) *Helicobacter pylori* sero-prevalence in asthma. Respir Med 94:756–759

Uemura N, Okamoto S, Yamamoto S, Matsumura N, Yamaguchi S, Yamakido M, Taniyama K, Sasaki N, Schlemper RJ (2001) *Helicobacter pylori* infection and the development of gastric cancer. N Engl J Med 345:784–789

Ünlüsoy Aksu A, Yılmaz G, Eğritaş Gürkan Ö, Sarı S, Dalgıç B (2018) The effect of *Helicobacter pylori* eradication on functional dyspepsia in Turkish children. Helicobacter 23:e12497. https://doi.org/10.1111/hel.12497

Villarreal-Calderon R, Luévano-González A, Aragón-Flores M, Zhu H, Yuan Y, Xiang Q, Yan B, Stoll KA, Cross JV, Iczkowski KA, Mackinnon AC Jr (2014) Antral atrophy, intestinal metaplasia, and preneoplastic markers in Mexican children with *Helicobacter pylori*-positive and *Helicobacter pylori*-negative gastritis. Ann Diagn Pathol 18:129–135. https://doi.org/10.1016/j.anndiagpath.2014.02.003

Wang C, Nishiyama T, Kikuchi S, Inoue M, Sawada N, Tsugane S, Lin Y (2017) Changing trends in the prevalence of *H. pylori* infection in Japan (1908-2003): a systematic review and meta-regression analysis of 170,752 individuals. Sci Rep 7:15491. https://doi.org/10.1038/s41598-017-15490-7

Wangda S, Richter JM, Kuenzang P, Wangchuk K, Choden T, Tenzin K, Malaty HM (2017) Epidemiology of *Helicobacter pylori infection* in asymptomatic schoolchildren in Bhutan. Helicobacter 22. https://doi.org/10.1111/hel.12439

Wu MC, Sung CH, Chang YC, Ho CL, Wu CC, Wu KH, Lee CY, Yang KD (2015) Seroprevalence of *Helicobacter pylori* and hepatitis A virus among children in rural central Taiwan. Jpn J Infect Dis 68:494–503

Yokota S, Konno M, Fujiwara S, Toita N, Takahashi M, Yamamoto S, Ogasawara N, Shiraishi T (2015) Intrafamilial, preferentially mother-to-child and intraspousal, *Helicobacter pylori* infection in Japan determined by mutilocus sequence typing and random amplified polymorphic DNA fingerprinting. Helicobacter 20:334–342. https://doi.org/10.1111/hel.12217

Yuan W, Li Yumin D, Yang L (2010) Iron deficiency anemia in *Helicobacter pylori* infection: meta-analysis of randomized controlled trials. Scand J Gastroenterol 45:665–676

Zabala Torrres B, Lucero Y, Lagomarcino AJ, Orellana-Manzano A, George S, Torres JP, O'Ryan M (2017) Review: prevalence and dynamics of *Helicobacter pylori* infection during childhood. Helicobacter 22. https://doi.org/10.1111/hel.12399

Zhou Y, Ye Z, Huang J, Huang Y, Yan W, Zhang Y (2018) High prevalence and low spontaneous eradication rate of *Helicobacter pylori* infection among schoolchildren aged 7-12 years. Acta Paediatr 107:1624–1628. https://doi.org/10.1111/apa.14387

Adv Exp Med Biol - Advances in Microbiology, Infectious Diseases and Public Health (2019) 1149: 121–134
https://doi.org/10.1007/5584_2019_362

Published online: 24 April 2019

Non-malignant *Helicobacter pylori*-Associated Diseases

Christina Falkeis-Veits and Michael Vieth

Abstract

Helicobacter pylori infection of the human stomach is associated with chronic gastritis, peptic ulcer disease or gastric carcinoma, and thus a high burden for the public health systems worldwide. Fortunately, only a small subfraction of up to 15–20% of infected individuals will develop serious complications. Unfortunately, it is not always known upfront, who will be affected by serious diesease outcome. For risk stratifications, it is therefore necessary to establish a common terminology and grading system, that can be applied worldwide to compare population data. The updated Sydney System for classification of gastritis with its semi-quantitative analogue scale is the system, that is currently used worldwide. Additionally, pathologists should always try to classify the etiology of the inflammatory infiltrates in the stomach to instruct the clinicians for choosing a proper treatment regime. Risk factors such as intestinal metaplasia, atrophy and scoring systems to classify these risk factors into a clinical context such as OLGA and OLGIM are discussed. Also, special forms of gastritis like lymphocytic gastritis, autoimmune gastritis and peptic ulcer disease are explained and discussed e.g. how to diagnose and how to treat. Extra-gastric sequelae of *H. pylori* infections inside and outside the stomach are shown in this chapter as well. Important host and bacterial risk factors such as pathogenicity islands are dicussed to draw a complete landscape around a *H. pylori* infection, that still can be diagnosed in patients. However, it needs to be noted that some countries have almost no *H. pylori* infection anymore, while others have still a very high frequency of infections with or without serious complications. The understanding and application of risk assessements may help to save money and quality of life. Extra-gastric *H. pylori* infections are rarely reported in the literature until today. The pathogenitiy is still under debate, but especially in the bile ducts and gallbladder, several pathological conditions may be also based on *H. pylori* infection, and will be also discussed.

Keywords

Helicobacter pylori · Gastritis · Sydney system · Atrophic gastritis

C. Falkeis-Veits and M. Vieth (✉)
Institute of Pathology, Klinikum Bayreuth, Bayreuth, Germany
e-mail: christina.falkeis@klinikum-bayreuth.de; vieth.lkpathol@uni-bayreuth.de

1 *Helicobacter pylori* Gastritis

1.1 The Sydney Classification System

H. pylori is a spiral-shaped, Gram-negative bacterium (Fig. 1a), that specifically colonizes

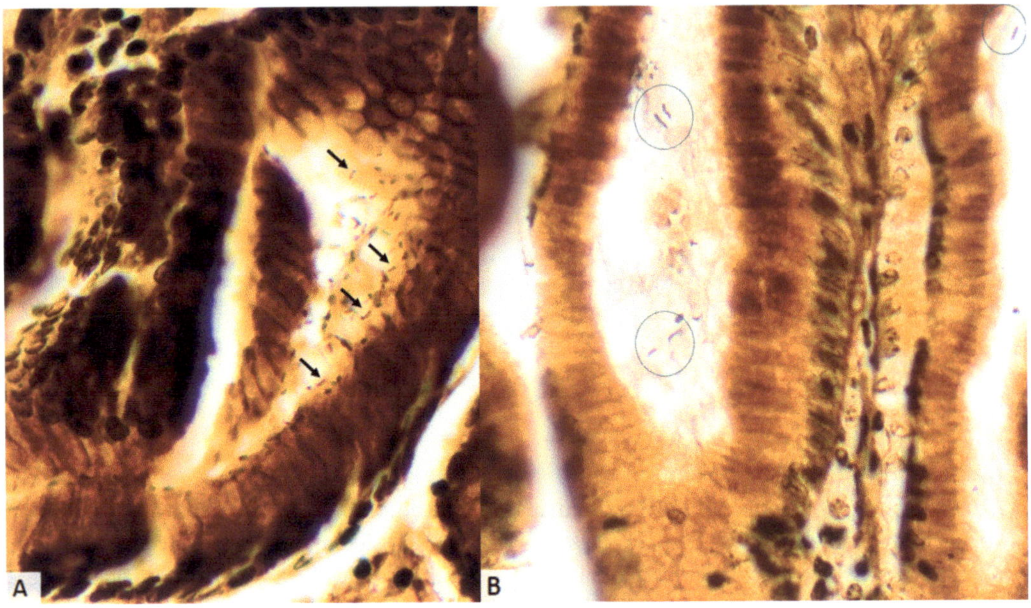

Fig. 1 Microscopic analysis of different *Helicobacter* species (Warthin-Starry stain, 400x). (**a**) *Helicobacter pylori* are small bacteria (3–5 μm), often found attached to the foveolar epithelium (arrows). (**b**) Non-pylori *Helicobacter* are formerly known as *Helicobacter heilmannii* (circles) are longer and more spiraled than *H. pylori* and tend to stay in the foveolar lumen

the gastric epithelium in about 50% of the human world population (Kusters et al. 2006). This infection induces a state of chronic inflammation due to highly sophisticated mechanisms by *H. pylori* crosstalk with the host immune system (Pachathundikandi et al. 2016). For a long time, the classification and grading of gastritis was sort of an open field. Until a meeting in Sydney in 1986, and another meeting in Houston in 1994, there were mainly two ideas how to classify gastritis (Dixon et al. 1996). One concept was just to detect and describe inflammatory reactions and cells in various versions, whereas especially in Germany a system was proposed that focused more on the etiology of the inflammatory infiltrates (Fig. 2). The idea behind the second system was that clinicians can use a scheme of diagnosis and etiology of inflammation better than just the knowledge, that there are a certain numbers of neutrophils somewhere in the samples. This approach is coming from a background that the strength of a pathologist is to give etiological causes for given pathological changes, and that this increases the precision of diagnostics, and thus clinical impact for potential treatments and the clinical course. When the original Sydney System was proposed in 1986, there was also an endoscopic part, but it turned out that the classification of gastritis is a sole histological diagnosis and endoscopy cannot reliably generate reproducible descriptions or diagnoses. Thus, the endoscopic approach was dropped over the years and classification of gastritis became a sole histological diagnosis.

The really new proposals of the updated Sydney System were the introduction of mandatory 4-tiered semiquantitative analog scales for antrum and corpus mucosa, scoring the number of lymphocytes/plasma cells as expression of chronic inflammation, the number of neutrophils as expression of acute inflammation, the density of *Helicobacter* bacteria on the mucosal surfaces, assessing intestinal metaplasia and grading atrophy (Dixon et al. 1996). Traditionally, the semiquantitative grades are stated as "none", "mild", "moderate" or "severe", respectively.

Bystanders such as lymphoid aggregates and follicles, mucous depletion and reactive epithelial

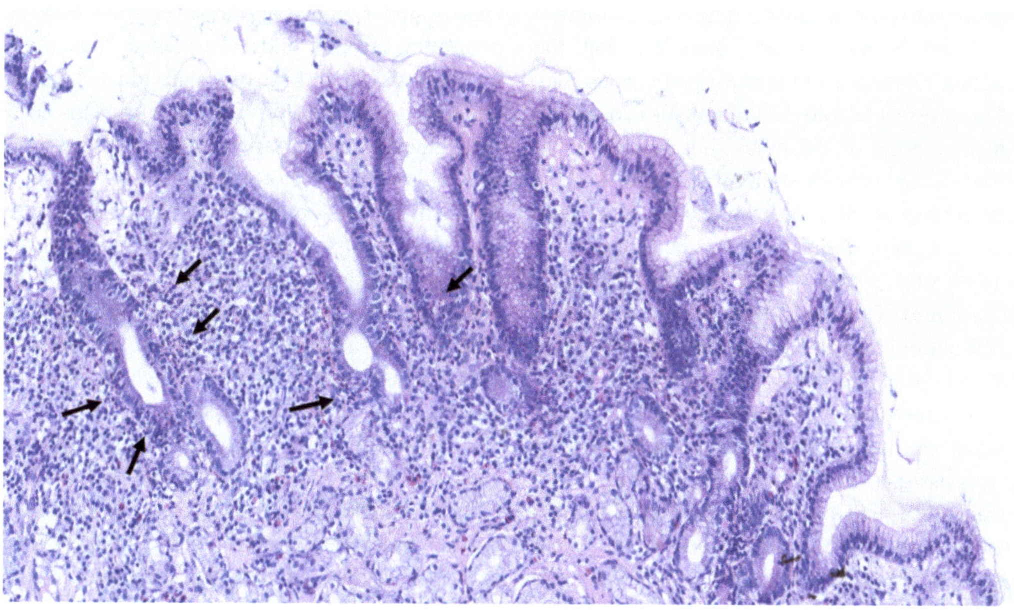

Fig. 2 Histology showing typical moderate, moderately active *H. pylori* gastritis. The band – like lymphoplasmocytic infiltrate can be found in the upper half of the mucosa. Neutrophilic infiltrates in the surface epithelium are marked with arrows

changes were seen as optional criteria. It was recommended to name the etiology of the scored criteria after grading of the chronic and acute inflammatory infiltrates. Thus, the fathers of the Sydney System managed to create a combination of the two main scoring classifications used at that time. This helped to spread the use of the Sydney System worldwide and led to a uniform way to classify gastritis within a very short time period.

1.2 Intestinal Metaplasia and Atrophy

Intestinal metaplasia (IM) in the stomach is defined as goblet cell bearing metaplasia either of colonic type or small intestinal type, the latter including Paneth cells (Giroux and Rustqi 2017). This is seen in sites of mucosal damage of any type including chemical-reactive gastritis and autoimmune gastritis, and can be found both in the antrum and corpus mucosa. In many cases it does accompany gastric atrophy, which is defined as a loss of gastric glands, a hallmark of the pathological pattern. The goblet cells in IM are known to produce Mucin-2, which hampers the ability of *H. pylori* to adhere to the epithelium, and thus in cases with severe IM often no bound *H. pylori* can be detected anymore. The odds ratios (ORs) for IM and atrophy are to be seen around 5–7, whereas first degree relatives of patients with gastric cancer have an OR of 10, individuals with pan-gastritis with equal inflammatory infiltrates in the antrum and the corpus show an OR of about 15 to develop a gastric carcinoma. The 10 years risk for developing cancer in the stomach with IM and/or atrophy is less than 1.8% (Reddy et al. 2016). In contrast, the OR of a so-called corpus predominant gastritis with more severe inflammatory infiltrates in the corpus is believed to reach almost 35 for malignant transformation. The OR numbers show how the risk for patients with IM and/or atrophy has to be compared with other risk factors, that are much higher, and it is well known that the frequency of IM and atrophy decreased enormously since the 1970s as compared to data from today's populations.

In 1994, an update of the Sydney System during the Houston workshop has shown that the biopsy protocol should be changed in a way that biopsies of the incisura are not obligate anymore (Dixon et al. 1996). The only shortcoming of the Sydney System may be that the requested biopsies from antrum and corpus may lead to a correct histological diagnosis in a certain high percentage of the cases, but patchiness of infiltrates can make it sometimes rather bothersome to identify the correct diagnosis, especially since the recommended sites for the biopsies are not those sites, where most of the IM and atrophy can be found. This leads to the idea that atrophy and IM should be graded separately, also because inter-and intraobserver variations of grading atrophy are somewhat poor with the Sydney System. Separate grading systems like operative link for gastritis assessment (OLGA) (Rugge et al. 2005) and operative link on IM assessment (OLGIM) (Capelle et al. 2010) were introduced in recent years, but made the situation somewhat more complicated, than it actually was. The shortcomings of poor inter-and intraobserver variations were not improved.

The observation that patients with multifocal atrophy and/or IM may have a higher risk for malignant transformation than those without is valid for *H. pylori* induced gastritis only. However, OLGA and OLGIM are applied to all various etiologies of gastritis leading to the situation that young patients with chemical reactive gastritis and foci of IM and/or atrophy get high scores in OLGA and/or OLGIM, suggesting a high risk for malignant transformation and regular endoscopic follow-up. It is also known for a long time that patients with duodenal ulcer during the course of *H. pylori* infection have a very low risk of malignant transformation, but can end up in high OLGA and OLGIM scores. A vice versa effect is known for patients with autoimmune gastritis in corpus and normal antrum mucosa showing low scores in OLGA and OLGIM leading to the recommendation that no follow-up is required, although they have higher prevalence of malignancy, especially for gastric neuroendocrine tumors (NET), squamous cell carcinoma of the esophagus and distal gastric adenocarcinoma (Ye and Nyren 2003).

Unfortunately, protagonists of OLGA and OLGIM seem to have dominated guideline meetings and forced OLGA and OLGIM into most European guidelines. The problem with OLGA and OLGIM is, however, that they are overrating atrophy and IM in a way that mixes up high risk and low risk individuals and are underrating the etiology of gastritis. This leads to frightened patients in the end and has been shown no benefit in form of number of saved lives so far, and probably never will. Besides this, there is also some criticism concerning the methods used against OLGA and OLGIM, evaluating the risk for gastric cancer development. It is also not clear why OLGA and OLGIM use four stages to conclude two risk groups, low risk and high risk.

Additionally, the widely used term "atrophic gastritis" is used in the literature for *H. pylori* induced gastric atrophy as well as autoimmune gastritis associated atrophy, in some papers there isn't even a distinction between the two. Sometimes even atrophy during the course of or due to chemical reactive gastritis may have been included in those studies. This situation leads to reports of cancer risks and survival rates of a mixed patient cohort since etiology and thus risk for malignant transformation of atrophy due to *H. pylori* infection and autoimmune gastritis may differ.

In recent studies, growing skepticism was raised towards the impact of IM on gastric cancer, due to the fact that it's not the IM that progresses to gastric cancer, but the destabilization of a gastric stem cell, which causes malignant transformation and growth of malignant cells (Graham and Zou 2018).

In conclusion, the Sydney System should still be used for all assessments of gastric specimen and the etiology of gastritis, always seen as integral part of the histological diagnosis. All further classifications such as OLGA and OLGIM should not be used in routine cases. The latter is against European and national guidelines, but makes a lot of sense due to the shortcomings of these additional scoring systems.

1.3 The Individuality of *H. pylori* Gastritis

H. pylori infection leads to a combined active and chronic inflammation in the gastric mucosa. The density of acute and chronic inflammatory infiltrates and their distribution throughout the stomach shows high variability of changes depending on *H. pylori* presence, duration of the infection, but also modifications due to medications like acetylsalicylic acids (ASA) or proton pump inhibitors (PPI). Both groups of drugs lower the number of *H. pylori* on the epithelium both in antrum and corpus. PPI medication often normalizes the antrum mucosa, whereas the status of the corpus mucosa switches to a corpus predominant gastritis with severe active inflammatory infiltrates within the corpus mucosa. This is the rationale why in long term PPI treatment a *H. pylori* infection should be excluded since there is an elevated risk for malignant transformation in individuals with predominant corpus gastritis, even when there is no such case described under PPI therapy until now in the literature (Sipponen and Stolte 1997). In ASA medication, the antrum can also normalize, but the corpus doesn't reveal more severe inflammatory infiltrates. This shows why always two biopsies from the antrum and the corpus are recommended by the updated Sydney System to ensure a high probability of a correct histological assessment of the status of the gastric mucosa with the correct given etiology in case there are any inflammatory infiltrates.

Besides the variability of immune reactions in different patients, the variability of the disease can also be explained by the presence of *H. pylori* strains that contain different factors of virulence. As discussed in this book, not all *H. pylori* strains possess the *cag*A gene, whereas all have the *vac*A gene, but not all isolates express the more virulent *vacA* s1/m1 allele product. The presence of the *cag*A gene usually is a marker for *cag* pathogenicity island, which leads to the translocation of the CagA protein (Backert et al. 2015). Injected CagA can then be tyrosine-phosphorylated by Src and Abl kinases, which induces a change in cell morphology and provokes pro-inflammatory and mitogenic responses (Naumann et al. 2017). These genes are expressed in differing combinations in the various *H. pylori* strains throughout the world (Mueller et al. 2012). In eastern Asia nearly all *Helicobacter* strains express the *cag*A gene, regardless of developing certain diseases. So this alone is not the solution to the individually different reactions to *H. pylori* infection.

There are also differences in the CagA protein, for example the number and configuration of the tyrosine phosphorylation motifs (EPIYA) at the carboxy-terminal end of the protein (Lind et al. 2014, 2016). Of the four known EPIYAs, the EPIYA-type C is predominantly found in Western countries, whereas EPIYA type D almost exclusively exists in East Asia. On the other hand, the *vacA* gene includes three alleles with slightly differing codes in different strains, forming proteins of varying virulence (Posselt et al. 2013).

Of course, this is only the tip of the iceberg. Many more genetic variations have been found in the different *H. pylori* strains across the world. Maybe these discriminations are leading to the eminently varying rates of gastric ulcer and gastric cancer and other complications of a *H. pylori* infection in different countries (da Costa et al. 2015), and should be investigated in more detail in future.

1.4 *H. pylori* vs. Other *Helicobacter* Species

It took until 1984 when Marshall and Warren discovered the pathogenic nature of *H. pylori* in the human stomach (Marshall and Warren 1984). Even that the WHO graded *H. pylori* as a carcinogen of class I, this didn't lead to a complete stop of the discussion how pathogenic *H. pylori* really is. The arguments and facts against its cancerogenic potential are, however, rather minor.

It is now known for around two decades that mixed infections of the human stomach can occur by different *Helicobacter* species, most of them usually colonizing also cats and dogs, and thus could be seen as a zoonotic disease. Besides *H. pylori*, more spiral-shaped *Helicobacter* species have been sporadically detected in the human stomach, and were named *Helicobacter heilmannii* (Fig. 1b). Gene sequencing revealed that *H. heilmannii* does not represent one, but several different species. *H. heilmanii* type I is similar to *Helicobacter suis*, whereas *H. heilmannii* type II represents different species like *Helicobacter felis*, *Helicobacter bizzozeronii*, *Helicobacter salomonis* and *Candidatus H. heilmannii* (Kubota-Aizawa et al. 2017). Other non-*pylori Helicobacter* species have also been found in the human stomach, like *Candidatus Helicobacter bovis*, *Candidatus H. suis*, *Helicobacter cinaedi* and many more (Bauwens et al. 2018; Smet et al. 2018). Mixed infections with more than one *Helicobacter* species have been found by RNA analysis in up to 10% of the infections (De Groote et al. 2005).

Morphologically, it is difficult to detect these co-infections since *Helicobacter* harbors an enormous capacity of adapting its shape. Sole "*Helicobacter heilmannii*" infections can be identified on the base of longer shape and a more pronounced spiral form. But as shown above, *H. heilmannii* is more a morphological description than a species diagnosis and should rather be stated as "*Helicobacter heilmannii*-like organisms (HHLO)" (Goji et al. 2015) or "non-*Helicobacter pylori Helicobacter* (NHPH) (De Groote et al. 2005)". Fortunately, this seems not to play any role for an eradication therapy, since all these *Helicobacter* species in the stomach are susceptible for the regular antibiotic treatment regimen, unless there are antibiotic resistances. One major difference between *H. heilmannii* and *H. pylori*, however, lies in the discussion whether *H. heilmannii* shows an increased risk for low grade MALT lymphoma compared to *H. pylori,* but this is still an ongoing discussion in the community.

It needs to be noted, however, that according to textbooks so called coccoid forms of *Helicobacter* should not be diagnosed without "normal" forms of *Helicobacter* bacteria in the gastric mucosa aside, since the coccoid forms are not believed to be viable especially with no "normal shaped" *Helicobacter* bacteria being around.

1.5 Special Forms of Gastritis

1.5.1 Atrophic Gastritis

As mentioned above, an ongoing *H. pylori* gastritis can lead to atrophy of both antrum and corpus mucosa, also known as gastritis with atrophy. The atrophic gastric mucosa can also show metaplastic changes such as IM and pancreatic-acinar metaplasia or fibrosis. In total corpus atrophy the elimination of the parietal cells leads to iron deficiency anemia first in the clinical course, hypochlorohydria and later within decades to Vitamin B12 deficiency. Multifocal atrophy of stomach mucosa is associated with higher risk for gastric adenocarcinoma (Rugge et al. 2005).

Eradication of *Helicobacter* can stop and sometimes reverse atrophy in the corpus mucosa, but not in antrum mucosa. IM also stays after eradication therapy and at least does not progress (Wang et al. 2011).

1.5.2 Autoimmune Gastritis

Autoimmune gastritis is caused by autoantibodies against parietal cells (PCAs) (Fig. 3). Those antibodies have been reported to be present in 19.5% of patients recruited during a general health check at a general practitioners' office. The numbers are increasing with *H. pylori* infection and age (Kulnigg-Dabsch 2016). PCAs are reactive against the proton pump mechanism in gastric parietal cells. A molecular similarity of the proton pump and *H. pylori* antigens has been reported, but it is unclear, which role the bacterium really plays in the pathogenesis of autoimmune gastritis. Destruction of the parietal cells leads to atrophy of the corpus mucosa with IM (Fig. 4). Individuals with autoimmune gastritis have not an elevated risk for gastric carcinoma, but reveal an increased rate of neuroendocrine tumors. Autoimmune gastritis has a different risk profile, etiology and even microbiome than *H. pylori* induced gastritis with atrophy and should be distinguished from the latter (Parsons et al. 2017). Many recent studies on atrophy and IM combine often all kinds of atrophy and IM, and do not distinguish between the etiologies. Thus, this is the problem with the reporting systems OLGA and OLGIM, as mentioned above.

1.5.3 Ex-*Helicobacter*-Gastritis

A very controversial form of gastritis is the so-called Ex-*H. pylori*-gastritis or Post-*H. pylori*-gastritis (Fig. 5), aiming to charcterize

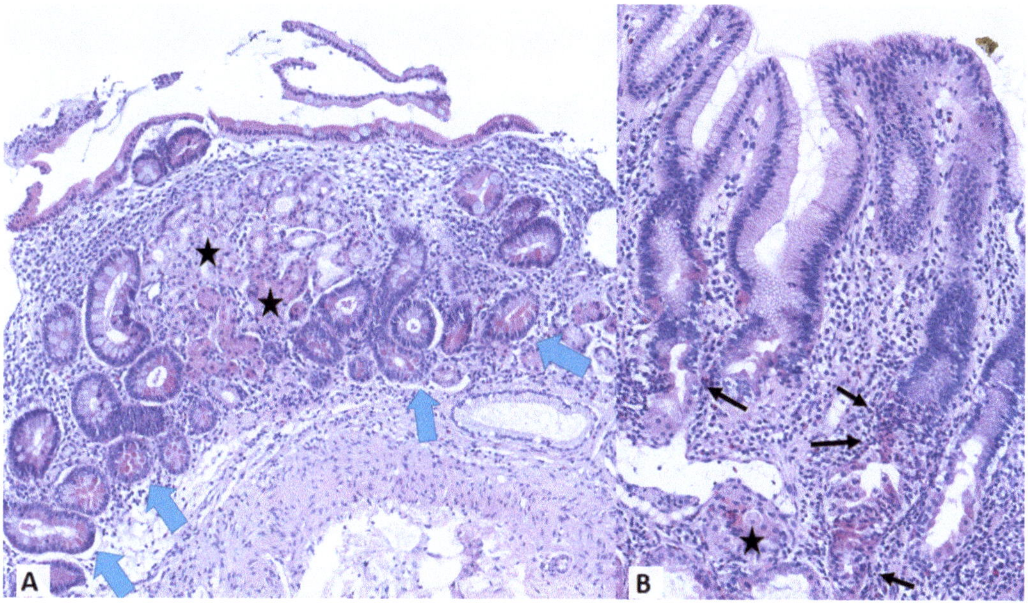

Fig. 3 Histology of autoimmune gastritis. (**a**) Autoimmune gastritis with severe atrophy and focal remnants of oxyntic mucosa are indicated (stars). Intestinal metaplasia with Paneth cell metaplasia (blue arrows). (**b**) Autoimmune gastritis with active inflammation (black arrows) and destruction of oxyntic glands (star)

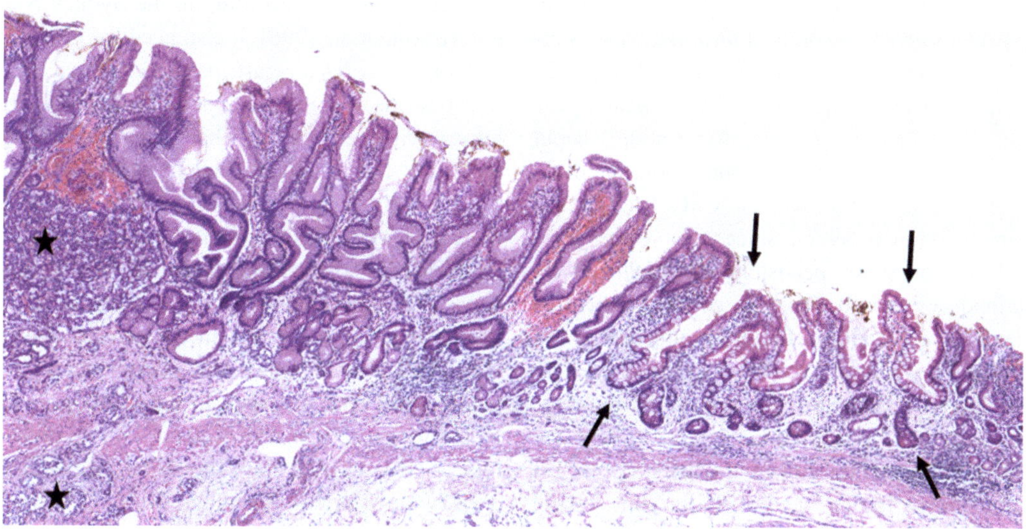

Fig. 4 Histology of autoimmune gastrisis with complete atrophy and intestinal metaplasia (arrows). Neuroendocrine tumor nests are indicated (stars)

the state after successful *Helicobacter* eradication therapy (Oberhuber and Haidenthaler 2000; Livzan et al. 2004). There are no convincing sets of histological datasets in the literature, giving precise criteria how to diagnose. Some pathologists even deny the existence of such a

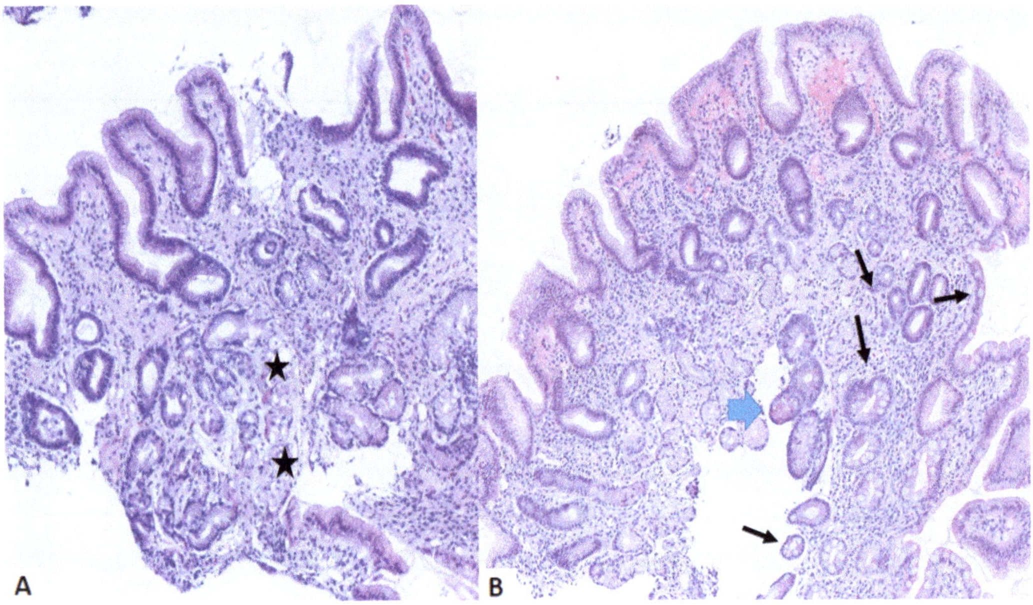

Fig. 5 Histology of post-*H. pylori* gastritis. (**a**) Nearly complete atrophy in corpus mucosa. Focal remnants of oxyntic glands are marked (stars).(**b**) Severe atrophy in the antrum mucosa with intestinal metaplasia (black arrows) including metaplastic Paneth cells (blue arrow)

gastritis form. Many colleagues refuse to use any specific term for a stomach, that has undergone successful eradication since for the vast majority of patients no precise clinical information is available regarding whether an eradication therapy has been undertaken and when. Most pathologists also like to have an information about *Helicobacter* serology to be sure about the diagnosis. But even this request will always leave some individuals behind, since the so-called serological scar (with persistent anti-*Heliocbacter* antibodies in serum after a successful eradication therapy) as a marker of a prior infection is not present in every patient after successful eradication therapy, since the duration of the presence may vary individually. It needs also to be taken into account, that it is not clear when a stomach is completetly "healed" and with no pathological changes anymore after eradication therapy. What most pathologists deny is, that one can diagnose such a form of gastritis. On the other hand with some experience, pathologists can very well diagnose an Ex-*Helicobacter*-gastritis: the number of lymphocytes and plasma cells is still mildly elevated with basal lymphoid aggregates or remnants of lymphoid follicles in antrum and corpus. Antrum and corpus biopsies show a mild not active gastritis according to the Sydney System (Dixon et al. 1996). It also needs to be noted that Ex-*Helicobacter*-gastritis and chemical reactive gastritis and even stomachs with no pathological changes may be hard to differentiate. Here, the corpus mucosa helps: when there are basal lymphoid aggregates or follicles present in the corpus of adults the pathologist could go for Ex-*Helicobacter*-gastritis, whenever the corpus shows no pathological changes most likely a chemical reactive gastritis in the antrum, which can be diagnosed unless the antrum shows no pathological changes, also. In the latter case, we would diagnose a gastric mucosa with no pathological changes. Attention should be paid in children with lymphocytic aggregates in corpus mucosa since these are considered physiological. From our daily routine we know, that there are some individuals in whom the stomach normalizes very fast after succuessful eradication, whereas in some others the picture of Ex-*Helicobacter*-gastritis stays for decades. Thus, it seems that there is a very individual variance of the duration of the histological picture.

The rationale why a pathologist should always try to diagnose a Ex-*Helicocater*-gastritis lies in the fact, that there is a point of no return for the development of a gastric carcinoma, and thus the risk of developing gastric malignancy even after successful eradication still persists to a somewhat smaller degree, and thus is different from a gastric mucosa that never harbored an *Helicobacter* infection. The argument that it cannot be diagnosed in all potential cases with certainty and with no serum titres or much clinical information about a possible eradication therapy does not really count since antibiotic treatment given for other reasons than *Helicobacter* may have led to eradication of the bacteria in the stomach, also.

1.5.4 Lymphocytic Gastritis

Lymphocytic gastritis shows an increase of lymphocytes in the lamina propria of gastric mucosa and additionally elevated counts of intraepithelial lymphocytes (IEL) above 25 per 100 epithelial cells (Fig. 6). In about a third of patients lymphocytic gastritis is associated with celiac disease, another third with *H. pylori* infection and the other patients have various causes for lymphocytic gastritis like varioliform gastritis or Crohn's disease. Unlike lymphocytic gastritis in celiac disease there is no antrum dominant inflammation reported in the *H. pylori* associated cases. The lymphocytes usually are positive for CD3 and CD8. Many patients with *H. pylori*-induced lymphocytic gastritis show only a low count of bacteria, in some there is only serological proof of the infection (Madisch et al. 2006).

Together with the lymphocytosis patients commonly develop atrophy in antrum and corpus mucosa and foveolar hyperplasia. Those also show increased proliferation in the foveolar epithelium. After eradication therapy even patients who were initially negative for *H. pylori* have been reported to improve serologically and histologically with improvement of mucosal atrophy in the corpus but not in the antrum mucosa and lower epithelial proliferation (Mäkinen et al. 2003). Thus, lympocytic gastritis is seen as a special form of *Helicobacter* infection that should be treated even if the bacteria cannot be detected morphologically.

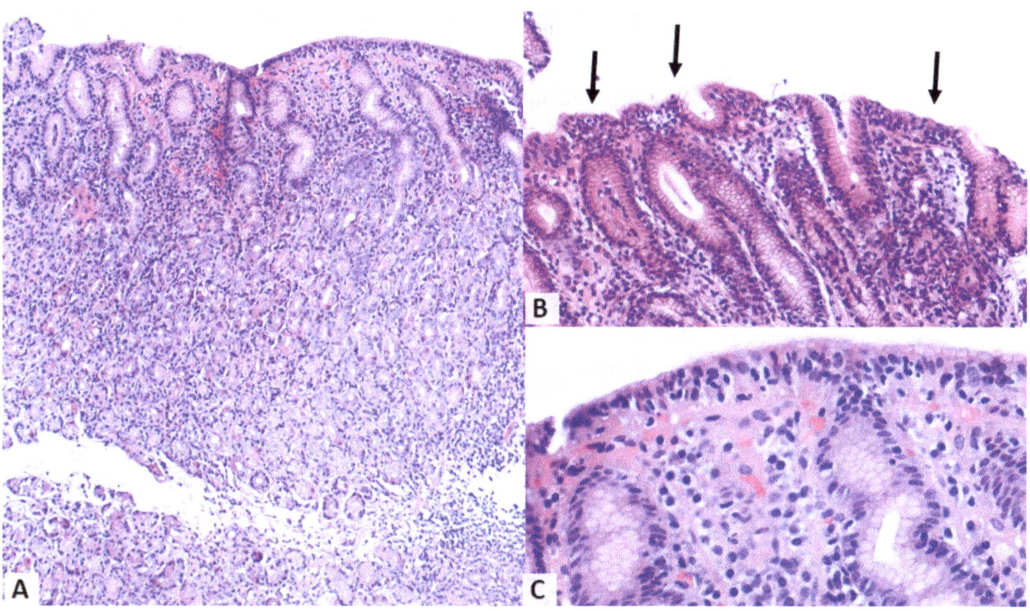

Fig. 6 Histology of lymphocytic gastritis. (**a**) Overview showing mononuclear cells in the mucosa. (**b**) Dense inflammatory infiltrates in the surface epithelium are marked with arrows. (**c**) Dense lymphocytic infiltration. The darker, variously shaped nuclei belong to lymphocytes, the lighter, oval shaped to epithelial cells. More than 25 lymphocytes per 100 epithelial cells are needed for the diagnosis

2 *H. pylori*-Induced Diseases Other Than Gastritis

2.1 Gastro-Esophageal Reflux Disease

It is known for a long time, that *H. pylori* is capable to buffer the gastric acid by its ammonium secretion leading to a higher pH than without *Helicobacter*. The whole discussion on positive effects of a *H. pylori* infection began, when Labenz and co-workers (1997) published a study that was never intended to answer the question whether eradication may provoke gastro-esophageal reflux disease. Basically, it was demonstrated that about 25.8% of the patients cured from *H. pylori* infection in case of a duodenal ulcer will develop reflux disease within 3 years and only 12.9% in case the infection was ongoing. Risk factors identified are male sex, severity of corpus gastritis, and weight gain. Several comments and studies were following, some of which were partly supporting, others partly denying it. Progressive studies even suggested not to treat any *H. pylori* infection since this may escape from the reflux disease-Barrett-adenocarcinoma-sequence. When looking unemotionally at the numbers, it becomes very clear, that the whole discussion makes no sense at all. It needs to be noted that *H. pylori* eradication doesn't induce reflux disease, but may unmask a prior existing reflux disease.

In Germany, with a population of 85 million inhabitants, there is an annual incidence of approximately 21,000 gastric carcinomas compared to 2,500 Barrett adenocarcinomas. Even if an eradication therapy against *H. pylori* would somehow trigger the gastro-esophageal reflux disease, we theoretically could "save" 2,500 lives from Barrett's carcinoma, but would rather lose 21,000 according to gastric carcinoma. As albeit gastro-esophageal reflux disease is only unmasked by eradication therapy, we theoretically have the chance to save not 21,000 people, only from gastric cancer but also 2,500 from Barrett's carcinoma, because unmasked reflux disease can be treated easily by standard medications.

2.2 Gastric and Duodenal Ulcer

In the 1960s, it was well known, that the risk of peptic ulcer in the stomach and the duodenum was still rising, but it was also known that there was a cohort phenomenon e.g. in the British population. Those born between 1870 and 1890 had the highest risk for ulcer disease. Susser and co-workers (1962) speculated that the affected individuals grew older and since their proportion in the population declined over time, and they were able to predict correctly from their birth-cohort analysis of peptic ulcer mortality in England and Wales between 1900 and 1950. The future decline of peptic ulcer disease correctly appeared 10–20 years prior to its actual occurrence. A birth cohort phenomenon suggests an influence by acquired relevant risk factors of a disease that occurs early in life (besides genetic bases). The marked changes of gastric and duodenal ulcer disease over time indicate that their presence was influenced by exogenous risk factors. Comparing frequencies, age and birth dates from numerous countries in 1987 showed, that the environmental factors must have taken effect prior to the age of 15 (Sonnenberg 1987). At that time, *H. pylori* was not yet discussed but could explain the findings very well. Thus, a lot of the disease burden and effects of a *Helicobacter* infection were already known before *H. pylori* was first described in 1984 (Warren and Marshall 1984).

2.3 Extra-Gastric *Helicobacter* Infections

Recently, there were cases in our routine biopsy practice showing no pathological changes of the gastric mucosa, but detection of *H. pylori* bacteria in a stool test was noted. It is known that the stool test is very sensitive (Iannone et al. 2018). Since this was seen independently across several laboratories, these observations lead to the idea that *Helicobacter* species may also colonize the human body outside the stomach. The clinical implications and whether there is a need to treat the infection, and if yes, how, still remains unclear. Interestingly, numerous case reports mainly in Asian populations, but also in European, have described that *H. plyori* can be found in the gallbladder, sometimes in conjunction with or without *Helicobacter* in the stomach. A reason for these sparse reports may be that gallbladders are not routinely checked for *H. pylori* or colonization by other bacteria. In addition, it is known that *Helicobacter* may colonize gastric heterotopias in the esophagus and small bowel (Meckel's diverticulum) and rectum (Dye et al. 1990), leading there to rare ulcerations and (very rarely) to malignant transformations (Pech et al. 2001).

Several reports on extra-gastric *Helicobacter* species associated infections, such as *Helicobacter hepaticus,* pointed out early to the possibility that *Helicobacter* species could also survive outside the stomach (Kawaguchi et al. 1996; Chen et al. 2007; Bansal et al. 2012; Zhou et al. 2013).

In recent years a preference for non-invasive *Helicobacter* tests, especially after *Helicobacter* eradication therapy, evolved. Serological tests harbor the disadvantage of the so-called serum scar. This means that positive antibodies in some individuals can be detected also after successful antibiotic treatment of a *H. pylori* infection (Backert et al. 2018). The false positive tests lead to the development of more sensitive tests, like *Helicobacter* stool tests (Vaira et al. 1999).

Finally, numerous cases for a successful *Helicobacter* eradication therapy have been documented in the stomach by means of histology, which showed positive findings in the stool test against *H. pylori.* It has to be noted that *Helicobacter* stool tests are rather sensitive and specific, and false positive tests are very rare (da Silva-Etto et al. 2017). The number of these cases is increasing to probably less than 3% of all cases with successful eradication therapy, where a stomach colonization by *H. pylori* can definitively be excluded (unpublished data). Various reports on *Helicobacter* species detection in the gallbladder led us also to screen cases with acute cholecystitis and subsequent cholecystectomy for *Helicobacter* species. Among more than 1,000 gallbladder specimen within the past 15 years we identified 1 case with *H. pylori* based on histology, immunohistochemistry, histochemistry, sequencing and positive culture among other concomitant bacteria in the gallbladder. In this specific case it turned out that a previously eradicated gastric *H. pylori* had the same fingerprint-pattern and *cagA* sequence compared to the strain found in the gallbladder (Backert et al. 2018). This finding is interesting for several reasons:

(a) *H. pylori* is able to survive without the acidic environment outside the stomach
(b) Extra-gastric *H. pylori* seems to be less sensitive for antibiotics eradication therapy than gastric *H. pylori*
(c) It needs to be studied whether *H. pylori* infection of the gallbladder can cause or fuel acute cholecystitis

It is striking that despite case reports have been published on this topic, the topic has not drawn any wider clinical interest. Thus, more systematic studies on large patient cohorts are required in the future.

A possible relation of gastric *H. pylori* infection to extra gastric diseases is based on more or less reliable up to obscure data on breast and prostate diseases (Kast 2007). Some seem to have rather strong evidence such as thrombocytic purpura and iron deficiency anemia (Malfertheiner and Selgrad 2010) but others as sudden infant death don't (Vieth et al. 2001). Dermatologists also have some interest in *H. pylori* due to the therapy of urticaria that can be caused by *H. pylori* infection (Rebora et al. 1995).

3 Concluding Remarks

In conclusion, there is strong evidence for the influence of *Helicobacter* infection on the course of gastric and certain extra-gastric diseases, but others lack such evidence and seem more or less like urban legends. In our present review, we gave deeper explanations and rationales behind risk factors for malignant transformation and the etiology of different forms of gastritis for the daily routine use. Up till now, substantial knowledge has been acquired for *H. pylori* effects on the gastric mucosa including precancerous conditions and risk factors. During the last decades, it became evident that also *Helicobacter* species other than *H. pylori* may colonize the stomach in humans and animals. Extra gastric colonization by *H. pylori* may gain more interest in the future since carcinomas may also arise on these colonization sites outside the stomach.

References

Backert S, Tegtmeyer N, Fischer W (2015) Composition, structure and function of the *Helicobacter pylori* cag pathogenicity island encoded type IV secretion system. Future Microbiol 10(6):955–965. https://doi.org/10.2217/fmb.15.32

Backert S, Tegtmeyer N, Oyarzabal OA, Osman D, Rohde M, Grützmann R, Vieth M. (2018)Unusual manifestation of live *Staphylococcus saprophyticus, Corynebacterium urinapleomorphum* and *Helicobacter pylori* in the gallbladder with cholecystitis. Int J Mol Sci 19(7) pii:E1826. https://doi.org/10.3390/ijms19071826

Bansal VK, Misra MC, Chaubal G, Datta Gupta S, Da B, Ahuja V, Sagar S (2012) *Helicobacter pylori* in gallbladder mucosa in patients with gallbladder disease. Indian J Gastroenterol 31:57–60. https://doi.org/10.1007/s12664-012-0162-8

Bauwens E, Joosten M, Taganna J, Rossi M, Debraekeleer A, Tay A, Peters F, Backert S, Fox J, Ducatelle R, Remaut H, Haesebrouck F, Smet A (2018) In silico proteomic and phylogenetic analysis of the outer membrane protein repertoire of gastric *Helicobacter* species. Sci Rep 8(1):15453. https://doi.org/10.1038/s41598-018-32476-1

Capelle LG, de Vries AC, Haringsma J, Ter Borg F, de Vries RA, Bruno MJ, van Dekken H, Meijer J, van Grieken NC, Kuipers EJ (2010) The staging of gastritis with the OLGA system by using intestinal metaplasia as an accurate alternative for atrophic gastritis. Gastrointest Endosc 71(7):1150–1158. https//doi.org/https://doi.org/10.1016/j.gie.2009.12.029

Chen DF, Hu L Yi P, Liu WW, Fang DC, Cao H (2007) *H pylori* exist in the gallbladder mucosa of patients with chronic cholecystitis. World J Gastroenterol 13 (10):1608–1611. https://doi.org/10.3748/wjg.v13.i10.1608

da Costa DM, Pereira Edos S, Rabenhorst SH (2015) What exists beyond *cag*A and *vac*A? *Helicobacter* genes in gastric diseases. World J Gastroenterol 21 (37):10563–10572. https://doi.org/10.3748/wjg.v21.i37.10563

da Silva-Etto JMK, Mattar R, Villares-Lopes CA, Marques SB, Carrilho FJ (2017) Evaluation of diagnostic accuracy of two rapid stool antigen tests using an immunochromatographic assay to detect *Helicobacter pylori*. Clin Biochem 50(16–17):959–962. https://doi.org/10.1016/jclinbiochem.2017.05.005

De Groote D, van Doorn LJ, van den Bulck K, Vandamme P, Vieth M, Stolte M, Debongnie JC, Burette A, Haesebrouck F, Ducatelle R (2005) Detection of non-*pylori* species in "*Helicobacter heimannii*" infected humans. Helicobacter 10(5):398–402. https://doi.org/10.1111/j.1523-5378.2005.00347.x

Dixon MF, Genta RM, Yardley JH, Correa P (1996) Classification and grading of gastritis. The updated Sydney System. International workshop on the histopathology of gastritis, Houston 1994. Am J Surg Pathol 20 (10):1161–1181

Dye KR, Marshall BJ, Frierson HF Jr, Pambianco DJ, McCallum RW (1990) *Camplyobacter pylori* colonizing heterotopic gastric tissue in the rectum. Am J Clin Pathol 93(1):144–147

Giroux V, Rustqi AK (2017) Metaplasia: tissue injury adaption and a precursor to the dysplasia-cancer sequence. Nat Rev Cancer 17(10):594–604. https://doi.org/10.1038/nrc.2017.68

Goji S, Tamujra Y, Sasaki M, Nakamura M, Matsui H, Murayama SY, Ebi M, Ogasawara N, Funaki Y, Kasugai K (2015) *Helicobacter suis*-infected nodular gastritis and a review of diagnostic sensitivity for *Helicobacter heilmanii*-like organisms. Case Rep Gastroenterol 9(2):179–197. https://doi.org/10.1159/000431169

Graham DY, Zou WY (2018) Guilt by association: intestinal metaplasia does not progress to gastric cancer. Curr Opin Gastroenterol 34(6):458–464. https://doi.org/10.1097/MOG.0000000000000472

Iannone A, Giorgio F, Russo F, Riezzo G, Girardi B, Pricci M, Palmer SC, Barone M, Principi M, Strippoli GF, Di Leo A, Ierardi E (2018) New fecal test for non-invasive *Helicobacter pylori* detection: a diagnostic accuracy study. World J Gastroenterol 24 (27):3021–3029. https://doi.org/10.3748/wjg.v24.i27.3021

Kast RE (2007) Some fibrocystic breast change may be caused by sexually transmitted *H. pylori* during oral nipple contact: supporting literature and case report of resolution after gut H. pylori eradication treatment. Med Hypotheses 68(5):1041–1046. https://doi.org/10.1016/j.mehy.2006.09.050

Kawaguchi M, Saito T, Ohno H, Midorikawa S, Sanji T, Handa Y, Morita S, Yoshida H, Tsurui M, Misaka R, Hirota T, Saito M, Minami K (1996) Bacteria closely resembling *Helicobacter pylori* detected immunohistologically and genetically in resected gallbladder mucosa. J Gastroenterol 31:294–298. https://doi.org/10.1007/BF02389534

Kubota-Aizawa S, Ohno K, Fukushima K, Kenemoto H, Nakashima K, Uchida K, Chambers JK, Goto-Koshino Y, Watanabe T, Sekizaki T, Mimuro H, Tsujimoto H (2017) Epidemiological study of gastric *Helicobacter* spp. in dogs with gastrointestinal disease in Japan and diversity of *Helicobacter heilmanii* sensu stricto. Vet J 225:56–62. https://doi.org/10.1016/j.tvjl.2017.04.004

Kulnigg-Dabsch S (2016) Autoimmune gastritis. Wien Med Wochenschr 166:424–430. https://doi.org/10.1007/s10354-016-0515-5

Kusters JG, van Vliet AH, Kuipers EJ (2006) Pathogenesis of *Helicobacter pylori* infection. Clin Microbiol Rev 19(3):449–490. https://doi.org/10.1128/CMR.00054-05

Labenz J, Blum AL, Bayerdörffer E, Meining A, Stolte M, Börsch G (1997) Curing *Helicobacter pylori* infection in patients with duodenal ulcer may provoke reflux esophagitis. Gastroenterology 112(5):1442–1447. https://doi.org/10.1016/S0016-5085(97)70024-6

Lind J, Backert S, Pfleiderer K, Berg DE, Yamaoka Y, Sticht H, Tegtmeyer N (2014) Systematic analysis of phosphotyrosine antibodies recognizing single phosphorylated EPIYA-motifs in CagA of Western-type *Helicobacter pylori* strains. PLoS One 9(5): e96488. https://doi.org/10.1371/journal.pone.0096488

Lind J, Backert S, Hoffmann R, Eichler J, Yamaoka Y, Perez-Perez GI, Torres J, Sticht H, Tegtmeyer N (2016) Systematic analysis of phosphotyrosine antibodies recognizing single phosphorylated EPIYA-motifs in CagA of East Asian-type *Helicobacter pylori* strains. BMC Microbiol 16(1):201. https://doi.org/10.1186/s12866-016-0820-6

Livzan MA, Kononov AV, Mozgovoĭ SI (2004) Ex-*Helicobacter* gastritis: is it a neologism or clinical reality? Eksp Klin Gastroenterol 148(5):55–59

Madisch A, Miehlke S, Neuber F, Morgner A, Kuhlisch E, Rappel S, Lehn N, Bayerdörffer E, Seitz G, Stolte M (2006) Healing of lymphocytic gastritis after helicobacter pylori eradication therapy–a randomized, double-blind, placebo-controlled multicentre trial. Aliment Pharmacol Ther 23(4):473–479

Mäkinen JM, Niemelä S, Kerola T, Lehtola J, Karttunen TJ (2003) Epithelial cell proliferation and glandular atrophy in lymphocytic gastritis: effect of *H. pylori* treatment. World J Gastroenterol 9(12):2706–2710

Malfertheiner P, Selgrad M (2010) H*elicobacter pylori* infection and current clinical areas of contention. Curr Opin Gastroenterol 26(6):618–623. https://doi.org/10.1097/MOG.0b013e32833efede

Marshall BJ, Warren JR (1984) Unidentified curved bacilli in the stomach of patients with gastritis and peptic ulceration. Lancet 1(8390):1311–1315

Mueller D, Tegtmeyer N, Brandt S, Yamaoka Y, De Poire E, Sgouras D, Wessler S, Torres J, Smolka A, Backert S (2012) c-Src and c-Abl kinases control hierarchic phosphorylation and function of the CagA effector protein in Western and East Asian *Helicobacter pylori* strains. J Clin Invest 122(4):1553–1566. https://doi.org/10.1172/JCI61143

Naumann M, Sokolova O, Tegtmeyer N, Backert S (2017) *Helicobacter pylori*: a paradigm pathogen for subverting host cell signal transmission. Trends Microbiol 25(4):316–328. https://doi.org/10.1016/j.tim.2016.12.004

Oberhuber G, Haidenthaler A (2000) Histopathology of *Helicobacter pylori* infections. Acta Med Austriaca 27 (4):100–103

Pachathundikandi SK, Muller A, Backert S (2016) Inflammasome activation by *Helicobacter pylori* and its implications for persistence and immunity. Curr Top Microbiol Immunol 397:117–131. https://doi.org/10.1007/978-3-319-41171-2_6

Parsons BN, Ijaz UZ, D'Amore R, Burkitt MD, Eccles R, Lenzi L, Duckworth CA, Moore AR, Tiszlavicz L, Varro A, Hall N, Pritchard DM (2017) Comparison of the human gastric microbiota in hypochlorhydric states arising as a result of *Helicobacter pylori*-induced atrophic gastritis, autoimmune gastritis and proton pump inhibitor use. PLoS Pathog 13(11): e1006653. https://doi.org/10.1371/journal.ppat.1006653

Pech O, May A, Gossner L, Vieth M, Trump F, Stolte S, Ell C (2001) Early stage adenocarcinoma of the esophagus arising in circular heterotopic gastric mucosa treated by endoscopic mucosal resection. Gastrointest Endosc 54(5):658–658

Posselt G, Backert S, Wessler S (2013) The functional interplay of *Helicobacter pylori* factors with gastric epithelial cells induces a multi-step process in pathogenesis. Cell Commun Signal 11:77. https://doi.org/10.1186/1478-811X-11-77

Rebora A, Drago F, Parodi A (1995) May *Helicobacter pylori* be important for dermatologists. Dermatology 191(1):6–8. https://doi.org/10.1159/000246470

Reddy KM, Chang JI, Shi JM, Wu BU (2016) Risk of gastric cancer among patients with intestinal metaplasia of the stomach in a US integrated health care system. Clin Gastroenterol Hepatol 14 (10):1420–1425. https://doi.org/10.1016/j.cgh.2016.05.045

Rugge M, Genta RM, OLGA Group (2005) Staging gastritis: an international proposal. Gastroenterology 129 (5):1807–1808. https://doi.org/10.1053/j.gastro.2005.09.056

Sipponen P, Stolte M (1997) Endoscopy 29(7):671–678. PMID: 9360882

Smet A, Yahara K, Rossi M, Tay A, Backert S, Armin E, Fox JG, Flahou B, Ducatelle R, Haesebrouck F, Corander J (2018) Macroevolution of gastric *Helicobacter* species unveils interspecies admixture and time of divergence. ISME J 12(10):2518–2531. https://doi.org/10.1038/s41396-018-0199-5

Sonnenberg A (1987) Causative factors in the etiology of peptic ulcer disease become effective before the age of

15 years. J Clin Epidemiol 40:193–202. https://doi.org/10.1016/0021-9681(87)90153-6

Susser M, Stein Z (1962) Civilisation and peptic ulcer. Lancet 1(7221):115–119

Vaira D, Malfertheiner P, Mégraud F, Axon AT, Deltenre M, Hirschl AM, Gasbarrini G, O'Morain C, Garcia JM, Quina M, Tytgat GN (1999) Diagnosis of *Helicobacter pylori* infection with a new non-invasive antigen-based assay. HpSA European study group. Lancet 354(9172):30–33. https://doi.org/10.1016/S0140-6736(98)08103-3

Vieth M, Stolte M, De Groote D, Deeg KH, Seitz G (2001) Death kisses for newborns? Arch Dis Child 84(6):525. PMID:11372081

Wang J, Xu L, Shi R, Huang X, Li SW, Huang Z, Zhang G (2011) Gastric atrophy and intestinal metaplasia before and after helicobacter pylori eradication: a meta-analysis. Digestion 83(4):253–260. https://doi.org/10.1159/000280318. Epub 2011 Feb 1. PMID: 21282951

Warren JR, Marshall B (1984) Unidentified curved bacilli on gastric epithelium in active chronic gastritis. Lancet 1(8336):1273–1275. https://doi.org/10.1016/S0140-6736(84)91816-6

Ye W, Nyren O (2003) Risk of cancers of the oesophagus and stomach by histology or subsite in patients hospitalised for pernicious anaemia. Gut 52(7):938–941

Zhou D, Guan WB, Wang JD, Zhang Y, Gong W, Quan ZW (2013) A comparative study of clinicopathological features between chronic cholecystitis patients with and without *Helicobacter pylori* infection in gallbladder mucosa. PLoS One 8(7):e70265. https://doi.org/10.1371/journal.pone.0070265

Adv Exp Med Biol - Advances in Microbiology, Infectious Diseases and Public Health (2019) 1149: 135–149
https://doi.org/10.1007/5584_2019_363

Published online: 24 April 2019

Malignant *Helicobacter pylori*-Associated Diseases: Gastric Cancer and MALT Lymphoma

Masanori Hatakeyama

Abstract

Helicobacter pylori is the first bacterium formally recognized to play a causative role in human malignancies, gastric cancer and gastric mucosa-associated lymphoid tissue (MALT) lymphoma. Evidence accumulates that *H. pylori cagA*-positive strains play a crucial role in the neoplastic transformation of mammalian cells. The *cagA*-encoded CagA protein is delivered into the host cells via bacterial type IV secretion, where it interacts with and thereby aberrantly activates pro-oncogenic phosphatase SHP2. The CagA-SHP2 interaction requires tyrosine phosphorylation of CagA at the Glu-Pro-Ile-Tyr-Ala (EPIYA) motif. The incidences of gastric cancer in East Asian countries such as Japan, China, and Korea are among the highest worldwide. A vast majority of *H. pylori* circulating in East Asia produce a CagA variant termed East Asian CagA, which possesses the SHP2-binding EPIYA motif (EPIYA-D) that is substantially diverged in sequence from the SHP2-binding EPIYA motif (EPIYA-C) of CagA isolated in the rest of the world (Western CagA). Tyrosine-phosphorylated EPIYA-D interacts with SHP2 approximately two orders of magnitude stronger than tyrosine-phosphorylated EPIYA-C does. The strong SHP2 binding of East Asian CagA is achieved by a cryptic interaction between the phenylalanine residue located at the +5 position from the phospho-tyrosine in EPIYA-D and a small hollow on the N-SH2 phosphopeptide-binding floor, the latter of which cannot be created by the corresponding aspartic acid in EPIYA-C. Thus, a variation in a single amino-acid residue determines the magnitude for the pathogenic/oncogenic action of CagA, which may influence the worldwide landscape in the incidence of *H. pylori*-associated malignancies, especially gastric cancer.

Keywords

H. pylori CagA · EPIYA motif · Tyrosine phosphorylation · SHP2 · Geographic polymorphism

1 Introduction

Cancers are deadly diseases characterized by deregulated proliferation and invasion of cells of clonal origin, made by accumulation of multiple genetic and epigenetic alterations in cells. Causative roles in the development of human cancers have been established for various viruses including HPV (human papillomavirus) in cervical carcinoma, HBV and HCV (hepatitis B and C viruses) in hepatocellular carcinoma, HTLV-1

M. Hatakeyama (✉)
Division of Microbiology, Graduate School of Medicine, The University of Tokyo, Tokyo, Japan
e-mail: mhata@m.u-tokyo.ac.jp

(human T-lymphotropic virus type 1) in ATL (adult T-cell leukemia), EBV (Epstein-Barr virus) in Burkitt lymphoma and Hodgkin disease, KSHV (Kaposi sarcoma herpesvirus) in Kaposi's sarcoma, and MCV (Merkel cell polyomavirus) in Merkel cell carcinoma (Chang and Moore 2012). Given that viral replication requires host cell machineries involved in DNA replication, RNA transcription, and protein translation, deregulation of those machineries through neoplastic transformation acts advantageously for a virus to produce its offsprings in large quantities. On the other hand, the contribution of a bacterium, another major microbial pathogen, in human carcinogenesis had remained ambiguous until quite recently. This was at least in part because of the difficulty in explaining the biological merit of neoplastic transformation from a bacterial standpoint. However, recent studies have gradually revealed a substantial contribution of bacteria to the development of several human cancers. Be that as it may, *Helicobacter pylori* is so far the only bacterium for which infection is widely accepted to play a major etiologic role in the development of human cancers.

H. pylori is a spiral-shaped Gram-negative bacterium that colonizes the stomach mucosa of approximately half of the world's human population (Warren and Marshall 1983). Infection with *H. pylori* is established during early childhood, mostly by the oral-oral or fecal-oral mode of transmission (Cave 1997). *H. pylori* infection lasts for the rest of one's life unless eradicated by antibiotics. There has been an accumulation of compelling evidence that chronic infection with *H. pylori* is causatively associated with the development of gastric cancer and gastric MALT (mucosa-associated lymphoid tissues) lymphoma. In turn, this gives *H. pylori*-directed carcinogenesis a paradigm of bacterial carcinogenesis. Here, I discuss molecular mechanisms underlying neoplastic transformation of host cells by the bacterial pathogen.

2 *H. pylori* Infection and Gastric Cancer

Gastric cancer is the fifth most-common cancer and the third leading cause of cancer deaths (Ferlay et al. 2015). Each year approximately 700,000 people are killed by this malignancy. This number corresponds to approximately 10% of total cancer cell deaths, rendering gastric cancer as a major health burden worldwide. Gastric cancer is also known for its unique geographic distribution in morbidity and mortality; it is highly prevalent in East Asian countries such as Japan, China and Korea, and about 50% of gastric cancer cases occur in East Asia (Ferlay et al. 2015), indicating the existence of an inherent risk factor for gastric cancer, either genetic or environmental, in East Asia.

The relationship between *H. pylori* infection and gastric cancer was first reported in 1991 (Parsonnet et al. 1991; Nomura et al. 1991; Forman et al. 1991). Subsequent large-scale epidemiological studies unequivocally demonstrated that chronic infection with *H. pylori* is intimately associated with the development of gastric cancer (Uemura et al. 2001). Those studies were followed by intervention studies showing that *H. pylori* eradication significantly reduced the incidence of gastric cancer, providing a rationale for mass screening and mass eradication of *H. pylori* infection in the conquest of gastric cancer (Wong et al. 2004; Wu et al. 2009; Fukase et al. 2008). In 1994, the International Agency for Research on Cancer (IARC), a specialized agency of the World Health Organization (WHO), classified *H. pylori* as a Group 1 carcinogen. In 2014, IARC reported that *H. pylori* is considered to be the cause of 80% of all stomach cancers (Herrero et al. 2014). However, this percentage may vary depending on geographic regions. Indeed, several reports showed that the incidence of *H. pylori*-negative gastric cancer is less than 5% in Japanese cases (Matsuo et al. 2011; Ono et al. 2012), suggesting that the vast majority of gastric cancers are associated with *H. pylori* infection in East Asia. This strong connection further points to the special importance of mass *H. pylori* eradication as a strategy for prevention of gastric cancer in this geographic area. Without effective *H. pylori* eradication, the current incidence of gastric cancer is projected to remain stable or even increase by 2030 (Herrero et al. 2014). Pathologically, gastric cancer is subclassified into intestinal type, which is more common in older adults, and diffuse type, which is more aggressive and occurs more frequently at younger ages. *H. pylori* infection has been considered to be associated with both types of gastric cancer.

3 *H. pylori* Infection and MALT Lymphoma

Lymphomas are hematological malignancies originating from lymphoid cells, most notably T and B lymphocytes that clonally proliferate in response to specific antigenic stimuli. MALT lymphoma is an extra-nodal low-grade lymphoma originating from B cells in the marginal zone of MALT. MALT lymphomas account for approximately 8% of all non-Hodgkin lymphomas (Troppan et al. 2015). Although MALT lymphomas can arise from various extra-nodal sites such as the lung, ocular adnexa, thyroid gland, and small intestine, many of them (~70%) originate from the stomach of patients infected chronically with *H. pylori* and approximately 13,000 peoples are diagnosed with gastric MALT lymphomas every year worldwide (Plummer et al. 2016). Under physiological conditions, the stomach mucosa does not contain lymphoid tissues. Chronic *H. pylori* infection induces the formation of MALT in the gastric mucosa as a result of persistent stimuli with *H. pylori* antigens. Importantly, eradication of *H. pylori* with antibiotics results in regression of MALT lymphoma in more than 75% of cases, indicating that continuous presence of the microbe is required to maintain malignant phenotypes that underlie the *in vivo* tumorigenicity in the majority of gastric MALT lymphomas (Zullo et al. 2014). Acquisition of resistance of MALT lymphoma to *H. pylori* eradication, which is associated with a more aggressive phenotype, appears to require chromosomal translocations such as t(1;14)(p22; q32) and t(11;18)(q(21;q21) translocations, which cause deregulated activation of the transcription factor NF-κB signaling pathway through overexpression of BCL10 and the generation of MALT1 fusion protein, respectively (Bautista-Quach et al. 2012).

4 Involvement of *H. pylori cagA*-Positive Strains in Human Malignancies

Highly virulent *H. pylori* strains possess a genomic DNA fragment, termed the *cag* pathogenicity island (*cag*PAI), which is a 40-kilobase segment that was thought to be acquired by a horizontal DNA transfer of unknown origin (Censini et al. 1996; Akopyants et al. 1998). In Western countries, approximately 30–40% of *H. pylori* strains isolated do not carry the *cag*PAI, whereas almost all of the *H. pylori* isolates from East Asian countries possess the *cag*PAI (Ito et al. 1997; Azuma et al. 2002). The *cag*PAI DNA contains 27–31 putative genes. Of those, at least 18 genes encode proteins serving as structural components of a type IV secretion system (T4SS), a syringe-like transmembrane apparatus that is capable of delivering macromolecules such as nucleic acids and proteins across the bacterial inner and outer membranes. The *cag*PAI DNA also contains the *cagA* (*cytotoxin-associated* gene *A*) gene, which encodes a 130~145-kilodalton (kDa) protein termed CagA (Covacci et al. 1993; Tummuru et al. 1993). CagA is the one and only known effector protein that is delivered into gastric epithelial cells via the *cag*PAI-encoded T4SS (Segal et al. 1999; Asahi et al. 2000; Backert et al. 2000; Odenbreit et al. 2000; Stein et al. 2000). Accordingly, *cagA*-positive *H. pylori* is used synonymously with *cag*PAI-positive *H. pylori,* and *cagA*-negative *H. pylori* is used synonymously with *cag*PAI-negative *H. pylori*. The association of *H. pylori cagA*-positive strains with increased risk of gastric cancer was first reported in mid 1990s (Kuipers et al. 1995; Blaser et al. 1995). The important role of infection with *cagA*-positive *H. pylori* in the development of gastric cancer, especially non-cardia gastric cancer, was subsequently consolidated by a number of clinico-epidemiological studies and by recent meta-analyses (Parsonnet et al. 1997; Huang et al. 2003; Park et al. 2018). The relationship was further supported by the results of an infection study of Mongolian gerbils with *cagA*-positive or isogenic *cagA*-negative *H. pylori* showing that a *cagA*-positive strain substantially facilitates atrophic corpus gastritis, a premalignant mucosal lesion of gastric cancer (Rieder et al. 2005).

Infection with *cagA*-positive *H. pylori* is also associated with the development of gastric MALT lymphoma. Serum anti-CagA antibody was positive in more than 95% of patients suffering from gastric MALT lymphoma (Eck et al. 1997). The CagA protein was detected in B lymphocytes in individuals infected with *cagA*-positive *H. pylori* (Lin et al. 2010) and in tumor cells of *cagA*-positive *H. pylori*-associated gastric MALT lymphoma (Kuo et al. 2013).

5 Molecular Structure of the *H. pylori* CagA Protein

The *cagA*-encoded CagA protein has a unique tertiary structure consisting of a solid N-terminal region (70% of the entire CagA) and an intrinsically disordered C-terminal tail (30% of the entire CagA) (Hayashi et al. 2012a). The crystal structure of N-terminal CagA predicts a square plate-like shape that contains three discrete domains (Fig. 1) (Hayashi et al. 2012a; Kaplan-Türköz et al. 2012). Domain I, the extreme N-terminal domain of CagA, has a small interacting surface area with Domain II or Domain III, indicating that Domain I is highly mobile. Domain II and Domain III comprise a structural core of CagA. The basic patch, a cluster of basic residues in Domain II, plays an important role in the interaction of CagA with phosphatidylserine, a membrane phospholipid specifically enriched in the inner leaflet of the plasma membrane (Murata-Kamiya et al. 2010). The intrinsically disordered CagA C-terminal tail is characterized by the presence of a variable number of EPIYA (Glu-Pro-Ile-Tyr-Ala) motifs, which serve as motifs for tyrosine phosphorylation of T4SS-delivered CagA by host cell kinases such as Src-family kinases (SFKs) and c-Abl (Selbach et al. 2002; Stein et al. 2002; Poppe et al. 2007; Tammer et al. 2007). Based on the sequences flanking each of the EPIYA motifs, four distinct EPIYA segments, EPIYA-A, -B -C and -D, have been identified (Fig. 1) (Hatakeyama 2004). The EPIYA-repeat region of CagA from *H. pylori* circulating worldwide except East Asian countries is in an arrangement of EPIYA-A, EPIYA-B and EPIYA-C segments (Western CagA). In addition, the EPIYA-C segment variably duplicates (mostly 1–3 times) in tandem among distinct Western CagA species. CagA from East Asian *H. pylori* isolates also contains the EPIYA-A and EPIYA–B segments but not the EPIYA-C segment. Instead, it has a distinct EPIYA-containing segment termed EPIYA-D, and the EPIYA-repeat region is in an arrangement of EPIYA-A, EPIYA-B and EPIYA-D segments (East Asian CagA).

The disordered C-terminal tail of CagA contains another repeatable sequence motif, originally designated as the CagA multimerization sequence motif (CM) (Fig. 1) (Ren et al. 2006). The CM motif, comprising 16 amino-acid residues, is located immediately distal to the last repeat of the EPIYA segments. Whereas the CM motif sequence is highly conserved, there are several amino-acid alterations between East Asian and Western CagA species (Lu et al. 2008). Based on this variation, the CM motif of Western CagA is termed CM^W and that of East Asian CagA is termed CM^E. Remarkably, the amino-terminal (N-terminal) 16-amino-acid stretch that constitutes the EPIYA-C segment is identical to CM^W. Multiplication of the EPIYA-C segment therefore increases the number of CM^W motifs in Western CagA.

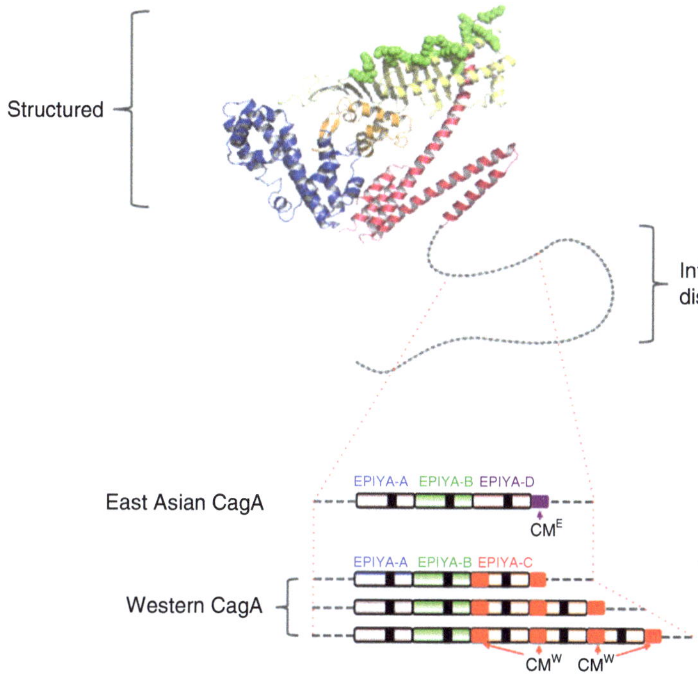

Fig. 1 Structural polymorphisms of *Helicobacter pylori* CagA
The CagA protein consists of the structured N-terminal region (70% of entire CagA) and the disordered C-terminal region (30% of entire CagA). The structured N-terminal region displays a heretofore-unidentified structure with three distinct domains, Domain I (blue), Domain II (yellow), and Domain III (red) (I-III). Domain II tethers CagA to the inner plasma membrane through electrostatic interaction between the basic patch (green) and the acidic phosphatidylserine that is primarily distributed to the inner face of the plasma membrane. The disordered C-terminal region of Western CagA contains the EPIYA-A, EPIYA-B and a variable number (mostly 1–3) of EPIYA-C segments. The disordered C-terminal region of East Asian CagA contains the EPIYA-A, EPIYA-B and EPIYA-D segments. Each of the EPIYA segments contains a single EPIYA tyrosine phosphorylation motif (shown as a black box), which is phosphorylated by SFKs and c-Abl. East Asian CagA has a single CM (CM^E; shown in purple) motif immediately downstream of the EPIYA-D segment. Western CagA possesses at least two CM (CM^W; shown in red) motifs, one within the EPIYA-C segment, and the other located distal to the last EPIYA-C segment. The CM^E and CM^W motif sequences are well conserved but have some amino acid differences

6 Pathobiological Actions of CagA Delivered into Host Cells

Expression of CagA in cultured gastric epithelial cells such as AGS cells, either by infection, transfection or transduction, induces a uniquely elongated cell-shape known as the "hummingbird" phenotype (Segal et al. 1999; Higashi et al. 2002b). Induction of the hummingbird phenotype is dependent on tyrosine phosphorylation of CagA on the EPIYA-C or EPIYA-D segment and requires the interaction of CagA with SHP2 (SH2 domain-containing protein tyrosine phosphatase 2), a cytoplasmic tyrosine phosphatase that possesses two related SH2 domains (N-SH2 and C-SH2) in tandem at the N-terminal region (Higashi et al. 2002a, b). CagA-SHP2 complex formation occurs through the interaction between the tyrosine-phosphorylated EPIYA-C or EPIYA-D segment and the SH2 domain(s) of SHP2. Physiologically, SHP2 is indispensable for full activation of the Ras-ERK pathway, which conveys a potent mitogenic signal, and is also involved in cell morphogenesis as well as cell motility (Neel et al. 2003). Consistent with this, expression of CagA in gastric epithelial cells induces sustained ERK activation, which provokes pro-mitogenic cellular response while morphologically inducing the hummingbird phenotype. Given that gain-of-function mutations of *PTPN11*, the gene encoding SHP2, have been found in a variety of human malignancies (Tartaglia et al. 2003; Bentires-Alj et al. 2004), aberrant activation of SHP2 by CagA is thought to play an important role in the neoplastic transformation of gastric epithelial cells. CagA is also capable of binding with the Crk adaptor protein in a tyrosine phosphorylation-dependent manner, though the responsible EPIYA segment remains unknown (Suzuki et al. 2005). Through the interaction, CagA disrupts adherens junctions by perturbing the adaptor function of Crk in cell signaling. Additionally, CagA binds to Grb2 via the EPIYA-containing region, which again deregulates the Ras signaling pathway to stimulate abnormal cell proliferation, in a tyrosine phosphorylation-independent manner (Mimuro et al. 2002). SHP1, the only SHP2 homologue in mammals, is primarily expressed in hematopoietic cells, where it negatively regulates immune cell activation. Although less abundant, it is also present in gastrointestinal linings. Like SHP2, SHP1 forms a physical complex with CagA in gastric epithelial cells. However, the interaction is independent of EPIYA tyrosine phosphorylation and, in striking contrast to SHP2, which acts as a downstream effector of CagA, SHP1 mediates tyrosine dephosphorylation of CagA on the EPIYA motifs. Accordingly, SHP1 is a tyrosine phosphatase that counteracts phosphorylation-dependent CagA biological actions (Saju et al. 2016). Further complexity was added to the phosphorylation–dependent interactions of CagA by the identification of five additional binding partners (PI3-kinase, Grb2, Grb7, RAS-GAP and Csk) obtained in a proteomics-based screen (Selbach et al. 2009), which need to be investigated in more detail in future studies.

The function of CagA as a pathogenic scaffold/hub in host cells into which it has been delivered does not always rely on the EPIYA motifs. The CM motif serves as a binding site for the polarity-regulating serine/threonine kinase PAR1 (Partitioning-defective 1) in a tyrosine phosphorylation-independent manner (Saadat et al. 2007). In mammals, PAR1 was independently identified as MARK (microtubule affinity-regulating kinase), which regulates the stability of microtubules by phosphorylating microtubule-associated proteins (MAPs) such as MAP1, MAP2, and tau (Matenia and Mandelkow 2009). Mammalian PAR1 comprises four PAR1/MARK isoforms, PAR1a/MARK3, PAR1b/MARK2, PAR1c/MARK1, and PAR1d/MARK4. Of these, PAR1b is a major PAR1 isoform in gastric epithelial cells and distributes to the basolateral membrane of polarized epithelial cells, where it is critically involved in the establishment and/or maintenance of epithelial apical-basal polarity (Hayashi et al. 2012b). CagA is capable of binding to all the PAR1 isoforms, with the highest affinity to PAR1b. In the CagA-PAR1 interaction, the CM motif of CagA directly binds to the kinase catalytic domain of PAR1 and thus strongly inhibits the kinase activity (Saadat et al. 2007; Nesić et al. 2010). As a result, the CagA-PAR1b interaction elicits junctional and polarity defects, making cells susceptible to various oncogenic insults (Saadat et al. 2007; Zeaiter et al. 2008). The CM motif also interacts with the activated HGF (hepatocyte growth factor) receptor c-Met, which in turn strengthens c-Met-dependent activation of the Ras and PI3K (phosphatidylinositol-3 kinase)/Akt signaling pathways (Suzuki et al. 2009). In addition, CagA associates with the cytoplasmic domain of E-cadherin via the CM motif and thereby disrupts the E-cadherin/β-catenin complex, leading to deregulated activation of E-cadherin-Wnt signaling (Murata-Kamiya et al. 2007). Furthermore, CagA associates with GSK-3β (glycogen synthase kinase-3β) via the C-terminal region, which in turn sequesters GSK-3β from the APC/β-catenin destruction complex and thereby potentiates Wnt signal activation (Lee et al. 2014).

The structured N-terminal region of CagA also interacts with a number of cellular proteins. For example, N-terminal CagA binds to the RUNX3 tumor suppressor to promote its degradation (Tsang et al. 2010). Furthermore, N-terminal

CagA enhances protease-dependent degradation of the p53 tumor suppressor by interacting with ASPP2 (apoptosis-stimulating protein of p53 2) (Buti et al. 2011). CagA-stimulated Akt phosphorylates and activates human double minute 2 (HDM2) and ARF-binding protein 1 (ARF-BP1) E3 ligases (Wei et al. 2015), which may further promote p53 degradation in conjunction with ASPP2. These findings collectively indicate that the N-terminal structured region of CagA promotes the survival of CagA-delivered cells by inactivating multiple tumor suppressors.

7 Oncogenic Role of *H. pylori* CagA *In Vivo*

Transgenic mice systemically expressing wild-type CagA show gastric epithelial hyperplasia and some of them spontaneously develop adenocarcinomas of the stomach and small intestine (Ohnishi et al. 2008). The mice also show leukocytosis that is associated with hypersensitivity to hematopoietic cytokines in bone-marrow cells, which may be due to deregulated SHP2 activation, and a fraction of them develop myeloid leukemias and B-cell lymphomas, hematological malignancies that can also be induced by gain-of-function mutations of SHP2. In contrast, no pathological abnormalities are observed in transgenic mice expressing phosphorylation-resistant CagA. These results provide the first direct evidence for the role of CagA as a bacterial protein that can promote carcinogenesis in mammals. The results further point to the importance of CagA tyrosine phosphorylation, which is an essential prerequisite for CagA to bind and deregulate SHP2, in *in vivo* tumorigenesis. In zebrafish, transgenic expression of CagA synergistically potentiates the induction of pre-cancerous intestinal metaplasia by *p53* inactivation (Neal et al. 2013), providing additional evidence for the *in vivo* oncogenic role of CagA in animals.

Approximately 10% of gastric cancer cases are comprised of cancer cells latently infected with Epstein Barr virus (EBV) (Akiba et al. 2008). Those EBV-positive gastric cancer cells are characterized by strong genome-wide CpG hypermethylation (Matsusaka et al. 2014), and *in vitro* infection of gastric epithelial cells with EBV induces promoter CpG hypermethylation of *PTPN6*, the gene encoding SHP1. Since EBV-positive gastric cancer also arises from the stomach infected with *cagA*-positive *H. pylori*, epigenetic silencing of SHP1, the CagA phosphatase, renders the level of CagA tyrosine phosphorylation high (Saju et al. 2016). Consistently, clinical specimens of EBV-positive gastric cancers exhibit *PTPN6* hypermethylation with reduced SHP1 expression. Augmented *H. pylori* CagA activity may therefore contribute to the development of EBV-positive gastric cancer. Thus EBV-positive gastric cancer provides for the first time the oncogenic collaboration between a virus (EBV) and a bacterium (*cagA*-positive *H. pylori*) in the development of human cancer.

H. pylori infection induces inflammation in the stomach mucosa by utilizing multiple distinct mechanisms. *H. pylori* peptidoglycan has been reported to enter gastric epithelial cells via the T4SS, where it stimulates the cytoplasmic pathogen recognition receptor Nod1 to activate NF-κB and thereby to induce inflammatory cytokines (Viala et al. 2004). *H. pylori* LPS (lipopolysaccharide), primarily recognized by toll-like receptor 2 (TLR2), also stimulates NF-κB (Nemati et al. 2017). More recent studies have revealed that βHBP (D-glycero-β-D-manno-heptose 1,7-bisphosphate), a metabolic precursor in LPS biosynthesis, is a non-protein effector of *H. pylori* T4SS. Upon delivery into gastric epithelial cells, βHBP stimulates the TIFA (TRAF-interacting protein with forkhead-associated domain)-dependent NF-κB activation pathway (Gall et al. 2017; Stein et al. 2017; Zimmermann et al. 2017). On the other hand, Zhou et al. showed that ADP-Hep (ADP-β-D-manno-heptose), but not βHBP, is a bacterial effector, which upon delivery via bacterial secretion system, directly binds to and thereby activates ALPK1 (alpha-kinase 1) to stimulate the TIFA-NF-kB axis, suggesting that ADP-heptose is a bacterial ligand that is recognized by the cytosolic innate immune receptor ALPK1 (Zhou et al. 2018). In any case, these studies indicate that the βHBP-dependent and/or ADP-heptose-dependent pathway is a major contributor of NF-κB activation by *H. pylori*. In CagA-transgenic mice, the cellular pool of IκB, which binds to and thereby sequesters NF-κB in the cytoplasm, is substantially diminished in gastrointestinal epithelial cells in which CagA is expressed (Suzuki et al. 2015). The CagA-mediated IκB reduction is, however, insufficient to spontaneously activate NF-κB, and thus no overt inflammation is induced in the gastrointestinal tract of CagA-transgenic mice. On the other hand, experimental colitis induced by administration of dextran sodium sulfate (DSS) is markedly deteriorated in CagA-transgenic mice due to the reduced cellular pool of IκB (Suzuki et al. 2015). Thus, CagA acts to lower the threshold of NF-κB activation that is triggered by other inflammogenic insults. The incidence of colonic dysplasia, the precursor of colon carcinoma, is also increased in CagA-transgenic mice treated with DSS. Accordingly, CagA deteriorates inflammation, whereas inflammation strengthens the oncogenic potential of CagA. Through this functional interplay, CagA and inflammation reinforce each other in creating a feed-forward regulatory loop that promotes neoplastic transformation of cells.

Recent studies have shown that the *H. pylori* CagA protein is detectable in B lymphoid cells *in vitro* infected with *cagA*-positive *H. pylori* and also in tumor cells of MALT lymphoma that have infiltrated the stomach epithelial linings, indicating a scenario in which CagA is directly involved in the neoplastic transformation of B cells (Lin et al. 2010). As in the case of gastric epithelial cells, CagA expressed in B cells undergoes tyrosine phosphorylation and binds to SHP2 (Kuo et al. 2013). CagA-deregulated SHP2 then aberrantly activates Erk, which in turn induces phosphorylation of Bad while up-regulating the anti-apoptotic proteins Bcl-2 and Bcl-xL. CagA also inhibits apoptosis of B cells by simultaneously perturbing the function of p53 and the JAK/STAT signaling pathway (Umehara et al. 2003). These CagA activities may promote accumulation of preneoplastic B cells by subverting their elimination through apoptosis, which may additionally contribute to the development of MALT lymphoma.

8 Geographic Polymorphisms of *H. pylori* CagA

Physical complex formation of CagA with SHP2, mediated via the tyrosine-phosphorylated EPIYA-C or EPIYA-D segment of CagA, is considered to be one of the key interactions through which CagA exerts its oncogenic action. The sequence flanking the phosphotyrosine (pTyr) residue of the EPIYA-D segment perfectly matches the consensus high-affinity binding sequence for the SHP2 SH2 domains, whereas the sequence flanking the pTyr residue of the EPIYA-C segment differs from the consensus sequence by a single amino acid at the pTyr+5 position: Phe in EPIYA-D and Asp in EPIYA-C (Higashi et al. 2002a). Comparison of the three-dimensional atomic structures of East Asian CagA (containing a single EPIYA-D segment) and Western CagA (containing a single EPIYA-C segment) complexed with SHP2, which were determined by using X-ray crystallography analysis, revealed that the high-affinity SHP2 binding of East Asian CagA is achieved via 1) an interaction of pTyr with its specific binding pocket in the N-SH2 domain of SHP2 and 2) a cryptic interaction between Phe at the +5 position from pTyr in EPIYA-D and a small hollow on the phosphopeptide-binding floor of N-SH2, the latter of which cannot be created by the corresponding Asp residue at the pTyr+5 position in the Western CagA EPIYA-C sequence (Fig. 2) (Hayashi et al. 2017). Reflecting the differences in their binding modes, East Asian CagA exhibits more than a hundred-fold greater SHP2-binding affinity than Western CagA does, as indicted by a comparison of binding dissociation constants (Nagase et al. 2015). Thus, East Asian CagA is far more potent than Western CagA in terms of SHP2 binding and catalytic perturbation of SHP2. Consistent with these biophysical/biochemical findings, clinical studies have revealed that gastric cancer is more closely associated with

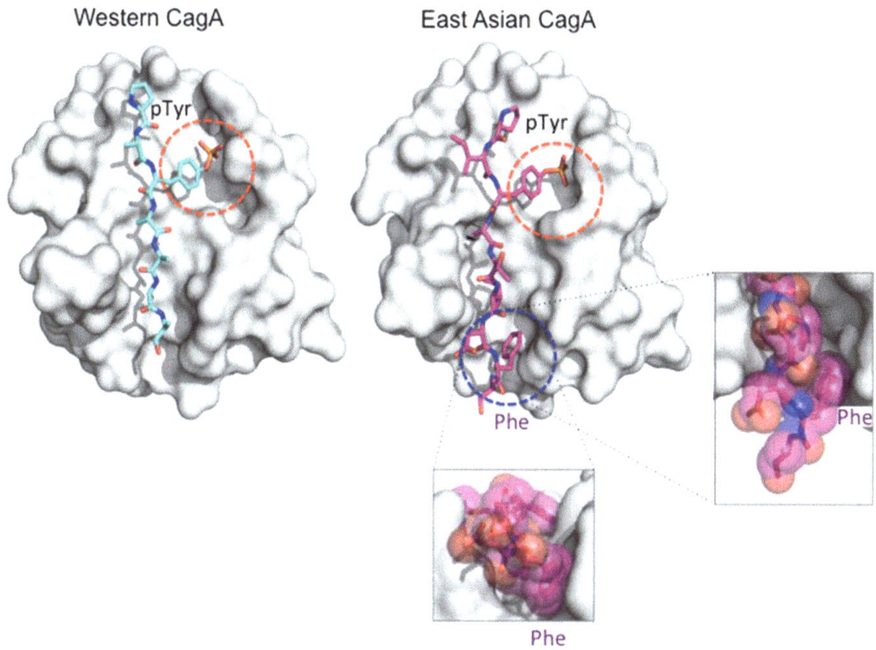

Fig. 2 Differential modes of SHP2 binding between East Asian and Western CagA species
Three-dimensional structures of the CagA EPIYA-C peptide (VSPEPIpYATIDDL) (cyan, left) and the CagA EPIYA-D peptide (ASPEPIpYATIDFD) (magenta, right) complexed with the N-SH2 domain of SHP2 (gray, molecular surface representation model) are shown. The specific interaction of phosphotyrosine (pTyr) on the EPIYA-C or EPIYA-D peptide with N-SH2 is indicated in the red dashed circle. The second interaction, which is specifically observed between Phe at the pTyr+5 position in the EPIYA-D peptide and a small hollow in the peptide-binding cleft of the SHP2 N-SH2 domain is shown in the blue dashed circle. The detailed molecular structures involved the second interaction depicted with the van der Waals surface of the EPIYA-D peptide (magenta) from the side and the bottom are presented in the boxes. The second interaction, which does not occur between EPIYA-C, robustly increases the SHP2-binding activity of CagA. The figure was modified from Hayashi et al. (2017) with permission

H. pylori strains producing East Asian CagA than with *H. pylori* strains producing Western CagA in geographical regions where the two distinct strains co-circulate (Azuma et al. 2004; Vilaichone et al. 2004; Satomi et al. 2006).

The EPIYA-C segment of Western CagA is unique in that it tandem-duplicates with a variable number, mostly from one to three. The proportions of Western CagA species containing one, two, and three EPIYA-C segments are approximately 60–70%, 20–30%, and < 5%, respectively (Xia et al. 2009). Within Western CagA species, those having a greater number of EPIYA-C segments exhibit stronger ability to interact with SHP2. A number of recent clinico-epidemiological studies have shown that infection with *H. pylori* strains carrying Western CagA with two or more EPIYA-C segments is a greater risk for the development of gastric carcinoma than is infection with *H. pylori* carrying CagA with a single EPIYA-C segment (Argent et al. 2004; Basso et al. 2008; Batista et al. 2011; Beltrán-Anaya et al. 2014; Ferreira Júnior et al. 2015; González et al. 2011; Sicinschi et al. 2010). Since SHP2 binding is the only known CagA activity for which the magnitude is correlated with the number of EPIYA-C segments, the degree of CagA-SHP2 interaction may link the EPIYA-C multiplication with gastric cancer risk. Indeed, a quantitative study revealed that the strength of CagA-SHP2 binding is dramatically elevated by a hundredfold upon duplication of the CagA EPIYA-C segment from one to two (Nagase et al. 2015). This is most probably due to monomeric interaction vs. dimeric interaction of CagA with SHP2, which possesses two CagA-binding SH2 domains in tandem. The robust increase in the SHP2-binding activity of CagA by EPIYA-C duplication is also associated with dramatic activation of SHP2 catalytic activity and marked enhancement of cell invasion phenotype into the extracellular matrix, a malignant cellular trait associated with SHP2 deregulation (Hayashi et al. 2017; Nagase et al. 2015). These observations provide a molecular basis for the role of multiple EPIYA-C segments as a distinct risk factor of gastric cancer.

Qualitative and quantitative variations in the CM motif also influence the pathobiological activity of individual CagA. Western CagA contains two or more copies of the CM^W motif as it usually possesses 1–3 repeats of the EPIYA-C segment, each of which contains a single CM^W motif (Ren et al. 2006). On the other hand, East Asian CagA carries only a single CM^E motif immediately downstream of the EPIYA-D segment. Consistent with the fact that the CM motif serves as the binding site for PAR1b, an increase in the number of CM^W motifs in a single Western CagA potentiates PAR1b-binding activity, which is in proportion to the magnitudes of disruption of tight junctions (Nishikawa et al. 2016). For instance, a Western CagA containing 4 CM^W motifs displays a PAR1b-binding affinity that is more than 30-fold higher than that of a Western CagA containing a single CM^W. Also notably, a single CM^E shows a binding affinity to PAR1b that is similar to that of two tandemly repeated CM^W motifs. The polymorphism in the CM motif is therefore another determinant for the magnitude of CagA pathobiological action that may additionally influence the clinical outcome of *cagA*-positive *H. pylori* infection.

9 Conclusions

Gastric cancer and gastric MALT lymphoma are the only human malignancies in which the etiological role of a particular bacterial infection has been widely accepted. In both malignancies, chronic infection with *H. pylori*, especially a *cagA*-positive strain, plays a key role in development of the malignancy. The *cagA*-encoded CagA protein is an oncogenic bacterial protein that is delivered into gastric epithelial cells (and B lymphocytes as well) by the *H. pylori* T4SS. In this sense, CagA is reminiscent of viral oncogene/oncoproteins that also act inside host cells. Oncogenic activity of CagA is primarily cell-autonomous and is based on its capability to promiscuously interact with host cell molecules and thereby to perturb multiple intracellular signaling pathways. Also notably, CagA on its own is not a potent inflammogen although it can deteriorate ongoing inflammation. Although sustained inflammation is known to play an important role in carcinogenesis and although *H. pylori* infection indeed induces chronic inflammation in the stomach mucosa, inflammation on its own may not be sufficient for the generation of cancerous cells at a high frequency (Hayashi et al. 2012c). Rather, the existence of a specific cancer-predisposing protein such as *H. pylori* CagA may be required for a steady progression of the multistep gastric carcinogenesis, which is further boosted in the presence of chronic inflammation. This notion is in line with the fact that most oncogenic viruses carry unique oncogenes/oncoproteins while inducing chronic inflammation at the infected organs and is indeed supported by the results of treatment of CagA-transgenic mice with a chemical inflammogen.

The prevalence of gastric cancer is extremely high in East Asian countries such as Japan, China, and Korea. Although there may be multiple factors underlying the massive accumulation of gastric cancer patients in East Asia, endemic circulation of *H. pylori* strains that produce pathobiologically more active CagA appears to be a crucial environmental factor. It is tempting to speculate that a nano-scale variation at a single amino acid residue in geographically distinct CagA, which determines the magnitude of the pathobiological action of CagA, plays a substantial role in shaping up the worldwide landscape of *H. pylori*-associated malignant diseases.

References

Akiba S, Koriyama C, Herrera-Goepfert R, Eizuru Y (2008) Epstein-Barr virus associated gastric carcinoma: epidemiological and clinicopathological features. Cancer Sci 99(2):195–201. https://doi.org/10.1111/j.1349-7006.2007.00674.x

Akopyants NS, Clifton SW, Kersulyte D, Crabtree JE, Youree BE, Reece CA, Bukanov NO, Drazek ES, Roe BA, Berg DE (1998) Analyses of the cag pathogenicity island of *Helicobacter pylori*. Mol Microbiol 28 (1):37–53

Argent RH, Kidd M, Owen RJ, Thomas RJ, Limb MC, Atherton JC (2004) Determinants and consequences of different levels of CagA phosphorylation for clinical isolates of *Helicobacter pylori*. Gastroenterology 127 (2):514–523. https://doi.org/10.1053/j.gastro.2004.06.006

Asahi M, Azuma T, Ito S, Ito Y, Suto H, Nagai Y, Tsubokawa M, Tohyama Y, Maeda S, Omata M, Suzuki T, Sasakawa C (2000) *Helicobacter pylori* CagA protein can be tyrosine phosphorylated in gastric epithelial cells. J Exp Med 191(4):593–602. PMC2195829

Azuma T, Yamakawa A, Yamazaki S, Fukuta K, Ohtani M, Ito Y, Dojo M, Yamazaki Y, Kuriyama M (2002) Correlation between variation of the 3′ region of the *cagA* gene in *Helicobacter pylori* and disease outcome in Japan. J Infect Dis 186(11):1621–1630. https://doi.org/10.1086/345374

Azuma T, Yamazaki S, Yamakawa A, Ohtani M, Muramatsu A, Suto H, Ito Y, Dojo M, Yamazaki Y, Kuriyama M, Keida Y, Higashi H, Hatakeyama M (2004) Association between diversity in the Src homology 2 domain-containing tyrosine phosphatase binding site of *Helicobacter pylori* CagA protein and gastric atrophy and cancer. J Infect Dis 189(5):820–827. https://doi.org/10.1086/381782

Backert S, Ziska E, Brinkmann V, Zimmy-Arndt U, Fauconnier A, Jungblut PR, Naumann M, Meyer T (2000) Translocation of the CagA protein in gastric epithelial cells by a type IV secretion apparatus. Cell Microbiol 2(2):155–164

Basso D, Zambon CF, Letley DP, Stranges A, Marchet A, Rhead JL, Schiavon S, Guariso G, Ceroti M, Nitti D, Rugge M, Plebani M, Atherton JC (2008) Clinical relevance of *Helicobacter pylori cagA* and *vacA* gene polymorphisms. Gastroenterology 135(1):91–99. https://doi.org/10.1053/j.gastro.2008.03.041

Batista SA, Rocha GA, Rocha AM, Saraiva IE, Cabral MM, Oliveira RC, Queiroz DM (2011) Higher number of *Helicobacter pylori* CagA EPIYA C phosphorylation sites increases the risk of gastric cancer, but not duodenal ulcer. BMC Microbiol 11:61. https://doi.org/10.1186/1471-2180-11-61

Bautista-Quach MA, Ake CD, Chen M, Wang J (2012) Gastrointestinal lymphomas: Morphology, immunophenotype and molecular features. J Gastrointest Oncol 3(3):209–225. https://doi.org/10.3978/j.issn.2078-6891.2012.024

Beltrán-Anaya FO, Poblete TM, Román-Román A, Reyes S, deSampedro J, Peralta-Zaragoza O, Rodríguez MÁ, del Moral-Hernández O, Illades-Aguiar B, Fernández-Tilapa G (2014) The EPIYA-ABCC motif pattern in CagA of *Helicobacter pylori* is associated with peptic ulcer and gastric cancer in Mexican population. BMC Gastroenterol 14(223). https://doi.org/10.1186/s12876-014-0223-9

Bentires-Alj M, Paez JG, David FS, Keilhack H, Halmos B, Naoki K, Maris JM, Richardson A, Bardelli A, Sugarbaker DJ, Richards WG, Du J, Girard L, Minna JD, Loh ML, Fisher DE, Velculescu VE, Vogelstein B, Meyerson M, Sellers WR, Neel BG (2004) Activating mutations of the noonan syndrome-associated *SHP2/PTPN11* gene in human solid tumors and adult acute myelogenous leukemia. Cancer Res 64 (24):8816–8820. https://doi.org/10.1158/0008-5472.CAN-04-1923

Blaser MJ, Perez-Perez GI, Kleanthous H, Cover TL, Peek RM, Chyou PH, Stemmermann GN, Nomura A (1995) Infection with *Helicobacter pylori* strains possessing cagA is associated with an increased risk of developing adenocarcinoma of the gastric cancer. Cancer Res 55 (10):2111–2115

Buti L, Spooner E, Van der Veen AG, Rappuoli R, Covacci A, Ploegh HL (2011) *Helicobacter pylori* cytotoxin-associated gene A (CagA) subverts the apoptosis-stimulating protein of p53 (ASPP2) tumor suppressor pathway of the host. Proc Natl Acad Sci U S A 108(22):9238–9243. https://doi.org/10.1073/pnas.1106200108

Cave DR (1997) How is *Helicobacter pylori* transmitted? Gastroenterology 113(6):S9–S14. https://doi.org/10.1016/S0016-5085(97)80004-2

Censini S, Lange C, Xiang Z, Crabtree JE, Ghiara P, Borodovsky M, Rappuoli R, Covacci A (1996) *cag*, a pathogenicity island of *Helicobacter pylori*, encodes type I-specific and disease-associated virulence factors. Proc Natl Acad Sci U S A 93(25):14648–14653. https://doi.org/PMC26189

Chang Y, Moore PS (2012) Merkel cell carcinoma: a virus-induced human cancer. Annu Rev Pathol 7:123–144. https://doi.org/10.1146/annurev-pathol-011110-130227

Covacci A, Censini S, Bugnoli M, Petracca R, Burroni D, Macchia G, Rappuoli R (1993) Molecular characterization of the 128-kDa immunodominant antigen of *Helicobacter pylori* associated with cytotoxicity and duodenal ulcer. Proc Natl Acad Sci U S A 90 (12):5791–5795. PMC46808

Eck M, Schmausser B, Haas R, Greiner A, Czub S, Müller-Hermelink HK (1997) MALT-type lymphoma of the stomach is associated with *Helicobacter pylori* strains expressing the CagA protein. Gastroenterology 112(5):1482–1486

Ferlay J, Soerjomataram I, Dikshit R, Eser S, Mathers C, Rebelo M, Parkin DM, Forman D, Bray F (2015) Cancer incidence and mortality worldwide: sources, methods and major patterns in GLOBOCAN 2012. Int J Cancer 136(5):E359–E386. https://doi.org/10.1002/ijc.29210

Ferreira Júnior M, Batista SA, Vidigal PV, Cordeiro AA, Oliveira FM, Prata LO, Diniz AE, Barral CM, Barbuto RC, Gomes AD, Araújo ID, Queiroz DM, Caliari MV (2015) Infection with CagA-positive *Helicobacter pylori* strain containing three EPIYA C phosphorylation sites is associated with more severe gastric lesions in experimentally infected Mongolian gerbils (Meriones unguiculatus). Eur J Histochem 59(2):2489. https://doi.org/10.4081/ejh.2015.2489

Forman D, Newell DG, Fullerton F, Yarnell JW, Stacey AR, Wald N, Sitas F (1991) Association between infection with *Helicobacter pylori* and risk of gastric cancer: evidence from a prospective investigation. BMJ 302(6788):1302–1305

Fukase K, Kato M, Kikuchi S, Inoue K, Uemura N, Okamoto S, Terao S, Amagai K, Hayashi S, Asaka M, Japan Gast Study Group (2008) Effect of eradication of *Helicobacter pylori* on incidence of metachronous gastric carcinoma after endoscopic resection of early gastric cancer: an open-label, randomised controlled trial. Lancet 372(9536):392–397. https://doi.org/10.1016/S0140-6736(08)61159-9

Gall A, Gaudet RG, Gray-Owen SD, Salama NR (2017) TIFA signaling in gastric epithelial cells initiates the cag type 4 secretion system-dependent innate immune response to *Helicobacter pylori* infection. MBio 8(4):e01168–e01117. https://doi.org/10.1128/mBio.01168-17

González I, Romero J, Rodríguez B, Llanos J, Morales E, Figueroa H, Perez-Castro R, Valdés E, Cofre C, Rojas A (2011) High prevalence of virulence-associated genotypes in *Helicobacter pylori* clinical isolates in the region del Maule, Chile. Scand J Infect Dis 43(8):652–655. https://doi.org/10.1016/j.micinf.2011.04.005

Hatakeyama M (2004) Oncogenic mechanisms of *Helicobacter pylori* CagA protein. Nat Rev Cancer 4(9):688–694. https://doi.org/10.1038/nrc1433

Hayashi T, Senda M, Morohashi H, Higashi H, Horio M, Kashiba Y, Nagase L, Sasaya D, Shimizu T, Venugopalan N, Kumeta H, Noda NN, Inagaki F, Senda T, Hatakeyama M (2012a) Tertiary structure-function analysis reveals the pathogenic signaling potentiation mechanism of *Helicobacter pylori* oncogenic effector CagA. Cell Host Microbe 12(1):20–33. https://doi.org/10.1016/j.chom.2012.05.010

Hayashi K, Suzuki A, Ohno S (2012b) PAR-1/MARK: a kinase essential for maintaining the dynamic state of microtubules. Cell Struct Funct 37(1):21–25

Hayashi D, Tamura A, Tanaka H, Yamazaki Y, Watanabe S, Suzuki K, Suzuki K, Sentani K, Yasui W, Rakugi H, Isaka Y, Tsukita S (2012c) Deficiency of claudin-18 causes paracellular H^+ leakage, up-regulation of interleukin-1β, and atrophic gastritis in mice. Gastroenterology 142(2):292–304. https://doi.org/10.1053/j.gastro.2011.10.040

Hayashi T, Senda M, Suzuki N, Nishikawa H, Ben C, Tang C, Nagase L, Inoue K, Senda T, Hatakeyama M (2017) Differential mechanisms for SHP2 binding and activation are exploited by geographically distinct *Helicobacter pylori* CagA oncoproteins. Cell Rep 20(12):2876–2890. https://doi.org/10.1016/j.celrep.2017.08.080

Herrero R, Parsonnet J, Greenberg ER (2014) Prevalence of gastric cancer. JAMA 312(12):1197–1198. https://doi.org/10.1001/jama.2014.10498

Higashi H, Tsutsumi R, Fujita A, Yamazaki S, Asaka M, Azuma T, Hatakeyama M (2002a) Biological activity of the *Helicobacter pylori* virulence factor CagA is determined by variation in the tyrosine phosphorylation sites. Proc Natl Acad Sci U S A 99(22):14428–14433. https://doi.org/10.1073/pnas.222375399

Higashi H, Tsutsumi R, Muto S, Sugiyama T, Azuma T, Asaka M, Hatakeyama M (2002b) SHP-2 tyrosine phosphatase as an intracellular target of *Helicobacter pylori* CagA protein. Science 295(5555):683–686. https://doi.org/10.1126/science.1067147

Huang JQ, Zheng GF, Sumanac K, Irvine EJ, Hunt RH (2003) Meta-analysis of the relationship between *cagA* seropositivity and gastric cancer. Gastroenterology 125(6):1636–1644

Ito Y, Azuma T, Ito S, Miyaji H, Hirai M, Yamazaki Y, Sato F, Kato T, Kohli Y, Kuriyama M (1997) Analysis and typing of the *vacA* gene from *cagA*-positive strains of *Helicobacter pylori* isolated in Japan. J Clin Microbiol 35(7):1710–1714

Kaplan-Türköz B, Jiménez-Soto LF, Dian C, Ertl C, Remaut H, Louche A, Tosi T, Haas R, Terradot L (2012) Structural insights into *Helicobacter pylori* oncoprotein CagA interaction with β1 integrin. Proc Natl Acad Sci U S A 109(36):14640–14645. https://doi.org/10.1073/pnas.1206098109

Kuipers EJ, Pérez-Pérez GI, Meuwissen SG, Blaser MJ (1995) *Helicobacter pylori* and atrophic gastritis: importance of the *cagA* status. J Natl Cancer Inst 87(23):1777–1780

Kuo SH, Chen LT, Lin CW, Wu MS, Hsu PN, Tsai HJ, Chu CY, Tzeng YS, Wang HP, Yeh KH, Cheng AL (2013) Detection of the *Helicobacter pylori* CagA protein in gastric mucosa-associated lymphoid tissue lymphoma cells: clinical and biological significance. Blood Cancer J 3:e125. https://doi.org/10.1038/bcj.2013.22

Lee DG, Kim HS, Lee YS, Kim S, Cha SY, Ota I, Kim NH, Cha YH, Yang DH, Lee Y, Park GJ, Yook JI, Lee YC (2014) *Helicobacter pylori* CagA promotes Snail-mediated epithelial-mesenchymal transition by reducing GSK-3 activity. Nat Commun 5(4423). https://doi.org/10.1038/ncomms5423

Lin WC, Tsai HF, Kuo SH, Wu MS, Lin CW, Hsu PI, Cheng AL, Hsu PN (2010) Translocation of *Helicobacter pylori* CagA into human B lymphocytes, the origin of mucosa-associated lymphoid tissue lymphoma. Cancer Res 70(14):5740–5748. https://doi.org/10.1158/0008-5472.CAN-09-4690

Lu H, Saito Y, Umeda M, Murata-Kamiya N, Zhang H, Higashi H, Hatakeyama M (2008) Structural and functional diversity in the PAR1b/MARK2-binding region of *Helicobacter pylori* CagA. Cancer Sci 99 (10):2004–2011. https://doi.org/10.1111/j.1349-7006.2008.00950.x

Matenia D, Mandelkow EM (2009) The tau of MARK: a polarized view of the cytoskeleton. Trends Biochem Sci 34(7):332–342. https://doi.org/10.1016/j.tibs.2009.03.008

Matsuo T, Ito M, Takata S, Tanaka S, Yoshihara M, Chayama K (2011) Low prevalence of *Helicobacter pylori*-negative gastric cancer among Japanese. Helicobacter 16(6):415–419. https://doi.org/10.1111/j.1523-5378.2011.00889.x

Matsusaka K, Funata S, Fukayama M, Kaneda A (2014) DNA methylation in gastric cancer, related to *Helicobacter pylori* and Epstein-Barr virus. World J Gastroenterol 20(14):3916–3926. https://doi.org/10.3748/wjg.v20.i14.3916

Mimuro H, Suzuki T, Tanaka J, Asahi M, Haas R, Sasakawa C (2002) Grb2 is a key mediator of *Helicobacter pylori* CagA protein activities. Mol Cell 10(4):745–755

Murata-Kamiya N, Kurashima Y, Teishikata Y, Yamahashi Y, Saito Y, Higashi H, Aburatani H, Akiyama T, Peek RMJR, Azuma T, Hatakeyama M (2007) *Helicobacter pylori* CagA interacts with E-cadherin and deregulates the β-catenin signal that promotes intestinal transdifferentiation in gastric epithelial cells. Oncogene 26(32):4617–4626. https://doi.org/10.1038/sj.onc.1210251

Murata-Kamiya N, Kikuchi K, Hayashi T, Higashi H, Hatakeyama M (2010) *Helicobacter pylori* exploits host membrane phosphatidylserine for delivery, localization and pathophysiological action of the CagA oncoprotein. Cell Host Microbe 7(5):399–411. https://doi.org/10.1016/j.chom.2010.04.005

Nagase L, Hayashi T, Senda T, Hatakeyama M (2015) Dramatic increase in SHP2 binding activity of *Helicobacter pylori* Western CagA by EPIYA-C duplication: its implications in gastric carcinogenesis. Sci Rep 5:15749. https://doi.org/10.1038/srep15749

Neal JT, Peterson TS, Kent ML, Guillemin K (2013) *H. pylori* virulence factor CagA increases intestinal cell proliferation by Wnt pathway activation in a transgenic zebrafish model. Dis Model Mech 6 (3):802–810. https://doi.org/10.1242/dmm.011163

Neel BG, Gu H, Pao L (2003) The 'Shp'ing news: SH2 domain-containing tyrosine phosphatases in cell signaling. Trends Biochem Sci 28(6):284–293. https://doi.org/10.1016/S0968-0004(03)00091-4

Nemati M, Larussa T, Khorramdelazad H, Mahmoodi M, Jafarzadeh A (2017) Toll-like receptor 2: An important immunomodulatory molecule during *Helicobacter pylori* infection. Life Sci 178:17–29. https://doi.org/10.1016/j.lfs.2017.04.006

Nesić D, Miller MC, Quinkert ZT, Stein M, Chait BT, Stebbins CE (2010) *Helicobacter pylori* CagA inhibits PAR1-MARK family kinases by mimicking host substrates. Nat Struct Mol Biol 17(1):130–132. https://doi.org/10.1038/nsmb.1705

Nishikawa H, Hayashi T, Arisaka F, Senda T, Hatakeyama M (2016) Impact of structural polymorphism for the *Helicobacter pylori* CagA oncoprotein on binding to polarity-regulating kinase PAR1b. Sci Rep 6:30031. https://doi.org/10.1038/srep30031

Nomura A, Stemmermann GN, Chyou PH, Kato I, Perez-Perez GI, Blaser MJ (1991) *Helicobacter pylori* infection and gastric carcinoma among Japanese Americans in Hawaii. N Engl J Med 325(16):1132–1136. https://doi.org/10.1056/NEJM199110173251604

Odenbreit S, Puls J, Sedlmaier B, Gerland E, Fischer W, Haas R (2000) Translocation of *Helicobacter pylori* CagA into gastric epithelial cells by type IV secretion. Science 287(5457):1497–1500

Ohnishi N, Yuasa H, Tanaka S, Sawa H, Miura M, Matsui A, Higashi H, Musashi M, Iwabuchi K, Suzuki M, Yamada G, Azuma T, Hatakeyama M (2008) Transgenic expression of *Helicobacter pylori* CagA induces gastrointestinal and hematopoietic neoplasms in mouse. Proc Natl Acad Sci U S A 105 (3):1003–1008. https://doi.org/10.1073/pnas.0711183105

Ono S, Kato M, Suzuki M, Ishigaki S, Takahashi M, Haneda M, Mabe K, Shimizu Y (2012) Frequency of *Helicobacter pylori*-negative gastric cancer and gastric mucosal atrophy in a Japanese endoscopic submucosal dissection series including histological, endoscopic and serological atrophy. Digestion 86(1):59–65. https://doi.org/10.1159/000339176

Park JY, Forman D, Waskito LA, Yamaoka Y, Crabtree JE (2018) Epidemiology of *Helicobacter pylori* and CagA-positive infections and global variations in gastric cancer. Toxins 10(4):163. https://doi.org/10.3390/toxins10040163

Parsonnet J, Friedman GD, Vandersteen DP, Chang Y, Vogelman JH, Orentreich N, Sibley RK (1991) *Helicobacter pylori* infection and the risk of gastric carcinoma. N Engl J Med 325(16):1127–1131. https://doi.org/10.1056/NEJM199110173251603

Parsonnet J, Friedman GD, Orentreich N, Vogelman H (1997) Risk for gastric cancer in people with CagA positive or CagA negative *Helicobacter pylori* infection. Gut 40(3):297–301. PMC1027076

Plummer M, de MArtel C, Vignat J, Ferlay J, Bray F, Franceschi S (2016) Global burden of cancers attributable to infections in 2012: a synthetic analysis. Lancet Glob Health 4(9):e609–e616. https://doi.org/10.1016/S2214-109X(16)30143-7

Poppe M, Feller SM, Römer G, Wessler S (2007) Phosphorylation of *Helicobacter pylori* CagA by c-Abl leads to cell motility. Oncogene 26(24):3462–3472. https://doi.org/10.1038/sj.onc.1210139

Ren S, Higashi H, Lu H, Azuma T, Hatakeyama M (2006) Structural basis and functional consequence of *Helicobacter pylori* CagA multimerization in Cells. J Biol Chem 281(43):32344–32352. https://doi.org/10.1074/jbc.M606172200

Rieder G, Merchant JL, Haas R (2005) *Helicobacter pylori cag*-type IV secretion system facilitates corpus colonization to induce precancerous conditions in Mongolian gerbils. Gastroenterology 128(5):1229–1242

Saadat I, Higashi H, Obuse C, Umeda M, Murata-Kamiya-N, Saito Y, Lu H, Ohnishi N, Azuma T, Suzuki A, Ohno S, Hatakeyama M (2007) *Helicobacter pylori* CagA targets PAR1/MARK kinase to disrupt epithelial cell polarity. Nature 447(7142):330–333. https://doi.org/10.1038/nature05765

Saju P, Murata-Kamiya N, Hayashi T, Senda Y, Nagase L, Noda S, Matsusaka K, Funata S, Kunita A, Urabe M, Seto Y, Fukayama M, Kaneda A, Hatakeyama M (2016) Host SHP1 phosphatase antagonizes *Helicobacter pylori* CagA and can be downregulated by Epstein-Barr virus. Nat Microbiol 1:16026. https://doi.org/10.1038/nmicrobiol.2016.26

Satomi S, Yamakawa A, Matsunaga S, Masaki R, Inagaki T, Okuda T, Suto H, Ito Y, Yamazaki Y, Kuriyama M, Keida Y, Kutsumi H, Azuma T (2006) Relationship between the diversity of the *cagA* gene of *Helicobacter pylori* and gastric cancer in Okinawa, Japan. J Gastroenterol 41(7):668–673. https://doi.org/10.1007/s00535-006-1838-6

Segal ED, Cha J, Lo J, Falkow S, Tompkins LS (1999) Altered states: Involvement of phosphorylated CagA in the induction of host cellular growth changes by *Helicobacter pylori*. Proc Natl Acad Sci U S A 96 (25):14559–14564. PMC24475

Selbach M, Moese S, Hauck CR, Meyer TF, Backert S (2002) Src is the kinase of the *Helicobacter pylori* CagA protein *in vitro* and *in vivo*. J Biol Chem 277 (9):6775–6778. https://doi.org/10.1074/jbc.C100754200

Selbach M, Paul FE, Brandt S, Guye P, Daumke O, Backert S, Dehio C, Mann M (2009) Host cell interactome of tyrosine-phosphorylated bacterial proteins. Cell Host Microbe 5(4):397–403. https://doi.org/10.1016/j.chom.2009.03.004

Sicinschi LA, Correa P, Peek RM, Camargo MC, Piazuelo MB, Romero-Gallo J, Hobbs SS, Krishna U, Delgado A, Mera R, Bravo LE, Schneider BG (2010) CagA C-terminal variations in *Helicobacter pylori* strains from Colombian patients with gastric precancerous lesions. Clin Microbiol Infect 16 (4):369–378. https://doi.org/10.1111/j.1469-0691.2009.02811.x

Stein M, Rappuoli R, Covacci A (2000) Tyrosine phosphorylation of the *Helicobacter pylori* CagA antigen after cag-driven host cell translocation. Proc Natl Acad Sci U S A 97(3):1263–1268. PMC15590

Stein M, Bagnoli F, Halenbeck R, Rappuoli R, Fantl WJ, Covacci A (2002) c-Src/Lyn kinases activate *Helicobacter pylori* CagA through tyrosine phosphorylation of the EPIYA motifs. Mol Microbiol 43 (4):971–980. https://doi.org/10.1046/j.1365-2958.2002.02781.x

Stein SC, Faber E, Bats SH, Murillo T, Speidel Y, Coombs N, Josenhans C (2017) *Helicobacter pylori* modulates host cell responses by CagT4SS-dependent translocation of an intermediate metabolite of LPS inner core heptose biosynthesis. PLoS Pathog 13(7): e1006514. https://doi.org/10.1371/journal.ppat.1006514

Suzuki M, Mimuro H, Suzuki T, Park M, Yamamoto T, Sasakawa C (2005) Interaction of CagA with Crk plays an important role in *Helicobacter pylori*-induced loss of gastric epithelial cell adhesion. J Exp Med 202 (9):1235–1247. https://doi.org/10.1084/jem.20051027

Suzuki M, Mimuro H, Kiga K, Fukumatsu M, Ishijima N, Morikawa H, Nagai S, Koyasu S, Gilman RH, Kersulyte D, Berg DE, Sasakawa C (2009) *Helicobacter pylori* CagA phosphorylation-independent function in epithelial proliferation and inflammation. Cell Host Microbe 5(1):23–34. https://doi.org/10.1016/j.chom.2008.11.010

Suzuki N, Murata-Kamiya N, Yanagiya K, Suda W, Hattori M, Kanda H, Bingo A, Fujii Y, Maeda S, Koike K, Hatakeyama M (2015) Mutual reinforcement of inflammation and carcinogenesis by the *Helicobacter pylori* CagA oncoprotein. Sci Rep 5:10024. https://doi.org/10.1038/srep10024

Tammer I, Brandt S, Hartig R, König W, Backert S (2007) Activation of Abl by *Helicobacter pylori*: a novel kinase for CagA and crucial mediator of host cell scattering. Gastroenterology 132(4):1309–1319. https://doi.org/10.1053/j.gastro.2007.01.050

Tartaglia M, Niemeyer CM, Fragale A, Song X, Buechner J, Jung A, Hählen K, Hasle H, Licht JD, Gelb BD (2003) Somatic mutations in *PTPN11* in juvenile myelomonocytic leukemia, myelodysplastic syndromes and acute myeloid leukemia. Nat Genet 34(2):148–150. https://doi.org/10.1038/ng1156

Troppan K, Wenzl K, Neumeister P, Deutsch A (2015) Molecular pathogenesis of MALT lymphoma. Gastroenterol Res Pract 2015:102656. https://doi.org/10.1155/2015/102656

Tsang YH, Lamb A, Romero-Gallo J, Huang B, Ito K, Peek RM Jr, Ito Y, Chen LF (2010) *Helicobacter pylori* CagA targets gastric tumor suppressor RUNX3 for proteasome-mediated degradation. Oncogene 29 (41):5643–5650. https://doi.org/10.1038/onc.2010.304

Tummuru MK, Cover TL, Blaser MJ (1993) Cloning and expression of a high-molecular-mass major antigen of *Helicobacter pylori*: evidence of linkage to cytotoxin production. Infect Immun 61(5):1799–1809. PMC280768

Uemura N, Okamoto S, Yamamoto S, Matsumura N, Yamaguchi S, Yamakido M, Taniyama K, Sasaki N, Schlemper RJ (2001) *Helicobacter pylori* infection and the development of gastric cancer. N Engl J Med 345 (11):784–789. https://doi.org/10.1056/NEJMoa001999

Umehara S, Higashi H, Ohnishi N, Asaka M, Hatakeyama M (2003) Effects of *Helicobacter pylori* CagA protein on the growth and survival of B lymphocytes, the origin of MALT lymphoma. Oncogene 22 (51):8337–8342. https://doi.org/10.1038/sj.onc.1207028

Viala J, Chaput C, Boneca IG, Cardona A, Girardin SE, Moran AP, Athman R, Mémet S, Huerre MR, Coyle AJ, DiStefano PS, Sansonetti PJ, Labigne A, Bertin J, Philpott DJ, Ferrero RL (2004) Nod1 responds to peptidoglycan delivered by the *Helicobacter pylori cag* pathogenicity island. Nat Immunol 5(11):1166–1174. https://doi.org/10.1038/ni1131

Vilaichone RK, Mahachai V, Tumwasorn S, Wu JY, Graham DY, Yamaoka Y (2004) Molecular epidemiology and outcome of *Helicobacter pylori* infection in Thailand: a cultural cross roads. Helicobacter 9 (5):453–459. https://doi.org/10.1111/j.1083-4389.2004.00260.x

Warren JR, Marshall B (1983) Unidentified curved bacilli on gastric epithelium in active chronic gastritis. Lancet 1(8390):1273–1275. https://doi.org/10.1016/S0140-6736(83)92719-8

Wei J, Noto JM, Zaika E, Romero-Gallo J, Piazuelo MB, Schneider B, El-Rifai W, Correa P, Peek RM, Zaika AI (2015) Bacterial CagA protein induces degradation of p53 protein in a p14^{ARF}-dependent manner. Gut 64 (7):1040–1048. https://doi.org/10.1136/gutjnl-2014-307295

Wong BC, Lam SK, Wong WM, Chen JS, Zheng TT, Feng RE, Lai KC, Hu WH, Yuen ST, Leung SY, Fong DY, Ho J, Ching CK, Chen JS, China Gastric Cancer Study Group (2004) *Helicobacter pylori* eradication to prevent gastric cancer in a high-risk region of China: a randomized controlled trial. JAMA 291 (12):187–194. https://doi.org/10.1001/jama.291.2.187

Wu CY, Kuo KN, Wu MS, Chen YJ, Wang CB, Lin JT (2009) Early *Helicobacter pylori* eradication decreases risk of gastric cancer in patients with peptic ulcer disease. Gastroenterology 137(5):1641–1648. https://doi.org/10.1053/j.gastro.2009.07.060

Xia Y, Yamaoka Y, Zhu Q, Matha I, Gao X (2009) A comprehensive sequence and disease correlation analyses for the C-terminal region of CagA protein of *Helicobacter pylori*. PLoS One 4(11):e7736. https://doi.org/10.1371/journal.pone.0007736

Zeaiter Z, Cohen D, Müsch A, Bagnoli F, Covacci A, Stein M (2008) Analysis of detergent-resistant membranes of *Helicobacter pylori* infected gastric adenocarcinoma cells reveals a role for MARK2/Par1b in CagA-mediated disruption of cellular polarity. Cell Microbiol 10(3):781–794. https://doi.org/10.1111/j.1462-5822.2007.01084.x

Zhou P, She Y, Dong N, Li P, He H, Borio A, Wu Q, Lu S, Ding X, Cao Y, Xu Y, Gao W, Dong M, Ding J, Wang DC, Zamyatina A, Shao F (2018) Alpha-kinase 1 is a cytosolic innate immune receptor for bacterial ADP-heptose. Nature 561(7721):122–126. https://doi.org/10.1038/s41586-018-0433-3

Zimmermann S, Pfannkuch L, Al-Zeer MA, Bartfeld S, Koch M, Liu J, Rechner C, Soerensen M, Sokolova O, Zamyatina A, Kosma P, Mäurer AP, Glowinski F, Pleissner KP, Schmid M, Brinkmann V, Karlas A, Naumann M, Rother M, Machuy N, Meyer TF (2017) ALPK1- and TIFA-dependent innate immune response triggered by the *Helicobacter pylori* Type IV secretion system. Cell Rep 20(10):2384–2395. https://doi.org/10.1016/j.celrep.2017.08.039

Zullo A, Hassan C, Ridola L, Repici A, Manta R, Andriani A (2014) Gastric MALT lymphoma: old and new insights. Ann Gastroenterol 27(1):27–33. PMC3959547

Adv Exp Med Biol - Advances in Microbiology, Infectious Diseases and Public Health (2019) 1149: 151–172
https://doi.org/10.1007/5584_2019_364

Published online: 24 April 2019

The Role of Host Genetic Polymorphisms in *Helicobacter pylori* Mediated Disease Outcome

Marguerite Clyne and Marion Rowland

Abstract

The clinical outcome of infection with the chronic gastric pathogen *Helicobacter pylori* is not the same for all individuals and also differs in different ethnic groups. Infection occurs in early life (<3 years of age), and while all infected persons mount an immune response and develop gastritis, the majority of individuals are asymptomatic. However, up to 10–15% develop duodenal ulceration, up to 1% develop gastric cancer (GC) and up to 0.1% can develop gastric mucosa-associated lymphoid tissue (MALT) lymphoma. The initial immune response fails to clear infection and *H. pylori* can persist for decades. *H. pylori* has been classified as a group one carcinogen by the WHO. Interestingly, development of duodenal ulceration protects against GC. Factors that determine the outcome of infection include the genotype of the infecting strains and the environment. Host genetic polymorphisms have also been identified as factors that play a role in mediating the clinical outcome of infection. Several studies present compelling evidence that polymorphisms in genes involved in the immune response such as pro and anti-inflammatory cytokines and pathogen recognition receptors (PRRs) play a role in modulating disease outcome. However, as the number of studies grows emerging confounding factors are small sample size and lack of appropriate controls, lack of consideration of environmental and bacterial factors and ethnicity of the population. This chapter is a review of current evidence that host genetic polymorphisms play a role in mediating persistent *H. pylori* infection and the consequences of the subsequent inflammatory response.

Keywords

Helicobacter pylori mediated disease · Host genetic polymorphism · Toll-like receptor · Interleukin 1 beta · NOD-like receptor · Autophagy

M. Clyne (✉)
School of Medicine and The Conway Institute of Biomolecular and Biomedical Research, University College Dublin, Dublin, Ireland
e-mail: marguerite.clyne@ucd.ie

M. Rowland
School of Medicine, University College Dublin, Dublin, Ireland

1 Introduction

Helicobacter pylori is one of the commonest infections of mankind with up to 50% of the global population harbouring the bacteria. However, there has been a rapid decline in the prevalence of infection since it was first reported in 1983 and an accompanying decline in the incidence of *H. pylori* associated diseases such as peptic ulcer disease (PUD) and GC (Lanas and

Chan 2017; Leow et al. 2016; McColl 2010). Despite this rapid decline in the incidence of *H. pylori* in developing and developed countries GC remains an important public health issue particularly in East Asia, and South America. A recent meta-analysis on the worldwide burden of cancer related to infection estimated that almost 90% of non-cardia GC is due to *H. pylori* and that those infected with *H. pylori* are a high-risk subgroup even in populations at low risk of GC (Lee et al. 2016).

H. pylori infection occurs in early childhood usually before the age of 3 years (Rowland et al. 2006), and similar to other infectious diseases is clustered in families (Drumm et al. 1990). Infection is life-long unless eradicated with antimicrobials. The response of the gastric mucosa to *H. pylori* infection is markedly different between children and adults (Serrano et al. 2013; Bontems et al. 2014). Children do not mount a strong humoral immune response to *H. pylori* (Crabtree et al. 1991b). They develop an antral predominant gastritis with minimal polymorphonuclear and mononuclear cell infiltrates. In contrast, the gastric mucosa of *H. pylori* infected adults demonstrates a dense polymorphorphonuclear and mononuclear cell infiltrate, lymphoid follicles and reactive epithelial changes. In two studies from Chile using two different cohorts of patients, it has been shown that children display a much milder histological pattern of gastritis compared to adults (Harris et al. 2008; Hernandez et al. 2014). The reduced inflammatory response to *H. pylori* in children may be explained by an enhanced regulatory T cell (Treg) response. *H. pylori* infected children have significantly more Treg cells in their antral mucosa and higher levels of protein and mRNA for Treg cytokines TGF-β1 and IL-10 compared to uninfected children or *H. pylori* infected adults. This was not due to a difference in either bacterial load or bacterial virulence factors between children and adults. Furthermore, this group also demonstrated that *H. pylori* infected children had significantly fewer gastric Th17 cells and significantly lower levels of IL-17 specific mRNA protein compared to similarly infected adults (Harris et al. 2008). These findings indicate that in adults, *H. pylori* gastritis is a consequence of both Th17 and Th1 immune-mediated inflammatory pathways, but in children both pathways are downregulated. This suggests the possibility that there is an upregulation of the immune regulatory response over time with a loss of the host Treg response, the reasons for which are unknown.

The clinical outcome of *H. pylori* infection is not the same in all individuals (Posselt et al. 2013). While all infected individuals have histological evidence of gastritis, those with duodenal ulcer (DU) disease have antral predominant gastritis with a high acid output (McColl et al. 1997) and GC is associated with corpus predominant gastritis, hypochlorhydria and gastric atrophy (Kuipers 1999). Long before the relationship between *H. pylori* and GC was established, it was known that duodenal ulceration was protective against GC (Howson et al. 1986; Susser and Stein 1962).

The outcome of *H. pylori* also differs throughout the world in different ethnic groups. There is a particularly high incidence of infection and GC in Eastern Asia, Eastern Europe and parts of Central and South America (Yamaoka et al. 2008). In Africa there is a high incidence of infection, but this does not result in high rates of either DU or GC (de Martel et al. 2013; Holcombe 1992). While the genotype of the infecting strain certainly plays a pivotal role in determining the outcome of infection, host and environmental factors are also involved. The gold standard for distinguishing genetic from environmental traits has been the study of twins. In twin studies the genetic influences have been shown to be of moderate importance for risk of PUD (Malaty et al. 1994) and heritability has been estimated to account for 28% (95% CI 0-51%, $P = \text{NS}$) of the susceptibility to GC (Lichtenstein et al. 2000). Shared environment accounted for 10%, and non-shared environment for 62% of the risk associated with GC. This suggests that while genetic factors may play a role, it is likely that the interaction between host and environmental factors is of pivotal importance for GC development.

Numerous studies have sought to determine the role of genetic host polymorphisms of genes encoding for immune response molecules such as the PRR molecules, Toll-like receptors (TLRs) and nucleotide-binding oligomerisation domain (NOD) like receptors (NLRs), and cytokines and the outcome of *H. pylori* infection (Pachathundikandi et al. 2013). This chapter focusses on reviewing host genetic polymorphisms and their potential relevance to understanding the divergent outcomes of *H. pylori* mediated disease in different individuals and in different ethnic populations.

2 Host Genetic Polymorphisms

2.1 Interleukin-1 Beta (IL-1β)

IL-1β is a potent inhibitor of gastric acid secretion and expression is upregulated upon *H. pylori* infection. The IL-1 receptor antagonist (IL-1RA), encoded by *IL-1RN*, competes with IL-1β for binding to the IL-1 receptor on cells, and thus can inhibit IL-1β production. Transgenic mice overproducing IL-1β in the stomach develop gastric inflammation and carcinoma (Tu et al. 2008) and IL-1β null mice have a decreased propensity to develop tumours in a model, in which GCs are induced by N-methyl-N-nitrosourea (MNU), a direct-acting alkylating agent (Shigematsu et al. 2013). Together, these data demonstrate the role of IL-1β in development of GC. The IL-1β gene is mapped to chromosome 2q14 and has 3 single nucleotide polymorphisms (SNPs): 31T/C, 511 C/T, and 3,954 C/T. These 3 SNPs are located in the promoter region of chromosome 2q14 and functional expression of them results in overexpression of IL-1β. Building on previous work that found a high prevalence of hypochlorhydria and gastric atrophy within a cohort of *H. pylori* infected first degree relatives of GC patients in Scotland, El-Omar and colleagues assessed the impact of polymorphisms in the *IL-1β* gene cluster, consisting of *IL-1β* and *IL-1RN*, on the development of GC in the same population and in Polish individuals (El-Omar et al. 2000). Specifically, the presence of two pro-inflammatory genotypes of the *IL-1* loci, *IL-1β -31T+* and *IL-1RN*2/*2* increased both the likelihood of a chronic hypochlorhydric response to *H. pylori* infection and the risk of GC development. In the absence of *H. pylori* infection this relationship did not exist. Machado and co-workers provided further evidence, which showed the association between *IL-1* gene cluster pro-inflammatory polymorphisms and increased risk of gastric carcinoma in a population from Northern Portugal, an area that has a particularly high incidence of GC (Machado et al. 2001). Statistical analysis showed an interaction between the *IL-1β-511T* and *IL-1RN*2* loci with the risk conferred by the *IL-1β*-511T allele substantially increased in individuals homozygous for the *IL*-1RN*2 allele (Machado et al. 2001).

Several studies have since looked at the effect of *IL-1β* polymorphism and susceptibility of different populations to development of GC. An association between *IL-1β* polymorphism and GC has been shown in both Asian and Latin American populations (Kumar et al. 2009; Sun et al. 2015). A recent meta-analysis, which included studies on both Caucasian and Asian populations, suggested that the *IL-1β*-31 polymorphism confers susceptibility to GC in the presence of *H. pylori* infection (Ying et al. 2016). A positive relationship between Asian ethnicity, *H. pylori* infection and the presence of the *IL-1β*-511 C/T SNP and stomach carcinoma susceptibility has also been described (Park et al. 2015). A meta-analysis of studies done in Chinese populations found there was a synergistic interaction between the IL-1β 511 C/T polymorphism and *H. pylori* in the development of GC (Chen et al. 2016). In the Brazilian population the *IL-1β*-511 CC and CT gene polymorphisms were associated with chronic gastritis and GC development in *H. pylori* infected individuals. However, there was no correlation found between *IL-1RN* gene polymorphisms and gastric disease (Santos et al. 2012). Other studies from Brazil found either no association with the *IL-1β*-511 and -31T SNPs (Gatti et al. 2004) or found a positive correlation between the *IL-1 RN**2 polymorphism and infection with CagA-positive strains of *H. pylori* with GC (Melo Barbosa et al. 2009). These contrasting results may be due to the large population in Brazil and the presence of different ethnic groups as the frequency of genetic polymorphisms can differ between different populations (Melo Barbosa et al. 2009).

The relationship of gastric mucosal IL-1β levels and development of gastroesophageal reflux disease (GERD) has been evaluated in a number of different populations. *IL-1β* and *IL-1RN* pro-inflammatory genotypes and infection with CagA-positive *H. pylori* were associated with a decreased risk of erosive oesophagitis in a Brazilian population (Queiroz et al. 2004). Ando and colleagues hypothesised that the pro-inflammatory *IL-1β* polymorphisms might decrease the risk of GERD symptoms and erosive oesophagitis by increasing the development of hypochlorhydria and gastric atrophy, especially among *H. pylori* infected subjects (Ando et al. 2006). They examined prospectively the effect of the pro-inflammatory cytokine polymorphisms, *IL-1β*-511 C/T, *IL-10*-819T/C, and *TNFα*-1,031T/C, on *H. pylori* induced gastric atrophy, reflux symptoms, and erosive oesophagitis in 320 Japanese patients. A pro-inflammatory *IL-1β* profile was associated with an increased risk of gastric atrophy. Among *H. pylori* positive patients, subjects homozygous for the pro-inflammatory allele *IL-1β*-511T had a significantly lower risk of erosive oesophagitis and GERD compared with those homozygous for the −511C allele indicating that *IL-1β* polymorphisms, *H. pylori*, and atrophy are independent markers of decreased risk of GERD (Ando et al. 2006). However, there was no statistically significant association found between genetic polymorphisms of *IL-1β*-511 and *IL-1RN* and GERD in a Korean population (Kim et al. 2013c) and *IL-1β*-511T/T and -31 C/C genotypes, were found to be associated with an increased risk of reflux esophagitis in Taiwan (Cheng et al. 2010).

2.2 Tumour Necrosis Factor Alpha (TNF-α)

Like *IL-1β,* the cytokine TNF-α is also pro-inflammatory and an inhibitor of acid production. Expression of TNF-α is upregulated in the gastric mucosa of *H. pylori* infected individuals (Crabtree et al. 1991a; Wilson and Crabtree 2007). Five biallelic SNPs within the promotor region of *TNF-α* have been identified. G23A, G308A, C857T, C863A and T1031C some of which affect the expression of TNF-α. Machado and co-workers found that carriers of the *TNF-α*-308A allele were at increased risk for GC development (Machado et al. 2003). However, no statistically significant increased risk of developing GC was found for any combination of bacterial and *TNF-α*-308 genotypes when compared with the risks conferred by the bacterial genotypes alone. El-Omar and co-workers reported that pro-inflammatory genotypes of *TNF-α* and *IL-10* were each associated with doubling the risk of noncardia GC (El-Omar et al. 2003). This risk increased further when combined with carriage of other pro-inflammatory polymorphisms in *IL-1β, IL-1 RN, TNF-α,* and *IL-10* with odds ratios (ORs) increasing from 2.8 for one polymorphism to 27.3 for up to 4 high-risk genotypes. Strikingly, these polymorphisms were not related to development of upper gastrointestinal tract cancers such as oesophageal or gastric cardia cancer. In a Mexican population, 278 individuals were studied to investigate the role of the *TNF-α*-308 and *IL-1β-31, IL-1RN* gene polymorphisms as risk factors for the development of GC. The *IL-1β* pro-inflammatory genotype increased the risk of distal GC, but there was no association with other polymorphisms (Garza-Gonzalez et al. 2005).

In a recent study in China, 47 families containing individuals diagnosed with GC (case family, n = 296), were matched with 47 families without GC (control family, n = 319). Haplotypes of *TNF-α*-308/-238 GA/GG, AA/GG and AA/GA were found to increase the risk of developing GC. First degree relatives of GC patients with *TNF-α* G308 polymorphisms positive for *H. pylori* were more likely to develop GC than second-degree relatives were (Xu et al. 2017b).

Polymorphisms in *TNF-α* have also been shown to be associated with PUD. In a study, which looked at 130 patients with DU, 50 with gastric ulcer and 102 ethnically-matched Spanish Caucasian healthy controls the *TNF-α*-G308A genotype was associated with an increased risk of DU (Lanas et al. 2001). For *H. pylori* positive patients from Japan that were classified according to disease type gastritis only (n = 164), gastric ulcers (n = 110), DU (n = 94), or GC (n = 105) polymorphisms in high-producer alleles of *TNF-α* (*TNF-α*-857T, *TNF-α*-863 A and *TNF-α*-1,031 C) were associated with an increased risk for both gastric ulcers and GC when compared with *H. pylori* negative controls (n = 172). This risk increased with carriage of more than one allele. However, no relationship was found between the development of *H. pylori*-related diseases and polymorphisms of *TNF-α*-308 and *IL-1β*-511/-31 (Sugimoto et al. 2007).

Meta-analysis studies have provided both contradictory and confirmatory results. Meta-analysis of 24 studies suggested that *TNF-α* -308G > A and − 1,031T > C polymorphisms may actually be protective factors against *H. pylori* infection, but that -863C > A may be a risk factor, especially in Asian populations (Sun et al. 2016). A total of 16 studies reporting *TNF-α* -308G/A, −1,031T/C, -863C/A, -857C/T, and -238G/A polymorphism were included in another meta-analysis. There was no statistically significant association found between -308G/A polymorphism and DU in the overall study population, as well as subgroup analyses by ethnicity, study design, and *H. pylori* status. Moreover, there was no statistical evidence of significant association between any of the studied *TNF-α* SNPs and DU (Zhang et al. 2013).

2.3 Other Cytokines

Altered expression of pro- and anti-inflammatory cytokines play an important role in mediating the immune response and development of gastritis upon infection with *H. pylori*. Therefore, genetic polymorphisms that might alter the production of such cytokines are likely to play a role in mediating the outcome of the infection.

Studies on the role of polymorphisms in *IL-2* have to date revealed controversial results. A study from Japan found that the *IL-2*T-330G polymorphism was associated with an increased risk of gastric atrophy, a precursor of GC, induced by *H. pylori* infection (Togawa et al. 2005). In a later study from Korea this polymorphism was not found to be associated with either an increased risk of GC or DU or gastric ulcer disease (Shin et al. 2008). Another study from Brazil found that carriers of the *IL-2*-330T/G polymorphism had a decreased risk of becoming infected with *H. pylori*. The polymorphism was associated with increased IL-2 and IFN-γ levels and it was suggested that these might protect against *H. pylori* infection in adulthood (Queiroz et al. 2009). A more recent study from Brazil where 80% of the subjects studied were of European descent found that, among patients with *H. pylori* infection, the *IL-2* -330 GG and + 114 TT genotypes are significantly associated with a high risk of developing GC, as is the -330G/+114T haplotype (Melchiades et al. 2017).

A case-control study in Chinese subjects investigated the association between polymorphisms in *IL-6/IL-6R* and GC risk. Three polymorphisms in the promoter region of *IL-6* (rs6949149, rs1800796, rs10499563) with the potential to effect IL-6 production, and one in exon 9 of *IL-6R* (rs2228145), all of which may alter the IL-6/IL-6R pathway, were examined. 473 GC patients and 474 age- and sex-matched healthy controls were included. In contrast to previous studies in this population (Du et al. 2015; Yin et al. 2012), *IL-6* rs1800796 CG genotype was associated with a decreased risk of GC in male subjects. In addition, *IL-6* rs10499563 was associated with a decreased risk of GC for *H. pylori* positive subjects, while, *IL-6R* rs2228145 was associated with decreased GC risk for *H. pylori* negative subjects (Zhang et al. 2017).

There are not many studies on genetic polymorphisms in children and the outcome of *H. pylori* infection. Such studies are of interest because it is as children that the majority of people are infected. A recent study from Romania found that the *IL-6*190C/T, *IL-6572*G/C, *TNF-α* 308G/A, and *ACE* I/D genetic polymorphisms were all associated with *H. pylori* infection and the risk for acquiring *H. pylori* infection was twice as high where 2 variant alleles coexisted

(Marginean et al. 2017). The role of polymorphisms in genes encoding for IL-6, IL-8 and IL-10 and the risk of PUD is reviewed elsewhere (Miftahussurur and Yamaoka 2015).

2.4 Toll-Like Receptors (TLRs)

TLRs are a class of PRRs expressed on the surface of immune cells such as monocytes, macrophages and dendritic cells and also on epithelial cells which act to recognise essential microbial products conserved in structure such as lipopolysaccharide (LPS) (TLR4), flagellin (TLR5) and unmethylated CPG motifs (TLR9). These conserved microbial products are referred to as pathogen associated molecular patterns (PAMPs). TLRs play a key role in innate immunity. *H. pylori* expresses several PAMPs that are ordinarily recognised by TLRs. While *H. pylori* LPS was initially thought to bind to TLR4 (Maeda et al. 2001; Su et al. 2003), subsequent studies have contradicted this and to date there is still some confusion in the literature. There is evidence that *H. pylori* LPS binds to TLR2 (Rad et al. 2009) and induces the expression of a discrete set of chemokines in epithelial cells (Smith et al. 2011). *H. pylori* LPS binding to TLR2 may induce cell proliferation which in turn upregulates expression of TLR4. Increased expression of TLR4 may increase interaction of TLR4 with LPS from other bacteria, and thus promote a strong inflammatory reaction (Yokota et al. 2010). Heat shock protein 60 (HSP60) has been shown to induce IL-8 production through TLR2 and TLR4 pathways in Kato-III cells, a gastric adenocarcinoma epithelial cell line (Takenaka et al. 2004) and through TLR2 recognition in human monocytes (Zhao et al. 2007). However, HSP 60 mediated induction of IL-6 by macrophages was shown to be via a TLR2- and TLR4-independent mechanism (Gobert et al. 2004).

Although flagellin of *H. pylori* does not activate TLR5 (Andersen-Nissen et al. 2005), both TLR2 and TLR5 can be activated by *H. pylori* by a yet unknown mechanism (Smith et al. 2006; Kumar Pachathundikandi et al. 2011). In addition, TLR9 on dendritic cells recognises *H. pylori* DNA (Smith et al. 2011). The gastric epithelia of children have been shown to respond to *H. pylori* infection by increasing the expression of TLR2, TLR4, TLR5 and TLR9 together with the cytokines IL-8, IL-10 and TNF-α (Lagunes-Servin et al. 2013), indicating that *H. pylori* is recognised by various TLRs upon infection in humans.

Chronic infection with *H. pylori* is a strong risk factor for development of GC (Schistosomes, liver flukes and *Helicobacter pylori*. IARC Working Group on the Evaluation of Carcinogenic Risks to Humans. Lyon, 7–14 June 1994 1994). *TLR* genes, which play a role in *H. pylori* induced chronic inflammation, have also been implicated in GC susceptibility. A SNP in the LPS recognition receptor complex TLR4, *TLR4*-896A/G, rs4986790 has been shown to be a risk factor at various stages of *H. pylori* induced gastric carcinogenesis and carriers have an 11 fold increased OR for hypochlorhydria (Hold et al. 2007). Castano-Rodriguez and co-workers (2013) conducted a case-control study and meta-analysis assessing *TLR2*–196 to −174del, *TLR4* Asp299Gly, *TLR4* Thr399Ile, *TLR4* rs11536889 and *CD14*–260 C/T and their association with gastric carcinogenesis (Castano-Rodriguez et al. 2013). Seventy Chinese individuals with primary non-cardia GC and 214 control patients with functional dyspepsia were included in the case control study. Individuals harbouring the *TLR4* rs11536889 T allele or the *TLR2*–196 to -174 deletion allele, who were also infected with *H. pylori,* were found to be at most risk of developing GC (OR: 9.75 and 3.10 respectively). Meta-analysis showed that the role of TLR signalling in development of GC is ethnic specific (Castano-Rodriguez et al. 2013). Another case-control study comprising of 310 ethnic Chinese individuals (87 non-cardia GC cases and 223 controls with functional dyspepsia) was conducted by the same group. Out of 25 polymorphisms detected by MALDI-TOF mass spectrometry, PCR, PCR-RFLP and real-time PCR, 7 polymorphisms showed significant associations with GC (*TLR4* rs11536889, *TLR4* rs10759931, *TLR4* rs1927911, *TLR4* rs10116253, *TLR4* rs10759932, *TLR4* rs2149356 and *CD14*–260 C/T). In multivariate analyses, *TLR4* rs11536889 remained a risk factor for GC (OR: 3.58) and *TLR4* rs10759932 decreased the risk of *H. pylori* infection (OR: 0.59) (Castano-Rodriguez et al. 2014b).

Another case control study recently evaluated genetic variations of *TLR5* in a Chinese population, and their interaction with *H. pylori* infection in GC patients. Minor alleles of three SNPs, including rs5744174 ($P = 0.001$), rs1640827 ($P = 0.005$), and rs17163737 ($P = 0.004$), were significantly associated with an increased risk of GC (ORs ranged from 1.20 to 1.24). Significant associations with *H. pylori* infection were identified for rs1640827 ($P = 0.009$) and rs17163737 ($P = 0.006$). Together the findings suggest that genetic variants in *TLR5* can alter the role of *H. pylori* infection in development of GC in a Chinese population (Xu et al. 2017a). Another study in Indian Tamils found the rs2072493, rs5744174, and rs5744168 polymorphisms within *TLR5* gene did not confer a significant risk for chronic *H. pylori* infection and the rs2072493 polymorphism conferred resistance to infection in the population. However, it should be noted that the authors highlighted that incidence of these polymorphisms was low in the population (Goda et al. 2017). In the same population individuals harbouring heterozygous and homozygous polymorphic variants of *TLR4* had a significantly higher risk of developing chronic *H. pylori* infection and PUD [rs4986790 AG, OR-2.7, GG, OR-9.8, rs4986791CT, OR-7.2, TT, OR-7.9] compared to control subjects. Also, the heterozygous variant of *TLR9* rs352140, was associated with persistence of infection (OR-1.87) (Loganathan et al. 2017).

The *TLR9*–1237 C allele has also been shown to be significantly associated with the development of *H. pylori* induced premalignant gastric changes in 168 healthy Caucasian first-degree relatives of GC patients from the West of Scotland. While the *TLR9* polymorphism was not associated with risk of *H. pylori* infection there was a significantly higher frequency of the variant C allele in *H. pylori* infected subjects with hypochlorhydria and gastric atrophy compared to infected subjects alone. Functional analysis of the SNP showed that carriage of the C allele increased *TLR9* transcriptional activity driven mainly by activation of the pro-inflammatory transcription factor NF-κB (Ng et al. 2010).

The association between single nucleotide polymorphisms in *TLR9* and *TLR5* and gastroduodenal diseases in 561 patients in Mexico was assessed. All patients were genotyped by allelic discrimination in regions 1174C > T and 1775A > G of TLR5 and -1,237T > C and 2848G > A of TLR9. The 2848A allele of *TLR9* was associated with a risk of DU formation but no polymorphisms in *TLR5* were associated with disease. Patients with polymorphisms in *TLR9* and *TLR5* expressed significantly lower levels of IL-1β and TNF-α, whereas polymorphisms in *TLR5* also decreased the expression of IL-6 and IL-10. Therefore 2848G > A polymorphism in *TLR9* may increase the risk of development of DU by modifying the inflammatory response to *H. pylori* infection (Trejo-de la et al. 2015).

2.5 TLR Signalling Pathway

Castano-Rodriquez and co-workers investigated the association between polymorphisms in genes involved in the TLR signalling pathway, including *TLR2, TLR4, LBP, MD-2, CD14* and *TIRAP*, and the risk of *H. pylori* infection and related development of GC in Chinese individuals (Castano-Rodriquez et al. 2014b). *TLR4* rs11536889 was identified as a risk factor for GC (OR: 3.58, 95% CI:1.20–10.65) and *TLR4* rs10759932 decreased the risk of *H. pylori* infection (OR: 0.59, 95% CI: 0.41–0.86). Strikingly statistical analyses revealed a strong association with *H. pylori* infection combined with polymorphisms in *TLR2, TLR4, MD-2, LBP* and

TIRAP and development of GC (Castano-Rodriguez et al. 2014b). Interestingly, another study examined the effect of polymorphisms in *MyD88* (rs6853) and *TIRAP* (rs8177374) in an Italian population and found that when the genes were analysed separately, TIRAP on its own was associated with protection from infection but analysis of the genes concurrently, revealed that certain combinations of *MyD88* and *TIRAP* protected the host against *H. pylori* colonisation more efficiently than could be done by *TIRAP* alone (Fulgione et al. 2016).

2.6 Nucleotide-Binding Oligomerisation Domain (NOD) Like Receptors (NLRs)

NLRs are PRRs that are activated and induce innate immune responses through cytosolic sensing of microbial components. There are over 20 members of the NLR family and several respond to various PAMPs, non-PAMP particles and cellular stresses to trigger pro-inflammatory responses, including the secretion of IL-1β (Franchi et al. 2009; Kanneganti et al. 2007). While TLRs are located on the surface of cells NLRs are located intracellularly and detect microbial components in the host cell cytosol. NLRs are components of inflammasomes, which are part of the innate immune system that act to induce maturation of inflammatory cytokines such as IL-1β and IL-18 in response to infection (Ghiringhelli et al. 2009; Gross et al. 2012; Pachathundikandi et al. 2016). Two NLR family members, NOD1 and NOD2, are activated by molecules produced during the synthesis and/or degradation of bacterial peptidoglycan. NOD1 is composed of a caspase activation and recruitment domain (CARD), a NOD and multiple leucine rich repeats (LRRs). NOD2 has an additional CARD domain. Both NOD1 and NOD2 can activate NF-κB in response to peptidoglycan fragments. NOD1 is activated by the dipeptide γ-D-glutamyl-*meso*-diaminopimelic acid (Chamaillard et al. 2003; Girardin et al. 2003a), which is produced by most Gram-negative bacteria and by some Gram-positive organisms. NOD2 is activated by muramyl di-peptide (Girardin et al. 2003b; Inohara et al. 2003), a component of peptidoglycan, found in nearly all bacteria. Kim and colleagues using murine derived dendritic cells identified the *H. pylori cag* PAI and the CagL protein, but not vacuolating cytotoxin A (VacA) or CagA, as playing a role in regulating the induction of pro-IL-1β and production of mature IL-1β in response to *H. pylori* infection (Kim et al. 2013a). Furthermore they showed that TLR2 and NOD2, but not NOD1, are required for induction of pro-IL-1β and NOD-like receptor pyrin domain containing 3 (NLRP3) in *H. pylori* infected dendritic cells (DCs) (Kim et al. 2013a), which was basically confirmed by follow up studies (Koch et al. 2015).

Both NOD1 and NOD2 protein expression is upregulated in epithelial cells isolated from gastric biopsy tissue of *H. pylori* infected individuals (Rosenstiel et al. 2006). The *H. pylori* T4SS delivers peptidoglycan fragments into host cells which results in NF-κB activation and mice deficient in NOD1 have an increased susceptibility to infection by *H. pylori* (Viala et al. 2004). While *H. pylori* is characterised by its helical shape coccoid forms of the organism also exist and the N-acetylmuramoyl-L-alanine amidase, AmiA, has been shown to be essential for this morphological transition and also for a modification of the cell wall peptidoglycan (Chaput et al. 2006). This change in cell wall peptidoglycan allows escape from detection by the immune system and therefore is a mechanism which could promote chronic infection.

Hofner and co-workers investigated SNPs of *NOD1* and *IL-8* in *H. pylori* infected patients from Hungary with gastritis and DU, compared with asymptomatic *H. pylori* infected individuals (Hofner et al. 2007). They found that carriage of the *NOD1* E266K polymorphism, encoding for a protein with a glutamic acid residue exchanged for a lysine increased the risk of DU in $CagA^{+}$ *H. pylori* positive patients. The same polymorphism was also associated with *H. pylori* associated gastric mucosal inflammation in a population from Korea (Kim et al. 2013b). Furthermore the *IL-8-251* polymorphism, associated with increased IL-8 production, was significantly associated with both gastritis and duodenal ulceration in *H. pylori* infected subjects (Hofner et al. 2007). Rosenstiel and co-workers examined the relationship between *H. pylori* related disease and five further SNPs in NOD1 but found no association with gastritis or gastric ulcer development (Rosenstiel et al. 2006). However, they did find in cells carrying the Crohn's disease associated *NOD2* variant, R702W, that NF-κB activation was significantly decreased and that there was an association between carriage of R702W and development of gastric lymphoma.

In a study looking at 574 Caucasian subjects, 114 with GC, 222 with high risk atrophic gastritis and 238 controls there was a tendency for the *NOD1* 796G/G genotype to be associated with an increased risk of atrophic gastritis, but there was no significant association found between the *NOD1* E266K (796G > A) polymorphism in relation to GC, atrophic gastritis or infection with *H. pylori* (Kupcinskas et al. 2011). In addition, Li and co-workers reported that there was no statistical significance observed between *NOD1* E266K and GC in a large Chinese population (Li et al. 2015).

Castano-Rodriguez and colleagues assessed polymorphisms in the NLR signaling pathway in *H. pylori* infection and GC in 310 Chinese subjects (Castano-Rodriguez et al. 2014a). Fifty-one polymorphisms in six genes, *NLRP3, NLRP12, NLRX1, CASP1, ASC* and *CARD8*, involved in the NLR signaling pathway and their association with GC was examined. Novel polymorphisms in *CARD8* (rs11672725) and in *NLRP12* (rs2866112) were found to increase the risk for GC and infection with *H. pylori,* respectively. Simultaneous carriage of polymorphisms in *CARD8, NLRP3, CASP1* and *NLRP12* together with *H. pylori* significantly increased the risk of GC in this population.

Li and co-workers performed a population based study of Chinese patients, including 132 with GC and, 1,198 with precancerous gastric lesions, and reported that *NOD2*rs718226 was associated with risk of precancerous gastric lesions and with GC (Li et al. 2015). There was a significant synergistic interaction between *NOD2*rs718226 and infection with *H. pylori* for the risk of dysplasia and for GC. They also found in a cohort of 766 patients with follow up data that two SNPs in *NOD2*, rs2111235 and rs7205423, and *H. pylori* infection correlated with a risk of development of gastric lesions.

2.7 Autophagy

Upon infection of gastric epithelial cells VacA of *H. pylori* has been shown to mediate autophagy, a self-degrading process whereby cytoplasmic components are recycled within the cell (Terebiznik et al. 2009). This was initially thought to be a mechanism to limit toxin derived host damage, however, the autophagosmes induced by VacA were subsequently shown to be defective due to absence of cathespsin D, a key hydrolase, required for degradation. Induction of autophagy is dependent on NOD2. Raju and co-workers studying two cohorts from both Scotland and Germany showed that infection with $VacA^{+}$ *H. pylori* was increased in individuals harboring a polymorphism in the *ATG16L1* autophagy gene, rs2241880, causing amino acid substitution T300A (Raju et al. 2012), which is associated with the risk of developing Crohn's disease (Hampe et al. 2007; Prescott et al. 2007). Carriage of this variant increased the risk of acquisition of $VacA^{+}$ *H. pylori* and chronic infection.

Gene expression studies showed that 28 molecules involved in vesicle trafficking, vesicle elongation and maturation were significantly down regulated in AGS gastric epithelial cells challenged with a highly virulent strain of *H. pylori* and core autophagy proteins and autophagy regulators were differentially expressed in infected THP-1 macrophage cells (Castano-Rodriguez et al. 2015). In addition carriage of the Crohn's disease associated polymorphism *ATG16L1* rs2241880 increased the risk of GC in *H. pylori* infected individuals in a Chinese population (Castano-Rodriguez et al. 2015). Autophagy related gene expression has recently been examined in *H. pylori* infected human gastric tissue taken from 136 Bhutanese volunteers with mild dyspeptic symptoms. mRNA expression of genes encoding for core molecules of autophagy ATH16L1 and ATG5 were significantly decreased compared with expression in *H. pylori* negative tissue (Tanaka et al. 2017). Interestingly, in this population an inverse relationship was found between the presence of the *ATG16L1* rs2241880 GG genotype and the density of *H. pylori* in the tissue (Tanaka et al. 2017). Although the majority of *H. pylori* organisms are extracellular and it is not thought to be an invasive pathogen down regulation of core autophagy machinery genes that inhibits autophagy may allow *H. pylori* to invade cells, and this may promote development of GC. Intracellular *H. pylori* have been observed in clinical specimens (Necchi et al. 2007; Tegtmeyer et al. 2017) and *in vitro* (Kwok et al. 2002), and the number of intracellular and interstitial *H. pylori* and clusters of $CagA^+$ bacteria are increased in patients with gastric intestinal metaplasia and cancer compared to control subjects with gastritis (Semino-Mora et al. 2003).

2.8 Prostate Stem Cell Antigen (PSCA)

PSCA acts to suppress cell growth and so it could have a preventative role against cancer in the gastric epithelium. *PSCA* rs2294008 C > T polymorphism influences *PSCA* expression and the fragment containing the T allele has lower transcriptional activity than the C allele (Yoshida et al. 2010). Therefore the presence of this allele may lead to greater risk of GC. A genome wide association (GWAS) study found a strong association between *PSCA* gene polymorphisms and the risk of GC in Japanese and Korean populations (Study Group of Millennium Genome Project 2008). The *PSCA* rs2294008 C > T polymorphism was also found to increase the risk of non-cardia GC and its precursors in Caucasian populations but to protect against proximal cancers (Lochhead et al. 2011). Matsuo and co-workers confirmed an association between *PSCA* polymorphism and risk of GC in Japanese subjects but found no association with *H. pylori* infection (Matsuo et al. 2009). Likewise, in a case control study that included three different nationalities living in the same area of China *PSCA* rs2294008 polymorphism was found to differentially contribute to GC among the different nationalities but its role was independent of *H. pylori* infection (Zhao et al. 2013). However, when the genotypes TT, TC and CC, of *PSCA* rs2294008 were compared among 339 *H. pylori* infected and uninfected Bhutanese, the PSCA TT genotype was associated with a greater than threefold increase in the prevalence and extent of intestinal metaplasia compared to C allele carriers among *H. pylori* infected individuals (Uotani et al. 2016). Analysis of genomic DNA from 603 Spanish patients with primary GC, 139 with DUs and 675 healthy controls determined that the *PSCA* rs2294008T allele was associated with reduced risk of developing DU disease and with increased risk of GC in the presence of *H. pylori* (Garcia-Gonzalez et al. 2015).

In a GWAS study of 7,035 DU cases and 25,323 controls from Japan, Tanikawa and colleagues demonstrated a role for two *PSCA* genetic variants rs2294008 ($P = 0.021$; $OR = 2.59$) and rs505922 ($P = 0.076$; $OR = 1.90$) in the development of DU disease in Japan (Tanikawa et al. 2012). Specifically, the C allele of *PSCA* rs2294008 was associated with increased risk of DU ($OR = 1.84$) and with a decreased risk of GC ($OR = 0.79$). The data also indicated that these SNPs are likely to be associated with DU development after *H. pylori* infection and not with susceptibility to persistent *H. pylori* infection *per se* (Tanikawa et al. 2012). Thus, the *PSCA* rs2294008 polymorphism

Table 1 Highlighted genetic polymorphisms and their role in *H. pylori* mediated disease outcome

Gene and SNP	Test population cohort studied	Findings	References
IL-1β -31T -511T	149 fist degree relatives of GC patients from West Scotland, 393 GC patients from Poland	IL-1β-31T and IL-1RN polymorphisms associated with increased risk of both hypochlorhydria induced by *H. pylori* and GC	El-Omar et al. (2000)
IL-1RN *2/*2	152 GC patients from Portugal	IL-1β -511T and IL-1RN polymorphisms associated with increased risk for intestinal-type GC.	Machado et al. (2001)
TNF-a -308A	221 Portugese individuals with chronic gastritis, and 287 GC patients.	Increased risk for development of GC	Machado et al. (2003)
	130 Spanish DU and 50 gastric ulcer patients	Increased risk for DU	Lanas et al. (2001)
TLR4 -896A/G	149 relatives of GC patients from Scotland, 312 Polish and 307 American GC patients.	Increased risk for hypochlorhydria	Hold et al. (2007)
rs11535889	310 ethnic Chinese	Risk factor for GC	Castano-Rodriguez et al. (2014b)
NOD1 E266K	85 *H. pylori* -positive DU and 136 chronic active gastritis patients from Hungary.	Increased risk of PUD	Hofner et al. (2007)
	412 healthy individuals in Korea from 2 healthcare clinics, 213 of whom were positive for *H. pylori*	Increase in gastric mucosal inflammation	Kim et al. (2013b)
NOD2 rs718226	132 Chinese patients with GC, 1,198 with precancerous gastric lesions	Increased risk of dysplasia and GC	Li et al. (2015)
ATG16L1	107 *H. pylori* positive individuals from Scotland and 273 from Germany	Increased risk of acquiring VacA^{+} *H. pylori*	Raju et al. (2012)
rs2241880 T300A	86 Chinese non cardia GC patients	Increased risk of GC	Castano-Rodriguez et al. (2015)
PSCA rs2294008 C allele	7,035 Japanese DU patient	Increased risk of DU and decreased risk of GC	Tanikawa et al. (2012)
rs2294008 T allele	603 GC and 139 DU patients from Spain	Increased risk of GC and decreased risk of DU.	Garcia-Gonzalez et al. (2015)

Abbreviations use: *GC* gastric cancer, *PUD* peptic ulcer disease, *DU* duodenal ulceration

appears to be involved in increased susceptibility to DU disease and decreased risk of GC.

Table 1 highlights some of the SNPs that have been discussed so far and the role they may play in mediating the outcome of infection with *H. pylori* in different ethnic groups.

2.9 DNA Repair

A nested case control study which included 246 gastric adenocarcinomas and 1,175 matched controls and 91 cases with chronic atrophic gastritis analysed 12 polymorphisms at DNA repair genes (*MSH2, MLH1, XRCC1, OGG1* and *ERCC2*) and seropositivity for *H. pylori*. No association was observed for any of the polymorphisms with a risk of GC. However, *ERCC2* K751Q polymorphism was associated with an increased risk for non-cardia neoplasm [OR = 1.78; 95% confidence interval (CI) 1.02–3.12], and *ERCC2* K751Q and D312N polymorphisms were associated with diffuse type GC. Furthermore *ERCC2* D312N (OR = 2.0; 95% CI 1.09–3.65) and K751Q alleles (OR = 1.82; 95% CI 1.01–3.30) and *XRCC1* R399Q (OR = 1.69; 95% CI 1.02–2.79) allele were associated with an increased risk for severe chronic atrophic gastritis (Capella et al. 2008). Other studies have suggested that genetic polymorphisms in *XRCC1*-Arg194Trp, *XRCC1*-Arg399Gln and *OGG1*-Ser326Cys may play an important role in the evolution of *H. pylori* -associated gastric lesions in the Chinese population (Li et al. 2009).

In order to clarify conflicting results on the role of *XRCC1*, a meta-analysis of published case-control and cohort studies which included 18 studies with 3,915 GC cases and 6,759 controls was performed. The results showed that there was a significant difference in genotype distribution between GC cases and controls among Asians (especially in the Chinese population), but not among Caucasians. Only the *XRCC1* Arg194Trp homozygous mutant genotype (Trp/Trp) was found to be associated with increased risk of GC (Chen et al. 2012).

Poly(adenosine diphosphate [ADP]-ribose) polymerase 1 (PARP-1) is important for base excision repair and maintenance of genomic integrity. A hospital-based, case–control study from China which included 556 individuals (236 GC cases and 320 controls with no evidence of gastrointestinal disease) assessed the association of *PARP-1* polymorphism and *H. pylori* infection in GC risk. Results indicated that the *PARP-1*762A/A genotype could be a risk factor for GC in China; and that the *PARP-1*762V/A polymorphism combined with infection by CagA$^+$ strains of *H. pylori* is associated with a higher risk for development of GC (Zhang et al. 2009). A study from Brazil examined 101 gastric tumours from *H. pylori* positive patients and assessed the interaction between the virulence of the infecting strain and the polymorphisms of host repair genes. The wild type *PARP-1* allele (A/A) was associated with intestinal type GC in older patients and in younger patients with this *PARP-1* genotype, low-virulence *H. pylori* strains contributed to gastric carcinogenesis when a polymorphic allele of *APE-1* was also present (Silva-Fernandes et al. 2012).

Excision repair cross-complementing group 8 (ERCC8) like PARP-1 also plays a critical role in DNA repair. Genetic polymorphisms in *ERCC8* together with *H. pylori* infection and environmental factors such as smoking and drinking were assessed in 394 controls, 394 atrophic gastritis, and 394 GC cases in northern China. Males and individuals over 50 years of age with *EGRCC8* rs158572 GA/GG/GG + GA genotypes had significantly increased risk of GC, while older individuals with rs158916 CT/CC + CT genotypes had a decreased risk of atrophic gastritis. Statistically significant interactions between *ERCC8* SNPs and *H. pylori* infection were observed for GC and atrophic gastritis risk ($P < 0.05$). Smokers and drinkers with *ERCC8* rs158572 GG + GA genotype were more susceptible to GC compared with non-smokers and non-drinkers homozygous for AA (Jing et al. 2015).

2.10 DNA Methylation

DNA methylation is an important epigenetic feature of DNA that plays a role in gene regulation and cellular differentiation mechanisms. Polymorphisms in genes that encode for DNA methylation enzymes may explain the susceptibility of individuals to GC. A study from China which included 447 patients with GC; 111 individuals with gastric atrophy and 961 healthy controls investigated associations between SNPs of the DNA methyltransferase-3a (*DNMT3a*) gene and risk of *H. pylori* infection, gastric atrophy and GC. Among healthy controls, the risk of *H. pylori* infection was significantly higher in subjects with the rs1550117 AA genotype, compared to those with GG/AG DNMT3a genotypes, but there was no significant correlation between the two SNPs and risk of developing gastric atrophy or GC (Cao et al. 2013). Another study examined associations between SNPs in the *DNMT1* gene and risks of *H. pylori* infection with development of gastric atrophy and GC in the Chinese population (Jiang et al. 2012). 447 patients with GC; 111 patients with gastric atrophy; and 961 healthy controls were included in the study. Carriers of rs10420321 GG and rs8111085 CC *DNMT1* genotype were associated with reduced risks for infection with *H. pylori* and a higher risk of gastric atrophy (Jiang et al. 2012).

MLH1 is a mismatch repair gene and O6-methylguanine DNA methyltransferase (MGMT) is a protein required for the repair of alkylated guanines in DNA that arise from exposures to environmental alkylation mutagens or through endogenous mechanisms. A study in Brazil looked at 239 patients which included 50 children between the ages of 2 and 18 years old (average age $= 8 \pm 4$ years) with dyspepsia, 97 patients with chronic gastritis (average age $= 35 \pm 13$ years), and 92 with GC (average age $= 60 \pm 12$ years). Methylation was not detected in the promoter regions of *MLH1* and *MGMT* in gastric biopsy samples from any of the children, regardless of *H. pylori* infection status. The *MGMT* promoter was methylated in 51% of chronic gastritis adult patients and was associated with *H. pylori* infection ($P < 0.05$); this region was methylated in 66% of GC patients, and the difference in the percentage of methylated samples between these patients and those from *H. pylori* infected chronic gastritis patients was statistically significant ($P < 0.05$). The frequency of promoter methylation for both genes was higher in GC samples than in *H. pylori* -positive chronic gastritis samples ($P < 0.05$) (Alvarez et al. 2013).

Methylenetetrahydrofolate reductase (MTHFR) catalyzes synthesis of 5-methylenetetrahydrofolate which leads to the production of S-adenosylmethionine, which is required for DNA methylation. Polymorphisms that alter MTHFR activity could influence the status of DNA methylation and susceptibility to GC. A C/T nucleotide change at 677 position of the *MTHFR* gene (*MTHFR* C677T), resulting in an alanine-to-valine substitution in the MTHFR protein, was found to produce a thermolabile variant of MTHFR with reduced enzyme activity. A recent meta-analysis has reported a significant positive correlation between GC and the *MTHFR* 677TT genotype (Boccia et al. 2008). In an analysis of 71 tumour samples from patients in Brazil the association between the methylation status of *CDKN2A, HMLH1,* and *COX-2,* the *MTHFR* C677T polymorphism and the influence of infection with pathogenic *H. pylori* strains in GC were analyzed. Hypermethylation of *CDKN2A* was associated with carriage of the *MTHFR* 677TT genotype and with infection by *H. pylori cagA/ vacA*s1m1 strains and this was a characteristic of distal tumours especially in older patients (> 60 years) (Neves Filho et al. 2010).

3 Gene Amplification and Gene Deletions

In addition to SNPs, increased gene copy number or gene deletions can also have functional consequences. The role of gene amplification and gene deletion in GC was investigated by examining the genomic profile of DNA from 193 primary GCs, 98 primary matched gastric normal samples and 40 GC cell lines using Affymetrix SNP6 microarrays containing approximately 1.8 million probes. Approximately 150 genomic aberrations per GC, comprising a mixture of broad and focally altered regions were seen. 22 focal genomic alterations (typically <100 kb) exhibiting high levels of copy number gain or loss were identified. Oncogenes, previously known to be amplified in GC, were identified, including *EGFR* and *ERBB2/HER2* (Deng et al. 2012). Focally deleted genes identified included *FHIT, RB1, CDKN2A/B*, and *WWOX* (Deng et al. 2012). In addition, novel genes not previously reported in GC were identified. These included genomic amplification of the transcription factors *GATA6* and *KLF5*, and somatic deletions in *PARK2, PDE4D, CSMD1* and *GMDS* (Deng et al. 2012).

Several of the genes identified have also been associated with the host response to *H. pylori* infection. Using *H. pylori* infected wild-type mice and mice with attenuated EGFR activity phosphorylated EGFR and ERBB2 were shown to be increased in tissues showing gastritis or atrophic gastritis (Chaturvedi et al. 2014). In addition, a specific EGFR inhibitor used in *H. pylori* infected transgenic INS-GAS mice and gerbils markedly reduced dysplasia and GC (Sierra et al. 2018). Furthermore, EGFR has been shown recently to play a crucial role in stimulating a potent IL-8 response in *H. pylori* infected endothelial cells (Tafreshi et al. 2018). Together these studies highlight EGFR as a potential therapeutic target for prevention of *H. pylori* induced inflammation and subsequent disease development.

FHIT encodes for a fragile histidine triad (FHIT) protein and is a tumour suppressor gene. It has been shown that decreased FHIT mRNA and protein expression is influenced by having a first degree relative with GC and by infection with *H. pylori* strains positive for both *vacA* and *cagA* (Stec-Michalska et al. 2009).

GATA6, which is crucial for gastrointestinal development and differentiation, is frequently amplified and/or overexpressed in human GC. *H. pylori* infected patients with DUs have been shown to have higher *GATA6* mRNA levels compared to infected patients without DU or with non-infected patients (Lario et al. 2012).

4 Conclusions

There is compelling evidence that host genetic polymorphisms especially those that predispose individuals to a risk of GC may go some way to explaining the different clinical outcomes of *H. pylori* infection in different populations. The initial immune response to infection in children fails to clear the infection and as children grow older *H. pylori* induces a complex inflammatory response and chronic gastritis, therefore it is not surprising that polymorphisms in immune response genes should alter the outcome of infection. However, despite the large number of studies it is difficult to draw definitive conclusions about the exact contribution of genetic polymorphisms to disease outcome (Fig. 1). Many studies are confounded by small sample size and lack appropriate control groups. In addition, few studies consider the interaction between bacterial, environmental and host factors. There is a need for appropriately powered studies which include participants with GC and DUs as well as uninfected participants and participants with different ethnic backgrounds. Determination of the functional consequences of gene polymorphisms and possible epistatic interactions of several host genetic polymorphisms are required to explain why DU disease protects against GC and the differences in outcome in different geographic locations. Conflicting results from studies around the world may be due to variation in allele frequencies among different populations. Future studies in large well defined ethnic populations that also take into account the bacterial genotypes present in that population and environmental factors combined with molecular cellular studies are required to comprehensively explain the effect of host genetics on *H. pylori* infection.

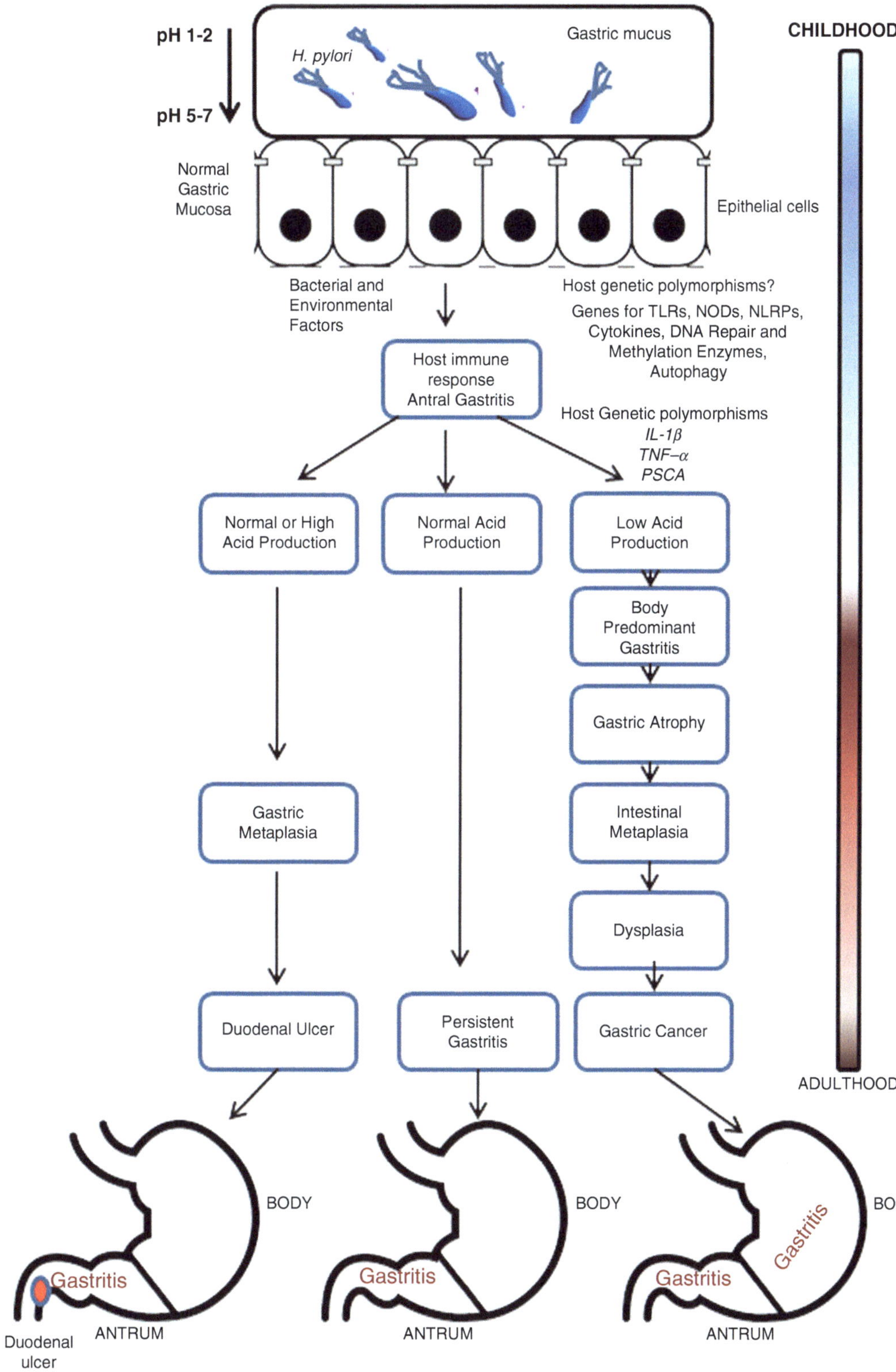

Fig. 1 Pathways involved in *H. pylori* mediated disease outcome and role of host genetic polymorphisms. Infection occurs in childhood and an immune response is mounted which does not clear the infection which can persist for decades. The functional consequences of some host SNPs such as those in genes encoding for the acid inhibitory cytokines IL-1β and TNF-α are well defined but for others it is less clear. The interaction of bacterial and environmental factors along with host genetic factors determine the outcome of *H. pylori* mediated disease

References

Alvarez MC, Santos JC, Maniezzo N, Ladeira MS, da Silva AL, Scaletsky IC, Pedrazzoli J Jr, Ribeiro ML (2013) MGMT and MLH1 methylation in *Helicobacter pylori* -infected children and adults. World J Gastroenterol 19(20):3043–3051. https://doi.org/10.3748/wjg.v19.i20.3043

Andersen-Nissen E, Smith KD, Strobe KL, Barrett SL, Cookson BT, Logan SM, Aderem A (2005) Evasion of Toll-like receptor 5 by flagellated bacteria. Proc Natl Acad Sci U S A 102(26):9247–9252. https://doi.org/10.1073/pnas.0502040102

Ando T, El-Omar EM, Goto Y, Nobata K, Watanabe O, Maeda O, Ishiguro K, Minami M, Hamajima N, Goto H (2006) Interleukin 1B proinflammatory genotypes protect against gastro-oesophageal reflux disease through induction of corpus atrophy. Gut 55 (2):158–164. https://doi.org/10.1136/gut.2005.072942

Boccia S, Hung R, Ricciardi G, Gianfagna F, Ebert MP, Fang JY, Gao CM, Gotze T, Graziano F, Lacasana-Navarro M, Lin D, Lopez-Carrillo L, Qiao YL, Shen H, Stolzenberg-Solomon R, Takezaki T, Weng YR, Zhang FF, van Duijn CM, Boffetta P, Taioli E (2008) Meta- and pooled analyses of the methylenetetrahydrofolate reductase C677T and A1298C polymorphisms and gastric cancer risk: a huge-GSEC review. Am J Epidemiol 167(5):505–516. https://doi.org/10.1093/aje/kwm344

Bontems P, Aksoy E, Burette A, Segers V, Deprez C, Mascart F, Cadranel S (2014) NF-kappaB activation and severity of gastritis in *Helicobacter pylori* -infected children and adults. Helicobacter 19 (3):157–167. https://doi.org/10.1111/hel.12118

Cao XY, Jia ZF, Cao DH, Kong F, Jin MS, Suo J, Jiang J (2013) DNMT3a rs1550117 polymorphism association with increased risk of *Helicobacter pylori* infection. Asian Pac J Cancer Prev 14(10):5713–5718

Capella G, Pera G, Sala N, Agudo A, Rico F, Del Giudicce G, Plebani M, Palli D, Boeing H, Bueno-de-Mesquita HB, Carneiro F, Berrino F, Vineis P, Tumino R, Panico S, Berglund G, Siman H, Nyren O, Hallmans G, Martinez C, Dorronsoro M, Barricarte A, Navarro C, Quiros JR, Allen N, Key T, Bingham S, Caldas C, Linseisen J, Nagel G, Overvad K, Tjonneland A, Boshuizen HC, Peeters PH, Numans ME, Clavel-Chapelon F, Trichopoulou A, Lund E, Jenab M, Kaaks R, Riboli E, Gonzalez CA (2008) DNA repair polymorphisms and the risk of stomach adenocarcinoma and severe chronic gastritis in the EPIC-EURGAST study. Int J Epidemiol 37(6):1316–1325. https://doi.org/10.1093/ije/dyn145

Castano-Rodriguez N, Kaakoush NO, Goh KL, Fock KM, Mitchell HM (2013) The role of TLR2, TLR4 and CD14 genetic polymorphisms in gastric carcinogenesis: a case-control study and meta-analysis. PLoS One 8(4):e60327. https://doi.org/10.1371/journal.pone.0060327

Castano-Rodriguez N, Kaakoush NO, Goh KL, Fock KM, Mitchell HM (2014a) The NOD-like receptor signalling pathway in *Helicobacter pylori* infection and related gastric cancer: a case-control study and gene expression analyses. PLoS One 9(6):e98899. https://doi.org/10.1371/journal.pone.0098899

Castano-Rodriguez N, Kaakoush NO, Pardo AL, Goh KL, Fock KM, Mitchell HM (2014b) Genetic polymorphisms in the Toll-like receptor signalling pathway in *Helicobacter pylori* infection and related gastric cancer. Hum Immunol 75(8):808–815. https://doi.org/10.1016/j.humimm.2014.06.001

Castano-Rodriguez N, Kaakoush NO, Goh KL, Fock KM, Mitchell HM (2015) Autophagy in *Helicobacter pylori* infection and related gastric cancer. Helicobacter 20 (5):353–369. https://doi.org/10.1111/hel.12211

Chamaillard M, Hashimoto M, Horie Y, Masumoto J, Qiu S, Saab L, Ogura Y, Kawasaki A, Fukase K, Kusumoto S, Valvano MA, Foster SJ, Mak TW, Nunez G, Inohara N (2003) An essential role for NOD1 in host recognition of bacterial peptidoglycan containing diaminopimelic acid. Nat Immunol 4 (7):702–707. https://doi.org/10.1038/ni945

Chaput C, Ecobichon C, Cayet N, Girardin SE, Werts C, Guadagnini S, Prevost MC, Mengin-Lecreulx D, Labigne A, Boneca IG (2006) Role of AmiA in the morphological transition of *Helicobacter pylori* and in immune escape. PLoS Pathog 2(9):e97. https://doi.org/10.1371/journal.ppat.0020097

Chaturvedi R, Asim M, Piazuelo MB, Yan F, Barry DP, Sierra JC, Delgado AG, Hill S, Casero RA Jr, Bravo LE, Dominguez RL, Correa P, Polk DB, Washington MK, Rose KL, Schey KL, Morgan DR, Peek RM Jr, Wilson KT (2014) Activation of EGFR and ERBB2 by *Helicobacter pylori* results in survival of gastric epithelial cells with DNA damage. Gastroenterology 146 (7):1739–1751 e1714. https://doi.org/10.1053/j.gastro.2014.02.005

Chen B, Zhou Y, Yang P, Wu XT (2012) Polymorphisms of XRCC1 and gastric cancer susceptibility: a meta-analysis. Mol Biol Rep 39(2):1305–1313. https://doi.org/10.1007/s11033-011-0863-6

Chen B, Luo MX, Zhou X, Lv Y, Su GQ (2016) Correlation between interleukin-1beta-511 C/T polymorphism and gastric cancer in chinese populations: a Meta-Analysis. Med Sci Monit 22:1742–1750

Cheng HH, Chang CS, Wang HJ, Wang WC (2010) Interleukin-1beta and −10 polymorphisms influence erosive reflux esophagitis and gastritis in Taiwanese patients. J Gastroenterol Hepatol 25(8):1443–1451. https://doi.org/10.1111/j.1440-1746.2010.06310.x

Crabtree JE, Shallcross TM, Heatley RV, Wyatt JI (1991a) Mucosal tumour necrosis factor alpha and interleukin-6 in patients with *Helicobacter pylori* associated gastritis. Gut 32(12):1473–1477

Crabtree JE, Shallcross TM, Wyatt JI, Taylor JD, Heatley RV, Rathbone BJ, Losowsky MS (1991b) Mucosal humoral immune response to *Helicobacter pylori* in patients with duodenitis. Dig Dis Sci 36(9):1266–1273 de Martel C, Forman D, Plummer M (2013) Gastric cancer: epidemiology and risk factors. Gastroenterol Clin N Am 42(2):219–240. https://doi.org/10.1016/j.gtc.2013.01.003

Deng N, Goh LK, Wang H, Das K, Tao J, Tan IB, Zhang S, Lee M, Wu J, Lim KH, Lei Z, Goh G, Lim QY, Tan AL, Sin Poh DY, Riahi S, Bell S, Shi MM, Linnartz R, Zhu F, Yeoh KG, Toh HC, Yong WP, Cheong HC, Rha SY, Boussioutas A, Grabsch H, Rozen S, Tan P (2012) A comprehensive survey of genomic alterations in gastric cancer reveals systematic patterns of molecular exclusivity and co-occurrence among distinct therapeutic targets. Gut 61 (5):673–684. https://doi.org/10.1136/gutjnl-2011-301839

Drumm B, Perez-Perez GI, Blaser MJ, Sherman PM (1990) Intrafamilial clustering of *Helicobacter pylori* infection. N Engl J Med 322(6):359–363. https://doi.org/10.1056/NEJM199002083220603

Du Y, Gao L, Zhang K, Wang J (2015) Association of the IL6 polymorphism rs1800796 with cancer risk: a meta-analysis. Genet Mol Res 14(4):13236–13246. https://doi.org/10.4238/2015.October.26.20

El-Omar EM, Carrington M, Chow WH, McColl KE, Bream JH, Young HA, Herrera J, Lissowska J, Yuan CC, Rothman N, Lanyon G, Martin M, Fraumeni JF Jr, Rabkin CS (2000) Interleukin-1 polymorphisms associated with increased risk of gastric cancer. Nature 404(6776):398–402. https://doi.org/10.1038/35006081

El-Omar EM, Rabkin CS, Gammon MD, Vaughan TL, Risch HA, Schoenberg JB, Stanford JL, Mayne ST, Goedert J, Blot WJ, Fraumeni JF Jr, Chow WH (2003) Increased risk of noncardia gastric cancer associated with proinflammatory cytokine gene polymorphisms. Gastroenterology 124(5):1193–1201

Franchi L, Eigenbrod T, Munoz-Planillo R, Nunez G (2009) The inflammasome: a caspase-1-activation platform that regulates immune responses and disease pathogenesis. Nat Immunol 10(3):241–247. https://doi.org/10.1038/ni.1703

Fulgione A, Di Matteo A, Contaldi F, Manco R, Ianniello F, Incerti G, De Seta M, Esposito N, Crasto A, Iannelli D, Capparelli R (2016) Epistatic interaction between MyD88 and TIRAP against *Helicobacter pylori*. FEBS Lett 590(14):2127–2137. https://doi.org/10.1002/1873-3468.12252

Garcia-Gonzalez MA, Bujanda L, Quintero E, Santolaria S, Benito R, Strunk M, Sopena F, Thomson C, Perez-Aisa A, Nicolas-Perez D, Hijona E, Carrera-Lasfuentes P, Piazuelo E, Jimenez P, Espinel J, Campo R, Manzano M, Geijo F, Pellise M, Zaballa M, Gonzalez-Huix F, Espinos J, Tito L, Barranco L, Pazo-Cid R, Lanas A (2015) Association of PSCA rs2294008 gene variants with poor prognosis and increased susceptibility to gastric cancer and decreased risk of duodenal ulcer disease. Int J Cancer 137(6):1362–1373. https://doi.org/10.1002/ijc.29500

Garza-Gonzalez E, Bosques-Padilla FJ, El-Omar E, Hold G, Tijerina-Menchaca R, Maldonado-Garza HJ, Perez-Perez GI (2005) Role of the polymorphic IL-1B, IL-1RN and TNF-A genes in distal gastric cancer in Mexico. Int J Cancer 114(2):237–241. https://doi.org/10.1002/ijc.20718

Gatti LL, Burbano RR, de Assumpcao PP, Smith Mde A, Payao SL (2004) Interleukin-1beta polymorphisms, *Helicobacter pylori* infection in individuals from Northern Brazil with gastric adenocarcinoma. Clin Exp Med 4(2):93–98

Ghiringhelli F, Apetoh L, Tesniere A, Aymeric L, Ma Y, Ortiz C, Vermaelen K, Panaretakis T, Mignot G, Ullrich E, Perfettini JL, Schlemmer F, Tasdemir E, Uhl M, Genin P, Civas A, Ryffel B, Kanellopoulos J, Tschopp J, Andre F, Lidereau R, McLaughlin NM, Haynes NM, Smyth MJ, Kroemer G, Zitvogel L (2009) Activation of the NLRP3 inflammasome in dendritic cells induces IL-1beta-dependent adaptive immunity against tumors. Nat Med 15 (10):1170–1178. https://doi.org/10.1038/nm.2028

Girardin SE, Boneca IG, Carneiro LA, Antignac A, Jehanno M, Viala J, Tedin K, Taha MK, Labigne A, Zahringer U, Coyle AJ, DiStefano PS, Bertin J, Sansonetti PJ, Philpott DJ (2003a) Nod1 detects a unique muropeptide from gram-negative bacterial peptidoglycan. Science 300(5625):1584–1587. https://doi.org/10.1126/science.1084677

Girardin SE, Boneca IG, Viala J, Chamaillard M, Labigne A, Thomas G, Philpott DJ, Sansonetti PJ (2003b) Nod2 is a general sensor of peptidoglycan through muramyl dipeptide (MDP) detection. J Biol Chem 278(11):8869–8872. https://doi.org/10.1074/jbc.C200651200

Gobert AP, Bambou JC, Werts C, Balloy V, Chignard M, Moran AP, Ferrero RL (2004) *Helicobacter pylori* heat shock protein 60 mediates interleukin-6 production by macrophages via a toll-like receptor (TLR)-2-, TLR-4-, and myeloid differentiation factor 88-independent mechanism. J Biol Chem 279(1):245–250. https://doi.org/10.1074/jbc.M307858200

Goda V, Jayaraman M, Loganathan R, Nazeer M, Ali M, Karunakaran P, Devaraju P (2017) TLR5 polymorphisms rs2072493, rs5744174, and rs5744168 are not genetic risk factors for chronic *Helicobacter pylori* infection in Indian tamils. Immunol Investig 46(6):537–543. https://doi.org/10.1080/08820139.2017.1319381

Gross O, Yazdi AS, Thomas CJ, Masin M, Heinz LX, Guarda G, Quadroni M, Drexler SK, Tschopp J (2012) Inflammasome activators induce interleukin-1alpha secretion via distinct pathways with differential requirement for the protease function of caspase-1. Immunity 36(3):388–400. https://doi.org/10.1016/j.immuni.2012.01.018

Hampe J, Franke A, Rosenstiel P, Till A, Teuber M, Huse K, Albrecht M, Mayr G, De La Vega FM, Briggs J, Gunther S, Prescott NJ, Onnie CM, Hasler R, Sipos B, Folsch UR, Lengauer T,

Platzer M, Mathew CG, Krawczak M, Schreiber S (2007) A genome-wide association scan of nonsynonymous SNPs identifies a susceptibility variant for Crohn disease in ATG16L1. Nat Genet 39 (2):207–211. https://doi.org/10.1038/ng1954

Harris PR, Wright SW, Serrano C, Riera F, Duarte I, Torres J, Pena A, Rollan A, Viviani P, Guiraldes E, Schmitz JM, Lorenz RG, Novak L, Smythies LE, Smith PD (2008) *Helicobacter pylori* gastritis in children is associated with a regulatory T-cell response. Gastroenterology 134(2):491–499. https://doi.org/10.1053/j.gastro.2007.11.006

Hernandez C, Serrano C, Einisman H, Villagran A, Pena A, Duarte I, Torres J, Riera F, Harris PR (2014) Peptic ulcer disease in *Helicobacter pylori* -infected children: clinical findings and mucosal immune response. J Pediatr Gastroenterol Nutr 59(6):773–778. https://doi.org/10.1097/MPG.0000000000000500

Hofner P, Gyulai Z, Kiss ZF, Tiszai A, Tiszlavicz L, Toth G, Szoke D, Molnar B, Lonovics J, Tulassay Z, Mandi Y (2007) Genetic polymorphisms of NOD1 and IL-8, but not polymorphisms of TLR4 genes, are associated with *Helicobacter pylori* -induced duodenal ulcer and gastritis. Helicobacter 12(2):124–131. https://doi.org/10.1111/j.1523-5378.2007.00481.x

Holcombe C (1992) *Helicobacter pylori*: the African enigma. Gut 33(4):429–431

Hold GL, Rabkin CS, Chow WH, Smith MG, Gammon MD, Risch HA, Vaughan TL, McColl KE, Lissowska J, Zatonski W, Schoenberg JB, Blot WJ, Mowat NA, Fraumeni JF Jr, El-Omar EM (2007) A functional polymorphism of toll-like receptor 4 gene increases risk of gastric carcinoma and its precursors. Gastroenterology 132(3):905–912. https://doi.org/10.1053/j.gastro.2006.12.026

Howson CP, Hiyama T, Wynder EL (1986) The decline in gastric cancer: epidemiology of an unplanned triumph. Epidemiol Rev 8:1–27

Inohara N, Ogura Y, Fontalba A, Gutierrez O, Pons F, Crespo J, Fukase K, Inamura S, Kusumoto S, Hashimoto M, Foster SJ, Moran AP, Fernandez-Luna JL, Nunez G (2003) Host recognition of bacterial muramyl dipeptide mediated through NOD2. Implications for Crohn's disease. J Biol Chem 278 (8):5509–5512. https://doi.org/10.1074/jbc.C200673200

Jiang J, Jia Z, Cao D, Jin MS, Kong F, Suo J, Cao X (2012) Polymorphisms of the DNA methyltransferase 1 associated with reduced risks of *Helicobacter pylori* infection and increased risks of gastric atrophy. PLoS One 7(9):e46058. https://doi.org/10.1371/journal.pone.0046058

Jing JJ, Sun LP, Xu Q, Yuan Y (2015) Effect of ERCC8 tagSNPs and their association with *H. pylori* infection, smoking, and alcohol consumption on gastric cancer and atrophic gastritis risk. Tumour Biol 36 (12):9525–9535. https://doi.org/10.1007/s13277-015-3703-9

Kanneganti TD, Lamkanfi M, Nunez G (2007) Intracellular NOD-like receptors in host defense and disease. Immunity 27(4):549–559. https://doi.org/10.1016/j.immuni.2007.10.002

Kim DJ, Park JH, Franchi L, Backert S, Nunez G (2013a) The Cag pathogenicity island and interaction between TLR2/NOD2 and NLRP3 regulate IL-1beta production in *Helicobacter pylori* infected dendritic cells. Eur J Immunol 43(10):2650–2658. https://doi.org/10.1002/eji.201243281

Kim EJ, Lee JR, Chung WC, Jung SH, Sung HJ, Lee YW, Oh YS, Kim SB, Paik CN, Lee KM, Noh SJ (2013b) Association between genetic polymorphisms of NOD 1 and *Helicobacter pylori* -induced gastric mucosal inflammation in healthy Korean population. Helicobacter 18(2):143–150. https://doi.org/10.1111/hel.12020

Kim JJ, Kim N, Hwang S, Kim JY, Kim JY, Choi YJ, Lee DH, Jung HC (2013c) Relationship of interleukin-1beta levels and gastroesophageal reflux disease in Korea. J Gastroenterol Hepatol 28(1):90–98. https://doi.org/10.1111/j.1440-1746.2012.07274.x

Koch KN, Hartung ML, Urban S, Kyburz A, Bahlmann AS, Lind J, Backert S, Taube C, Müller A (2015) *Helicobacter* urease-induced activation of the TLR2/NLRP3/IL-18 axis protects against asthma. J Clin Invest 125:3297–3302. https://doi.org/10.1172/JCI79337

Kuipers EJ (1999) Review article: exploring the link between *Helicobacter pylori* and gastric cancer. Aliment Pharmacol Ther 13(Suppl 1):3–11

Kumar Pachathundikandi S, Brandt S, Madassery J, Backert S (2011) Induction of TLR-2 and TLR-5 expression by *Helicobacter pylori* switches cagPAI-dependent signalling leading to the secretion of IL-8 and TNF-alpha. PLoS One 6(5):e19614. https://doi.org/10.1371/journal.pone.0019614

Kumar S, Kumar A, Dixit VK (2009) Evidences showing association of interleukin-1B polymorphisms with increased risk of gastric cancer in an Indian population. Biochem Biophys Res Commun 387(3):456–460. https://doi.org/10.1016/j.bbrc.2009.07.033

Kupcinskas J, Wex T, Bornschein J, Selgrad M, Leja M, Juozaityte E, Kiudelis G, Jonaitis L, Malfertheiner P (2011) Lack of association between gene polymorphisms of Angiotensin converting enzyme, Nod-like receptor 1, Toll-like receptor 4, FAS/FASL and the presence of *Helicobacter pylori* -induced premalignant gastric lesions and gastric cancer in Caucasians. BMC Med Genet 12:112. https://doi.org/10.1186/1471-2350-12-112

Kwok T, Backert S, Schwarz H, Berger J, Meyer TF (2002) Specific entry of *Helicobacter pylori* into cultured gastric epithelial cells via a zipper-like mechanism. Infect Immun 70(4):2108–2120

Lagunes-Servin H, Torres J, Maldonado-Bernal C, Perez-Rodriguez M, Huerta-Yepez S, Madrazo de la Garza A, Munoz-Perez L, Flores-Luna L, Ramon-Garcia G, Camorlinga-Ponce M (2013) Toll-like receptors and cytokines are upregulated during *Helicobacter pylori* infection in children. Helicobacter 18(6):423–432. https://doi.org/10.1111/hel.12067

Lanas A, Chan FK (2017) Peptic ulcer disease. Lancet 390 (10094):613–624. https://doi.org/10.1016/S0140-6736 (16)32404-7

Lanas A, Garcia-Gonzalez MA, Santolaria S, Crusius JB, Serrano MT, Benito R, Pena AS (2001) TNF and LTA gene polymorphisms reveal different risk in gastric and duodenal ulcer patients. Genes Immun 2(8):415–421. https://doi.org/10.1038/sj.gene.6363798

Lario S, Ramirez-Lazaro MJ, Aransay AM, Lozano JJ, Montserrat A, Casalots A, Junquera F, Alvarez J, Segura F, Campo R, Calvet X (2012) microRNA profiling in duodenal ulcer disease caused by *Helicobacter pylori* infection in a Western population. Clin Microbiol Infect 18(8):E273–E282. https://doi.org/10.1111/j.1469-0691.2012.03849.x

Lee YC, Chiang TH, Chou CK, Tu YK, Liao WC, Wu MS, Graham DY (2016) Association between *Helicobacter pylori* eradication and gastric cancer incidence: a systematic review and meta-analysis. Gastroenterology 150(5):1113–1124 e1115. https://doi.org/10.1053/j.gastro.2016.01.028

Leow AH, Lim YY, Liew WC, Goh KL (2016) Time trends in upper gastrointestinal diseases and *Helicobacter pylori* infection in a multiracial Asian population–a 20-year experience over three time periods. Aliment Pharmacol Ther 43(7):831–837. https://doi.org/10.1111/apt.13550

Li WQ, Zhang L, Ma JL, Zhang Y, Li JY, Pan KF, You WC (2009) Association between genetic polymorphisms of DNA base excision repair genes and evolution of precancerous gastric lesions in a Chinese population. Carcinogenesis 30(3):500–505. https://doi.org/10.1093/carcin/bgp018

Li ZX, Wang YM, Tang FB, Zhang L, Zhang Y, Ma JL, Zhou T, You WC, Pan KF (2015) NOD1 and NOD2 genetic variants in association with risk of gastric cancer and its precursors in a Chinese population. PLoS One 10(5):e0124949. https://doi.org/10.1371/journal.pone.0124949

Lichtenstein P, Holm NV, Verkasalo PK, Iliadou A, Kaprio J, Koskenvuo M, Pukkala E, Skytthe A, Hemminki K (2000) Environmental and heritable factors in the causation of cancer–analyses of cohorts of twins from Sweden, Denmark, and Finland. N Engl J Med 343(2):78–85. https://doi.org/10.1056/NEJM200007133430201

Lochhead P, Frank B, Hold GL, Rabkin CS, Ng MT, Vaughan TL, Risch HA, Gammon MD, Lissowska J, Weck MN, Raum E, Muller H, Illig T, Klopp N, Dawson A, McColl KE, Brenner H, Chow WH, El-Omar EM (2011) Genetic variation in the prostate stem cell antigen gene and upper gastrointestinal cancer in white individuals. Gastroenterology 140 (2):435–441. https://doi.org/10.1053/j.gastro.2010.11.001

Loganathan R, Nazeer M, Goda V, Devaraju P, Ali M, Karunakaran P, Jayaraman M (2017) Genetic variants of TLR4 and TLR9 are risk factors for chronic *Helicobacter pylori* infection in South Indian Tamils. Hum Immunol 78(2):216–220. https://doi.org/10.1016/j.humimm.2016.12.002

Machado JC, Pharoah P, Sousa S, Carvalho R, Oliveira C, Figueiredo C, Amorim A, Seruca R, Caldas C, Carneiro F, Sobrinho-Simoes M (2001) Interleukin 1B and interleukin 1RN polymorphisms are associated with increased risk of gastric carcinoma. Gastroenterology 121(4):823–829

Machado JC, Figueiredo C, Canedo P, Pharoah P, Carvalho R, Nabais S, Castro Alves C, Campos ML, Van Doorn LJ, Caldas C, Seruca R, Carneiro F, Sobrinho-Simoes M (2003) A proinflammatory genetic profile increases the risk for chronic atrophic gastritis and gastric carcinoma. Gastroenterology 125 (2):364–371

Maeda S, Akanuma M, Mitsuno Y, Hirata Y, Ogura K, Yoshida H, Shiratori Y, Omata M (2001) Distinct mechanism of *Helicobacter pylori* -mediated NF-kappa B activation between gastric cancer cells and monocytic cells. J Biol Chem 276 (48):44856–44864. https://doi.org/10.1074/jbc.M105381200

Malaty HM, Engstrand L, Pedersen NL, Graham DY (1994) *Helicobacter pylori* infection: genetic and environmental influences. A study of twins. Ann Intern Med 120(12):982–986

Marginean MO, Marginean CO, Melit LE, Voidazan S, Moldovan V, Banescu C (2017) The impact of host's genetic susceptibility on *Helicobacter pylori* infection in children. Medicine (Baltimore) 96(30):e7612. https://doi.org/10.1097/MD.0000000000007612

Matsuo K, Tajima K, Suzuki T, Kawase T, Watanabe M, Shitara K, Misawa K, Ito S, Sawaki A, Muro K, Nakamura T, Yamao K, Yamamura Y, Hamajima N, Hiraki A, Tanaka H (2009) Association of prostate stem cell antigen gene polymorphisms with the risk of stomach cancer in Japanese. Int J Cancer 125 (8):1961–1964. https://doi.org/10.1002/ijc.24519

McColl KE (2010) Clinical practice. *Helicobacter pylori* infection. N Engl J Med 362(17):1597–1604. https://doi.org/10.1056/NEJMcp1001110

McColl KE, El-Omar EM, Gillen D (1997) The role of *H. pylori* infection in the pathophysiology of duodenal ulcer disease. J Physiol Pharmacol 48(3):287–295

Melchiades JL, Zabaglia LM, Sallas ML, Orcini WA, Chen E, Smith MAC, Payao SLM, Rasmussen LT (2017) Polymorphisms and haplotypes of the interleukin 2 gene are associated with an increased risk of gastric cancer. The possible involvement of *Helicobacter pylori*. Cytokine 96:203–207. https://doi.org/10.1016/j.cyto.2017.04.020

Melo Barbosa HP, Martins LC, Dos Santos SE, Demachki S, Assumpcao MB, Aragao CD, de Oliveira Corvelo TC (2009) Interleukin-1 and TNF-alpha polymorphisms and *Helicobacter pylori* in a Brazilian

Amazon population. World J Gastroenterol 15 (12):1465–1471
Miftahussurur M, Yamaoka Y (2015) *Helicobacter pylori* virulence genes and host genetic polymorphisms as risk factors for peptic ulcer disease. Expert Rev Gastroenterol Hepatol 9(12):1535–1547. https://doi.org/10.1586/17474124.2015.1095089
Necchi V, Candusso ME, Tava F, Luinetti O, Ventura U, Fiocca R, Ricci V, Solcia E (2007) Intracellular, intercellular, and stromal invasion of gastric mucosa, preneoplastic lesions, and cancer by *Helicobacter pylori*. Gastroenterology 132(3):1009–1023. https://doi.org/10.1053/j.gastro.2007.01.049
Neves Filho EH, Alves MK, Lima VP, Rabenhorst SH (2010) MTHFR C677T polymorphism and differential methylation status in gastric cancer: an association with *Helicobacter pylori* infection. Virchows Arch 457(6):627–633. https://doi.org/10.1007/s00428-010-0996-3
Ng MT, Van't Hof R, Crockett JC, Hope ME, Berry S, Thomson J, McLean MH, McColl KE, El-Omar EM, Hold GL (2010) Increase in NF-kappaB binding affinity of the variant C allele of the toll-like receptor 9 -1237T/C polymorphism is associated with *Helicobacter pylori* -induced gastric disease. Infect Immun 78(3):1345–1352. https://doi.org/10.1128/IAI.01226-09
Pachathundikandi SK, Tegtmeyer N, Backert S (2013) Signal transduction of *Helicobacter pylori* during interaction with host cell protein receptors of epithelial and immune cells. Gut Microbes 4:454–474. https://doi.org/10.4161/gmic.27001
Pachathundikandi SK, Müller A, Backert S (2016) Inflammasome Activation by *Helicobacter pylori* and its implications for persistence and immunity. Curr Top Microbiol Immunol 397:117–131. https://doi.org/10.1007/978-3-319-41171-2_6
Park MJ, Hyun MH, Yang JP, Yoon JM, Park S (2015) Effects of the interleukin-1beta-511 C/T gene polymorphism on the risk of gastric cancer in the context of the relationship between race and *H. pylori* infection: a meta-analysis of 20,000 subjects. Mol Biol Rep 42(1):119–134. https://doi.org/10.1007/s11033-014-3748-7
Posselt G, Backert S, Wessler S (2013) The functional interplay of *Helicobacter pylori* factors with gastric epithelial cells induces a multi-step process in pathogenesis. Cell Commun Signal 11:77. https://doi.org/10.1186/1478-811X-11-77
Prescott NJ, Fisher SA, Franke A, Hampe J, Onnie CM, Soars D, Bagnall R, Mirza MM, Sanderson J, Forbes A, Mansfield JC, Lewis CM, Schreiber S, Mathew CG (2007) A nonsynonymous SNP in ATG16L1 predisposes to ileal Crohn's disease and is independent of CARD15 and IBD5. Gastroenterology 132(5):1665–1671. https://doi.org/10.1053/j.gastro.2007.03.034
Queiroz DM, Guerra JB, Rocha GA, Rocha AM, Santos A, De Oliveira AG, Cabral MM, Nogueira AM, De Oliveira CA (2004) IL1B and IL1RN polymorphic genes and *Helicobacter pylori cagA* strains decrease the risk of reflux esophagitis. Gastroenterology 127(1):73–79
Queiroz DM, Saraiva IE, Rocha GA, Rocha AM, Gomes LI, Melo FF, Bittencourt PF (2009) IL2-330G polymorphic allele is associated with decreased risk of *Helicobacter pylori* infection in adulthood. Microbes Infect 11(12):980–987. https://doi.org/10.1016/j.micinf.2009.07.008
Rad R, Ballhorn W, Voland P, Eisenacher K, Mages J, Rad L, Ferstl R, Lang R, Wagner H, Schmid RM, Bauer S, Prinz C, Kirschning CJ, Krug A (2009) Extracellular and intracellular pattern recognition receptors cooperate in the recognition of *Helicobacter pylori*. Gastroenterology 136(7):2247–2257. https://doi.org/10.1053/j.gastro.2009.02.066
Raju D, Hussey S, Ang M, Terebiznik MR, Sibony M, Galindo-Mata E, Gupta V, Blanke SR, Delgado A, Romero-Gallo J, Ramjeet MS, Mascarenhas H, Peek RM, Correa P, Streutker C, Hold G, Kunstmann E, Yoshimori T, Silverberg MS, Girardin SE, Philpott DJ, El Omar E, Jones NL (2012) Vacuolating cytotoxin and variants in Atg16L1 that disrupt autophagy promote *Helicobacter pylori* infection in humans. Gastroenterology 142(5):1160–1171. https://doi.org/10.1053/j.gastro.2012.01.043
Rosenstiel P, Hellmig S, Hampe J, Ott S, Till A, Fischbach W, Sahly H, Lucius R, Folsch UR, Philpott D, Schreiber S (2006) Influence of polymorphisms in the NOD1/CARD4 and NOD2/CARD15 genes on the clinical outcome of *Helicobacter pylori* infection. Cell Microbiol 8 (7):1188–1198. https://doi.org/10.1111/j.1462-5822.2006.00701.x
Rowland M, Daly L, Vaughan M, Higgins A, Bourke B, Drumm B (2006) Age-specific incidence of *Helicobacter pylori*. Gastroenterology 130(1):65–72.; quiz 211. https://doi.org/10.1053/j.gastro.2005.11.004
Santos JC, Ladeira MS, Pedrazzoli J Jr, Ribeiro ML (2012) Relationship of IL-1 and TNF-alpha polymorphisms with *Helicobacter pylori* in gastric diseases in a Brazilian population. Braz J Med Biol Res 45 (9):811–817
Schistosomes, liver flukes and *Helicobacter pylori* . IARC Working Group on the Evaluation of Carcinogenic Risks to Humans. Lyon, 7-14 June 1994 (1994) IARC Monogr Eval Carcinog Risks Hum 61:1–241
Semino-Mora C, Doi SQ, Marty A, Simko V, Carlstedt I, Dubois A (2003) Intracellular and interstitial expression of *Helicobacter pylori* virulence genes in gastric precancerous intestinal metaplasia and adenocarcinoma. J Infect Dis 187(8):1165–1177. https://doi.org/10.1086/368133
Serrano C, Wright SW, Bimczok D, Shaffer CL, Cover TL, Venegas A, Salazar MG, Smythies LE, Harris PR, Smith PD (2013) Downregulated Th17 responses are associated with reduced gastritis in *Helicobacter pylori*

-infected children. Mucosal Immunol 6(5):950–959. https://doi.org/10.1038/mi.2012.133

Shigematsu Y, Niwa T, Rehnberg E, Toyoda T, Yoshida S, Mori A, Wakabayashi M, Iwakura Y, Ichinose M, Kim YJ, Ushijima T (2013) Interleukin-1beta induced by *Helicobacter pylori* infection enhances mouse gastric carcinogenesis. Cancer Lett 340(1):141–147. https://doi.org/10.1016/j.canlet.2013.07.034

Shin WG, Jang JS, Kim HS, Kim SJ, Kim KH, Jang MK, Lee JH, Kim HJ, Kim HY (2008) Polymorphisms of interleukin-1 and interleukin-2 genes in patients with gastric cancer in Korea. J Gastroenterol Hepatol 23 (10):1567–1573. https://doi.org/10.1111/j.1440-1746.2008.05479.x

Sierra JC, Asim M, Verriere TG, Piazuelo MB, Suarez G, Romero-Gallo J, Delgado AG, Wroblewski LE, Barry DP, Peek RM Jr, Gobert AP, Wilson KT (2018) Epidermal growth factor receptor inhibition downregulates *Helicobacter pylori* -induced epithelial inflammatory responses, DNA damage and gastric carcinogenesis. Gut 67(7):1247–1260. https://doi.org/10.1136/gutjnl-2016-312888

Silva-Fernandes IJ, da Silva TA, Agnez-Lima LF, Ferreira MV, Rabenhorst SH (2012) *Helicobacter pylori* genotype and polymorphisms in DNA repair enzymes: where do they correlate in gastric cancer? J Surg Oncol 106(4):448–455. https://doi.org/10.1002/jso.23077

Smith MF Jr, Novotny J, Carl VS, Comeau LD (2006) *Helicobacter pylori* and toll-like receptor agonists induce syndecan-4 expression in an NF-kappaB-dependent manner. Glycobiology 16(3):221–229. https://doi.org/10.1093/glycob/cwj061

Smith SM, Moran AP, Duggan SP, Ahmed SE, Mohamed AS, Windle HJ, O'Neill LA, Kelleher DP (2011) Tribbles 3: a novel regulator of TLR2-mediated signaling in response to *Helicobacter pylori* lipopolysaccharide. J Immunol 186(4):2462–2471. https://doi.org/10.4049/jimmunol.1000864

Stec-Michalska K, Peczek L, Michalski B, Wisniewska-Jarosinska M, Krakowiak A, Nawrot B (2009) *Helicobacter pylori* infection and family history of gastric cancer decrease expression of FHIT tumor suppressor gene in gastric mucosa of dyspeptic patients. Helicobacter 14(5):126–134. https://doi.org/10.1111/j.1523-5378.2009.00715.x

Study Group of Millennium Genome Project for C, Sakamoto H, Yoshimura K, Saeki N, Katai H, Shimoda T, Matsuno Y, Saito D, Sugimura H, Tanioka F, Kato S, Matsukura N, Matsuda N, Nakamura T, Hyodo I, Nishina T, Yasui W, Hirose H, Hayashi M, Toshiro E, Ohnami S, Sekine A, Sato Y, Totsuka H, Ando M, Takemura R, Takahashi Y, Ohdaira M, Aoki K, Honmyo I, Chiku S, Aoyagi K, Sasaki H, Ohnami S, Yanagihara K, Yoon KA, Kook MC, Lee YS, Park SR, Kim CG, Choi IJ, Yoshida T, Nakamura Y, Hirohashi S (2008) Genetic variation in PSCA is associated with susceptibility to diffuse-type gastric cancer. Nat Genet 40(6):730–740. https://doi.org/10.1038/ng.152

Su B, Ceponis PJ, Lebel S, Huynh H, Sherman PM (2003) *Helicobacter pylori* activates toll-like receptor 4 expression in gastrointestinal epithelial cells. Infect Immun 71(6):3496–3502

Sugimoto M, Furuta T, Shirai N, Nakamura A, Xiao F, Kajimura M, Sugimura H, Hishida A (2007) Different effects of polymorphisms of tumor necrosis factor-alpha and interleukin-1 beta on development of peptic ulcer and gastric cancer. J Gastroenterol Hepatol 22 (1):51–59. https://doi.org/10.1111/j.1440-1746.2006.04442.x

Sun X, Xu Y, Zhang F, Jing T, Han J, Zhang J (2015) Association between the IL1B -31C > T polymorphism and *Helicobacter pylori* infection in Asian and Latin American population: a meta-analysis. Microb Pathog 86:45–52. https://doi.org/10.1016/j.micpath.2015.07.010

Sun X, Xu Y, Wang L, Zhang F, Zhang J, Fu X, Jing T, Han J (2016) Association between TNFA Gene Polymorphisms and *Helicobacter pylori* Infection: A Meta-Analysis. PLoS One 11(1):e0147410. https://doi.org/10.1371/journal.pone.0147410

Susser M, Stein Z (1962) Civilisation and peptic ulcer. Lancet 1(7221):115–119

Tafreshi M, Guan J, Gorrell RJ, Chew N, Xin Y, Deswaerte V, Rohde M, Daly RJ, Peek RM Jr, Jenkins BJ, Davies EM, Kwok T (2018) *Helicobacter pylori* type IV secretion system and its adhesin subunit, CagL, mediate potent inflammatory responses in primary human endothelial cells. Front Cell Infect Microbiol 8:22. https://doi.org/10.3389/fcimb.2018.00022

Takenaka R, Yokota K, Ayada K, Mizuno M, Zhao Y, Fujinami Y, Lin SN, Toyokawa T, Okada H, Shiratori Y, Oguma K (2004) *Helicobacter pylori* heat-shock protein 60 induces inflammatory responses through the Toll-like receptor-triggered pathway in cultured human gastric epithelial cells. Microbiology 150(Pt 12):3913–3922. https://doi.org/10.1099/mic.0.27527-0

Tanaka S, Nagashima H, Uotani T, Graham DY, Yamaoka Y (2017) Autophagy-related genes in *Helicobacter pylori* infection. Helicobacter 22(3). https://doi.org/10.1111/hel.12376

Tanikawa C, Urabe Y, Matsuo K, Kubo M, Takahashi A, Ito H, Tajima K, Kamatani N, Nakamura Y, Matsuda K (2012) A genome-wide association study identifies two susceptibility loci for duodenal ulcer in the Japanese population. Nat Genet 44(4):430–434., S431–432. https://doi.org/10.1038/ng.1109

Tegtmeyer N, Wessler S, Necchi V, Rohde M, Harrer A, Rau TT, Asche CI, Boehm M, Loessner H, Figueiredo C, Naumann M, Palmisano R, Solcia E, Ricci V, Backert S (2017) *Helicobacter pylori* employs a unique basolateral type IV secretion mechanism for CagA delivery. Cell Host Microbe 22:552–560.e555. https://doi.org/10.1016/j.chom.2017.09.005

Terebiznik MR, Raju D, Vazquez CL, Torbricki K, Kulkarni R, Blanke SR, Yoshimori T, Colombo MI, Jones NL (2009) Effect of *Helicobacter pylori* 's vacuolating cytotoxin on the autophagy pathway in gastric epithelial cells. Autophagy 5(3):370–379

Togawa S, Joh T, Itoh M, Katsuda N, Ito H, Matsuo K, Tajima K, Hamajima N (2005) Interleukin-2 gene polymorphisms associated with increased risk of gastric atrophy from *Helicobacter pylori* infection. Helicobacter 10(3):172–178. https://doi.org/10.1111/j.1523-5378.2005.00308.x

Trejo-de la OA, Torres J, Sanchez-Zauco N, Perez-Rodriguez M, Camorlinga-Ponce M, Flores-Luna L, Lazcano-Ponce E, Maldonado-Bernal C (2015) Polymorphisms in TLR9 but not in TLR5 increase the risk for duodenal ulcer and alter cytokine expression in the gastric mucosa. Innate Immun 21 (7):706–713. https://doi.org/10.1177/1753425915587130

Tu S, Bhagat G, Cui G, Takaishi S, Kurt-Jones EA, Rickman B, Betz KS, Penz-Oesterreicher M, Bjorkdahl O, Fox JG, Wang TC (2008) Overexpression of interleukin-1beta induces gastric inflammation and cancer and mobilizes myeloid-derived suppressor cells in mice. Cancer Cell 14 (5):408–419. https://doi.org/10.1016/j.ccr.2008.10.011

Uotani T, Sugimoto M, Ichikawa H, Tanaka S, Nagashima H, Uchida T, Graham DY, Yamaoka Y (2016) Prostate stem cell antigen gene TT genotype and development of intestinal metaplasia in *Helicobacter pylori* infection. J Dig Dis 17(1):20–27. https://doi.org/10.1111/1751-2980.12309

Viala J, Chaput C, Boneca IG, Cardona A, Girardin SE, Moran AP, Athman R, Memet S, Huerre MR, Coyle AJ, DiStefano PS, Sansonetti PJ, Labigne A, Bertin J, Philpott DJ, Ferrero RL (2004) Nod1 responds to peptidoglycan delivered by the *Helicobacter pylori* cag pathogenicity island. Nat Immunol 5(11):1166–1174. https://doi.org/10.1038/ni1131

Wilson KT, Crabtree JE (2007) Immunology of *Helicobacter pylori*: insights into the failure of the immune response and perspectives on vaccine studies. Gastroenterology 133(1):288–308. https://doi.org/10.1053/j.gastro.2007.05.008

Xu T, Fu D, Ren Y, Dai Y, Lin J, Tang L, Ji J (2017a) Genetic variations of TLR5 gene interacted with *Helicobacter pylori* infection among carcinogenesis of gastric cancer. Oncotarget 8(19):31016–31022. https://doi.org/10.18632/oncotarget.16050

Xu Y, Cao X, Jiang J, Chen Y, Wang K (2017b) TNF-alpha-308/−238 polymorphisms are associated with gastric cancer: a case-control family study in China. Clin Res Hepatol Gastroenterol 41 (1):103–109. https://doi.org/10.1016/j.clinre.2016.05.014

Yamaoka Y, Kato M, Asaka M (2008) Geographic differences in gastric cancer incidence can be explained by differences between *Helicobacter pylori* strains. Intern Med 47(12):1077–1083

Yin YW, Sun QQ, Hu AM, Wang Q, Liu HL, Hou ZZ, Zeng YH, Xu RJ, Shi LB, Ma JB (2012) Associations between interleukin-6 gene −174 C/G and −572 C/G polymorphisms and the risk of gastric cancer: a meta-analysis. J Surg Oncol 106(8):987–993. https://doi.org/10.1002/jso.23199

Ying HY, Yu BW, Yang Z, Yang SS, Bo LH, Shan XY, Wang HJ, Zhu YJ, Wu XS (2016) Interleukin-1B 31 C>T polymorphism combined with *Helicobacter pylori* -modified gastric cancer susceptibility: evidence from 37 studies. J Cell Mol Med 20(3):526–536. https://doi.org/10.1111/jcmm.12737

Yokota S, Okabayashi T, Rehli M, Fujii N, Amano K (2010) *Helicobacter pylori* lipopolysaccharides upregulate toll-like receptor 4 expression and proliferation of gastric epithelial cells via the MEK1/2-ERK1/2 mitogen-activated protein kinase pathway. Infect Immun 78(1):468–476. https://doi.org/10.1128/IAI.00903-09

Yoshida T, Ono H, Kuchiba A, Saeki N, Sakamoto H (2010) Genome-wide germline analyses on cancer susceptibility and GeMDBJ database: gastric cancer as an example. Cancer Sci 101(7):1582–1589. https://doi.org/10.1111/j.1349-7006.2010.01590.x

Zhang Q, Li Y, Li X, Zhou W, Shi B, Chen H, Yuan W (2009) PARP-1 Val762Ala polymorphism, CagA+ *H. pylori* infection and risk for gastric cancer in Han Chinese population. Mol Biol Rep 36(6):1461–1467. https://doi.org/10.1007/s11033-008-9336-y

Zhang BB, Liu XZ, Sun J, Yin YW, Sun QQ (2013) Association between TNF alpha gene polymorphisms and the risk of duodenal ulcer: a meta-analysis. PLoS One 8(2):e57167. https://doi.org/10.1371/journal.pone.0057167

Zhang JZ, Liu CM, Peng HP, Zhang Y (2017) Association of genetic variations in IL-6/IL-6R pathway genes with gastric cancer risk in a Chinese population. Gene 623:1–4. https://doi.org/10.1016/j.gene.2017.04.038

Zhao Y, Yokota K, Ayada K, Yamamoto Y, Okada T, Shen L, Oguma K (2007) *Helicobacter pylori* heat-shock protein 60 induces interleukin-8 via a Toll-like receptor (TLR)2 and mitogen-activated protein (MAP) kinase pathway in human monocytes. J Med Microbiol 56(Pt 2):154–164. https://doi.org/10.1099/jmm.0.46882-0

Zhao J, Geng P, Li Z, Cui S, Zhao J, Wang L, Li J, Ji F, Li G, Shen G, Lin M, Shen C (2013) Prostate stem cell antigen rs2294008 polymorphism differentially contributes to *Helicobacter pylori* -negative gastric cancer among various populations in China. Mol Clin Oncol 1(3):493–498. https://doi.org/10.3892/mco.2013.70

Adv Exp Med Biol - Advances in Microbiology, Infectious Diseases and Public Health (2019) 1149: 173–194
https://doi.org/10.1007/5584_2019_365

Published online: 24 April 2019

Helicobacter pylori Genetic Polymorphisms in Gastric Disease Development

Jeannette M. Whitmire and D. Scott Merrell

Abstract

Infecting half of the world's population, *Helicobacter pylori* is a medically important bacterium that induces a variety of gastric diseases, including gastritis, peptic ulcer disease, and gastric cancer. Sequencing of almost 1000 *H. pylori* isolates has revealed a diverse genome that contains abundant polymorphic genetic elements; many of these lie in factors likely to be associated with virulence. To ascertain the effect of these varying genetic elements, numerous epidemiological studies have investigated the contribution of the various polymorphisms to gastric disease development; particular focus has been placed on polymorphisms in the outer membrane proteins (OMPs), an effector protein, and a toxin produced by *H. pylori.* These studies have revealed geographic variation in the prevalence of various polymorphisms as well as in the associations between particular polymorphisms and gastric disease development. Furthermore, researchers have identified polymorphisms in multiple genes that frequently occur in combination. Though no single polymorphic genetic factor alone can fully account for gastric disease development in a population, the evaluation of multiple polymorphisms in a colonizing *H. pylori* strain can aid in the assessment of the pathogenic potential of the strain. Here we review specific *H. pylori* genetic polymorphisms (Bab proteins, Hom proteins, HopQ, SabA, SabB, OipA, IceA, VacA and CagA) that have been linked to disease outcome and discuss how geographic location and virulence factor polymorphisms together contribute to *H. pylori*-induced disease.

Keywords

Genetic polymorphisms · Virulence factors · *H. pylori* · Gastric disease · Toxin

J. M. Whitmire and D. S. Merrell (✉)
Department of Microbiology and Immunology, Uniformed Services University of the Health Sciences, Bethesda, MD, USA
e-mail: jeannette.whitmire@usuhs.edu; douglas.merrell@usuhs.edu

1 Introduction

Helicobacter pylori is a Gram-negative, microaerophilic, spiral bacterium that is capable of colonizing the inhospitable environment of the human stomach. Though infection rates vary greatly by region, the bacterium infects half of the human population across the globe, imparting a huge medical burden (Dunn et al. 1997). Despite its profound importance, the recognition and acceptance of this bacterium as medically important required many years of research and observation. Indeed, as far back as 1892, Giulio Bizzozero observed and noted the presence of spirilli in the stomachs of healthy dogs (as reviewed in (Figura and Oderda 1996; Marshall 2001). Throughout the twentieth century, researchers reported observations of spiral organisms in the stomachs of dogs, cats, macaques (Doenges 1938, 1939; Kasai and Kobayashi 1919), and humans presenting with gastric disturbances (Freedberg and Barron 1940). These reports led to the production of electron micrographs of *Helicobacter* inside a human parietal cell (Ito 1967) and in the stomachs of dogs (Lockard and Boler 1970). Finally, the groundbreaking advances of Marshall and Warren enabled the isolation and culture of this organism from human biopsies (Marshall et al. 1984; Marshall and Warren 1984; Warren and Marshall 1983), and ultimately, their research defined a link between *H. pylori* infection and development of gastritis and peptic ulcer disease (Marshall et al. 1985a, b).

The unique niche of *H. pylori* within the acidic environment of the stomach is a testament to its co-evolution with and adaptation to its human host. The publication of the first *H. pylori* chromosome sequence in 1997 uncovered a small 1.6 Mb genome characterized by a paucity of regulatory factors, limited metabolic abilities, and minimal biosynthetic capabilities compared to similar organisms (Tomb et al. 1997). This compact, restricted genome resulted from countless years of adaptations to its host and the accompanying acidic mucosal environment. Comparative genomic analyses have revealed that *H. pylori* strains display a great deal of genome-wide diversity (de Reuse and Bereswill 2007; Salama et al. 2000). Indeed, as detailed by Yamaoka and co-workers in Chapter 1, phylogenetic analyses have revealed several genetically distinct groups of *H. pylori* that can be linked to human migration (de Reuse and Bereswill 2007; Yamaoka et al. 2008). Furthermore, these various groups of *H. pylori* exhibit different polymorphisms within multiple genes that have been further linked to severity of the gastric diseases within the host (Chen et al. 2016; Yamaoka et al. 2008). Additionally, the plasticity of the *H. pylori* genome is further highlighted through the micro-evolution and genetic variability of strains identified within a single human host (de Reuse and Bereswill 2007), as well as through the passage and isolation of strains using established animal models (Asim et al. 2015; Draper et al. 2017; Franco et al. 2005).

As genomic analyses of bacterial species have become more accessible, studies that investigate genomic variations of *H. pylori* strains have abounded, and these studies have identified associations between particular genetic polymorphisms, geographic location of the host, and ultimate disease outcomes. Indeed, *H. pylori* can cause a range of gastric maladies within the human host and occurrence rates of these various diseases vary widely throughout the world (Kusters et al. 2006). Though a large portion of infected individuals remain asymptomatic, the primary presentation of *H. pylori* infection is through chronic gastritis and peptic ulcer disease (Atherton 2006). More severe manifestations of disease include gastric adenocarcinoma and MALT lymphoma (Atherton 2006; Kusters et al. 2006). More detailed information on *H. pylori*-associated diseases is presented in Chapters 7 and 8 by Vieth and colleagues and Hatakeyama and co-authors, respectively. Interestingly, multiple studies have recently, yet controversially, indicated a protective effect conferred by *H. pylori* infection against several diseases (Arnold et al. 2012; Eross et al. 2018; Kyburz and Muller 2017; Robinson 2015; White et al. 2015). These studies have led to a larger debate about whether a bacterium that colonizes 50% of the world's population should be considered a part of the normal flora and whether infection should be left untreated in the young and in asymptomatic patients (Cover and Blaser 2009; Kyburz and Muller 2017; Robinson 2015). However, on the other side of this debate lies the fact that gastric cancer is the third most common form of cancer-related death in the world (International Agency for Research on Cancer 2012). Thus, the risk of disease may outweigh the potential benefits of this organism. Overall, the seemingly contradictory findings underscore the need to investigate and understand the differences between *H. pylori* strains that may contribute to various disease outcomes; such an understanding may reveal those patients that are most at risk and help to define those that should receive treatment (Fig. 1).

The persistent colonization of the harsh environment of the stomach is accomplished through a delicate interplay between the host and the bacterium. Indeed, polymorphic genetic factors in both the bacterium and human host influence the course of infection and contribute to the ultimate disease pathology experienced by the host. An increasing number of epidemiological studies, in conjunction with molecular studies, have revealed roles for specific polymorphisms in both the host and bacterium that directly contribute to the various gastric pathologies caused by *H. pylori.* While Clyne and colleagues discuss polymorphic factors within the host that influence disease outcome in Chapter 9, the current chapter will focus on specific genetic polymorphisms within *H. pylori* that impact disease pathology in some populations.

2 Virulence Factors

To survive the gastric milieu and to persistently colonize the human stomach, *H. pylori* possesses multiple virulence factors that contribute to its infectious strategy. First, following ingestion, *H. pylori* must transit through the acidic gastric lumen, during which it employs multiple sheathed flagella for motility and the urease enzyme to combat acid stress (Kao et al. 2016) and follows a urea gradient to locate the epithelium (Huang et al. 2015). Next, it must colonize the gastric mucosa and adhere to the gastric epithelial cells, which it expertly accomplishes with an arsenal of adhesins and outer membrane proteins (OMPs) (Kao et al. 2016; Wroblewski et al. 2010). Lastly, it directly affects the gastric mucosal cells through the injection of CagA into the host cells and the secretion of VacA into the mucosal layer (Kao et al. 2016; Tegtmeyer et al. 2009; Wroblewski et al. 2010). These virulence factors can cause extensive damage to gastric cells, leading to the inflammation that characterizes the clinical diseases many patients experience.

Initial epidemiological studies of *H. pylori* have unveiled a correlation between disease severity and the presence of specific virulence factors. Excitingly, the expansion of genomic sequencing tools has led to the identification of polymorphisms within several virulence factors, and these polymorphisms also appear to be linked to the varying disease outcomes within infected individuals. However, large differences exist in the prevalence of various polymorphisms among strains from different geographic regions across the globe. Consequently, associations between the many polymorphisms exhibited by *H. pylori* and the various clinical disease manifestations vary tremendously depending on the geographic location of the research study and associated strains. For example, East Asian strains tend to exhibit less genotypic heterogeneity while harboring more virulent genotypes (Kim et al. 2015); this region of the world also has the highest global incidence of gastric cancer (Yamaoka et al. 2008). Conversely, Western strains exhibit a greater diversity in genotype distributions (Kim et al. 2015), contributing to a greater variation in the manifestation of clinical disease (Yamaoka et al. 2008). Thus, the geographic origin of strains used in studies must be taken into account when analyzing data from the numerous available epidemiological studies.

As described above, disease development within the human host results from a complex interaction between host polymorphisms, *H. pylori* polymorphisms, and environmental factors. The prevalence of these polymorphisms and environmental factors varies throughout the world, which adds another geographic layer of complexity to the study of *H. pylori* infection. Our approach in this chapter is to highlight findings for certain genes that have been linked to disease outcome. In particular, we focus on the polymorphisms identified in the OMPs, VacA and CagA, since these factors substantially contribute to the course of disease progression. Though certainly not exhaustive, we compare and contrast a selection of studies for each factor, since results from different studies are often not in

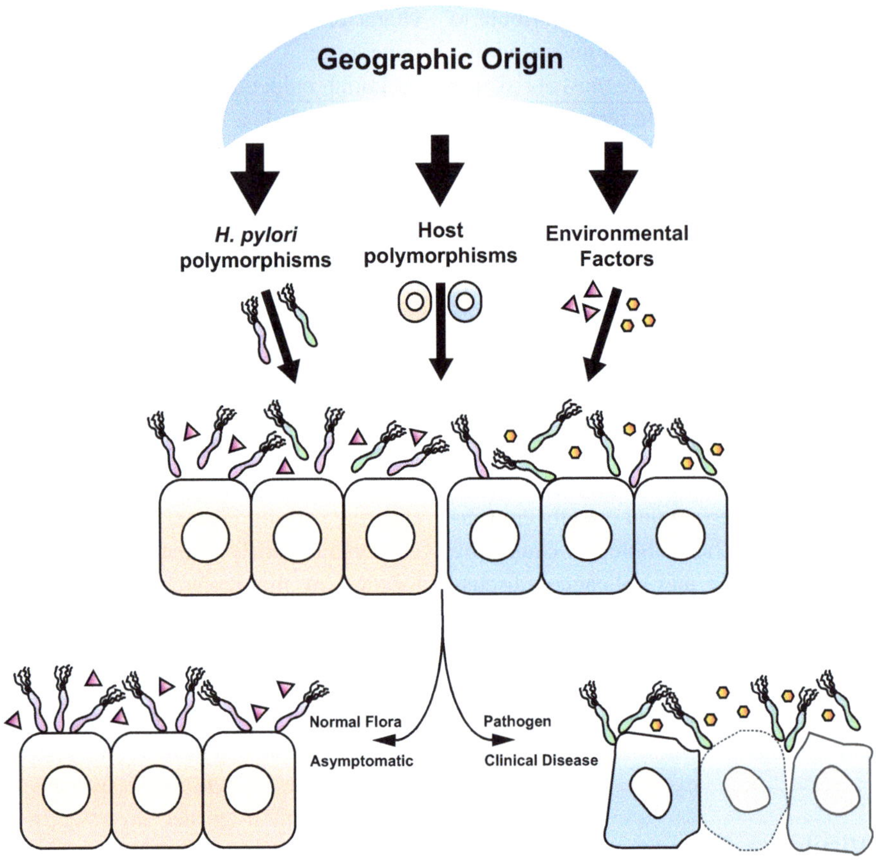

Fig. 1 Factors influencing *H. pylori*'s disease potential in the gastric mucosa
Geography plays an important role in the ultimate outcome of *H. pylori* infection; this is due to geographical influences on *H. pylori* polymorphisms, host polymorphisms, and environmental factors that contribute to pathogenicity. Possible gene polymorphisms of *H. pylori* are represented by two different colored bacteria in the diagram and host polymorphisms are represented by two different colored host cells. Environmental factors to which a host may be exposed are identified by two separate colored shapes. An individual has the potential to be infected with different *H. pylori* variants and to be exposed to various environmental factors. As such, the complex interplay between the polymorphisms within *H. pylori*, the polymorphisms of the host, and the environmental factors will determine the ultimate disease potential of *H. pylori* within the host. As illustrated, *H. pylori* can remain a member of the normal gastric flora and cause no symptoms in the infected individual, or *H. pylori* can become pathogenic and induce clinical symptoms within the host

agreement. The lack of unity among published studies underscores the complexity of disease development during the course of *H. pylori* infection.

2.1 Adhesins and OMPs

Within the compact genome of *H. pylori*, an astonishing 4% of the predicted genes are believed to encode OMPs (Alm et al. 2000; Chmiela et al. 2017; Oleastro and Menard 2013). This large number of OMPs suggests that these factors are a critical component of the *H. pylori* lifecycle. As such, efforts to define the expression and function of the many OMPs encoded by the *H. pylori* genome are a major focus of many ongoing research studies. Interestingly, unlike many other Gram-negative microorganisms, *H. pylori* does not possess a predominant OMP, instead it employs its repertoire of diverse OMPs as needed (Alm et al. 2000;

Oleastro and Menard 2013). Broadly speaking, the *H. pylori* OMPs are classified into several families of proteins, which include the *Helicobacter* outer membrane protein (Hop) family, hop-related (Hor) protein, *Helicobacter* OMP family (Hof), and the *Helicobacter* outer membrane (Hom) family (Matsuo et al. 2017). Sequencing of the first *H. pylori* genome revealed approximately 21 Hop proteins, some of which are involved in adherence to the gastric epithelium. In particular, subsequent studies have identified Bab proteins (HopS/HopT), Sab proteins (HopO/HopP), the OipA protein (HopH), and the HopQ protein, as important components for interaction with host cells during the course of infection (Matsuo et al. 2017). While BabA is the blood-group-antigen-binding adhesin that binds to Lewis B (Le^b) blood group antigens (Ilver et al. 1998), SabA is a sialic acid-binding adhesin that interacts with sialyl-Lewis^a, sialyl-Lewis^x, and Lewis^X receptors on host cells (Mahdavi et al. 2002). OipA is the outer inflammatory protein that is capable of inducing inflammatory cytokine production (Yamaoka et al. 2002). Finally, HopQ interacts with various proteins of the carcinoembryonic antigen-related cell adhesion molecule (CEACAM) family (Matsuo et al. 2017; Moonens et al. 2018). These various proteins and their polymorphisms are described in detail below.

2.1.1 Bab Proteins

The *bab* genes consist of three paralogues that contain extremely similar sequences: *babA*, *babB*, and *babC* (Alm et al. 2000; Ansari and Yamaoka 2017; Colbeck et al. 2006; Kim et al. 2015) (Fig. 2). Although the three *bab* genes contain highly similar sequences, BabA is the only protein encoded by these genes that has been shown to bind to Le^b antigens (Oleastro and Menard 2013); BabA was the first identified *H. pylori* adhesin (Ilver et al. 1998). In addition, two primary allelic variants of *babA* have been identified in clinical isolates: *babA1* and *babA2* (Fig. 2). The *babA2* allele contains the entire coding sequence of the *babA* gene, which results in the production of the BabA protein. In contrast, the *babA1* variant contains a deletion of 10 base pairs, which includes the ATG translation initiation codon; this deletion results in a loss of translation of this protein in *H. pylori* strains containing this gene variant (Ilver et al. 1998). Additionally, the 5′ end of the *babA* gene contains a string of cytosine-thymine (CT) repeats, which leads to deletion or insertion of nucleotides during DNA replication (Salaun et al. 2004; Solnick et al. 2004). This variability in CT repeats causes phase variation, whereby subsequent cells carry *babA* genes that encode for truncated non-functional proteins. The *bab* status of *H. pylori* is further complicated by the existence of 3 distinct genomic loci in which the *bab* genes can be found (Colbeck et al. 2006; Oleastro and Menard 2013; Pride and Blaser 2002) (Fig. 2).

Varying results have been obtained from the multitude of epidemiological studies that have been conducted to investigate the relationship between *babA* gene status and disease outcome. As noted earlier, genetic variations within *H. pylori* are frequently observed in conjunction with the geographic origin of the clinical isolate under investigation. As such, studies that investigated the influence of the *babA* genotype on disease status report differing associations that depend on the origin of the strain. In general, studies that employ isolates from Western countries tend to identify a significant association between the *babA2* genotype and disease outcome, whereas studies that utilize strains from East Asian countries frequently fail to detect a significant correlation between the *babA2* genotype and disease. Specifically, studies conducted using strains that originate from Western countries identify a correlation between the *babA2* genotype and development of gastritis (Homan et al. 2014), duodenal ulcers (Gerhard et al. 1999; Olfat et al. 2005; Torres et al. 2009), intestinal metaplasia (Zambon et al. 2003), and adenocarcinoma (Gerhard et al. 1999) (Fig. 2). Furthermore, a meta-analysis of 25 research articles also reveals a correlation between the *babA2* genotype and peptic ulcer disease and duodenal ulcers within Western countries (Chen et al. 2013). Interestingly, a recent genome-wide association study, which compares the entire genomes of 173 European *H. pylori* clinical

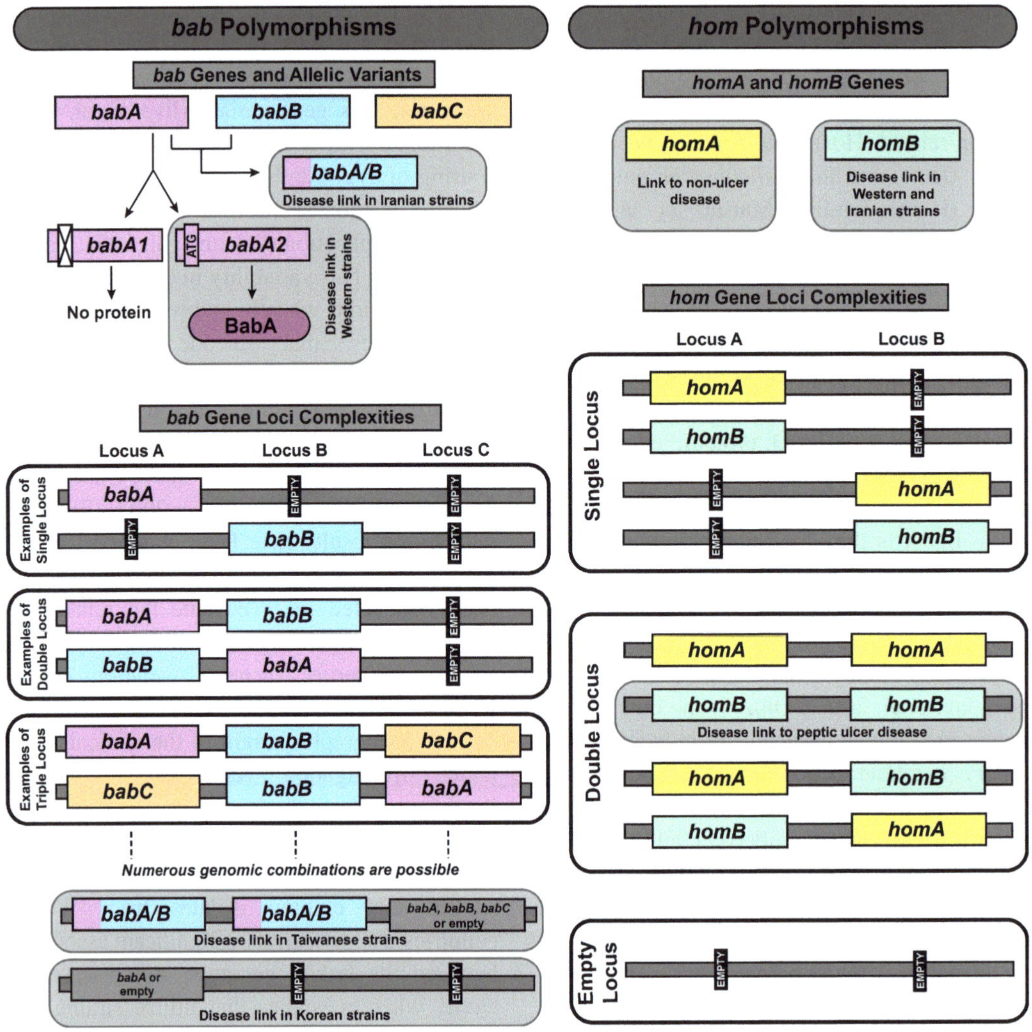

Fig. 2 OMP polymorphisms in *bab* and *hom* genes

***bab* polymorphisms** occur in both the gene sequences and the genomic loci in which the *bab* genes are located. The three *bab* genes, *babA, babB,* and *babC*, are depicted as three different colors. The *babA* gene has two alleles. The *babA1* allele contains a 10 bp deletion and lacks the ATG translation initiation codon; thus, cells with this allele produce no BabA protein. The *babA2* allele contains the ATG codon necessary for translation, which permits cells to produce the BabA protein. As noted, strains with the *babA2* allele have been linked to disease in Western countries. Additionally, the *bab* genes can recombine to form chimeric *bab* sequences; the *babA/B* gene is depicted with the combined colors of the *babA* and *babB* genes. The presence of the *babA/B* gene has been linked to disease in Iranian strains. Polymorphisms in the *bab* genes also appear in the genomic loci occupied by the genes. As depicted, there are three loci that can be occupied by the *bab* genes. Specific examples of *bab* genes occurring at a one, two, or three loci are shown; numerous unshown genomic combinations of the *bab* genes are possible. In addition, the diagram displays the *babA/B* gene in locus A and locus B, which has been linked to disease in Taiwanese strains. Furthermore, the genotype at locus B has been linked to disease in Korean strains; identified combinations shown to occur with the empty locus B are indicated

***hom* polymorphisms** also occur at the level of both the gene and the genomic loci. Two of the *hom* genes, *homA* and *homB,* are displayed as two separate colors. *homA* has been linked to non-ulcer disease, and the presence of *homB* has been correlated with disease in Western and Iranian strains. The two genomic loci that can be occupied by the *homA* and *homB* genes are depicted with the various combinations of the genes that can be found at these loci. A double copy of the *homB* gene has been linked to peptic ulcer disease isolates, investigates the different frequencies in the occurrence of genes and single nucleotide polymorphisms (SNPs) that occur in strains isolated from gastritis patients versus those obtained from gastric cancer patients. Of note, one of the observed differences is a significant association between the presence of the *babA* gene and gastric cancer in these European clinical isolates (Berthenet et al. 2018). Although multiple studies describe a link between *babA* and disease, some studies do not confirm a broad link across all Western countries (Oleastro et al. 2009b) and actually report variations in *babA* association that depend on the Western country from which the isolates originate (Olfat et al. 2005).

Unlike the association between *babA* and disease outcome frequently observed in strains from Western countries, studies of isolates from East Asian countries have largely been unable to find a correlation between the *babA2* genotype and disease status (Chomvarin et al. 2008; Mizushima et al. 2001; Oleastro et al. 2009b). However, one study did show an association of *babA2* strains with pre-neoplastic lesions in Chinese patients (Yu et al. 2002). It is worth noting that traditional PCR genotyping of *babA* in strains may not provide the best information on *babA* status or the correlation of *bab* genes and disease (Fujimoto et al. 2007). Indeed, a study in Taiwan highlights the potential role played by the particular genomic loci occupied by the *bab* genes; the *babA/B* genotype, which occurs through the recombination of the *babA* and *babB* genes, associates with pre-cancerous lesions and gastric cancer when present in both locus A and B (Sheu et al. 2012) (Fig. 2). Similarly, the study of Korean isolates reveals that the *bab* genotype at locus B associates with disease type (Kim et al. 2015). In particular, strains lacking a *bab* gene in locus B are more likely to originate from gastric ulcer or gastric cancer patients (Kim et al. 2015).

Additional data that indicate a link between *bab* gene status and disease outcome arises from the study of Asian strains, specifically from Iraq and Iran. In fact, the study of clinical isolates from Iraq shows an association between the *babA2* genotype and peptic ulcer disease, which is similar to the correlations observed in Western countries (Abdullah et al. 2012). Moreover, as chronic infection leads to the formation of chimeric *bab* genotypes (Matteo et al. 2011), a study using Iranian strains reveals an increased risk for duodenal ulcer development with a *babA/B* genotype as well as a low Leb binding (Le^b) phenotype (Saberi et al. 2016) (Fig. 2). This more complicated association between *bab* genotype and disease status is similar to studies in East Asian countries.

To further confound study of the *bab* genes, phase variation of *babA* and homologous recombination among the *bab* genes and their genomic loci allow strains to lose BabA protein production throughout the course of infection in animal models (Hansen et al. 2017; Liu et al. 2015; Ohno et al. 2011; Solnick et al. 2004; Styer et al. 2010). Indeed, a study involving isolates from Western and East Asian countries correlates increased disease severity with low production of BabA (Fujimoto et al. 2007). The varying characteristics of the *bab* genes are also hypothesized to contribute to *H. pylori*'s ability to adapt to the gastric niche through the modulation of its adherence profile (Liu et al. 2015; Solnick et al. 2004; Styer et al. 2010). Thus, the complexity of *bab* polymorphisms perhaps hinders the identification of a clinical link between *babA* genotype and disease outcome in more populations.

As described above, the multitude of epidemiological studies that attempt to define a link between *bab* and disease outcome produce varied results predominantly influenced by geography. Collectively, studies of strains from Western countries tend to identify a link between the *babA2* genotype and disease status, whereas research conducted on strains from East Asian countries reveal a more complicated relationship between *bab* and disease outcome.

2.1.2 Hom Proteins

H. pylori carries a small family of four OMPs called the Hom family (Alm et al. 2000). Most published studies focus specifically on *homA* and *homB*, both of which can be found at two distinct but overlapping genomic loci (Oleastro et al.

2008; Oleastro et al. 2009a) (Fig. 2). The two genes are 90% identical (Alm et al. 2000), and each gene has allelic variants (Oleastro et al. 2009a). Functional and epidemiological studies focus on the presence and copy number of *homA* and *homB*. *In vitro*, HomB has been shown to stimulate IL-8 secretion and to promote *H. pylori* adherence (Oleastro et al. 2008). Furthermore, the presence of *homB* is associated with peptic ulcer disease (Oleastro et al. 2008, 2010) in strains from Western countries but not strains from East Asian countries (Oleastro et al. 2009b) (Fig. 2). Comparatively, *homA* correlates with non-ulcer disease in children and adults (Oleastro et al. 2008, 2009b). In terms of more severe disease, *homB* is associated with gastric cancer (Talebi Bezmin Abadi et al. 2011), though the association is geographically-dependent (Hussein 2011; Kang et al. 2012). This geographic disparity is hypothesized to be a result of the variation in *homA* and *homB* gene status as well as the genomic loci at which the genes are carried in various strains (Kang et al. 2012). Indeed, *homA* and *homB* gene profiles appear to vary greatly based on the geographic origin of the strain (Servetas et al. 2018). Copy number also appears to be relevant as there is a correlation between strains carrying two copies of *homB* and the occurrence of peptic ulcer disease (Oleastro et al. 2009b) (Fig. 2). Thus, the complicated association between the *hom* genes and disease status depends not only on *homA* versus *homB* genotype but also on geographic location, genomic locus, and gene copy number.

2.1.3 HopQ

HopQ is yet another *H. pylori* adhesin. This protein binds to carcinoembryonic antigen-related cell adhesion molecule (CEACAM) receptors on gastric epithelial cells and aids in the translocation of CagA (Belogolova et al. 2013; Javaheri et al. 2016; Koniger et al. 2016; Moonens et al. 2018; Tegtmeyer et al. 2019). Evaluation of multiple *H. pylori* strains reveals that two predominant alleles exist for *hopQ*, named type I and type II (Cao and Cover 2002). Both of these allelic variants are capable of binding CEACAMs (Moonens et al. 2018), and strains can carry both variants within their genomes (Cao and Cover 2002; Chiarini et al. 2009). Strains from East Asian patients primarily carry *hopQ* type I, while strains from Western countries can carry either type (Cao et al. 2005). Several studies indicate an association between the presence of the *hopQ* type I allele and peptic ulcer disease (Cao and Cover 2002; Leylabadlo et al. 2016; Oleastro et al. 2010). Additionally, *hopQ* I, carried as either a single allele or in combination with a type II allele (*hopQ* I/II), correlates with gastritis and peptic ulcers (Chiarini et al. 2009). While the *hopQ* type I allele is also linked to gastric cancer (Talebi Bezmin Abadi and Mohabbati Mobarez 2014; Yakoob et al. 2016), an additional study correlates both *hopQ* I and II with gastric cancer in Iran (Leylabadlo et al. 2016). However, another study fails to find an association between *hopQ* allele and disease outcome (Ohno et al. 2009). Taken together, the data primarily indicate an association between the *hopQ* type I allele and gastric disease. Its role in translocation of CagA, however, is yet unclear and requires further study.

2.1.4 Additional OMPs

As previously discussed regarding the *babA* gene, other OMPs undergo phase variation through the presence of dinucleotide repeats within the 5′ region of their coding sequence (Salaun et al. 2004). This phase variation results in the classification of genes as being either "on" or "off" depending on the insertion of a premature stop codon. This polymorphic characteristic has been utilized in studies to identify associations between the "on" or "off" status of particular genes and the disease status of the patient; such studies have included the OMPs encoded by *sabA*, *sabB*, and *oipA*. As with other studies attempting to link specific polymorphisms to disease state, conflicting and controversial results have been reported, some of which are described below.

While its homolog, SabB, does not bind these glycoproteins, SabA is an adhesin that binds sialyl-Lewis^{a} and sialyl-Lewis^{x} antigens (de Jonge et al. 2004; Mahdavi et al. 2002). Like the *bab* genes, the *sab* genes can undergo recombination with one another. However, studies suggest that selective pressure to maintain *sabA* exists in the host (Talarico et al. 2012). Investigation of clinical isolates shows that the "off" status of *sabB* is associated with duodenal ulcers (de Jonge et al. 2004). That same study, along with other studies, fails to identify any link between *sabA* and disease status (Chiarini et al. 2009; de Jonge et al. 2004; Oleastro et al. 2010; Yadegar et al. 2014; Yanai et al. 2007). In contrast, *sabA* status is correlated with intestinal metaplasia and gastric cancer, but negatively associated with duodenal ulcers (Yamaoka et al. 2006). It is worth noting that the identification of the "on" status for *sabA*, which is often determined by sequencing of the gene, does not always reflect the actual production of SabA protein by individual isolates (Sheu et al. 2006). This finding likely affects the interpretation of all sequencing-based studies that have looked at the "on/off" status of *sabA* and *sabB*. Thus, the data linking *sabA* and *sabB* to disease outcome remain ambiguous.

Outer inflammatory protein A (OipA/HopH) is encoded by the phase-variable *oipA* gene (Yamaoka et al. 2000) and is involved in adhesion (Dossumbekova et al. 2006) and stimulation of IL-8 secretion (Yamaoka et al. 2002). While *oipA* "on" status is linked to peptic ulcers (Markovska et al. 2011; Oleastro et al. 2010), duodenal ulcers (Yamaoka et al. 2002, 2006) and gastric cancer (Yamaoka et al. 2006), other studies do not identify an association between *oipA* status and gastric pathologies (Torres et al. 2014) or disease outcome (Chiarini et al. 2009; de Jonge et al. 2004; Farzi et al. 2018; Zambon et al. 2003). Interestingly, the *oipA* "on" status associates with peptic ulcer disease in children but not in adults (Oleastro et al. 2008). Furthermore, a meta-analysis study identifies an association between *oipA* status and increased risk for peptic ulcer disease and gastric cancer (Liu et al. 2013). As described for other *H. pylori* factors, a selection of published studies identifies a link between *oipA* and disease status (Liu et al. 2013; Markovska et al. 2011; Oleastro et al. 2008, 2010; Yamaoka et al. 2002, 2006), while several others do not (Chiarini et al. 2009; de Jonge et al. 2004; Farzi et al. 2018; Zambon et al. 2003). However, the actual host cell receptor for OipA is still unclear and needs to be identified.

2.2 *iceA*

Although it is neither an OMP, an effector nor a toxin, the induced by contact with epithelium (*iceA*) locus is identified as a gene that correlates with gastric disease state (Peek Jr. et al. 1998). The *iceA* gene has two different alleles: *iceA1* and *iceA2* (Figueiredo et al. 2000; Peek Jr. et al. 1998). The *iceA1* allele exhibits sequence homology to an endonuclease found in *Neisseria* (Figueiredo et al. 2000). The initial identification of *iceA* reveals an association between *iceA1* and peptic ulcers as well as increased IL-8 in the mucosa (Peek Jr. et al. 1998; van Doorn et al. 1998). Subsequently, in a study of South Africa strains, *iceA1* associates with gastric cancer, while *iceA2* variants correlate with peptic ulcer disease (*iceA2D*) and gastritis (*iceA2C*) (Kidd et al. 2001). In Venezuelan strains, the *iceA2* genotype is a marker for atrophic gastritis, especially in combination with particular host genetic polymorphisms (Chiurillo et al. 2010); *iceA2* is also correlated with chronic gastritis (Caner et al. 2007). In contrast, a larger analysis of several published studies reveals an association between *iceA1* and peptic ulcer disease in China (Huang et al. 2016), while a study of Iranian isolates indicates a correlation between gastric cancer and *iceA1* (Dadashzadeh et al. 2017). The *iceA1* allele is also associated with duodenal ulcers (Caner et al. 2007). However, several studies fail to identify a definitive link between *iceA* genotype and clinical disease outcome (Abdullah et al. 2012; Chomvarin et al. 2008; Ito et al. 2000; Miehlke et al. 2001; Ribeiro et al. 2003; Yamaoka et al. 2002). Similar to the epidemiological data published for OMPs, conflicting results have accumulated regarding the role of *iceA* genotype in gastric disease development.

2.3 Secreted and Injected Proteins

While adhesins and OMPs allow the bacterium to interact with gastric cells, *H. pylori* also produces two proteins, VacA and CagA, which enter host cells and greatly perturb cellular processes. The vacuolating cytotoxin (VacA) is secreted by *H. pylori* and first generated interest for its ability to induce vacuole formation in epithelial cells (Cover and Blaser 1992; Leunk et al. 1988). The product of the cytotoxin-associated gene A (CagA) is injected into gastric cells and has been shown to affect many cellular processes through both phosphorylation-dependent and independent mechanisms (Bridge and Merrell 2013; Jones et al. 2010; Nishikawa and Hatakeyama 2017). The intimate interactions of these proteins with host cells contribute to the gastric disturbances caused by *H. pylori* infection. As such, these two virulence factors have been the subject of a multitude of epidemiological studies to identify allelic variants that affect disease progression.

2.3.1 *vacA*

VacA is a pore-forming toxin that is unique to *H. pylori*; it does not show significant homology to the sequence or structure of other known bacterial toxins (Cover and Blaser 1992; Leunk et al. 1988) (see review (Foegeding et al. 2016)). Following secretion by *H. pylori,* VacA can be internalized by host cells and then cause an accumulation of vesicles and the formation of anion channels. These toxic effects induce swelling, vacuolation, and death of the host cells (Foegeding et al. 2016; Thi Huyen Trang et al. 2016). At a molecular level, the VacA toxin is first produced as a 140 kDa protein that undergoes proteolytic cleavage to form the mature 88 kDa protein, comprised of 33 kDa (p33) and 55 kDa (p55) subunits from the N-terminal and C-terminal domains, respectively (Cover et al. 1994). Experimentally, only the p33 subunit in conjunction with a small portion of the N-terminus of the p55 subunit is required for vacuolation (de Bernard et al. 1998; Ye et al. 1999). More in-depth information on the function of VacA is described by Sgouras and colleagues in Chapter 3 and by Hatakeyama and co-workers in Chap. 8.

Current analysis shows that the *vacA* allelic variations are observed in five distinct regions of the *vacA* gene, resulting in strains that produce VacA proteins with varying toxicities (Thi Huyen Trang et al. 2016) (Fig. 3). The first region of variability is located within the signal peptide and the N-terminal region of the p33 subunit; this has been designated the s-region and the two primary variants are referred to as s1 and s2 (Atherton et al. 1995). The s1 variant is able to create vacuoles in host cells, whereas the s2 allele encodes a VacA protein that is unable to induce vacuolation (Atherton et al. 1995; McClain et al. 2001). In addition to functional differences in the produced VacA protein, there appears to be a transcriptional difference in the various *vacA* alleles; transcripts of the s2 forms are found at lower levels than the s1 forms (Forsyth et al. 1998). The second region of variability is also observed in the p33 subunit, but lies closer to the C-terminus of the subunit; this region is called the i-region (Rhead et al. 2007). The i-region has 3 predominant alleles, i1, i2, and i3. Of these, the i1 allele results in the most active form of the toxin (Rhead et al. 2007; Thi Huyen Trang et al. 2016). The third region of *vacA* diversity is found within the p55 subunit and has two primary variants, m1 and m2. These variants differ in channel-forming and cell-binding abilities (Atherton et al. 1995; Pagliaccia et al. 1998; Tombola et al. 2001; Wang et al. 2001). The fourth region of variability is termed the d-region; this variation lies at the junction of the p33 and p55 subunits, and the resulting d1 and d2 variants differ based on the presence or absence of nucleotides (Ogiwara et al. 2009). Finally, a recently identified fifth region of polymorphism has been termed the c-region and is defined by the inclusion (c2) or exclusion (c1) of a 15 bp nucleotide sequence (Bakhti et al. 2016). Clinical isolates have been found with virtually all possible combinations of these variable regions, though some combinations are observed more frequently than others.

An enormous number of epidemiological studies have been conducted to define the importance

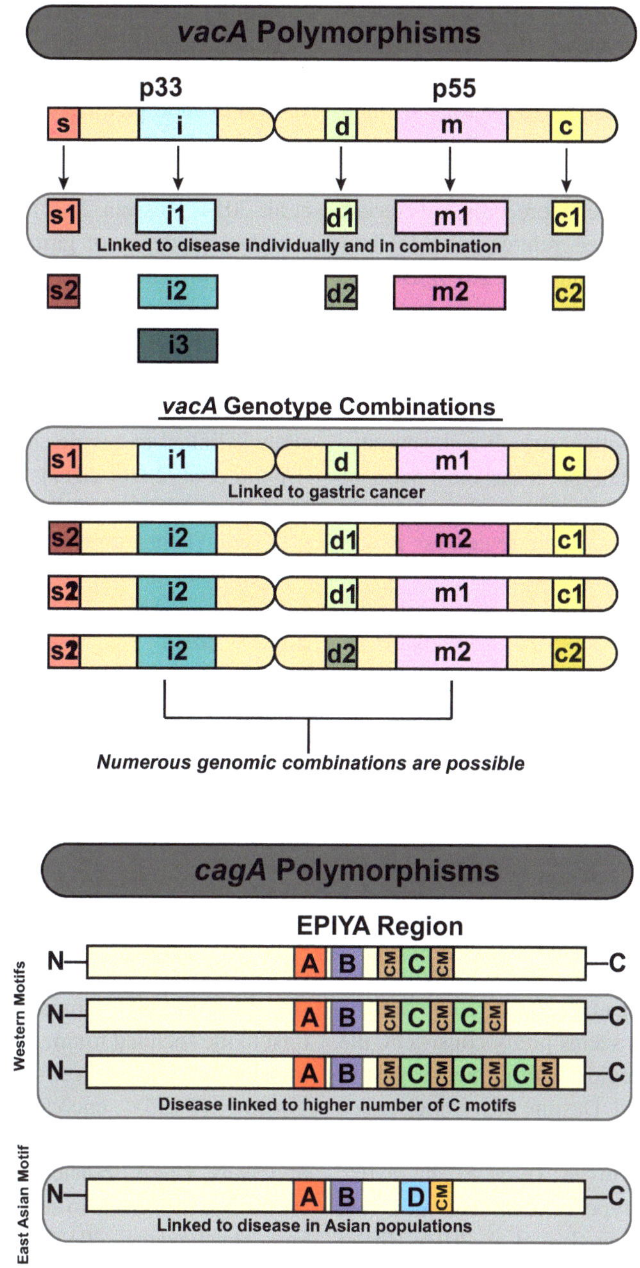

Fig. 3 CagA and VacA polymorphisms
Polymorphisms within the five regions of the *vacA* gene are represented by different colors. These regions include the s- and i-regions found in the p33 subunit of the protein encoded by *vacA* and the d-, m-, and c-regions located within the p55 subunit. Additionally, the allelic variations for each region are depicted in varying shades of the color that represents each region. The s1, i1, d1, m1, and c1 alleles have all been linked to disease. Representative combinations of the alleles are depicted; numerous unshown genomic combinations are possible. The combined s1/i1/m1 *vacA* allele has been linked to gastric cancer
Polymorphisms within the EPIYA region of the CagA protein are represented by different colors. The Western EPIYA motifs include an A, B, and C motif with varying number of C motifs possible. Additionally, the Western EPIYA sequence includes CM motifs found upstream and downstream of the EPIYA-C motif sequence. Disease has been linked to an increasing number of C motifs. The East Asian EPIYA motif includes an A, B, and D motif, which has been linked to disease in Asian populations. The East Asian motif contains a single CM motif downstream of the EPIYA-D motif sequence

of the various *vacA* alleles to disease development/progression. Of these, the vast majority focus primarily on the s- and m-regions of the *vacA* allele; more recent investigations include the i- and d-regions. While we will summarize some of the major findings in this area, the reader is encouraged to consider more thorough recent reviews of this topic-see references (McClain et al. 2017; Thi Huyen Trang et al. 2016). Briefly, initial studies of *vacA* reveal an association between the s1 allele and gastric inflammation (Atherton et al. 1997; Gunn et al. 1998; Zambon et al. 2003). Additional studies correlate the presence of the m1 allele with gastric disease severity. In fact, the s1/m1 genotype is the most virulent combination of these two alleles and toxicity is not induced with the m2 allele regardless if it is found in combination with the s1 or s2 polymorphisms (Atherton et al. 1995; Atherton et al. 1997; McClain et al. 2017; Yamaoka et al. 2008). Recent meta-analyses of published studies reveal links between the s1/m1 genotype of *vacA* and peptic ulcer disease (Matos et al. 2013) and gastric cancer (Matos et al. 2013; Pormohammad et al. 2018). Another meta-analysis of publications involving strains from Southeast Asia identifies an association between the *vacA* m1 allele and peptic ulcer disease (Sahara et al. 2012), while a study of Korean isolates confirms an association between the m-region, *cagA* allele, and disease state (Jang et al. 2010). While the role of the s and m alleles seems pretty consistent, the association of the i-region with particular gastric diseases is more varied. Despite this, some studies identify a link between the i-region genotype and disease status (Thi Huyen Trang et al. 2016). Indeed, the i1 genotype is identified as a risk factor for gastric cancer (Liu et al. 2016; Rhead et al. 2007) as well as is associated with peptic ulcers (Yordanov et al. 2012). A more in-depth study finds that specific amino acid positions within the i-region are linked to severe disease outcomes, particularly in isolates not carrying an EPIYA-ABD motif in the *cagA* allele (Jones et al. 2011); the *cagA* alleles are discussed in more detail in the subsequent section. In terms of the d-region, the d1 allele correlates with the s1, m1, and i1 alleles in Western strains, and strains containing all 4 of these variants display an increased risk of gastric cancer (Ogiwara et al. 2009). Additionally, the d1 polymorphism associates with peptic ulcers (Basiri et al. 2014) and gastric adenocarcinoma in Iranian strains (Abdi et al. 2017; Bakhti et al. 2016; Basiri et al. 2014; Hussein 2014). In areas of Iran with a higher incidence of gastric cancer, there is an increased prevalence of i1 and d1 polymorphisms (Latifi-Navid et al. 2013), and the i1 and d1 alleles increase the risk for the intestinal-type and diffuse-type of adenocarcinoma, respectively (Abdi et al. 2017). Finally, the presence of the c1 allele is linked to gastric cancer in Iranian strains (Bakhti et al. 2016, 2017). Taken *en masse*, gastric cancer risk is primarily associated with the s1, m1, and i1 genotypes (McClain et al. 2017; Foegeding et al. 2016).

2.3.2 *cagA*

The widely-studied CagA effector is a 120–145 kDa protein that is encoded on the *cag* pathogenicity island (*cag*PAI) (Censini et al. 1996); initial studies revealed a link between the presence of *cagA* and gastric diseases, including peptic ulcer disease and gastric cancer (Blaser et al. 1995; Covacci et al. 1993; Parsonnet et al. 1997; Weel et al. 1996). Indeed, CagA is classified as an oncoprotein due to its involvement in cancer development in a transgenic mouse model (Miura et al. 2009; Ohnishi et al. 2008). In contrast to the secreted toxin, VacA, CagA is directly translocated into host cells by a type IV secretion system (T4SS) encoded on the *cag*PAI (Odenbreit et al. 2000). For a more thorough review of CagA translocation and activity, we encourage the reader to see reference (Nishikawa and Hatakeyama 2017). Briefly, translocated CagA disrupts a variety of host cell signaling pathways through direct interaction with host proteins. More extensive discussion on the activity of CagA is provided by Sgouras and colleagues in Chapter 3 and by Hatakeyama and co-workers in Chapter 8.

The C-terminus of CagA contains two major polymorphic regions that are responsible for CagA's downstream effects: the EPIYA (Glu-Pro-Ile-Tyr-Ala) motif and the CagA multimerization (CM) motif (Ren et al. 2006; Stein et al. 2002) (Fig. 3). Polymorphisms within both the EPIYA and CM motifs have been extensively reported and further investigated for their *in vitro* effects on cells as well their contribution to disease progression (see reviews (Backert et al. 2010; Bridge and Merrell 2013; Jones et al. 2010; Nishikawa and Hatakeyama 2017)). First, the EPIYA motifs vary among clinical isolates based on the number of EPIYA repeats within the C-terminus as well as the amino acid sequence flanking the EPIYA sequence (Covacci et al. 1993; Hatakeyama 2006). The sequence of the polymorphic flanking regions is used to denote the motifs as EPIYA-A, -B, -C, and -D; CagA is further classified into Western (containing EPIYA-A, -B, and -C, where -C can be repeated multiple times) and East Asian (containing EPIYA-A, -B, and -D) strains due to the prevalence of the particular EPIYA motifs in these geographic regions (Hatakeyama 2006; Higashi et al. 2002; Miehlke et al. 1996). Similar to the EPIYA types, the polymorphisms in the CM sequences differ between the Western and East Asian strains. Western CagA carries the CM motif upstream of the EPIYA-C motif and also at the distal end of the last EPIYA-C motif; this repeated CM motif allows for the recombination that results in the repeated EPIYA-C motifs that are sometimes observed in Western strains (Furuta et al. 2011; Ren et al. 2006). In contrast, the East Asian EPIYA-D motif contains only one CM motif that is found downstream of the EPIYA-D motif (Ren et al. 2006). Accordingly, only one copy of the EPIYA-D motif is typically observed in East Asian strains.

Epidemiological studies have repeatedly sought to identify a link between the polymorphisms within the *cagA* gene and gastric disease development; however, as with the other factors discussed in this chapter, conflicting results are reported across the vast amount of literature on this topic. For example, when investigating Western strains, some studies identify an association between increasing numbers of EPIYA-C motifs and the development of intestinal metaplasia or dysplasia (Sicinschi et al. 2010) or gastric cancer (Basso et al. 2008; Batista et al. 2011; Ferreira et al. 2012; Yamaoka et al. 1999). However, other studies that also utilize Western strains find no correlation between increasing numbers of EPIYA-C motifs and disease state (Acosta et al. 2010; Figura et al. 2012; Rizzato et al. 2012; Shokrzadeh et al. 2010). Similarly, studies in Korea fail to establish an association between EPIYA-C motifs and disease (Choi et al. 2007); however, most strains in that region carry the EPIYA-D motif, making it difficult to evaluate the effect of the EPIYA-C polymorphism on disease (Argent et al. 2008a). In fact, a study in Japan describes only strains containing EPIYA-D in their gastric cancer patient population (Azuma et al. 2004), and a larger molecular epidemiological study uncovers a link between gastric carcinoma and strains carrying the EPIYA-D polymorphism (Jones et al. 2009). Furthermore, a meta-analysis study that evaluates publications that assess *H. pylori* isolates from Southeast Asia finds that possessing the EPIYA-D motif increases the risk of developing peptic ulcer disease and gastric cancer (Sahara et al. 2012). Another recent meta-analysis evaluates 23 published studies and reports correlations between the EPIYA-D polymorphism and gastric cancer in Asia, multiple EPIYA-C motifs and peptic ulcer disease and duodenal ulcers in Asia, and multiple EPIYA-C motifs and gastric cancer in Europe and the United States (Li et al. 2017). Finally, when investigating the association of the CM motif to gastric disease, a study of New York City hospital strains uncovers a link between strains harboring one or two Western CM motifs and peptic ulcer disease and gastric cancer in comparison to strains with an Eastern CM motif (Ogorodnik and Raffaniello 2013). Thus, both EPIYA type and CM motif appear to contribute to disease progression.

3 Combinations of Genotypes

Even though many studies focus on individual bacterial genes, it is worth noting that there are certainly higher order associations and interactions among the various factors that ultimately influence disease. Indeed, when thinking about combinations of various alleles, many epidemiological studies reveal that certain alleles of different virulence factors are more frequently found in combination and that these combinations are associated with particular gastric diseases. As a few examples, strains possessing *cagA*, *vacA* s1, and *babA2* associate with duodenal ulcer (Gerhard et al. 1999; Torres et al. 2009) and adenocarcinoma (Gerhard et al. 1999). Also, the *cagA*, *vacA* s1 m1, and *babA2* genotype creates a higher risk for intestinal metaplasia (Zambon et al. 2003). Chronic active gastritis links to a genotype of *cagA*, *vacA* s1 m1, *babA2*, and *hopQ* I or *hopQ* I/II (Chiarini et al. 2009). In addition, a study of strains isolated from German patients experiencing chronic gastritis reveals an association between the *oipA* (*hopH*) "on" status and the *vacA* s1, *vacA* m1, *babA2* and *cagA* genotypes (Dossumbekova et al. 2006). In a study of South African isolates, strains harboring the *iceA1* and *vacA* s1 genotypes are associated with gastric cancer (Kidd et al. 2001). Furthermore, an association between the presence of the *homB* gene and *cagA*, *babA2*, *vacA* s1, *hopQ* type I, and *oipA* "on" status has been identified (Oleastro et al. 2008). An additional study involving Western strains indicates that *homB* is associated with the presence of *cagA* and *vacA* s1, which together correlate with peptic ulcer disease (Oleastro et al. 2009b). *En masse*, epidemiological research has primarily linked gastric cancer risk to strains possessing *vacA* s1, *vacA* i1, *vacA* m1, *cagA*, the T4SS of *cag*, and OMPs (McClain et al. 2017). In contrast, a recent review by Floch *et al.* summarizes the epidemiological studies conducted using strains specifically from MALT lymphoma patients; those authors indicate that less virulent polymorphisms are associated with this disease (Floch et al. 2017). We do note that with all of these epidemiological studies, caution must be exercised when making concrete conclusions concerning the associations between specific genotypes and disease. One reason for this is the finding that independent strains isolated from the stomach of the same patient may have different genotypes (Lopez-Vidal et al. 2008). Due to the cost and labor involved, few studies have independently assessed large numbers of strains from individual patients.

In general, genotypes that are frequently found in association with each other are believed to work synergistically in order to enhance *H. pylori*'s persistence within the gastric mucosal layer. However, VacA and CagA exhibit an antagonistic relationship, whereby each protein dampens the effect of the other (Argent et al. 2008b; McClain et al. 2017). Thus, strains with the most virulent genotypes of both proteins harbor the best ability to control the effect of the opposing protein.

4 Conclusions

Many studies have attempted to identify particular polymorphisms that are linked to the most virulent *H. pylori* clinical isolates. Unfortunately, a single predictive factor has remained elusive; however, combinations of particular polymorphic virulence factors have repeatedly been linked to clinical disease outcome. Moreover, these combinations of virulence genotypes show variation based on the geographic origin of the strains being studied. Additionally, the complex contribution of polymorphisms within both the host and bacterium along with environmental factors influences *H. pylori*'s ultimate fate within the host (Fig. 1). Indeed, reports of *H. pylori* as a beneficial gastric organism underscore the importance of identifying factors that determine whether the bacterium survives as a commensal organism or becomes a pathogen within its human host. Despite the plethora of data gathered to date, more research is needed to adequately define the risk of gastric disease development attributable to various *H. pylori* genotypes. This information will ultimately help to inform patients and physicians as they weigh the benefits and risks of applying antimicrobial therapy to *H. pylori* infections.

References

Abdi E, Latifi-Navid S, Zahri S, Yazdanbod A, Safaralizadeh R (2017) *Helicobacter pylori* genotypes determine risk of non-cardia gastric cancer and intestinal- or diffuse-type GC in Ardabil: a very high-risk area in northwestern Iran. Microb Pathog 107:287–292. https://doi.org/10.1016/j.micpath.2017.04.007

Abdullah SM, Hussein NR, Salih AM, Merza MA, Goreal AA, Odeesh OY, Majed HS, Assafi MA, Hawrami K (2012) Infection with *Helicobacter pylori* strains carrying *babA2* and *cagA* is associated with an increased risk of peptic ulcer disease development in Iraq. Arab J Gastroenterol 13(4):166–169. https://doi.org/10.1016/j.ajg.2012.12.001

Acosta N, Quiroga A, Delgado P, Bravo MM, Jaramillo C (2010) *Helicobacter pylori* CagA protein polymorphisms and their lack of association with pathogenesis. World J Gastroenterol 16(31):3936–3943

Alm RA, Bina J, Andrews BM, Doig P, Hancock RE, Trust TJ (2000) Comparative genomics of *Helicobacter pylori*: analysis of the outer membrane protein families. Infect Immun 68(7):4155–4168

Ansari S, Yamaoka Y (2017) *Helicobacter pylori* BabA in adaptation for gastric colonization. World J Gastroenterol 23(23):4158–4169. https://doi.org/10.3748/wjg.v23.i23.4158

Argent RH, Hale JL, El-Omar EM, Atherton JC (2008a) Differences in *Helicobacter pylori* CagA tyrosine phosphorylation motif patterns between western and East Asian strains, and influences on interleukin-8-secretion. J Med Microbiol 57(Pt 9):1062–1067. https://doi.org/10.1099/jmm.0.2008/001818-0

Argent RH, Thomas RJ, Letley DP, Rittig MG, Hardie KR, Atherton JC (2008b) Functional association between the *Helicobacter pylori* virulence factors VacA and CagA. J Med Microbiol 57(Pt 2):145–150. https://doi.org/10.1099/jmm.0.47465-0

Arnold IC, Hitzler I, Muller A (2012) The immunomodulatory properties of *Helicobacter pylori* confer protection against allergic and chronic inflammatory disorders. Front Cell Infect Microbiol 2:10. https://doi.org/10.3389/fcimb.2012.00010

Asim M, Chikara SK, Ghosh A, Vudathala S, Romero-Gallo J, Krishna US, Wilson KT, Israel DA, Peek RM Jr, Chaturvedi R (2015) Draft genome sequence of gerbil-adapted carcinogenic *Helicobacter pylori* strain 7.13. Genome Announc 3(3). https://doi.org/10.1128/genomeA.00641-15

Atherton JC (2006) The pathogenesis of *Helicobacter pylori*-induced gastro-duodenal diseases. Annu Rev Pathol 1:63–96. https://doi.org/10.1146/annurev.pathol.1.110304.100125

Atherton JC, Cao P, Peek RM Jr, Tummuru MK, Blaser MJ, Cover TL (1995) Mosaicism in vacuolating cytotoxin alleles of *Helicobacter pylori*. Association of specific *vacA* types with cytotoxin production and peptic ulceration. J Biol Chem 270(30):17771–17777

Atherton JC, Peek RM Jr, Tham KT, Cover TL, Blaser MJ (1997) Clinical and pathological importance of heterogeneity in *vacA*, the vacuolating cytotoxin gene of *Helicobacter pylori*. Gastroenterology 112(1):92–99

Azuma T, Yamazaki S, Yamakawa A, Ohtani M, Muramatsu A, Suto H, Ito Y, Dojo M, Yamazaki Y, Kuriyama M, Keida Y, Higashi H, Hatakeyama M (2004) Association between diversity in the Src homology 2 domain--containing tyrosine phosphatase binding site of *Helicobacter pylori* CagA protein and gastric atrophy and cancer. J Infect Dis 189(5):820–827. https://doi.org/10.1086/381782

Backert S, Tegtmeyer N, Selbach M (2010) The versatility of *Helicobacter pylori* CagA effector protein functions: the master key hypothesis. Helicobacter 15(3):163–176. https://doi.org/10.1111/j.1523-5378.2010.00759.x

Bakhti SZ, Latifi-Navid S, Mohammadi S, Zahri S, Bakhti FS, Feizi F, Yazdanbod A, Siavoshi F (2016) Relevance of *Helicobacter pylori vacA* 3′-end region polymorphism to gastric cancer. Helicobacter 21(4):305–316. https://doi.org/10.1111/hel.12284

Bakhti SZ, Latifi-Navid S, Zahri S, Bakhti FS, Hajavi N, Yazdanbod A (2017) Are *Helicobacter pylori* highly cytotoxic genotypes and cardia gastric adenocarcinoma linked? Lessons from Iran. Cancer Biomark 21(1):235–246. https://doi.org/10.3233/cbm-170701

Basiri Z, Safaralizadeh R, Bonyadi MJ, Somi MH, Mahdavi M, Latifi-Navid S (2014) *Helicobacter pylori vacA* d1 genotype predicts risk of gastric adenocarcinoma and peptic ulcers in northwestern Iran. Asian Pac J Cancer Prev 15(4):1575–1579

Basso D, Zambon CF, Letley DP, Stranges A, Marchet A, Rhead JL, Schiavon S, Guariso G, Ceroti M, Nitti D, Rugge M, Plebani M, Atherton JC (2008) Clinical relevance of *Helicobacter pylori cagA* and *vacA* gene polymorphisms. Gastroenterology 135(1):91–99. https://doi.org/10.1053/j.gastro.2008.03.041

Batista SA, Rocha GA, Rocha AM, Saraiva IE, Cabral MM, Oliveira RC, Queiroz DM (2011) Higher number of *Helicobacter pylori* CagA EPIYA C phosphorylation sites increases the risk of gastric cancer, but not duodenal ulcer. BMC Microbiol 11:61. https://doi.org/10.1186/1471-2180-11-61

Belogolova E, Bauer B, Pompaiah M, Asakura H, Brinkman V, Ertl C, Bartfeld S, Nechitaylo TY, Haas R, Machuy N, Salama N, Churin Y, Meyer TF (2013) *Helicobacter pylori* outer membrane protein HopQ identified as a novel T4SS-associated virulence factor. Cell Microbiol 15(11):1896–1912. https://doi.org/10.1111/cmi.12158

Berthenet E, Yahara K, Thorell K, Pascoe B, Meric G, Mikhail JM, Engstrand L, Enroth H, Burette A, Megraud F, Varon C, Atherton JC, Smith S, Wilkinson TS, Hitchings MD, Falush D, Sheppard SK (2018) A GWAS on *Helicobacter pylori* strains points to genetic variants associated with gastric cancer risk. BMC Biol 16(1):84. https://doi.org/10.1186/s12915-018-0550-3

Blaser MJ, Perez-Perez GI, Kleanthous H, Cover TL, Peek RM, Chyou PH, Stemmermann GN, Nomura A (1995) Infection with *Helicobacter pylori* strains possessing *cagA* is associated with an increased risk of developing adenocarcinoma of the stomach. Cancer Res 55 (10):2111–2115

Bridge DR, Merrell DS (2013) Polymorphism in the *Helicobacter pylori* CagA and VacA toxins and disease. Gut Microbes 4(2):101–117. https://doi.org/10.4161/gmic.23797

Caner V, Yilmaz M, Yonetci N, Zencir S, Karagenc N, Kaleli I, Bagci H (2007) *H. pylori iceA* alleles are disease-specific virulence factors. World J Gastroenterol 13(18):2581–2585

Cao P, Cover TL (2002) Two different families of *hopQ* alleles in *Helicobacter pylori*. J Clin Microbiol 40 (12):4504–4511

Cao P, Lee KJ, Blaser MJ, Cover TL (2005) Analysis of *hopQ* alleles in East Asian and Western strains of *Helicobacter pylori*. FEMS Microbiol Lett 251 (1):37–43. https://doi.org/10.1016/j.femsle.2005.07.023

Censini S, Lange C, Xiang Z, Crabtree JE, Ghiara P, Borodovsky M, Rappuoli R, Covacci A (1996) *cag*, a pathogenicity island of *Helicobacter pylori*, encodes type I-specific and disease-associated virulence factors. Proc Natl Acad Sci U S A 93(25):14648–14653

Chen MY, He CY, Meng X, Yuan Y (2013) Association of *Helicobacter pylori babA2* with peptic ulcer disease and gastric cancer. World J Gastroenterol 19 (26):4242–4251. https://doi.org/10.3748/wjg.v19.i26.4242

Chen YL, Mo XQ, Huang GR, Huang YQ, Xiao J, Zhao LJ, Wei HY, Liang Q (2016) Gene polymorphisms of pathogenic *Helicobacter pylori* in patients with different types of gastrointestinal diseases. World J Gastroenterol 22(44):9718–9726. https://doi.org/10.3748/wjg.v22.i44.9718

Chiarini A, Cala C, Bonura C, Gullo A, Giuliana G, Peralta S, D'Arpa F, Giammanco A (2009) Prevalence of virulence-associated genotypes of *Helicobacter pylori* and correlation with severity of gastric pathology in patients from western Sicily, Italy. Eur J Clin Microbiol Infect Dis 28(5):437–446. https://doi.org/10.1007/s10096-008-0644-x

Chiurillo MA, Moran Y, Canas M, Valderrama E, Alvarez A, Armanie E (2010) Combination of *Helicobacter pylori-iceA2* and proinflammatory interleukin-1 polymorphisms is associated with the severity of histological changes in Venezuelan chronic gastritis patients. FEMS Immunol Med Microbiol 59 (2):170–176. https://doi.org/10.1111/j.1574-695X.2010.00675.x

Chmiela M, Karwowska Z, Gonciarz W, Allushi B, Staczek P (2017) Host pathogen interactions in *Helicobacter pylori* related gastric cancer. World J Gastroenterol 23(9):1521–1540. https://doi.org/10.3748/wjg.v23.i9.1521

Choi KD, Kim N, Lee DH, Kim JM, Kim JS, Jung HC, Song IS (2007) Analysis of the 3′ variable region of the *cagA* gene of *Helicobacter pylori* isolated in Koreans. Dig Dis Sci 52(4):960–966. https://doi.org/10.1007/s10620-005-9030-z

Chomvarin C, Namwat W, Chaicumpar K, Mairiang P, Sangchan A, Sripa B, Tor-Udom S, Vilaichone RK (2008) Prevalence of *Helicobacter pylori vacA, cagA, cagE, iceA* and *babA2* genotypes in Thai dyspeptic patients. Int J Infect Dis 12(1):30–36. https://doi.org/10.1016/j.ijid.2007.03.012

Colbeck JC, Hansen LM, Fong JM, Solnick JV (2006) Genotypic profile of the outer membrane proteins BabA and BabB in clinical isolates of *Helicobacter pylori*. Infect Immun 74(7):4375–4378. https://doi.org/10.1128/iai.00485-06

Covacci A, Censini S, Bugnoli M, Petracca R, Burroni D, Macchia G, Massone A, Papini E, Xiang Z, Figura N (1993) Molecular characterization of the 128-kDa immunodominant antigen of *Helicobacter pylori* associated with cytotoxicity and duodenal ulcer. Proc Natl Acad Sci U S A 90(12):5791–5795

Cover TL, Blaser MJ (1992) Purification and characterization of the vacuolating toxin from *Helicobacter pylori*. J Biol Chem 267(15):10570–10575

Cover TL, Blaser MJ (2009) *Helicobacter pylori* in health and disease. Gastroenterology 136(6):1863–1873. https://doi.org/10.1053/j.gastro.2009.01.073

Cover TL, Tummuru MK, Cao P, Thompson SA, Blaser MJ (1994) Divergence of genetic sequences for the vacuolating cytotoxin among *Helicobacter pylori* strains. J Biol Chem 269(14):10566–10573

Dadashzadeh K, Peppelenbosch MP, Adamu AI (2017) *Helicobacter pylori* pathogenicity factors related to gastric cancer. Can J Gastroenterol Hepatol 2017:7942489. https://doi.org/10.1155/2017/7942489

de Bernard M, Burroni D, Papini E, Rappuoli R, Telford J, Montecucco C (1998) Identification of the *Helicobacter pylori* VacA toxin domain active in the cell cytosol. Infect Immun 66(12):6014–6016

de Jonge R, Pot RG, Loffeld RJ, van Vliet AH, Kuipers EJ, Kusters JG (2004) The functional status of the *Helicobacter pylori sabB* adhesin gene as a putative marker for disease outcome. Helicobacter 9 (2):158–164. https://doi.org/10.1111/j.1083-4389.2004.00213.x

de Reuse H, Bereswill S (2007) Ten years after the first *Helicobacter pylori* genome: comparative and functional genomics provide new insights in the variability and adaptability of a persistent pathogen. FEMS Immunol Med Microbiol 50(2):165–176. https://doi.org/10.1111/j.1574-695X.2007.00244.x

Doenges JL (1938) Spirochetes in gastric glands of *Macacus rhesus* and humans without definite history of related disease. Proc Soc Exp Biol Med 38 (4):536–538

Doenges JL (1939) Spirochetes in the gastric glands of *Macacus rhesus* and of man without related diseases. Arch Pathol 27:469–477

Dossumbekova A, Prinz C, Mages J, Lang R, Kusters JG, Van Vliet AH, Reindl W, Backert S, Saur D, Schmid RM, Rad R (2006) *Helicobacter pylori* HopH (OipA) and bacterial pathogenicity: genetic and functional genomic analysis of *hopH* gene polymorphisms. J Infect Dis 194(10):1346–1355. https://doi.org/10.1086/508426

Draper JL, Hansen LM, Bernick DL, Abedrabbo S, Underwood JG, Kong N, Huang BC, Weis AM, Weimer BC, van Vliet AH, Pourmand N, Solnick JV, Karplus K, Ottemann KM (2017) Fallacy of the unique genome: sequence diversity within single *Helicobacter pylori* strains. MBio 8(1). https://doi.org/10.1128/mBio.02321-16

Dunn BE, Cohen H, Blaser MJ (1997) *Helicobacter pylori*. Clin Microbiol Rev 10(4):720–741

Eross B, Farkas N, Vincze A, Tinusz B, Szapary L, Garami A, Balasko M, Sarlos P, Czopf L, Alizadeh H, Rakonczay Z Jr, Habon T, Hegyi P (2018) *Helicobacter pylori* infection reduces the risk of Barrett's esophagus: a meta-analysis and systematic review. Helicobacter 23(4):e12504. https://doi.org/10.1111/hel.12504

Farzi N, Yadegar A, Aghdaei HA, Yamaoka Y, Zali MR (2018) Genetic diversity and functional analysis of *oipA* gene in association with other virulence factors among *Helicobacter pylori* isolates from Iranian patients with different gastric diseases. Infect Genet Evol 60:26–34. https://doi.org/10.1016/j.meegid.2018.02.017

Ferreira RM, Machado JC, Leite M, Carneiro F, Figueiredo C (2012) The number of *Helicobacter pylori* CagA EPIYA C tyrosine phosphorylation motifs influences the pattern of gastritis and the development of gastric carcinoma. Histopathology 60(6):992–998. https://doi.org/10.1111/j.1365-2559.2012.04190.x

Figueiredo C, Quint WG, Sanna R, Sablon E, Donahue JP, Xu Q, Miller GG, Peek RM Jr, Blaser MJ, van Doorn LJ (2000) Genetic organization and heterogeneity of the *iceA* locus of *Helicobacter pylori*. Gene 246(1–2):59–68

Figura N, Oderda G (1996) Reflections on the first description of the presence of *Helicobacter* species in the stomach of mammals. Helicobacter 1(1):4–5

Figura N, Moretti E, Roviello F, Papini F, Marrelli D (2012) *cagA* structural types of *Helicobacter pylori* strains isolated from patients with gastric carcinoma and chronic gastritis only. Intern Emerg Med 7(Suppl 2):S103–S105. https://doi.org/10.1007/s11739-012-0759-z

Floch P, Megraud F, Lehours P (2017) *Helicobacter pylori* strains and gastric MALT lymphoma. Toxins (Basel) 9(4). https://doi.org/10.3390/toxins9040132

Foegeding NJ, Caston RR, McClain MS, Ohi MD, Cover TL (2016) An overview of *Helicobacter pylori* VacA toxin biology. Toxins (Basel) 8(6). https://doi.org/10.3390/toxins8060173

Forsyth MH, Atherton JC, Blaser MJ, Cover TL (1998) Heterogeneity in levels of vacuolating cytotoxin gene (*vacA*) transcription among *Helicobacter pylori* strains. Infect Immun 66(7):3088–3094

Franco AT, Israel DA, Washington MK, Krishna U, Fox JG, Rogers AB, Neish AS, Collier-Hyams L, Perez-Perez GI, Hatakeyama M, Whitehead R, Gaus K, O'Brien DP, Romero-Gallo J, Peek RM Jr (2005) Activation of beta-catenin by carcinogenic *Helicobacter pylori*. Proc Natl Acad Sci U S A 102(30):10646–10651. https://doi.org/10.1073/pnas.0504927102

Freedberg A, Barron L (1940) The presence of spirochetes in human gastric mucosa. Am J Dig Dis 7(10):443–445. https://doi.org/10.1007/bf02997393

Fujimoto S, Olaniyi Ojo O, Arnqvist A, Wu JY, Odenbreit S, Haas R, Graham DY, Yamaoka Y (2007) *Helicobacter pylori* BabA expression, gastric mucosal injury, and clinical outcome. Clin Gastroenterol Hepatol 5(1):49–58. https://doi.org/10.1016/j.cgh.2006.09.015

Furuta Y, Yahara K, Hatakeyama M, Kobayashi I (2011) Evolution of *cagA* oncogene of *Helicobacter pylori* through recombination. PLoS One 6(8):e23499. https://doi.org/10.1371/journal.pone.0023499

Gerhard M, Lehn N, Neumayer N, Boren T, Rad R, Schepp W, Miehlke S, Classen M, Prinz C (1999) Clinical relevance of the *Helicobacter pylori* gene for blood-group antigen-binding adhesin. Proc Natl Acad Sci U S A 96(22):12778–12783

Gunn MC, Stephens JC, Stewart JA, Rathbone BJ, West KP (1998) The significance of *cagA* and *vacA* subtypes of *Helicobacter pylori* in the pathogenesis of inflammation and peptic ulceration. J Clin Pathol 51(10):761–764

Hansen LM, Gideonsson P, Canfield DR, Boren T, Solnick JV (2017) Dynamic expression of the BabA adhesin and its BabB paralog during *Helicobacter pylori* infection in rhesus macaques. Infect Immun 85(6). https://doi.org/10.1128/iai.00094-17

Hatakeyama M (2006) *Helicobacter pylori* CagA – a bacterial intruder conspiring gastric carcinogenesis. Int J Cancer 119(6):1217–1223. https://doi.org/10.1002/ijc.21831

Higashi H, Tsutsumi R, Fujita A, Yamazaki S, Asaka M, Azuma T, Hatakeyama M (2002) Biological activity of the *Helicobacter pylori* virulence factor CagA is determined by variation in the tyrosine phosphorylation sites. Proc Natl Acad Sci U S A 99(22):14428–14433. https://doi.org/10.1073/pnas.222375399

Homan M, Sterbenc A, Kocjan BJ, Luzar B, Zidar N, Orel R, Poljak M (2014) Prevalence of the *Helicobacter pylori babA2* gene and correlation with the degree of gastritis in infected Slovenian children. Antonie Van Leeuwenhoek 106(4):637–645. https://doi.org/10.1007/s10482-014-0234-0

Huang JY, Sweeney EG, Sigal M, Zhang HC, Remington SJ, Cantrell MA, Kuo CJ, Guillemin K, Amieva MR (2015) Chemodetection and destruction of host urea allows *Helicobacter pylori* to locate the epithelium.

Cell Host Microbe 18(2):147–156. https://doi.org/10.1016/j.chom.2015.07.002

Huang X, Deng Z, Zhang Q, Li W, Wang B, Li M (2016) Relationship between the *iceA* gene of *Helicobacter pylori* and clinical outcomes. Ther Clin Risk Manag 12:1085–1092. https://doi.org/10.2147/tcrm.s107991

Hussein NR (2011) A study of *Helicobacter pylori* outer-membrane proteins (*hom*) A and B in Iraq and Turkey. J Infect Public Health 4(3):135–139. https://doi.org/10.1016/j.jiph.2011.03.004

Hussein NR (2014) *Helicobacter pylori vacA* d1 genotype is associated with gastric cancer but not peptic ulcers in Kurdistan region, northern Iraq. Asian Pac J Cancer Prev 15(14):5965–5966

Ilver D, Arnqvist A, Ogren J, Frick IM, Kersulyte D, Incecik ET, Berg DE, Covacci A, Engstrand L, Boren T (1998) *Helicobacter pylori* adhesin binding fucosylated histo-blood group antigens revealed by retagging. Science 279(5349):373–377

International Agency for Research on Cancer (2012) Cancer fact sheets: Stomach Cancer. World Health Organization. http://gco.iarc.fr/today/fact-sheets-cancers?cancer=5&type=0&sex=0. Accessed 9-6-2018

Ito S (1967) Anatomic structure of the gastric mucosa. In: Code CF (ed) Alimentary Canal, vol II. American Physiological Society, Washington, DC, pp 705–741

Ito Y, Azuma T, Ito S, Suto H, Miyaji H, Yamazaki Y, Kato T, Kohli Y, Keida Y, Kuriyama M (2000) Sequence analysis and clinical significance of the *iceA* gene from *Helicobacter pylori* strains in Japan. J Clin Microbiol 38(2):483–488

Jang S, Jones KR, Olsen CH, Joo YM, Yoo YJ, Chung IS, Cha JH, Merrell DS (2010) Epidemiological link between gastric disease and polymorphisms in VacA and CagA. J Clin Microbiol 48(2):559–567. https://doi.org/10.1128/jcm.01501-09

Javaheri A, Kruse T, Moonens K, Mejias-Luque R, Debraekeleer A, Asche CI, Tegtmeyer N, Kalali B, Bach NC, Sieber SA, Hill DJ, Koniger V, Hauck CR, Moskalenko R, Haas R, Busch DH, Klaile E, Slevogt H, Schmidt A, Backert S, Remaut H, Singer BB, Gerhard M (2016) *Helicobacter pylori* adhesin HopQ engages in a virulence-enhancing interaction with human CEACAMs. Nat Microbiol 2:16189. https://doi.org/10.1038/nmicrobiol.2016.189

Jones KR, Joo YM, Jang S, Yoo YJ, Lee HS, Chung IS, Olsen CH, Whitmire JM, Merrell DS, Cha JH (2009) Polymorphism in the CagA EPIYA motif impacts development of gastric cancer. J Clin Microbiol 47(4):959–968. https://doi.org/10.1128/jcm.02330-08

Jones KR, Whitmire JM, Merrell DS (2010) A tale of two toxins: *Helicobacter pylori* CagA and VacA modulate host pathways that impact disease. Front Microbiol 1:115. https://doi.org/10.3389/fmicb.2010.00115

Jones KR, Jang S, Chang JY, Kim J, Chung IS, Olsen CH, Merrell DS, Cha JH (2011) Polymorphisms in the intermediate region of VacA impact *Helicobacter pylori*-induced disease development. J Clin Microbiol 49(1):101–110. https://doi.org/10.1128/jcm.01782-10

Kang J, Jones KR, Jang S, Olsen CH, Yoo YJ, Merrell DS, Cha JH (2012) The geographic origin of *Helicobacter pylori* influences the association of the *homB* gene with gastric cancer. J Clin Microbiol 50(3):1082–1085. https://doi.org/10.1128/jcm.06293-11

Kao CY, Sheu BS, Wu JJ (2016) *Helicobacter pylori* infection: an overview of bacterial virulence factors and pathogenesis. Biom J 39(1):14–23. https://doi.org/10.1016/j.bj.2015.06.002

Kasai K, Kobayashi R (1919) The stomach spirochete occurring in mammals. J Parasitol 6:1–10

Kidd M, Peek RM, Lastovica AJ, Israel DA, Kummer AF, Louw JA (2001) Analysis of *iceA* genotypes in south African *Helicobacter pylori* strains and relationship to clinically significant disease. Gut 49(5):629–635

Kim A, Servetas SL, Kang J, Kim J, Jang S, Cha HJ, Lee WJ, Kim J, Romero-Gallo J, Peek RM Jr, Merrell DS, Cha JH (2015) *Helicobacter pylori bab* paralog distribution and association with *cagA, vacA*, and *homA/B* genotypes in American and South Korean clinical isolates. PLoS One 10(8):e0137078. https://doi.org/10.1371/journal.pone.0137078

Koniger V, Holsten L, Harrison U, Busch B, Loell E, Zhao Q, Bonsor DA, Roth A, Kengmo-Tchoupa A, Smith SI, Mueller S, Sundberg EJ, Zimmermann W, Fischer W, Hauck CR, Haas R (2016) *Helicobacter pylori* exploits human CEACAMs via HopQ for adherence and translocation of CagA. Nat Microbiol 2:16188. https://doi.org/10.1038/nmicrobiol.2016.188

Kusters JG, van Vliet AH, Kuipers EJ (2006) Pathogenesis of *Helicobacter pylori* infection. Clin Microbiol Rev 19(3):449–490. https://doi.org/10.1128/cmr.00054-05

Kyburz A, Muller A (2017) *Helicobacter pylori* and extragastric diseases. Curr Top Microbiol Immunol 400:325–347. https://doi.org/10.1007/978-3-319-50520-6_14

Latifi-Navid S, Mohammadi S, Maleki P, Zahri S, Yazdanbod A, Siavoshi F, Massarrat S (2013) *Helicobacter pylori vacA* d1/−i1 genotypes and geographic differentiation between high and low incidence areas of gastric cancer in Iran. Arch Iran Med 16(6):330–337. https://doi.org/10.13166/aim.005

Leunk RD, Johnson PT, David BC, Kraft WG, Morgan DR (1988) Cytotoxic activity in broth-culture filtrates of *Campylobacter pylori*. J Med Microbiol 26(2):93–99. https://doi.org/10.1099/00222615-26-2-93

Leylabadlo HE, Yekani M, Ghotaslou R (2016) *Helicobacter pylori hopQ* alleles (type I and II) in gastric cancer. Biomed Rep 4(5):601–604. https://doi.org/10.3892/br.2016.634

Li Q, Liu J, Gong Y, Yuan Y (2017) Association of CagA EPIYA-D or EPIYA-C phosphorylation sites with peptic ulcer and gastric cancer risks: a meta-analysis. Medicine (Baltimore) 96(17):e6620. https://doi.org/10.1097/md.0000000000006620

Liu J, He C, Chen M, Wang Z, Xing C, Yuan Y (2013) Association of presence/absence and on/off patterns of *Helicobacter pylori oipA* gene with peptic ulcer disease and gastric cancer risks: a meta-analysis. BMC Infect Dis 13:555. https://doi.org/10.1186/1471-2334-13-555

Liu H, Fero JB, Mendez M, Carpenter BM, Servetas SL, Rahman A, Goldman MD, Boren T, Salama NR, Merrell DS, Dubois A (2015) Analysis of a single *Helicobacter pylori* strain over a 10-year period in a primate model. Int J Med Microbiol 305(3):392–403. https://doi.org/10.1016/j.ijmm.2015.03.002

Liu X, He B, Cho WC, Pan Y, Chen J, Ying H, Wang F, Lin K, Peng H, Wang S (2016) A systematic review on the association between the *Helicobacter pylori vacA* i genotype and gastric disease. FEBS Open Bio 6 (5):409–417. https://doi.org/10.1002/2211-5463.12046

Lockard VG, Boler RK (1970) Ultrastructure of a spiraled microorganism in the gastric mucosa of dogs. Am J Vet Res 31(8):1453–1462

Lopez-Vidal Y, Ponce-de-Leon S, Castillo-Rojas G, Barreto-Zuniga R, Torre-Delgadillo A (2008) High diversity of *vacA* and *cagA Helicobacter pylori* genotypes in patients with and without gastric cancer. PLoS One 3(12):e3849. https://doi.org/10.1371/journal.pone.0003849

Mahdavi J, Sonden B, Hurtig M, Olfat FO, Forsberg L, Roche N, Angstrom J, Larsson T, Teneberg S, Karlsson KA, Altraja S, Wadstrom T, Kersulyte D, Berg DE, Dubois A, Petersson C, Magnusson KE, Norberg T, Lindh F, Lundskog BB, Arnqvist A, Hammarstrom L, Boren T (2002) *Helicobacter pylori* SabA adhesin in persistent infection and chronic inflammation. Science 297(5581):573–578. https://doi.org/10.1126/science.1069076

Markovska R, Boyanova L, Yordanov D, Gergova G, Mitov I (2011) *Helicobacter pylori oipA* genetic diversity and its associations with both disease and *cagA, vacA* s, m, and i alleles among Bulgarian patients. Diagn Microbiol Infect Dis 71(4):335–340. https://doi.org/10.1016/j.diagmicrobio.2011.08.008

Marshall BJ (2001) One hundred years of discovery and rediscovery of *Helicobacter pylori* and its association with peptic ulcer disease. In: Mobley HLT, Mendz GL, Hazell SL (eds) *Helicobacter pylori*: physiology and genetics. ASM Press, Washington, DC, pp 19–24

Marshall BJ, Warren JR (1984) Unidentified curved bacilli in the stomach of patients with gastritis and peptic ulceration. Lancet 323(8390):1311–1315

Marshall BJ, Royce H, Annear DI, Goodwin CS, Pearman JW, Warren JR, Armstrong JA (1984) Original isolation of *Campylobacter pyloridis* from human gastric mucosa. Microbios Lett 25:83–88

Marshall BJ, Armstrong JA, McGechie DB, Glancy RJ (1985a) Attempt to fulfil Koch's postulates for pyloric *Campylobacter*. Med J Aust 142(8):436–439

Marshall BJ, McGechie DB, Rogers PA, Glancy RJ (1985b) Pyloric *Campylobacter* infection and gastroduodenal disease. Med J Aust 142(8):439–444

Matos JI, de Sousa HA, Marcos-Pinto R, Dinis-Ribeiro M (2013) *Helicobacter pylori* CagA and VacA genotypes and gastric phenotype: a meta-analysis. Eur J Gastroenterol Hepatol 25(12):1431–1441. https://doi.org/10.1097/MEG.0b013e328364b53e

Matsuo Y, Kido Y, Yamaoka Y (2017) *Helicobacter pylori* outer membrane protein-related pathogenesis. Toxins (Basel) 9(3). https://doi.org/10.3390/toxins9030101

Matteo MJ, Armitano RI, Romeo M, Wonaga A, Olmos M, Catalano M (2011) *Helicobacter pylori bab* genes during chronic colonization. Int J Mol Epidemiol Genet 2(3):286–291

McClain MS, Cao P, Iwamoto H, Vinion-Dubiel AD, Szabo G, Shao Z, Cover TL (2001) A 12-amino-acid segment, present in type s2 but not type s1 *Helicobacter pylori* VacA proteins, abolishes cytotoxin activity and alters membrane channel formation. J Bacteriol 183(22):6499–6508. https://doi.org/10.1128/jb.183.22.6499-6508.2001

McClain MS, Beckett AC, Cover TL (2017) *Helicobacter pylori* Vacuolating toxin and gastric Cancer. Toxins (Basel) 9(10). https://doi.org/10.3390/toxins9100316

Miehlke S, Kibler K, Kim JG, Figura N, Small SM, Graham DY, Go MF (1996) Allelic variation in the *cagA* gene of *Helicobacter pylori* obtained from Korea compared to the United States. Am J Gastroenterol 91 (7):1322–1325

Miehlke S, Yu J, Schuppler M, Frings C, Kirsch C, Negraszus N, Morgner A, Stolte M, Ehninger G, Bayerdorffer E (2001) *Helicobacter pylori vacA, iceA,* and *cagA* status and pattern of gastritis in patients with malignant and benign gastroduodenal disease. Am J Gastroenterol 96(4):1008–1013. https://doi.org/10.1111/j.1572-0241.2001.03685.x

Miura M, Ohnishi N, Tanaka S, Yanagiya K, Hatakeyama M (2009) Differential oncogenic potential of geographically distinct *Helicobacter pylori* CagA isoforms in mice. Int J Cancer 125(11):2497–2504. https://doi.org/10.1002/ijc.24740

Mizushima T, Sugiyama T, Komatsu Y, Ishizuka J, Kato M, Asaka M (2001) Clinical relevance of the *babA2* genotype of *Helicobacter pylori* in Japanese clinical isolates. J Clin Microbiol 39(7):2463–2465. https://doi.org/10.1128/jcm.39.7.2463-2465.2001

Moonens K, Hamway Y, Neddermann M, Reschke M, Tegtmeyer N, Kruse T, Kammerer R, Mejias-Luque R, Singer BB, Backert S, Gerhard M, Remaut H (2018) *Helicobacter pylori* adhesin HopQ disrupts trans dimerization in human CEACAMs. EMBO J 37(13). https://doi.org/10.15252/embj.201798665

Nishikawa H, Hatakeyama M (2017) Sequence polymorphism and intrinsic structural disorder as related to pathobiological performance of the *Helicobacter pylori* CagA oncoprotein. Toxins (Basel) 9(4). https://doi.org/10.3390/toxins9040136

Odenbreit S, Puls J, Sedlmaier B, Gerland E, Fischer W, Haas R (2000) Translocation of *Helicobacter pylori* CagA into gastric epithelial cells by type IV secretion. Science 287(5457):1497–1500

Ogiwara H, Sugimoto M, Ohno T, Vilaichone RK, Mahachai V, Graham DY, Yamaoka Y (2009) Role of deletion located between the intermediate and middle regions of the *Helicobacter pylori vacA* gene in cases of gastroduodenal diseases. J Clin Microbiol 47 (11):3493–3500. https://doi.org/10.1128/jcm.00887-09

Ogorodnik E, Raffaniello RD (2013) Analysis of the 3-′-variable region of the *cagA* gene from *Helicobacter pylori* strains infecting patients at New York City hospitals. Microb Pathog 56:29–34. https://doi.org/10.1016/j.micpath.2012.10.003

Ohnishi N, Yuasa H, Tanaka S, Sawa H, Miura M, Matsui A, Higashi H, Musashi M, Iwabuchi K, Suzuki M, Yamada G, Azuma T, Hatakeyama M (2008) Transgenic expression of *Helicobacter pylori* CagA induces gastrointestinal and hematopoietic neoplasms in mouse. Proc Natl Acad Sci U S A 105 (3):1003–1008. https://doi.org/10.1073/pnas.0711183105

Ohno T, Sugimoto M, Nagashima A, Ogiwara H, Vilaichone RK, Mahachai V, Graham DY, Yamaoka Y (2009) Relationship between *Helicobacter pylori hopQ* genotype and clinical outcome in Asian and Western populations. J Gastroenterol Hepatol 24 (3):462–468. https://doi.org/10.1111/j.1440-1746.2008.05762.x

Ohno T, Vallstrom A, Rugge M, Ota H, Graham DY, Arnqvist A, Yamaoka Y (2011) Effects of blood group antigen-binding adhesin expression during *Helicobacter pylori* infection of Mongolian gerbils. J Infect Dis 203(5):726–735. https://doi.org/10.1093/infdis/jiq090

Oleastro M, Menard A (2013) The role of *Helicobacter pylori* outer membrane proteins in adherence and pathogenesis. Biology (Basel) 2(3):1110–1134. https://doi.org/10.3390/biology2031110

Oleastro M, Cordeiro R, Ferrand J, Nunes B, Lehours P, Carvalho-Oliveira I, Mendes AI, Penque D, Monteiro L, Megraud F, Menard A (2008) Evaluation of the clinical significance of *homB*, a novel candidate marker of *Helicobacter pylori* strains associated with peptic ulcer disease. J Infect Dis 198(9):1379–1387. https://doi.org/10.1086/592166

Oleastro M, Cordeiro R, Menard A, Yamaoka Y, Queiroz D, Megraud F, Monteiro L (2009a) Allelic diversity and phylogeny of *homB*, a novel co-virulence marker of *Helicobacter pylori*. BMC Microbiol 9:248. https://doi.org/10.1186/1471-2180-9-248

Oleastro M, Cordeiro R, Yamaoka Y, Queiroz D, Megraud F, Monteiro L, Menard A (2009b) Disease association with two *Helicobacter pylori* duplicate outer membrane protein genes, *homB* and *homA*. Gut Pathog 1(1):12. https://doi.org/10.1186/1757-4749-1-12

Oleastro M, Santos A, Cordeiro R, Nunes B, Megraud F, Menard A (2010) Clinical relevance and diversity of two homologous genes encoding glycosyltransferases in *Helicobacter pylori*. J Clin Microbiol 48 (8):2885–2891. https://doi.org/10.1128/jcm.00401-10

Olfat FO, Zheng Q, Oleastro M, Voland P, Boren T, Karttunen R, Engstrand L, Rad R, Prinz C, Gerhard M (2005) Correlation of the *Helicobacter pylori* adherence factor BabA with duodenal ulcer disease in four European countries. FEMS Immunol Med Microbiol 44(2):151–156. https://doi.org/10.1016/j.femsim.2004.10.010

Pagliaccia C, de Bernard M, Lupetti P, Ji X, Burroni D, Cover TL, Papini E, Rappuoli R, Telford JL, Reyrat JM (1998) The m2 form of the *Helicobacter pylori* cytotoxin has cell type-specific vacuolating activity. Proc Natl Acad Sci U S A 95(17):10212–10217

Parsonnet J, Friedman GD, Orentreich N, Vogelman H (1997) Risk for gastric cancer in people with CagA positive or CagA negative *Helicobacter pylori* infection. Gut 40(3):297–301

Peek RM Jr, Thompson SA, Donahue JP, Tham KT, Atherton JC, Blaser MJ, Miller GG (1998) Adherence to gastric epithelial cells induces expression of a *Helicobacter pylori* gene, *iceA*, that is associated with clinical outcome. Proc Assoc Am Physicians 110 (6):531–544

Pormohammad A, Ghotaslou R, Leylabadlo HE, Nasiri MJ, Dabiri H, Hashemi A (2018) Risk of gastric cancer in association with *Helicobacter pylori* different virulence factors: a systematic review and meta-analysis. Microb Pathog 118:214–219. https://doi.org/10.1016/j.micpath.2018.03.004

Pride DT, Blaser MJ (2002) Concerted evolution between duplicated genetic elements in *Helicobacter pylori*. J Mol Biol 316(3):629–642. https://doi.org/10.1006/jmbi.2001.5311

Ren S, Higashi H, Lu H, Azuma T, Hatakeyama M (2006) Structural basis and functional consequence of *Helicobacter pylori* CagA multimerization in cells. J Biol Chem 281(43):32344–32352. https://doi.org/10.1074/jbc.M606172200

Rhead JL, Letley DP, Mohammadi M, Hussein N, Mohagheghi MA, Eshagh Hosseini M, Atherton JC (2007) A new *Helicobacter pylori* vacuolating cytotoxin determinant, the intermediate region, is associated with gastric cancer. Gastroenterology 133 (3):926–936. https://doi.org/10.1053/j.gastro.2007.06.056

Ribeiro ML, Godoy AP, Benvengo YH, Mendonca S, Pedrazzoli J Jr (2003) Clinical relevance of the *cagA*, *vacA* and *iceA* genotypes of *Helicobacter pylori* in Brazilian clinical isolates. FEMS Immunol Med Microbiol 36(3):181–185

Rizzato C, Torres J, Plummer M, Munoz N, Franceschi S, Camorlinga-Ponce M, Fuentes-Panana EM, Canzian F, Kato I (2012) Variations in *Helicobacter pylori* cytotoxin-associated genes and their influence in progression to gastric cancer: implications for prevention. PLoS One 7(1):e29605. https://doi.org/10.1371/journal.pone.0029605

Robinson K (2015) *Helicobacter pylori*-mediated protection against extra-gastric immune and inflammatory disorders: the evidence and controversies. Diseases 3 (2):34–55. https://doi.org/10.3390/diseases3020034

Saberi S, Schmidt A, Eybpoosh S, Esmaili M, Talebkhan Y, Mohajerani N, Oghalaie A, Eshagh Hosseini M, Mohagheghi MA, Bugaytova J, Boren T, Mohammadi M (2016) *Helicobacter pylori* strains from duodenal ulcer patients exhibit mixed *babA/B* genotypes with low levels of BabA adhesin and Lewis b binding. Dig Dis Sci 61(10):2868–2877. https://doi.org/10.1007/s10620-016-4217-z

Sahara S, Sugimoto M, Vilaichone RK, Mahachai V, Miyajima H, Furuta T, Yamaoka Y (2012) Role of *Helicobacter pylori cagA* EPIYA motif and *vacA* genotypes for the development of gastrointestinal diseases in southeast Asian countries: a meta-analysis. BMC Infect Dis 12:223. https://doi.org/10.1186/1471-2334-12-223

Salama N, Guillemin K, McDaniel TK, Sherlock G, Tompkins L, Falkow S (2000) A whole-genome microarray reveals genetic diversity among *Helicobacter pylori* strains. Proc Natl Acad Sci U S A 97(26):14668–14673. https://doi.org/10.1073/pnas.97.26.14668

Salaun L, Linz B, Suerbaum S, Saunders NJ (2004) The diversity within an expanded and redefined repertoire of phase-variable genes in *Helicobacter pylori*. Microbiology 150(Pt 4):817–830. https://doi.org/10.1099/mic.0.26993-0

Servetas SL, Kim A, Su H, Cha JH, Merrell DS (2018) Comparative analysis of the Hom family of outer membrane proteins in isolates from two geographically distinct regions: the United States and South Korea. Helicobacter 23(2):e12461. https://doi.org/10.1111/hel.12461

Sheu BS, Odenbreit S, Hung KH, Liu CP, Sheu SM, Yang HB, Wu JJ (2006) Interaction between host gastric Sialyl-Lewis X and *H. pylori* SabA enhances *H. pylori* density in patients lacking gastric Lewis B antigen. Am J Gastroenterol 101(1):36–44. https://doi.org/10.1111/j.1572-0241.2006.00358.x

Sheu SM, Sheu BS, Chiang WC, Kao CY, Wu HM, Yang HB, Wu JJ (2012) *H. pylori* clinical isolates have diverse *babAB* genotype distributions over different topographic sites of stomach with correlation to clinical disease outcomes. BMC Microbiol 12:89. https://doi.org/10.1186/1471-2180-12-89

Shokrzadeh L, Baghaei K, Yamaoka Y, Dabiri H, Jafari F, Sahebekhtiari N, Tahami A, Sugimoto M, Zojaji H, Zali MR (2010) Analysis of 3′-end variable region of the *cagA* gene in *Helicobacter pylori* isolated from Iranian population. J Gastroenterol Hepatol 25 (1):172–177. https://doi.org/10.1111/j.1440-1746.2009.05979.x

Sicinschi LA, Correa P, Peek RM, Camargo MC, Piazuelo MB, Romero-Gallo J, Hobbs SS, Krishna U, Delgado A, Mera R, Bravo LE, Schneider BG (2010) CagA C-terminal variations in *Helicobacter pylori* strains from Colombian patients with gastric precancerous lesions. Clin Microbiol Infect 16 (4):369–378. https://doi.org/10.1111/j.1469-0691.2009.02811.x

Solnick JV, Hansen LM, Salama NR, Boonjakuakul JK, Syvanen M (2004) Modification of *Helicobacter pylori* outer membrane protein expression during experimental infection of rhesus macaques. Proc Natl Acad Sci U S A 101(7):2106–2111. https://doi.org/10.1073/pnas.0308573100

Stein M, Bagnoli F, Halenbeck R, Rappuoli R, Fantl WJ, Covacci A (2002) c-Src/Lyn kinases activate *Helicobacter pylori* CagA through tyrosine phosphorylation of the EPIYA motifs. Mol Microbiol 43 (4):971–980

Styer CM, Hansen LM, Cooke CL, Gundersen AM, Choi SS, Berg DE, Benghezal M, Marshall BJ, Peek RM Jr, Boren T, Solnick JV (2010) Expression of the BabA adhesin during experimental infection with *Helicobacter pylori*. Infect Immun 78(4):1593–1600. https://doi.org/10.1128/iai.01297-09

Talarico S, Whitefield SE, Fero J, Haas R, Salama NR (2012) Regulation of *Helicobacter pylori* adherence by gene conversion. Mol Microbiol 84(6):1050–1061. https://doi.org/10.1111/j.1365-2958.2012.08073.x

Talebi Bezmin Abadi A, Mohabbati Mobarez A (2014) High prevalence of *Helicobacter pylori hopQ* II genotype isolated from Iranian patients with gastroduodenal disorders. J Pathog 2014:842469. https://doi.org/10.1155/2014/842469

Talebi Bezmin Abadi A, Rafiei A, Ajami A, Hosseini V, Taghvaei T, Jones KR, Merrell DS (2011) *Helicobacter pylori homB*, but not *cagA*, is associated with gastric cancer in Iran. J Clin Microbiol 49 (9):3191–3197. https://doi.org/10.1128/jcm.00947-11

Tegtmeyer N, Zabler D, Schmidt D, Hartig R, Brandt S, Backert S (2009) Importance of EGF receptor, HER2/Neu and Erk1/2 kinase signalling for host cell elongation and scattering induced by the *Helicobacter pylori* CagA protein: antagonistic effects of the vacuolating cytotoxin VacA. Cell Microbiol 11(3):488–505. https://doi.org/10.1111/j.1462-5822.2008.01269.x

Tegtmeyer N, Harrer A, Schmitt V, Singer BB, Backert S (2019) Expression of CEACAM1 or CEACAM5 in AZ-521 cells restores the type IV secretion deficiency for translocation of CagA by *Helicobacter pylori*. Cell Microbiol 21:e12965. https://doi.org/10.1111/cmi.12965

Thi Huyen Trang T, Thanh Binh T, Yamaoka Y (2016) Relationship between *vacA* types and development of gastroduodenal diseases. Toxins (Basel) 8(6). https://doi.org/10.3390/toxins8060182

Tomb JF, White O, Kerlavage AR, Clayton RA, Sutton GG, Fleischmann RD, Ketchum KA, Klenk HP, Gill S, Dougherty BA, Nelson K, Quackenbush J, Zhou L, Kirkness EF, Peterson S, Loftus B, Richardson D, Dodson R, Khalak HG, Glodek A, McKenney K, Fitzegerald LM, Lee N, Adams MD, Hickey EK, Berg DE, Gocayne JD, Utterback TR, Peterson JD, Kelley JM, Cotton MD, Weidman JM, Fujii C,

Bowman C, Watthey L, Wallin E, Hayes WS, Borodovsky M, Karp PD, Smith HO, Fraser CM, Venter JC (1997) The complete genome sequence of the gastric pathogen *Helicobacter pylori*. Nature 388 (6642):539–547. https://doi.org/10.1038/41483

Tombola F, Pagliaccia C, Campello S, Telford JL, Montecucco C, Papini E, Zoratti M (2001) How the loop and middle regions influence the properties of *Helicobacter pylori* VacA channels. Biophys J 81 (6):3204–3215. https://doi.org/10.1016/s0006-3495(01)75956-5

Torres LE, Melian K, Moreno A, Alonso J, Sabatier CA, Hernandez M, Bermudez L, Rodriguez BL (2009) Prevalence of *vacA, cagA* and *babA2* genes in Cuban *Helicobacter pylori* isolates. World J Gastroenterol 15 (2):204–210

Torres K, Valderrama E, Sayegh M, Ramirez JL, Chiurillo MA (2014) Study of the *oipA* genetic diversity and EPIYA motif patterns in *cagA*-positive *Helicobacter pylori* strains from Venezuelan patients with chronic gastritis. Microb Pathog 76:26–32. https://doi.org/10.1016/j.micpath.2014.09.006

van Doorn LJ, Figueiredo C, Sanna R, Plaisier A, Schneeberger P, de Boer W, Quint W (1998) Clinical relevance of the *cagA, vacA,* and *iceA* status of *Helicobacter pylori*. Gastroenterology 115(1):58–66

Wang WC, Wang HJ, Kuo CH (2001) Two distinctive cell binding patterns by vacuolating toxin fused with glutathione S-transferase: one high-affinity m1-specific binding and the other lower-affinity binding for variant m forms. Biochemistry 40(39):11887–11896

Warren JR, Marshall B (1983) Unidentified curved bacilli on gastric epithelium in active chronic gastritis. Lancet 1(8336):1273–1275

Weel JF, van der Hulst RW, Gerrits Y, Roorda P, Feller M, Dankert J, Tytgat GN, van der Ende A (1996) The interrelationship between cytotoxin-associated gene A, vacuolating cytotoxin, and *Helicobacter pylori*-related diseases. J Infect Dis 173(5):1171–1175

White JR, Winter JA, Robinson K (2015) Differential inflammatory response to *Helicobacter pylori* infection: etiology and clinical outcomes. J Inflamm Res 8:137–147. https://doi.org/10.2147/jir.s64888

Wroblewski LE, Peek RM Jr, Wilson KT (2010) *Helicobacter pylori* and gastric cancer: factors that modulate disease risk. Clin Microbiol Rev 23 (4):713–739. https://doi.org/10.1128/cmr.00011-10

Yadegar A, Mobarez AM, Alebouyeh M, Mirzaei T, Kwok T, Zali MR (2014) Clinical relevance of *cagL* gene and virulence genotypes with disease outcomes in a *Helicobacter pylori* infected population from Iran. World J Microbiol Biotechnol 30(9):2481–2490. https://doi.org/10.1007/s11274-014-1673-5

Yakoob J, Abbas Z, Khan R, Salim SA, Awan S, Abrar A, Jafri W (2016) *Helicobacter pylori* outer membrane protein Q allele distribution is associated with distinct pathologies in Pakistan. Infect Genet Evol 37:57–62. https://doi.org/10.1016/j.meegid.2015.10.027

Yamaoka Y, El-Zimaity HM, Gutierrez O, Figura N, Kim JG, Kodama T, Kashima K, Graham DY (1999) Relationship between the *cagA* 3′ repeat region of *Helicobacter pylori*, gastric histology, and susceptibility to low pH. Gastroenterology 117(2):342–349

Yamaoka Y, Kwon DH, Graham DY (2000) A M (r) 34,000 proinflammatory outer membrane protein (*oipA*) of *Helicobacter pylori*. Proc Natl Acad Sci U S A 97(13):7533–7538. https://doi.org/10.1073/pnas.130079797

Yamaoka Y, Kikuchi S, el-Zimaity HM, Gutierrez O, Osato MS, Graham DY (2002) Importance of *Helicobacter pylori oipA* in clinical presentation, gastric inflammation, and mucosal interleukin 8 production. Gastroenterology 123(2):414–424

Yamaoka Y, Ojo O, Fujimoto S, Odenbreit S, Haas R, Gutierrez O, El-Zimaity HM, Reddy R, Arnqvist A, Graham DY (2006) *Helicobacter pylori* outer membrane proteins and gastroduodenal disease. Gut 55 (6):775–781. https://doi.org/10.1136/gut.2005.083014

Yamaoka Y, Kato M, Asaka M (2008) Geographic differences in gastric cancer incidence can be explained by differences between *Helicobacter pylori* strains. Intern Med 47(12):1077–1083

Yanai A, Maeda S, Hikiba Y, Shibata W, Ohmae T, Hirata Y, Ogura K, Yoshida H, Omata M (2007) Clinical relevance of *Helicobacter pylori sabA* genotype in Japanese clinical isolates. J Gastroenterol Hepatol 22 (12):2228–2232. https://doi.org/10.1111/j.1440-1746.2007.04831.x

Ye D, Willhite DC, Blanke SR (1999) Identification of the minimal intracellular vacuolating domain of the *Helicobacter pylori* vacuolating toxin. J Biol Chem 274(14):9277–9282

Yordanov D, Boyanova L, Markovska R, Gergova G, Mitov I (2012) Significance of *Helicobacter pylori vacA* intermediate region genotyping-a Bulgarian study. Diagn Microbiol Infect Dis 74(3):253–257. https://doi.org/10.1016/j.diagmicrobio.2012.07.008

Yu J, Leung WK, Go MY, Chan MC, To KF, Ng EK, Chan FK, Ling TK, Chung SC, Sung JJ (2002) Relationship between *Helicobacter pylori babA2* status with gastric epithelial cell turnover and premalignant gastric lesions. Gut 51(4):480–484

Zambon CF, Navaglia F, Basso D, Rugge M, Plebani M (2003) *Helicobacter pylori babA2, cagA*, and s1 *vacA* genes work synergistically in causing intestinal metaplasia. J Clin Pathol 56(4):287–291

Adv Exp Med Biol - Advances in Microbiology, Infectious Diseases and Public Health (2019) 1149: 195–210
https://doi.org/10.1007/5584_2019_366

Published online: 24 April 2019

Helicobacter pylori Infection, the Gastric Microbiome and Gastric Cancer

Joana Pereira-Marques, Rui M. Ferreira, Ines Pinto-Ribeiro, and Ceu Figueiredo

Abstract

After a long period during which the stomach was considered as an organ where microorganisms could not thrive, *Helicobacter pylori* was isolated *in vitro* from gastric biopsies, revolutionising the fields of Microbiology and Gastroenterology. Since then, and with the introduction of high-throughput sequencing technologies that allowed deep characterization of microbial communities, a growing body of knowledge has shown that the stomach contains a diverse microbial community, which is different from that of the oral cavity and of the intestine. Gastric cancer is a heterogeneous disease that is the end result of a cascade of events arising in a small fraction of patients colonized with *H. pylori.* In addition to *H. pylori* infection and to multiple host and environmental factors that influence disease development, alterations to the composition and function of the normal gastric microbiome, also known as dysbiosis, may also contribute to malignancy. Chronic inflammation of the mucosa in response to *H. pylori* may alter the gastric environment, paving the way to the growth of a dysbiotic gastric bacterial community. This dysbiotic microbiome may promote the development of gastric cancer by sustaining inflammation and/or inducing genotoxicity. This chapter summarizes what is known about the gastric microbiome in the context of *H. pylori*-associated gastric cancer, introducing the emerging dimension of the microbiome into the pathogenesis of this highly incident and deadly disease.

Keywords

Helicobacter pylori · Gastric microbiome · Gastric microbiota · Gastric cancer · Microbial dysbiosis

Author contributed equally with all other contributors. Joana Pereira-Marques and Rui M. Ferreira

J. Pereira-Marques
i3S – Instituto de Investigação e Inovação em Saúde, Universidade do Porto, Porto, Portugal

Ipatimup – Institute of Molecular Pathology and Immunology of the University of Porto, Porto, Portugal

ICBAS – Institute of Biomedical Sciences Abel Salazar, University of Porto, Porto, Portugal

R. M. Ferreira
i3S – Instituto de Investigação e Inovação em Saúde, Universidade do Porto, Porto, Portugal

Ipatimup – Institute of Molecular Pathology and Immunology of the University of Porto, Porto, Portugal

I. Pinto-Ribeiro and C. Figueiredo (✉)
i3S – Instituto de Investigação e Inovação em Saúde, Universidade do Porto, Porto, Portugal

Ipatimup – Institute of Molecular Pathology and Immunology of the University of Porto, Porto, Portugal

Faculty of Medicine, University of Porto, Porto, Portugal
e-mail: cfigueiredo@ipatimup.pt

1 Introduction

The human body is inhabited in its different niches by a vast collection of microbes, generally known as the microbiota. These microorganisms, their genetic information, as well as the information of the niche in which they interact, are usually referred to as the microbiome (Cho and Blaser 2012). Currently, the term microbiome is also used to refer to the microorganisms themselves, i.e. the microbiota (Knight et al. 2017). The number of microbial cells was commonly thought to outnumber the quantity of human cells by a ten-fold ratio, but recent assessments propose a 1:1 ratio as a better estimate (Sender et al. 2016).

Bacteria constitute so far the best explored component of the microbiome. Progress in this research area had been hampered by the fact that only a very small fraction of the microbial species can be cultured *in vitro*. The advent of high-throughput sequencing technologies, together with the emergence of large international and interdisciplinary projects, have strongly contributed to expand our understanding of the microbiome structure and functions (Turnbaugh et al. 2007; Qin et al. 2010; Arnold et al. 2016).

It is currently accepted that the microbiome plays a major role in the maintenance of the normal physiology and health of the host, being involved in a wide variety of metabolic functions and participating in the normal maturation of the immune system (Gilbert et al. 2018). The composition of the normal microbiome varies between individuals and is influenced by local conditions inherent to the anatomic site, host genetics, diet, and antibiotic consumption (Lloyd-Price et al. 2017; Gilbert et al. 2018). Disruption of the balance that exists between the microbiome and the host, called dysbiosis, may promote numerous diseases, including cancer (Gilbert et al. 2018). For example, members of the gut microbiome such as *Fusobacterium nucleatum*, *Escherichia coli,* and *Bacteroides fragilis*, have been found enriched in colorectal cancer (Goodwin et al. 2011; Ahn et al. 2013; Bonnet et al. 2014). Nevertheless, and although the exact mechanisms linking microbial dysbiosis and cancer are still largely unknown, it can be anticipated that bacterial metabolites and toxins, as well as inflammation triggered by the microbiome contribute to the promotion of cancer. Here, we discuss in detail the current knowledge on the human gastric microbiome in the context of health and disease, and provide insights into the potential impact of microbial dysbiosis in the development of *H. pylori*-associated gastric cancer, by revisiting Correa's hypothesis of gastric carcinogenesis (Correa 1992).

2 Gastric Cancer

Gastric cancer is the fifth most incident cancer worldwide, with almost 1 million new cases per year (Ferlay et al. 2015). Gastric cancer is also the third cause of cancer-related death worldwide, with about 750,000 deaths estimated to having occurred in 2012. The incidence and mortality of gastric cancer show wide geographic variation, with East Asian countries registering the highest rates (Ferlay et al. 2015). Gastric cancer is a heterogeneous disease in what concerns morphology, genetics, and context. Histologically, gastric cancer heterogeneity is reflected by the diversity in classifications. The most commonly used histological classification systems are the one of the World Health Organization, comprising five main types – tubular, papillary, mucinous, poorly cohesive, and rare histological variants – and Lauren's, comprising two main types – diffuse and intestinal (Fenoglio-Preiser et al. 2010; Lauren 1965). Lauren's classification remains the most widely used and each cancer type has distinct epidemiologic and pathophysiological characteristics (Carneiro 1997; Spoto et al. 2018). Gastric cancer of the diffuse type occurs more frequently in females and at earlier ages, and is characterised by isolated or small groups of neoplastic cells that do not form glandular structures. In contrast, gastric cancer of the intestinal type is more prevalent at advanced ages, mainly in males, and is characterized by the presence of glandular structures and a higher to moderate degree of cell differentiation (Lauren 1965;

Carneiro 1997; Van Cutsem et al. 2016). The sequence of histological changes that culminate in intestinal type gastric cancer is better characterized than the one leading to diffuse type cancer, despite both types being associated with chronic gastritis as a consequence of *H. pylori* infection. Intestinal type gastric cancer is the result of a long, multifactorial and multistep process, which starts with *H. pylori* chronic gastritis, followed by atrophic gastritis, intestinal metaplasia, dysplasia, and cancer (Correa et al. 1975; Correa 1992).

Gastric cancer heterogeneity is also manifested at the molecular level (Ottini et al. 2006). Comprehensive analyses of gastric cancer tissues from large cohorts of patients recently emphasized the complexity of this disease and led to the proposal of different molecular classifications (Lei et al. 2013; Cancer Genome Atlas Research 2014; Cristescu et al. 2015). For example, the Cancer Genome Atlas research network classification proposed four main gastric cancer types (Cancer Genome Atlas Research 2014): chromosomally unstable tumours, which have marked aneuploidy, frequent mutations in *TP53*, amplification of receptor tyrosine kinases and *RAS*; microsatellite unstable tumours, which are characterised by *MLH1* promoter hypermethylation and a high mutational rate of genes including *TP53*, *KRAS*, *ARID1A*, *PIK3CA,* and *PTEN*; genomically stable tumours that have mutations in *CDH1*, encoding E-cadherin, *ARID1A* and *RHOA*; and Epstein-Barr virus-positive tumours, that show recurrent *PI3KCA* and *ARID1A*, but very rare *TP53* mutations, *CDKN2A* promoter hypermethylation, and amplification of *JAK2*, and of PD-L1- and PD-L2-encoding genes.

It is important to acknowledge that the great majority of gastric cancers occur in a sporadic setting, with about 10% of the cases having familial clustering, and 1–3% occurring in a hereditary setting (Oliveira et al. 2015; Van Cutsem et al. 2016). Hereditary diffuse gastric cancer (HDGC) is the most common and best-studied hereditary gastric cancer syndrome, where about 40% of the affected families have germline mutations in the *CDH1* gene, encoding the cell-cell adhesion protein E-cadherin (Oliveira et al. 2015). The other two syndromes, gastric adenocarcinoma and proximal polyposis of the stomach (GAPPS) and familial intestinal gastric cancer (FIGC), are very rare, and while in the former germline point mutations in the *APC* promoter have been identified (Worthley et al. 2012; Li et al. 2016), in the latter no aetiological genetic alterations are known (Oliveira et al. 2015). The exact extent to which *H. pylori* infection and/or the microbiome of the stomach may contribute to the different molecular profiles and contexts of gastric cancer, however, remains largely unexplored.

3 *H. pylori* Infection and Gastric Cancer

H. pylori is considered as the major risk factor for the development of gastric cancer, being categorized as a class I carcinogen by the International Agency for Research on Cancer (IARC 1994). It has been estimated that at least 90% of all non-cardia gastric cancers worldwide are attributable to *H. pylori* (Plummer et al. 2015). The estimated worldwide prevalence of *H. pylori* is 44.3%, with considerable variation according to the geographic region (Zamani et al. 2018). There is a major geographic overlap between *H. pylori* prevalence and gastric cancer incidence, and in general countries with highest cancer incidence have high infection rates (Ferlay et al. 2015; Zamani et al. 2018). Since the initial collection of epidemiological and functional data that provided grounds for the classification of *H. pylori* as a class I carcinogen, numerous studies have been published demonstrating the causal relationship between chronic *H. pylori* infection and gastric cancer (IARC 2011). The magnitude of the risk of gastric cancer associated with *H. pylori* infection has now been estimated in different populations, and varies with the type of assay used to detect the infection, being about three-fold if serology is used (Helicobacter and Cancer Collaborative Group 2001) and reaching over 20-fold when more sensitive assays are used (Gonzalez et al. 2012). As an additional piece of evidence that links *H. pylori* infection and gastric cancer, the eradication of the infection has an impact in reducing the incidence of this malignancy (Ford et al. 2015).

Although the association between *H. pylori* and gastric cancer is extensively recognized, the majority of the infected patients do not develop this malignancy, which arguments in favour of the multifactorial nature of this disease. Host genetic susceptibility, namely polymorphisms in genes that are involved in the inflammatory response to *H. pylori* infection have been associated with the risk of gastric cancer. Among the best studied are those that encode interleukin (IL)-1β, IL-1 receptor antagonist, tumour necrosis factor (TNF)-α pro-inflammatory cytokines and the anti-inflammatory IL-10. Genetic variation in the promoters or in non-coding regions of these genes are associated with increased risk for the development of gastric cancer (El-Omar et al. 2001; Machado et al. 2003; Persson et al. 2011). Remarkably, in genetically susceptible hosts, infection with more virulent *H. pylori* strains markedly enhances gastric cancer risk (Figueiredo et al. 2002).

Cigarette smoking, alcohol intake, and salt consumption are recognized environmental factors that influence the risk of gastric cancer. Indeed, ever and current smokers have higher risk to develop gastric cancer compared with never smokers, and among current smokers the risk increases with number of cigarettes per day (Praud et al. 2018). Heavy and very heavy alcohol drinkers have higher risks for developing gastric cancer in comparison with abstainers, and these associations are independent of the *H. pylori* infection status (Rota et al. 2017). Dietary salt intake is also associated with gastric cancer risk, the risk being gradually increased for higher consumption levels (D'Elia et al. 2012). Accordingly, in an animal model of infection, a diet with high salt content accelerated the development of gastric cancer, in particular in animals infected with *cagA*-positive *H. pylori* strains (Gaddy et al. 2013). On the other hand, the consumption of fruit and white vegetables, which are rich sources of vitamin C, are inversely associated with gastric cancer risk (Fang et al. 2015).

Adding to the influence of host and environmental factors in gastric cancer, the genetic diversity of *H. pylori*, and in particular variation in virulence genes associated with the pathogenicity of strains, also impact gastric cancer risk (Ferreira et al. 2014). CagA is the best-documented *H. pylori* virulence factor influencing gastric cancer. CagA is encoded by a pathogenicity island that is present in about 60–70% of *H. pylori* strains worldwide. The same pathogenicity island also encodes a type IV secretion system, which functions as a molecular syringe and allows CagA to be delivered into the host cells (Backert et al. 2015). Once in the host cell cytoplasm, CagA can be phosphorylated by host kinases within EPIYA motifs. Both phosphorylated and non-phosphorylated CagA are capable of activating signalling pathways that influence host responses, including inflammation, proliferation, and cell polarity (Backert et al. 2010). CagA phosphorylation, however, appears to be important in gastric cancer development, as transgenic mice expressing wild-type CagA, but not phosphorylation-resistant CagA, develop gastric tumours (Ohnishi et al. 2008). Patients who are infected with *H. pylori cagA*-positive strains, and with strains with CagA harbouring higher number of phosphorylation motifs, are associated with increased risk for gastric premalignant lesions and for gastric cancer (Ferreira et al. 2014). Additionally, CagA influences host disease progression, and infection with *H. pylori cagA*-positive strains increases the risk of progression of preneoplastic lesions (Plummer et al. 2007; Gonzalez et al. 2011). Variation in other *H. pylori* virulence factors, such as the VacA toxin, has also been associated with gastric precancerous lesions and cancer (Gonzalez et al. 2011; Ferreira et al. 2014). This and other virulence factors of *H. pylori* and their relationship with disease are discussed in Chap. 3 of this volume. Additionally, the molecular mechanisms that underlie *H. pylori*-mediated malignant transformation are discussed in Chap. 8.

4 The Gastric Microbiota, Is There More Than *H. pylori*?

For many years, the human stomach was assumed to be sterile, given its high acidic pH, gastric peristalsis, and the presence of digestive enzymes, among other protective and antimicrobial factors (Martinsen et al. 2005). With the discovery and isolation of *H. pylori* (Warren and Marshall 1983) this dogma was broken, and more recently the idea that the stomach harbours a complex bacterial community became accepted. Initial analyses of the bacteria present in the stomach relied on microbiological cultures. These have identified *Firmicutes* as the most common phylum, followed by *Proteobacteria*, *Bacteroidetes*, and *Actinobacteria*, and genera that were most commonly isolated included *Streptococcus*, *Lactobacillus*, *Bacteroides*, *Staphylococcus*, *Veillonella*, *Corynebacterium*, *Clostridium*, and *Neisseria* (Stockbruegger 1985; Thorens et al. 1996; Adamsson et al. 1999; Mowat et al. 2000; Zilberstein et al. 2007). This type of approach, however, yielded an incomplete and biased landscape of the gastric microbiota, since most of the bacteria are difficult to culture or are uncultivable. The development of culture-independent methods revealed that the human gastric ecosystem has a more diverse and complex microbiota than initially anticipated (Monstein et al. 2000; Bik et al. 2006; Andersson et al. 2008; Li et al. 2009; Delgado et al. 2013; Schulz et al. 2018).

The bacterial community of the normal stomach has not been extensively characterised, probably due to difficulties in recruiting normal individuals for upper endoscopy. A 16S rRNA gene cloning and sequencing-based approach was undertaken to analyse the gastric microbial communities of five individuals with normal gastric mucosa and five patients with non-*H. pylori* and non-NSAID (non-steroidal anti-inflammatory drug) (NHNN) gastritis, all Chinese from Hong-Kong (Li et al. 2009). *Firmicutes* and *Proteobacteria* were the most represented phyla, and while in the normal stomach the *Proteobacteria* was the most abundant, in the NHNN gastritis the most abundant phylum was the *Firmicutes*. The five most common genera were *Streptococcus*, *Prevotella*, *Neisseria*, *Haemophilus*, and *Porphyromonas*; together, *Streptococcus* and *Prevotella* represented over 40% of all sequences.

Following studies exposed the diversity and the inter-individual variability of the gastric microbiota derived from the analysis of populations from distinct origins, but also from different sample types, and using various technical approaches. Overall, the most common gastric bacteria can be assigned to five major phyla – *Proteobacteria*, *Firmicutes*, *Bacteroidetes*, *Actinobacteria* and *Fusobacteria*, and the two most prominent genera of the non-*H. pylori* infected stomach are *Streptococcus* and *Prevotella* (Bik et al. 2006; Andersson et al. 2008; Li et al. 2009; Delgado et al. 2013). A more recent study that included 20 Caucasians from the UK with a normal stomach, without evidence of *H. pylori* infection, concurred that the bacterial family *Prevotellaceae* was the most abundant (23%), followed by *Streptococcaceae* (10%). In fact, the microbiota of these stomachs had the highest levels of microbial diversity and bacterial richness in comparison with other groups of patients infected with *H. pylori* (Parsons et al. 2017).

According to the great majority of reports, when *H. pylori* is present, this bacterium is the most abundant microbial component, representing between 40% to over 95% of the gastric microbiota (Bik et al. 2006; Andersson et al. 2008; Li et al. 2017; Klymiuk et al. 2017; Schulz et al. 2018; Ferreira et al. 2018; Parsons et al. 2017). In addition to finding *H. pylori* as the most abundant bacterium in the stomach of patients who test positive for *H. pylori*, it has been shown that the microbiota of *H. pylori*-positive subjects has lower diversity than that of *H. pylori*-negative subjects (Bik et al. 2006; Andersson et al. 2008; Schulz et al. 2018). Our analysis of the gastric microbiota of 81 chronic gastritis cases from Portugal that were 99% *H. pylori*-positive, revealed that as *H. pylori* abundance increases, there is a significant decrease in diversity (data not shown). Accordingly, a study that evaluated the gastric microbiota before and after *H. pylori* eradication treatment showed that the eradication of *H. pylori* resulted in an increase in bacterial diversity (Li et al. 2017).

The influence of *H. pylori* on the composition and dynamics of the gastric microbiota is still not fully understood. Difficulties may in part relate to the differences in methods to diagnose *H. pylori* infection and various studies using sequencing-based methods have demonstrated that *H. pylori* could be detected at low levels in samples of subjects that were diagnosed as *H. pylori*-negative by conventional methods (histopathology, rapid urease test, serology, and PCR) (Bik et al. 2006; Maldonado-Contreras et al. 2011; Delgado et al. 2013; Thorell et al. 2017).

The majority of reports show no major alterations on the pattern of distribution of phyla between *H. pylori*-positive and *H. pylori*-negative patients (Bik et al. 2006; Maldonado-Contreras et al. 2011; Schulz et al. 2018). Using the PhyloChip microarray, Maldonado-Contreras *et al.* reported a similar representation of the four dominant phyla between *H. pylori*-infected and -uninfected rural Amerindians (Maldonado-Contreras et al. 2011). In regression analyses, authors were able to identify an association between *H. pylori* positivity and decreased relative abundance of *Actinobacteria*, *Bacteroidetes*, and *Firmicutes*. These results are sustained by our data on Portuguese patients with chronic gastritis, in which we found an inverse correlation between the relative abundance of *H. pylori* and non-*Helicobacter* α- and β-*Proteobacteria*, *Actinobacteria*, *Firmicutes*, and *Bacteroidetes* (Ferreira et al. 2018).

Experimental infections of the rhesus macaque model were used to assess the impact of *H. pylori* challenge upon the pre-existing gastric microbiota (Martin et al. 2013). Data showed that although *Helicobacter* became dominant in challenged animals, the removal of the *Helicobacter* reads from the libraries did not significantly alter the relative abundance of taxa between challenged and unchallenged animals. Nevertheless, the impact of *H. pylori* on relatively rare taxa was not determined. In contrast, in a mouse model of infection, challenge of animals with *H. pylori* significantly and consistently affected the abundance of several species, suggesting that *H. pylori* influences the gastric microbiota composition at lower taxonomic levels (Kienesberger et al. 2016).

It has been a matter of debate whether bacteria found in the stomach represent transient swallowed bacteria or active members of a resident microbiota colonizing the gastric mucosa. Comparisons of the microbial communities along different sites of the gastrointestinal (GI) tract have shown that the gastric microbiota is different from that at other sites. Although some proximity with the microbiota of the oral cavity and throat exists, the stomach microbial communities cluster together (Andersson et al. 2008; Stearns et al. 2011; Delgado et al. 2013). Recent data aiming to evaluate the metabolically active microbial communities in different regions of the GI tract found that the transient luminal microbiota present in gastric juice is closely related with that of saliva and of duodenal aspirates and significantly different from that of gastric biopsies, supporting the idea that the stomach has a local mucosa-associated microbiota (Schulz et al. 2018).

5 The Gastric Microbiota in Gastric Carcinogenesis

While *H. pylori* is recognized as being fundamental in gastric carcinogenesis, the role of non-*H. pylori* microbiota has not yet been established. The majority of the publications so far included low number of patients and/or had limitations in sensitivity and depth of coverage, which in general did not allow producing statistically based conclusions. One of the first DNA-based descriptions of the gastric bacterial community in patients with gastric cancer, used terminal restriction fragment length polymorphism (T-RFLP) in combination with 16S rRNA gene cloning and sequencing to characterize 10 patients with gastric cancer and five *H. pylori*-negative dyspeptics with normal gastric mucosa (Dicksved et al. 2009). A complex bacterial community dominated by different species of *Streptococcus*, *Lactobacillus*, *Veilonella* and *Prevotella*, and with low abundance of *H. pylori* was reported in the stomach of cancer patients.

A study of 15 patients from Mexico with non-atrophic gastritis, intestinal metaplasia, or gastric cancer, using the PhyloChip, showed a gastric microbiota profile separation between non-atrophic gastritis and gastric cancer based on the presence/absence of taxa. This analysis could neither separate non-atrophic gastritis and intestinal metaplasia, nor metaplasia and cancer (Aviles-Jimenez et al. 2014). Taxa with differences in abundance between non-atrophic gastritis and gastric cancer were identified, with significant decreases in the abundance of *Porphyromonas*, *Neisseria* and bacteria from the *TM7* phylum, and increases in the abundance of *Lactobacillus* and *Lachnospiraceae* observed in gastric cancer. Diversity, as measured by bacterial richness, was statistically significantly decreased from non-atrophic gastritis to gastric cancer. In contrast, a survey of the metabolic active bacteria of the stomach of 12 gastric cancer and 20 functional dyspepsia patients of Chinese ethnicity from Singapore and Malaysia, detected an increase in species richness and in phylogenetic diversity in cancer (Castano-Rodriguez et al. 2017). An earlier study of 10 chronic gastritis, 10 intestinal metaplasia and 11 gastric cancer patients from Korea, also suggested an increase in bacterial diversity from gastritis to cancer, but without supporting statistical analysis (Eun et al. 2014). Still, the majority of publications so far report a decrease in bacteria diversity and richness from non-atrophic gastritis to gastric cancer (Aviles-Jimenez et al. 2014; Li et al. 2017; Coker et al. 2018; Ferreira et al. 2018).

The two most complete gastric microbiota studies in the gastric cancer field using 16S rRNA gene sequencing were published in the beginning of 2018 (Coker et al. 2018; Ferreira et al. 2018). Coker and colleagues studied the gastric mucosal microbiota in different histological stages of gastric carcinogenesis in 81 patients from Xi'an in China (Coker et al. 2018). The analysis of 21 superficial gastritis, 23 atrophic gastritis, 17 intestinal metaplasia, and 20 gastric cancer patients, demonstrated that the gastric microbiota of patients with intestinal metaplasia and with gastric cancer had significantly reduced microbial richness in comparison with that of superficial gastritis patients. Although no significant differences were found in microbiota profiles between superficial gastritis, atrophic gastritis and intestinal metaplasia, the microbiota of these stages were significantly different from that of the gastric cancer. The screen for differentially abundant taxa revealed 21 taxa enriched and 10 taxa depleted in gastric cancer in comparison with superficial gastritis, with increasing strengths of interactions among them along the progression of disease. Among the cancer-enriched bacteria were members of the human oral microbiome *Peptostreptococcus*, *Streptococcus*, *Parvimonas*, *Slackia*, and *Dialister*, which were the most significant in network interaction analysis. These bacteria were able to distinguish gastric cancer from superficial gastritis in receiver-operating characteristic (ROC) analysis. The authors validated their results in a Chinese Inner Mongolian cohort of patients (Coker et al. 2018).

Our own studies analysing 135 Portuguese patients, showed significant differences in the structure as well as in the composition of the gastric microbial communities between chronic gastritis and gastric cancer patients (Ferreira et al. 2018). Overall, patients with cancer had significantly decreased gastric microbial diversity, as assessed by the Shannon index, in comparison with patients with chronic gastritis. The gastric microbiota profiles of the two patient groups could be separated based on both the presence/absence and the relative abundance of taxa. In our series, *Proteobacteria*, *Firmicutes*, *Bacteroidetes*, *Actinobacteria* and *Fusobacteria* were identified as the five most abundant phyla in the stomach, in agreement with earlier descriptions (Bik et al. 2006; Aviles-Jimenez et al. 2014; Jo et al. 2016). Phyla ranked in the same relative abundance in the two patient groups, with significantly increased abundance of non-*Helicobacter Proteobacteria*, *Actinobacteria* and *Firmicutes* and lower abundance of *Bacteroidetes* and *Fusobacteria* in the gastric cancer microbiota. While being the major genus in chronic gastritis with a mean relative abundance of 42% (varying from 0.01–95%), *Helicobacter* had a significant reduction in abundance in gastric cancer. In fact, and despite 87% of the gastric cancer patients were *H. pylori*-positive, the mean relative abundance of reads was just 6% (Ferreira et al. 2018). Actually, the gastric microbiota profiles of the two clinical settings could be distinguished based on *Helicobacter* abundance.

Overall, we have identified 29 microbial taxa, including 10 differentially abundant genera that best explain the differences between patient groups. Differential abundances in the great majority of these genera were further validated using quantitative polymerase chain reaction in the discovery cohort, and additionally confirmed in validation cohorts comprising patients from Portugal, China and Mexico (Ferreira et al. 2018). *Helicobacter*, *Neisseria*, *Prevotella*, and *Streptococcus* were enriched in the microbiota of chronic gastritis patients. *Streptococcus*, *Prevotella* and *Neisseria* are among the most abundant commensals of the oral cavity (Bik et al. 2010) and among the most frequently detected bacteria in the non-neoplastic stomach, having been cultured or identified in gastric juice and/or biopsies from *H. pylori*-positive and -negative gastritis and in the normal stomach (Thorens et al. 1996; Bik et al. 2006; Li et al. 2009; Delgado et al. 2013; Parsons et al. 2017; Schulz et al. 2018). Interestingly, in a comparison of the gastric microbiota of Colombian inhabitants from two regions with divergent gastric cancer risks, *Streptococcus* and *Neisseria* were identified only in individuals from the low risk, but not in those from the high risk gastric cancer region (Yang et al. 2016).

Genera that were enriched in the gastric cancer microbiota, and significantly more prevalent in patients with gastric cancer than in patients with chronic gastritis, were *Achromobacter*, *Citrobacter*, *Lactobacillus*, *Clostridium*, *Rhodococcus*, and *Phyllobacterium* (Ferreira et al. 2018). These bacteria comprise several intestinal residents that may become opportunistic pathogens (Kelly and LaMont 2008; Rajilic-Stojanovic and de Vos 2014), and indeed *Lactobacillus*, *Clostridium*, and *Citrobacter* have been detected in the gastric juice or gastric biopsies from patients taking acid suppressive drugs, and patients with intestinal metaplasia and gastric cancer (Sjostedt et al. 1985; Mowat et al. 2000; Dicksved et al. 2009; Aviles-Jimenez et al. 2014). In a recent study of nine gastritis and 11 gastric cancer patients from Taiwan, species of *Clostridium* and *Lactobacillus* were also found enriched in the gastric cancer microbiota (Hsieh et al. 2018).

Microbial dysbiosis was inversely correlated with the microbial diversity and was significantly higher in cancer than in gastritis, a finding that was validated in additional patient cohorts (Ferreira et al. 2018). Actually, microbial dysbiosis could distinguish gastric cancer and chronic gastritis patients in ROC analysis. Interestingly, microbial dysbiosis could discriminate gastric cancer better than individual genera, suggesting that alterations to the microbial community as a whole rather than particular bacteria contribute to the development of gastric cancer.

The role of the microbiota in the promotion of neoplasia is supported by data obtained in the insulin-gastrin (INS-GAS) transgenic mouse model. In comparison with germ-free INS-GAS mice, those harbouring a complex microbiota had higher levels of gastric inflammation, epithelial damage, oxyntic gland atrophy, hyperplasia, metaplasia, and dysplasia. When infected with *H. pylori*, INS-GAS mice that harboured a complex microbiota had more severe gastric lesions and an earlier development of gastrointestinal intraepithelial neoplasia (GIN) in comparison to *H. pylori*-infected germ-free INS-GAS mice (Lofgren et al. 2011). Furthermore, progression towards GIN occurred to a similar extent in *H. pylori*-infected INS-GAS mice with a complex microbiota and in *H. pylori*-infected INS-GAS mice colonized with a restricted microbiota consisting of only three species of commensal murine bacteria (*Clostridium* sp., *Lactobacillus murinus*, and *Bacteroides* sp.) (Lertpiriyapong et al. 2014). These results suggest that colonization of the stomach with commensal bacteria from other locations of the GI tract may promote *H. pylori*-associated gastric cancer. Altogether, these studies highlight that there is a shift in the composition of the stomach microbiome from gastritis to gastric cancer, with a likely reduction of bacterial diversity, and with increased microbial dysbiosis in the cancerous stomach.

6 Revisiting Correa's Hypothesis of Gastric Carcinogenesis

In the multistep model of gastric carcinogenesis proposed by Pelayo Correa, persistent infection of the gastric mucosa with *H. pylori* initiates and perpetuates an inflammatory process that can progress to atrophic gastritis, intestinal metaplasia, dysplasia, and gastric cancer (Correa 1992). In this model, *H. pylori* infection plays an important role in the initial phases of the cascade. Indeed, *H. pylori* scarcely colonizes the severe atrophic stomach and may progressively disappear in gastric tissues at later steps of carcinogenesis (Correa 1992; Kuipers 1998). Analyses of the gastric microbiome have also described decreased relative abundance of *H. pylori* in gastric cancer (Dicksved et al. 2009; Eun et al. 2014; Ferreira et al. 2018; Hsieh et al. 2018), although this was not consistently observed or not reported (Yu et al. 2017; Coker et al. 2018).

The hypothesis of Correa contemplated that the loss of acid-secreting parietal cells in *H. pylori*-induced atrophic gastritis leads to higher gastric pH, and to proliferation in the stomach of bacteria that are capable of reducing nitrate to nitrite, to form N-nitroso compounds that are mutagenic (Correa et al. 1975; Correa 1992). Actually, significant intragastric bacterial overgrowth has been demonstrated in patients on long-term acid suppression by the use of proton pump inhibitors (PPIs) or histamine-2 receptor antagonists (Stockbruegger 1985; Sanduleanu et al. 2001). A recent investigation of 24 dyspeptic Italian patients, showed that although PPI treatment did not have a major influence in the gastric microbiota composition, an increase in the relative abundance of *Firmicutes*, namely *Streptococcus* was reported (Paroni Sterbini et al. 2016). In accordance with these findings, in a study analysing the metabolically active gastric microbial communities of 19 patients from the UK receiving PPI therapy and 20 individuals with normal stomach, relatively few alterations in the gastric microbiota were detected, but *Streptococcus* was significantly enriched in PPI-treated patients (Parsons et al. 2017). An enrichment in *Streptococcaceae* in the gut microbiota of PPI users has also been reported in two large studies (Imhann et al. 2016; Jackson et al. 2016). The enrichment of upper GI tract commensals observed in the stomach and in the gut, may be related with the disruption of the highly acidic barrier of the stomach induced by the acid suppressive therapy.

Likewise, the increase of the pH of the stomach due to decreased acid production as a result of parietal cell loss in *H. pylori*-associated atrophy, may generate a niche that becomes suitable to the establishment of a different microbiome (Plottel and Blaser 2011). One may speculate that this altered gastric microbiome, where *H. pylori* is less abundant or absent, and where commensal bacteria from other locations of the GI tract thrive, would act as continuous stimuli by maintaining the inflammatory process and/or inducing genotoxicity, thus promoting gastric carcinogenesis (Fig.1). This would in part explain the lack of success of *H. pylori* eradication in preventing progression of preneoplastic lesions and gastric cancer in patients with atrophy or intestinal metaplasia at baseline (Wong et al. 2004; Mera et al. 2018).

The microbiome of the cancerous stomach is functionally different from that of the stomach without cancer (Coker et al. 2018; Ferreira et al. 2018). Although only a very limited number of studies have addressed this aspect, predictive functional analyses have revealed that gastric cancer patients have an enrichment of several microbial pathways, including those related with membrane transport, carbohydrate digestion and absorption, carbohydrate metabolism, xenobiotics biodegradation and metabolism, and

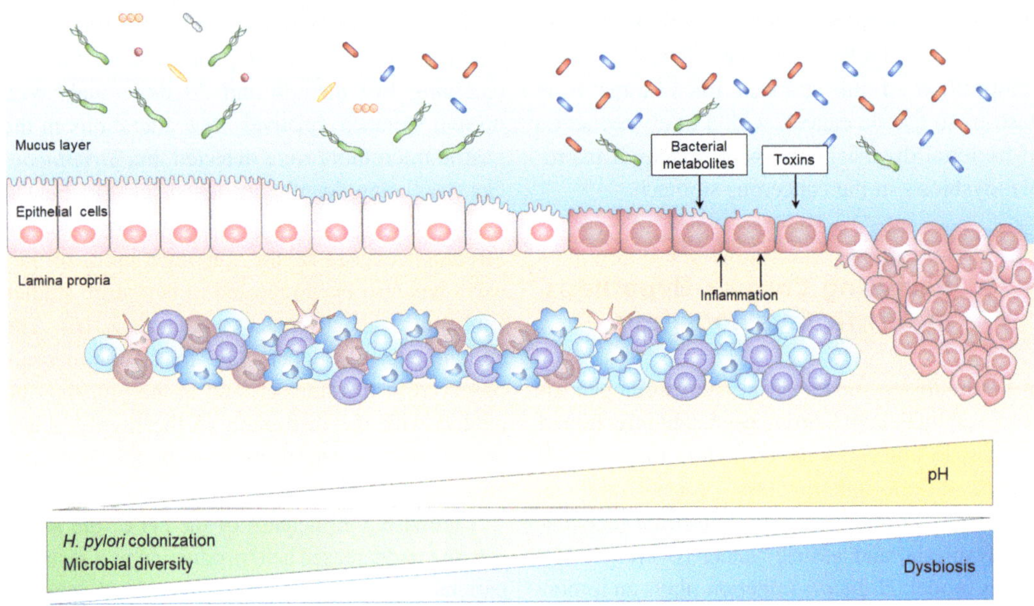

Fig. 1 Model for microbial dysbiosis in gastric cancer development. *H. pylori* infection triggers and perpetuates an inflammatory response in the gastric mucosa that, in some of the infected individuals, leads to loss of acid-secreting parietal cells with increase of the gastric pH. In this altered environment, *H. pylori* colonization decreases, and bacteria from other locations of the GI tract establish in the gastric niche, resulting in dysbiosis. This dysbiotic microbiome, characterized by reduced microbial diversity, may promote the development of gastric cancer by sustaining inflammation and/or inducing genotoxicity. Bacteria: green, *H. pylori;* orange, pink and grey, resident mucosa-associated microbiota; blue and red, dysbiotic microbiota; Inflammatory cells: dark blue, macrophages; pink, dendritic cells; dark pink, monocytes; light blue, CD4 T-lymphocytes; violet, CD8 T-lymphocytes

lipid metabolism (Tseng et al. 2016; Castano-Rodriguez et al. 2017; Coker et al. 2018; Ferreira et al. 2018). Findings are, however, relatively divergent between studies and results should therefore be interpreted with caution.

To revisit Correa's hypothesis that nitrate-reducing bacteria contribute to malignant transformation of the atrophic stomach by increasing the concentrations of nitrite and of N-nitroso compounds, we have assessed the functional features of the microbiome involved in these reactions (Ferreira et al. 2018). By fully reconstituting the metagenomes, based on the profiles obtained from the 16S rRNA gene sequences, we showed that in comparison with chronic gastritis, the gastric cancer microbiome had an increased representation of nitrate reductase and of nitrite reductase functions, the enzymes that respectively reduce nitrate to nitrite and nitrite to nitric oxide. The four genera *Citrobacter*, *Achromobacter*, *Clostridium* and *Phyllobacterium* were identified as the major contributors to these functions (Ferreira et al. 2018). Interestingly, and in agreement with our observations, are those of a follow-up study conducted in Taiwan to evaluate the effects of subtotal gastrectomy as a treatment for early-stage gastric cancer. The alteration of the gastric environment by the surgery led to significant changes in the gastric microbial community, and nitrate reductase, nitrite reductase, and other functions related to nitrosation were enriched in the gastric microbiome before, but not after subtotal gastrectomy (Tseng et al. 2016). These data suggest that the gastric cancer microbiome has the potential to produce carcinogenic N-nitroso compounds. Additional features linked to the dysbiotic microbiome may be involved in the promotion of a carcinogenic environment in the stomach. Microbial metabolites and toxins, as well as inflammation by-products generated by the dysbiotic microbiome, may directly induce host cell damage or interfere with host signalling pathways that influence cell turnover and survival, thus increasing the risk for gastric malignant transformation (Fig.1).

7 Conclusions

Despite the recent advances in the investigation of the human gastric microbiome, research in this area remains limited. Although a number of papers about the microbiome of the stomach in the context of gastric carcinogenesis have been published, caution should be taken with the interpretation of the results of very distinct technical approaches. Additionally, differences in the geographic origin, genetic background, and environmental exposures of the populations should be taken into consideration.

While it is clear that the microbial community present in gastric cancer is distinct from that present in chronic gastritis, research conducted on the microbiome of the histological stages that precede gastric cancer is still lacking. Studies in large and clinically well-defined patient populations will be key to determine the role of microbial dysbiosis in progression to cancer. The shift from descriptive to functionally based studies that investigate the effects of specific taxa and/or bacterial derived-metabolites in the gastric mucosa, will allow gaining insights into the mechanisms that lead to dysbiosis-associated genotoxicity and inflammation. Uncovering these mechanisms will create the grounds for translating microbiome research into prevention, diagnosis, and treatment improvements to control and decrease gastric cancer burden.

Acknowledgements JPM, RMF and IPR have fellowships from Fundação para a Ciência e a Tecnologia (FCT; PD/BD/114014/2015, SFRH/BPD/84084/2012, and SFRH/BD/110803/2015, respectively) through Programa Operacional Capital Humano (POCH) and the European Union. JPM's fellowship is in the framework of FCT's PhD Programme BiotechHealth (Ref PD/0016/2012). i3S-Instituto de Investigação e Inovação em Saúde is funded by Fundo Europeu de Desenvolvimento Regional (FEDER) funds through the COMPETE 2020-Operacional Programme for Competitiveness and Internationalisation (POCI), Portugal 2020, and by Portuguese funds through Fundação para a Ciência e a Tecnologia (FCT)/Ministério da Ciência, Tecnologia e Inovação (POCI-01-0145-FEDER-007274).

References

Adamsson I, Nord CE, Lundquist P, Sjostedt S, Edlund C (1999) Comparative effects of omeprazole, amoxycillin plus metronidazole versus omeprazole, clarithromycin plus metronidazole on the oral, gastric and intestinal microflora in *Helicobacter pylori*-infected patients. J Antimicrob Chemother 44 (5):629–640

Ahn J, Sinha R, Pei Z, Dominianni C, Wu J, Shi J, Goedert JJ, Hayes RB, Yang L (2013) Human gut microbiome and risk for colorectal cancer. J Natl Cancer Inst 105 (24):1907–1911. https://doi.org/10.1093/jnci/djt300

Andersson AF, Lindberg M, Jakobsson H, Backhed F, Nyren P, Engstrand L (2008) Comparative analysis of human gut microbiota by barcoded pyrosequencing. PLoS One 3(7):e2836. https://doi.org/10.1371/journal.pone.0002836

Arnold JW, Roach J, Azcarate-Peril MA (2016) Emerging technologies for Gut microbiome research. Trends Microbiol 24(11):887–901. https://doi.org/10.1016/j.tim.2016.06.008

Aviles-Jimenez F, Vazquez-Jimenez F, Medrano-Guzman R, Mantilla A, Torres J (2014) Stomach microbiota composition varies between patients with non-atrophic gastritis and patients with intestinal type of gastric cancer. Sci Rep 4:4202. https://doi.org/10.1038/srep04202

Backert S, Tegtmeyer N, Selbach M (2010) The versatility of *Helicobacter pylori* CagA effector protein functions: the master key hypothesis. Helicobacter 15 (3):163–176. https://doi.org/10.1111/j.1523-5378.2010.00759.x

Backert S, Tegtmeyer N, Fischer W (2015) Composition, structure and function of the *Helicobacter pylori* cag pathogenicity island encoded type IV secretion system. Future Microbiol 10(6):955–965. https://doi.org/10.2217/fmb.15.32

Bik EM, Eckburg PB, Gill SR, Nelson KE, Purdom EA, Francois F, Perez-Perez G, Blaser MJ, Relman DA (2006) Molecular analysis of the bacterial microbiota in the human stomach. Proc Natl Acad Sci U S A 103 (3):732–737. https://doi.org/10.1073/pnas.0506655103

Bik EM, Long CD, Armitage GC, Loomer P, Emerson J, Mongodin EF, Nelson KE, Gill SR, Fraser-Liggett CM, Relman DA (2010) Bacterial diversity in the oral cavity of 10 healthy individuals. ISME J 4 (8):962–974. https://doi.org/10.1038/ismej.2010.30

Bonnet M, Buc E, Sauvanet P, Darcha C, Dubois D, Pereira B, Dechelotte P, Bonnet R, Pezet D, Darfeuille-Michaud A (2014) Colonization of the human gut by *E. coli* and colorectal cancer risk. Clin Cancer Res 20(4):859–867. https://doi.org/10.1158/1078-0432.CCR-13-1343

Cancer Genome Atlas Research (2014) Comprehensive molecular characterization of gastric adenocarcinoma. Nature 513(7517):202–209. https://doi.org/10.1038/nature13480

Carneiro F (1997) Classification of gastric carcinomas. Curr Diagn Pathol 4(1):51–59. https://doi.org/10.1016/S0968-6053(97)80008-7

Castano-Rodriguez N, Goh KL, Fock KM, Mitchell HM, Kaakoush NO (2017) Dysbiosis of the microbiome in gastric carcinogenesis. Sci Rep 7(1):15957. https://doi.org/10.1038/s41598-017-16289-2

Cho I, Blaser MJ (2012) The human microbiome: at the interface of health and disease. Nat Rev Genet 13 (4):260–270. https://doi.org/10.1038/nrg3182

Coker OO, Dai Z, Nie Y, Zhao G, Cao L, Nakatsu G, Wu WK, Wong SH, Chen Z, Sung JJY, Yu J (2018) Mucosal microbiome dysbiosis in gastric carcinogenesis. Gut 67(6):1024–1032. https://doi.org/10.1136/gutjnl-2017-314281

Correa P (1992) Human gastric carcinogenesis: a multistep and multifactorial process–First American Cancer Society Award Lecture on cancer epidemiology and prevention. Cancer Res 52(24):6735–6740

Correa P, Haenszel W, Cuello C, Tannenbaum S, Archer M (1975) A model for gastric cancer epidemiology. Lancet 2(7924):58–60

Cristescu R, Lee J, Nebozhyn M, Kim KM, Ting JC, Wong SS, Liu J, Yue YG, Wang J, Yu K, Ye XS, Do IG, Liu S, Gong L, Fu J, Jin JG, Choi MG, Sohn TS, Lee JH, Bae JM, Kim ST, Park SH, Sohn I, Jung SH, Tan P, Chen R, Hardwick J, Kang WK, Ayers M, Hongyue D, Reinhard C, Loboda A, Kim S, Aggarwal A (2015) Molecular analysis of gastric cancer identifies subtypes associated with distinct clinical outcomes. Nat Med 21(5):449–456. https://doi.org/10.1038/nm.3850

D'Elia L, Rossi G, Ippolito R, Cappuccio FP, Strazzullo P (2012) Habitual salt intake and risk of gastric cancer: a meta-analysis of prospective studies. Clin Nutr 31 (4):489–498. https://doi.org/10.1016/j.clnu.2012.01.003

Delgado S, Cabrera-Rubio R, Mira A, Suarez A, Mayo B (2013) Microbiological survey of the human gastric ecosystem using culturing and pyrosequencing methods. Microb Ecol 65(3):763–772. https://doi.org/10.1007/s00248-013-0192-5

Dicksved J, Lindberg M, Rosenquist M, Enroth H, Jansson JK, Engstrand L (2009) Molecular characterization of the stomach microbiota in patients with gastric cancer and in controls. J Med Microbiol 58(Pt 4):509–516. https://doi.org/10.1099/jmm.0.007302-0

El-Omar EM, Carrington M, Chow WH, McColl KE, Bream JH, Young HA, Herrera J, Lissowska J, Yuan CC, Rothman N, Lanyon G, Martin M, Fraumeni JF Jr, Rabkin CS (2001) The role of interleukin-1 polymorphisms in the pathogenesis of gastric cancer. Nature 412(6842):99. https://doi.org/10.1038/35083631

Eun CS, Kim BK, Han DS, Kim SY, Kim KM, Choi BY, Song KS, Kim YS, Kim JF (2014) Differences in gastric mucosal microbiota profiling in patients with chronic gastritis, intestinal metaplasia, and gastric cancer using pyrosequencing methods. Helicobacter 19 (6):407–416. https://doi.org/10.1111/hel.12145

Fang X, Wei J, He X, An P, Wang H, Jiang L, Shao D, Liang H, Li Y, Wang F, Min J (2015) Landscape of dietary factors associated with risk of gastric cancer: a systematic review and dose-response meta-analysis of prospective cohort studies. Eur J Cancer 51 (18):2820–2832. https://doi.org/10.1016/j.ejca.2015.09.010

Fenoglio-Preiser C, Carneiro F, Correa P, Guilford P, Lambert R, Magraud F, Muñoz N, Powell SM, Ruggae M, Sasako M, Stolte M, Watanabe H (2010) Tumours of the stomach: gastric carcinoma. In: Bosman FT, Carneiro F, Hruban RH, Theise ND (eds) WHO classification of tumours of the digestive system, 4th edn. World Health Organization, Geneva, pp 37–67

Ferlay J, Soerjomataram I, Dikshit R, Eser S, Mathers C, Rebelo M, Parkin DM, Forman D, Bray F (2015) Cancer incidence and mortality worldwide: sources, methods and major patterns in GLOBOCAN 2012. Int J Cancer 136(5):E359–E386. https://doi.org/10.1002/ijc.29210

Ferreira RM, Machado JC, Figueiredo C (2014) Clinical relevance of *Helicobacter pylori* vacA and cagA genotypes in gastric carcinoma. Best Pract Res Clin Gastroenterol 28(6):1003–1015. https://doi.org/10.1016/j.bpg.2014.09.004

Ferreira RM, Pereira-Marques J, Pinto-Ribeiro I, Costa JL, Carneiro F, Machado JC, Figueiredo C (2018) Gastric microbial community profiling reveals a dysbiotic cancer-associated microbiota. Gut 67(2):226–236. https://doi.org/10.1136/gutjnl-2017-314205

Figueiredo C, Machado JC, Pharoah P, Seruca R, Sousa S, Carvalho R, Capelinha AF, Quint W, Caldas C, van Doorn LJ, Carneiro F, Sobrinho-Simoes M (2002) *Helicobacter pylori* and interleukin 1 genotyping: an opportunity to identify high-risk individuals for gastric carcinoma. J Natl Cancer Inst 94(22):1680–1687

Ford AC, Forman D, Hunt R, Yuan Y, Moayyedi P (2015) *Helicobacter pylori* eradication for the prevention of gastric neoplasia. Cochrane Database Syst Rev (7): CD005583. https://doi.org/10.1002/14651858.CD005583.pub2

Gaddy JA, Radin JN, Loh JT, Zhang F, Washington MK, Peek RM Jr, Algood HM, Cover TL (2013) High dietary salt intake exacerbates *Helicobacter pylori*-induced gastric carcinogenesis. Infect Immun 81 (6):2258–2267. https://doi.org/10.1128/IAI.01271-12

Gilbert JA, Blaser MJ, Caporaso JG, Jansson JK, Lynch SV, Knight R (2018) Current understanding of the human microbiome. Nat Med 24(4):392–400. https://doi.org/10.1038/nm.4517

Gonzalez CA, Figueiredo C, Lic CB, Ferreira RM, Pardo ML, Ruiz Liso JM, Alonso P, Sala N, Capella G, Sanz-Anquela JM (2011) *Helicobacter pylori* cagA and vacA genotypes as predictors of progression of gastric preneoplastic lesions: a long-term follow-up in a high-risk area in Spain. Am J Gastroenterol 106 (5):867–874. https://doi.org/10.1038/ajg.2011.1

Gonzalez CA, Megraud F, Buissonniere A, Lujan Barroso L, Agudo A, Duell EJ, Boutron-Ruault MC, Clavel-Chapelon F, Palli D, Krogh V, Mattiello A, Tumino R, Sacerdote C, Quiros JR, Sanchez-Cantalejo E, Navarro C, Barricarte A, Dorronsoro M, Khaw KT, Wareham N, Allen NE, Tsilidis KK, Bas Bueno-de-Mesquita H, Jeurnink SM, Numans ME, Peeters PH, Lagiou P, Valanou E, Trichopoulou A, Kaaks R, Lukanova-McGregor A, Bergman MM, Boeing H, Manjer J, Lindkvist B, Stenling R, Hallmans G, Mortensen LM, Overvad K, Olsen A, Tjonneland A, Bakken K, Dumeaux V, Lund E, Jenab M, Romieu I, Michaud D, Mouw T, Carneiro F, Fenge C, Riboli E (2012) *Helicobacter pylori* infection assessed by ELISA and by immunoblot and noncardia gastric cancer risk in a prospective study: the Eurgast-EPIC project. Ann Oncol 23 (5):1320–1324. https://doi.org/10.1093/annonc/mdr384

Goodwin AC, Destefano Shields CE, Wu S, Huso DL, Wu X, Murray-Stewart TR, Hacker-Prietz A, Rabizadeh S, Woster PM, Sears CL, Casero RA, Jr. (2011) Polyamine catabolism contributes to enterotoxigenic *Bacteroides fragilis*-induced colon tumorigenesis. Proc Natl Acad Sci U S A 108 (37):15354–15359. doi:https://doi.org/10.1073/pnas.1010203108

Helicobacter and Cancer Collaborative Group (2001) Gastric cancer and *Helicobacter pylori*: a combined analysis of 12 case control studies nested within prospective cohorts. Gut 49(3):347–353

Hsieh YY, Tung SY, Pan HY, Yen CW, Xu HW, Lin YJ, Deng YF, Hsu WT, Wu CS, Li C (2018) Increased abundance of *Clostridium* and *Fusobacterium* in gastric microbiota of patients with gastric cancer in Taiwan. Sci Rep 8(1):158. https://doi.org/10.1038/s41598-017-18596-0

IARC (1994) Schistosomes, liver flukes and *Helicobacter pylori*. IARC working group on the evaluation of carcinogenic risks to humans. Lyon, 7–14 June 1994. IARC Monogr Eval Carcinog Risks Hum 61:1–241

IARC (2011) IARC. Monographs on the evaluation of carcinogenic risks to humans, volume 100. A review of carcinogen—part B: biological agents. International Agency for Research on Cancer, Lyon

Imhann F, Bonder MJ, Vich Vila A, Fu J, Mujagic Z, Vork L, Tigchelaar EF, Jankipersadsing SA, Cenit MC, Harmsen HJ, Dijkstra G, Franke L, Xavier RJ, Jonkers D, Wijmenga C, Weersma RK, Zhernakova A (2016) Proton pump inhibitors affect the gut microbiome. Gut 65(5):740–748. https://doi.org/10.1136/gutjnl-2015-310376

Jackson MA, Goodrich JK, Maxan ME, Freedberg DE, Abrams JA, Poole AC, Sutter JL, Welter D, Ley RE, Bell JT, Spector TD, Steves CJ (2016) Proton pump inhibitors alter the composition of the gut microbiota. Gut 65(5):749–756. https://doi.org/10.1136/gutjnl-2015-310861

Jo HJ, Kim J, Kim N, Park JH, Nam RH, Seok YJ, Kim YR, Kim JS, Kim JM, Kim JM, Lee DH, Jung HC (2016) Analysis of gastric microbiota by pyrosequencing: minor role of bacteria other than *Helicobacter pylori* in the gastric carcinogenesis. Helicobacter 21(5):364–374. https://doi.org/10.1111/hel.12293

Kelly CP, LaMont JT (2008) *Clostridium difficile*–more difficult than ever. N Engl J Med 359(18):1932–1940. https://doi.org/10.1056/NEJMra0707500

Kienesberger S, Cox LM, Livanos A, Zhang XS, Chung J, Perez-Perez GI, Gorkiewicz G, Zechner EL, Blaser MJ (2016) Gastric *Helicobacter pylori* infection affects local and distant microbial populations and host responses. Cell Rep 14(6):1395–1407. https://doi.org/10.1016/j.celrep.2016.01.017

Klymiuk I, Bilgilier C, Stadlmann A, Thannesberger J, Kastner MT, Hogenauer C, Puspok A, Biowski-Frotz-S, Schrutka-Kolbl C, Thallinger GG, Steininger C (2017) The human gastric microbiome is predicated upon infection with *Helicobacter pylori*. Front Microbiol 8:2508. https://doi.org/10.3389/fmicb.2017.02508

Knight R, Callewaert C, Marotz C, Hyde ER, Debelius JW, McDonald D, Sogin ML (2017) The microbiome and human biology. Annu Rev Genomics Hum Genet 18:65–86. https://doi.org/10.1146/annurev-genom-083115-022438

Kuipers EJ (1998) Review article: relationship between *Helicobacter pylori*, atrophic gastritis and gastric cancer. Aliment Pharmacol Ther 12(Suppl 1):25–36

Lauren P (1965) The two histological main types of gastric carcinoma: diffuse and so-called intestinal-type carcinoma. An attempt at a histo-clinical classification. Acta Pathol Microbiol Scand 64:31–49

Lei Z, Tan IB, Das K, Deng N, Zouridis H, Pattison S, Chua C, Feng Z, Guan YK, Ooi CH, Ivanova T, Zhang S, Lee M, Wu J, Ngo A, Manesh S, Tan E, Teh BT, So JB, Goh LK, Boussioutas A, Lim TK, Flotow H, Tan P, Rozen SG (2013) Identification of molecular subtypes of gastric cancer with different responses to PI3-kinase inhibitors and 5-fluorouracil. Gastroenterology 145(3):554–565. https://doi.org/10.1053/j.gastro.2013.05.010

Lertpiriyapong K, Whary MT, Muthupalani S, Lofgren JL, Gamazon ER, Feng Y, Ge Z, Wang TC, Fox JG (2014) Gastric colonisation with a restricted commensal microbiota replicates the promotion of neoplastic lesions by diverse intestinal microbiota in the *Helicobacter pylori* INS-GAS mouse model of gastric carcinogenesis. Gut 63(1):54–63. https://doi.org/10.1136/gutjnl-2013-305178

Li XX, Wong GL, To KF, Wong VW, Lai LH, Chow DK, Lau JY, Sung JJ, Ding C (2009) Bacterial microbiota profiling in gastritis without *Helicobacter pylori* infection or non-steroidal anti-inflammatory drug use. PLoS One 4(11):e7985. https://doi.org/10.1371/journal.pone.0007985

Li J, Woods SL, Healey S, Beesley J, Chen X, Lee JS, Sivakumaran H, Wayte N, Nones K, Waterfall JJ, Pearson J, Patch AM, Senz J, Ferreira MA, Kaurah P, Mackenzie R, Heravi-Moussavi A, Hansford S, Lannagan TRM, Spurdle AB, Simpson PT, da Silva L, Lakhani SR, Clouston AD, Bettington M, Grimpen F, Busuttil RA, Di Costanzo N, Boussioutas A, Jeanjean M, Chong G, Fabre A, Olschwang S, Faulkner GJ, Bellos E, Coin L, Rioux K, Bathe OF, Wen X, Martin HC, Neklason DW, Davis SR, Walker RL, Calzone KA, Avital I, Heller T, Koh C, Pineda M, Rudloff U, Quezado M, Pichurin PN, Hulick PJ, Weissman SM, Newlin A, Rubinstein WS, Sampson JE, Hamman K, Goldgar D, Poplawski N, Phillips K, Schofield L, Armstrong J, Kiraly-Borri C, Suthers GK, Huntsman DG, Foulkes WD, Carneiro F, Lindor NM, Edwards SL, French JD, Waddell N, Meltzer PS, Worthley DL, Schrader KA, Chenevix-Trench G (2016) Point mutations in exon 1B of APC reveal gastric adenocarcinoma and proximal polyposis of the stomach as a familial adenomatous polyposis variant. Am J Hum Genet 98(5):830–842. https://doi.org/10.1016/j.ajhg.2016.03.001

Li TH, Qin Y, Sham PC, Lau KS, Chu KM, Leung WK (2017) Alterations in gastric microbiota after *H. pylori* eradication and in different histological stages of gastric carcinogenesis. Sci Rep 7:44935. https://doi.org/10.1038/srep44935

Lloyd-Price J, Mahurkar A, Rahnavard G, Crabtree J, Orvis J, Hall AB, Brady A, Creasy HH, McCracken C, Giglio MG, McDonald D, Franzosa EA, Knight R, White O, Huttenhower C (2017) Strains, functions and dynamics in the expanded human microbiome project. Nature 550(7674):61–66. https://doi.org/10.1038/nature23889

Lofgren JL, Whary MT, Ge Z, Muthupalani S, Taylor NS, Mobley M, Potter A, Varro A, Eibach D, Suerbaum S, Wang TC, Fox JG (2011) Lack of commensal flora in *Helicobacter pylori*-infected INS-GAS mice reduces gastritis and delays intraepithelial neoplasia. Gastroenterology 140(1):210–220. https://doi.org/10.1053/j.gastro.2010.09.048

Machado JC, Figueiredo C, Canedo P, Pharoah P, Carvalho R, Nabais S, Castro Alves C, Campos ML, Van Doorn LJ, Caldas C, Seruca R, Carneiro F, Sobrinho-Simoes M (2003) A proinflammatory genetic profile increases the risk for chronic atrophic gastritis and gastric carcinoma. Gastroenterology 125(2):364–371

Maldonado-Contreras A, Goldfarb KC, Godoy-Vitorino F, Karaoz U, Contreras M, Blaser MJ, Brodie EL, Dominguez-Bello MG (2011) Structure of the human gastric bacterial community in relation to *Helicobacter pylori* status. ISME J 5(4):574–579. https://doi.org/10.1038/ismej.2010.149

Martin ME, Bhatnagar S, George MD, Paster BJ, Canfield DR, Eisen JA, Solnick JV (2013) The impact of *Helicobacter pylori* infection on the gastric microbiota of the rhesus macaque. PLoS One 8(10):e76375. https://doi.org/10.1371/journal.pone.0076375

Martinsen TC, Bergh K, Waldum HL (2005) Gastric juice: a barrier against infectious diseases. Basic Clin Pharmacol Toxicol 96(2):94–102. https://doi.org/10.1111/j.1742-7843.2005.pto960202.x

Mera RM, Bravo LE, Camargo MC, Bravo JC, Delgado AG, Romero-Gallo J, Yepez MC, Realpe JL, Schneider BG, Morgan DR, Peek RM Jr, Correa P, Wilson KT, Piazuelo MB (2018) Dynamics of *Helicobacter pylori* infection as a determinant of progression of gastric precancerous lesions: 16-year follow-up of an eradication trial. Gut 67(7):1239–1246. https://doi.org/10.1136/gutjnl-2016-311685

Monstein HJ, Tiveljung A, Kraft CH, Borch K, Jonasson J (2000) Profiling of bacterial flora in gastric biopsies from patients with *Helicobacter pylori*-associated gastritis and histologically normal control individuals by temperature gradient gel electrophoresis and 16S rDNA sequence analysis. J Med Microbiol 49(9):817–822. https://doi.org/10.1099/0022-1317-49-9-817

Mowat C, Williams C, Gillen D, Hossack M, Gilmour D, Carswell A, Wirz A, Preston T, McColl KE (2000) Omeprazole, *Helicobacter pylori* status, and alterations in the intragastric milieu facilitating bacterial N-nitrosation. Gastroenterology 119(2):339–347

Ohnishi N, Yuasa H, Tanaka S, Sawa H, Miura M, Matsui A, Higashi H, Musashi M, Iwabuchi K, Suzuki M, Yamada G, Azuma T, Hatakeyama M (2008) Transgenic expression of *Helicobacter pylori* CagA induces gastrointestinal and hematopoietic neoplasms in mouse. Proc Natl Acad Sci U S A 105(3):1003–1008. https://doi.org/10.1073/pnas.0711183105

Oliveira C, Pinheiro H, Figueiredo J, Seruca R, Carneiro F (2015) Familial gastric cancer: genetic susceptibility, pathology, and implications for management. Lancet Oncol 16(2):e60–e70. https://doi.org/10.1016/S1470-2045(14)71016-2

Ottini L, Falchetti M, Lupi R, Rizzolo P, Agnese V, Colucci G, Bazan V, Russo A (2006) Patterns of genomic instability in gastric cancer: clinical implications and perspectives. Ann Oncol 17(Suppl 7):vii97–vi102. https://doi.org/10.1093/annonc/mdl960

Paroni Sterbini F, Palladini A, Masucci L, Cannistraci CV, Pastorino R, Ianiro G, Bugli F, Martini C, Ricciardi W, Gasbarrini A, Sanguinetti M, Cammarota G, Posteraro B (2016) Effects of proton pump inhibitors on the gastric mucosa-associated microbiota in dyspeptic patients. Appl Environ Microbiol 82(22):6633–6644. https://doi.org/10.1128/AEM.01437-16

Parsons BN, Ijaz UZ, D'Amore R, Burkitt MD, Eccles R, Lenzi L, Duckworth CA, Moore AR, Tiszlavicz L, Varro A, Hall N, Pritchard DM (2017) Comparison of the human gastric microbiota in hypochlorhydric states arising as a result of *Helicobacter pylori*-induced atrophic gastritis, autoimmune atrophic gastritis and proton pump inhibitor use. PLoS Pathog 13(11): e1006653. https://doi.org/10.1371/journal.ppat.1006653

Persson C, Canedo P, Machado JC, El-Omar EM, Forman D (2011) Polymorphisms in inflammatory response genes and their association with gastric cancer: a HuGE systematic review and meta-analyses. Am J Epidemiol 173(3):259–270. https://doi.org/10.1093/aje/kwq370

Plottel CS, Blaser MJ (2011) Microbiome and malignancy. Cell Host Microbe 10(4):324–335. https://doi.org/10.1016/j.chom.2011.10.003

Plummer M, van Doorn LJ, Franceschi S, Kleter B, Canzian F, Vivas J, Lopez G, Colin D, Munoz N, Kato I (2007) *Helicobacter pylori* cytotoxin-associated genotype and gastric precancerous lesions. J Natl Cancer Inst 99(17):1328–1334. https://doi.org/10.1093/jnci/djm120

Plummer M, Franceschi S, Vignat J, Forman D, de Martel C (2015) Global burden of gastric cancer attributable to *Helicobacter pylori*. Int J Cancer 136(2):487–490. https://doi.org/10.1002/ijc.28999

Praud D, Rota M, Pelucchi C, Bertuccio P, Rosso T, Galeone C, Zhang ZF, Matsuo K, Ito H, Hu J, Johnson KC, Yu GP, Palli D, Ferraroni M, Muscat J, Lunet N, Peleteiro B, Malekzadeh R, Ye W, Song H, Zaridze D, Maximovitch D, Aragones N, Castano-Vinyals G, Vioque J, Navarrete-Munoz EM, Pakseresht M, Pourfarzi F, Wolk A, Orsini N, Bellavia A, Hakansson N, Mu L, Pastorino R, Kurtz RC, Derakhshan MH, Lagiou A, Lagiou P, Boffetta P, Boccia S, Negri E, La Vecchia C (2018) Cigarette smoking and gastric cancer in the Stomach Cancer Pooling (StoP) project. Eur J Cancer Prev 27 (2):124–133. https://doi.org/10.1097/CEJ.0000000000000290

Qin J, Li R, Raes J, Arumugam M, Burgdorf KS, Manichanh C, Nielsen T, Pons N, Levenez F, Yamada T, Mende DR, Li J, Xu J, Li S, Li D, Cao J, Wang B, Liang H, Zheng H, Xie Y, Tap J, Lepage P, Bertalan M, Batto JM, Hansen T, Le Paslier D, Linneberg A, Nielsen HB, Pelletier E, Renault P, Sicheritz-Ponten T, Turner K, Zhu H, Yu C, Li S, Jian M, Zhou Y, Li Y, Zhang X, Li S, Qin N, Yang H, Wang J, Brunak S, Dore J, Guarner F, Kristiansen K, Pedersen O, Parkhill J, Weissenbach J, Meta HITC, Bork P, Ehrlich SD, Wang J (2010) A human gut microbial gene catalogue established by metagenomic sequencing. Nature 464(7285):59–65. https://doi.org/10.1038/nature08821

Rajilic-Stojanovic M, de Vos WM (2014) The first 1000 cultured species of the human gastrointestinal microbiota. FEMS Microbiol Rev 38(5):996–1047. https://doi.org/10.1111/1574-6976.12075

Rota M, Pelucchi C, Bertuccio P, Matsuo K, Zhang ZF, Ito H, Hu J, Johnson KC, Palli D, Ferraroni M, Yu GP, Muscat J, Lunet N, Peleteiro B, Ye W, Song H, Zaridze D, Maximovitch D, Guevara M, Fernandez-Villa T, Vioque J, Navarrete-Munoz EM, Wolk A, Orsini N, Bellavia A, Hakansson N, Mu L, Persiani R, Kurtz RC, Lagiou A, Lagiou P, Galeone C, Bonzi R, Boffetta P, Boccia S, Negri E, La Vecchia C (2017) Alcohol consumption and gastric cancer risk-a pooled analysis within the StoP project consortium. Int J Cancer 141(10):1950–1962. https://doi.org/10.1002/ijc.30891

Sanduleanu S, Jonkers D, De Bruine A, Hameeteman W, Stockbrugger RW (2001) Non-*Helicobacter pylori* bacterial flora during acid-suppressive therapy: differential findings in gastric juice and gastric mucosa. Aliment Pharmacol Ther 15(3):379–388

Schulz C, Schutte K, Koch N, Vilchez-Vargas R, Wos-Oxley ML, Oxley APA, Vital M, Malfertheiner P, Pieper DH (2018) The active bacterial assemblages of the upper GI tract in individuals with and without *Helicobacter* infection. Gut 67 (2):216–225. https://doi.org/10.1136/gutjnl-2016-312904

Sender R, Fuchs S, Milo R (2016) Are we really vastly outnumbered? Revisiting the ratio of bacterial to host cells in humans. Cell 164(3):337–340. https://doi.org/10.1016/j.cell.2016.01.013

Sjostedt S, Heimdahl A, Kager L, Nord CE (1985) Microbial colonization of the oropharynx, esophagus and stomach in patients with gastric diseases. Eur J Clin Microbiol 4(1):49–51

Spoto CPE, Gullo I, Carneiro F, Montgomery EA, Brosens LAA (2018) Hereditary gastrointestinal carcinomas and their precursors: An algorithm for genetic testing. Semin Diagn Pathol 35(3):170–183. https://doi.org/10.1053/j.semdp.2018.01.004

Stearns JC, Lynch MD, Senadheera DB, Tenenbaum HC, Goldberg MB, Cvitkovitch DG, Croitoru K, Moreno-Hagelsieb G, Neufeld JD (2011) Bacterial biogeography of the human digestive tract. Sci Rep 1:170. https://doi.org/10.1038/srep00170

Stockbruegger RW (1985) Bacterial overgrowth as a consequence of reduced gastric acidity. Scand J Gastroenterol Suppl 111:7–16

Thorell K, Bengtsson-Palme J, Liu OH, Palacios Gonzales RV, Nookaew I, Rabeneck L, Paszat L, Graham DY, Nielsen J, Lundin SB, Sjoling A (2017) In vivo analysis of the viable microbiota and *Helicobacter pylori* transcriptome in gastric infection and early stages of carcinogenesis. Infect Immun 85(10). https://doi.org/10.1128/IAI.00031-17

Thorens J, Froehlich F, Schwizer W, Saraga E, Bille J, Gyr K, Duroux P, Nicolet M, Pignatelli B, Blum AL, Gonvers JJ, Fried M (1996) Bacterial overgrowth during treatment with omeprazole compared with cimetidine: a prospective randomised double blind study. Gut 39(1):54–59

Tseng CH, Lin JT, Ho HJ, Lai ZL, Wang CB, Tang SL, Wu CY (2016) Gastric microbiota and predicted gene functions are altered after subtotal gastrectomy in patients with gastric cancer. Sci Rep 6:20701. https://doi.org/10.1038/srep20701

Turnbaugh PJ, Ley RE, Hamady M, Fraser-Liggett CM, Knight R, Gordon JI (2007) The human microbiome project. Nature 449(7164):804–810. https://doi.org/10.1038/nature06244

Van Cutsem E, Sagaert X, Topal B, Haustermans K, Prenen H (2016) Gastric cancer. Lancet 388 (10060):2654–2664. https://doi.org/10.1016/S0140-6736(16)30354-3

Warren JR, Marshall B (1983) Unidentified curved bacilli on gastric epithelium in active chronic gastritis. Lancet 1(8336):1273–1275

Wong BC, Lam SK, Wong WM, Chen JS, Zheng TT, Feng RE, Lai KC, Hu WH, Yuen ST, Leung SY, Fong DY, Ho J, Ching CK, Chen JS, China Gastric Cancer Study G (2004) *Helicobacter pylori* eradication to prevent gastric cancer in a high-risk region of China: a randomized controlled trial. JAMA 291(2):187–194. https://doi.org/10.1001/jama.291.2.187

Worthley DL, Phillips KD, Wayte N, Schrader KA, Healey S, Kaurah P, Shulkes A, Grimpen F, Clouston A, Moore D, Cullen D, Ormonde D, Mounkley D, Wen X, Lindor N, Carneiro F, Huntsman DG, Chenevix-Trench G, Suthers GK (2012) Gastric adenocarcinoma and proximal polyposis of the stomach (GAPPS): a new autosomal dominant syndrome. Gut 61(5):774–779. https://doi.org/10.1136/gutjnl-2011-300348

Yang I, Woltemate S, Piazuelo MB, Bravo LE, Yepez MC, Romero-Gallo J, Delgado AG, Wilson KT, Peek RM, Correa P, Josenhans C, Fox JG, Suerbaum S (2016) Different gastric microbiota compositions in two human populations with high and low gastric cancer risk in Colombia. Sci Rep 6:18594. https://doi.org/10.1038/srep18594

Yu G, Torres J, Hu N, Medrano-Guzman R, Herrera-Goepfert R, Humphrys MS, Wang L, Wang C, Ding T, Ravel J, Taylor PR, Abnet CC, Goldstein AM (2017) Molecular characterization of the human stomach microbiota in gastric cancer patients. Front Cell Infect Microbiol 7:302. https://doi.org/10.3389/fcimb.2017.00302

Zamani M, Ebrahimtabar F, Zamani V, Miller WH, Alizadeh-Navaei R, Shokri-Shirvani J, Derakhshan MH (2018) Systematic review with meta-analysis: the worldwide prevalence of *Helicobacter pylori* infection. Aliment Pharmacol Ther 47(7):868–876. https://doi.org/10.1111/apt.14561

Zilberstein B, Quintanilha AG, Santos MA, Pajecki D, Moura EG, Alves PR, Maluf Filho F, de Souza JA, Gama-Rodrigues J (2007) Digestive tract microbiota in healthy volunteers. Clinics (Sao Paulo) 62(1):47–54. https://doi.org/10.1590/S1807-59322007000100008

Adv Exp Med Biol - Advances in Microbiology, Infectious Diseases and Public Health (2019) 1149: 211–225
https://doi.org/10.1007/5584_2019_367

Published online: 24 April 2019

Current and Future Treatment of *Helicobacter pylori* Infections

Hiroshi Matsumoto, Akiko Shiotani, and David Y. Graham

Abstract

Helicobacter pylori is one of the most common human pathogens and it has been estimated that about 50% of the world's population is currently infected. The present consensus is that, unless there are compelling reasons, all *H. pylori* infections should be cured. Since the 1990s, different national and international guidelines for the management of *H. pylori*-related diseases have been published and periodically updated regarding indications for treatment, diagnostic procedures, and preferred treatment regimens. Most guidelines provide sophisticated meta-analyses examining the outcome of different regimens done in regions with variable, often high rates of resistance to antibiotics, for which the prevalence and effects of resistance was often ignored. Although successful antimicrobial therapy must be susceptibility-based, increasing antimicrobial resistance and general unavailability of susceptibility testing have required clinicians to generally rely on empiric regimens. Antibiotics resistance of *H. pylori* has reached alarming high levels worldwide, which has an effect to efficacy of treatment. The recommendations should provide regimes for multi-resistant infections or for those where susceptibility testing is unavailable or refused. The first rule is to use only proven locally effective therapies. Because of patient intolerances, drug allergies, and local experiences, the clinicians should have at least two options for first-line therapy. As with any antimicrobial therapy, a thorough review of prior antibiotic use is invaluable to identify the presence of probably resistance. The second key is patient education regarding potential and expected side-effects and the importance of completing the course of antibiotics. We also review here triple therapies, sequential-concomitant, hybrid therapies, bismuth therapies, dual therapy, vonoprazan, modern antibiotic treatments, probiotics and vaccination.

Keywords

Helicobacter pylori · Triple therapy · Sequential therapy · Concomitant therapy · Vonoprazan

H. Matsumoto (✉) and A. Shiotani
Division of Gastroenterology, Department of Internal medicine, Kawasaki Medical School, Kurashiki City, Okayama Prefecture, Japan
e-mail: hmatsumoto0311@gmail.com

D. Y. Graham
Michael E. DeBakey VA Medical Center and Baylor College of Medicine, Houston, TX, USA

1 Introduction

Helicobacter pylori is a Gram-negative spiral-shaped bacterium and human pathogen. *H. pylori* infection causes gastric inflammation and its related to diseases: peptic ulcer, gastric cancer, mucosa-associated lymphoid tissue (MALT) lymphoma and a variety of other conditions such as vitamin B12 deficiency, iron deficiency and idiopathic thrombocytopenia (Malfertheiner et al. 2012). *H. pylori* is one of the most common human pathogens and it has been estimated that about 50% of the world's population is currently infected (Williams and Pounder 1999). However, there are substantial geographic differences in the prevalence of infection, being most common in developing countries and infrequent in countries with advanced economies (Hunt et al. 2011). Living in, or birth in, a developing country and low socioeconomic status are associated with increased risk of *H. pylori* infection (Bastos et al. 2013; Bruce and Maaroos 2008). Humans are the primary host, and the transmission requires contact with an infected person either directly or through contaminated food or water. Transmission typically occurs within families and most often in childhood. Improvements in household hygiene and living conditions reduce transmission and such improvements are likely the most important factors in decreasing the prevalence of the *H. pylori* infection worldwide (Vale and Vitor 2010). The decline in transmission among children has resulted in a series of age-related cohorts with the prevalence of the infection at age 20, generally describing the lifetime prevalence for that age-cohort (Bastos et al. 2013; Fujimoto et al. 2007). Even within Europe, *H. pylori* prevalence ranges from 11% in Sweden to 60% in Spain (Hunt et al. 2011; Sanchez Ceballos et al. 2007). In China, *H. pylori* prevalence has been reported as high as 83%, but is decreasing rapidly in most areas (Zhang et al. 2014). In the USA, cross-sectional studies of the participants in the National Health and Nutrition Examination Survey (NHANES) III and NHANES 1999–2000 found an overall seropositivity of approximately 30%, which was similar to the prevalence in Canada (Chen and Blaser 2012). In Japan, there has been a progressive and a rapid decline in *H. pylori* prevalence, which is reflected in an overall decrease from 75% in the 1970s to 35% by 2010 (Kamada et al. 2015). The infection is becoming rare in Japanese children with an overall prevalence of <4% and falling.

1.1 *H. pylori* as a Bacterial Infection

H. pylori infection is a bacterial infection, and its growth and survival was shown to be susceptible to a variety of antimicrobials. However, *in vitro* susceptibility did not reliably predict *in vivo* effectiveness and *H. pylori* antimicrobial treatment proved to be more complicated than treatment of other common infections such as bacterial pneumonia, in part because *H. pylori* resides in the human stomach. The stomach presents a hostile environment as it is acidic and to be effective most antibiotics require neutral or near neutral environments. *H. pylori* can also be found deep in the mucosa and even within gastric epithelial cells (Kwok et al. 2002). This wide range of environments complicate delivering active antimicrobial drugs to ensure killing all of the organisms. Other problems have been caused by doctors in that most regimens have been developed using a trial-and-error approach rather than by a systematic strategy.

As with all antimicrobial therapies, treatment success depends on the details of therapy including: susceptibility, doses, formulations, frequency of dosing in relation to meals, use of adjuvants such as anti-secretory drugs, antacids, or probiotics, as well as treatment duration. Few of these important factors have been subjected to detailed clinical assessment and the optimum regimens still remain unknown. The effectiveness of some regimens may also vary in relation to host differences, for example polymorphisms in drug-metabolizing enzyme such as cytochrome P450 2C9 (CYP2C19) can greatly affect effectiveness (Yang and Lin 2010).

2 History of *H. pylori* Therapies

Early experiments showed that, although *H. pylori* was susceptible *in vitro* to many antimicrobials, only a few appeared useful *in vivo*. The first effective regimen, defined as reliably able to produce cure rates of >90%, was a triple therapy of largely acid-independent antimicrobials: bismuth, metronidazole (MTZ), and tetracycline (Graham and Lee 2015). It was soon discovered that its effectiveness was markedly reduced by MTZ resistance and that problem could be partially or completely overcome by increasing the dosage of MTZ to 1.5 g or greater and adding a proton pump inhibitor (PPI). These changes resulted in a quadruple therapy, now commonly called bismuth quadruple therapy. When given for 5–7 days it proved highly effective in MTZ susceptible infections, but required 14 day therapy in the presence of resistance. The second widely used regimen to be introduced, utilized the pH sensitive agents amoxicillin (AMX) and clarithromycin(CLR) and a PPI all given twice a day. To be effective, these pH sensitive antibiotics require bacterial replication. Effective regimens have been described using 200–500 mg of CLR twice a day. Both bismuth quadruple therapy and CAM therapy can reliably achieve 95% or greater cure rates in susceptible infections and adherent patients (Hsu et al. 2017; Macias-Garcia et al. 2018).

Because of convenience, tolerability, and vast marketing support from Pharma, CLR triple therapy became, and still is, one of the most widely prescribed treatment regime worldwide. Best results are obtained with 14-day therapy. However, to obtain marketing advantages, pharmaceutical companies introduced shorter duration CLR triple therapies (7 or 10 instead of 14 days), which were generally associated with a reduction in cure rates from >95% to between 88% and 94%, respectively. This decline was partially obscured by increasing CLR resistance, which was rapidly undermining the effectiveness of the regimen (Thung et al. 2016). In contrast to MTZ resistance, CLR resistance is all-or-none meaning that the presence of CLR resistance effectively reduces the three drug regimen to only AMX and the PPI, which at the doses given resulted in an overall marked fall in cure rates. The rapid rise in CLR resistance was a bystander effect related to high use of macrolides worldwide for respiratory infections. By the year 2000, the *H. pylori* cure rates with CLR were often between 70% and 75%. However, in many countries this regimen was the only one approved which left physicians with few options. Treatment guidelines from various groups and countries continued to recommend short duration CLR triple therapy long after its cure rate had become unacceptably low. In 2012 the Maastricht IV guideline (Malfertheiner et al. 2012) recommended CLR not be used if the prevalence of resistance was 15% or greater, but as clinicians had no access to local resistance rates data, this failed to reduce the continued popularity and use of this obsolete regimen.

2.1 Antibiotic Resistance

Since at least the year 2000, the *H. pylori* eradication rates have been decreasing because of increasing resistance to one or more of the antibiotics (Grad et al. 2012; Graham and Fischbach 2010; Malfertheiner et al. 2012; Savoldi et al. 2018; Selgrad and Malfertheiner 2011). The World Health Organization groups antimicrobial resistance data by region and among east Asian countries the prevalence of CLR resistance is high. MTZ resistance is low only in Japan. Besides, high prevalence resistance to both CLR and MTZ is recognized in Italy, Vietnam, and Mexico as well as China. In northern Europe, there is generally low resistance to CLR also because of restricted use. The prevalence of bacterial resistance is related to the consumption rates of these antibiotics (Graham 2015; Meyer et al. 2002). A 2017 retrospective review from the Netherlands reported increasing resistance rates for CLR (from 9.8% to 18.1%), MTZ (20.7–23.2%), and AMX (6.3–10%) over 10 years (Ruiter et al. 2017). The first systemic reviews of primary antibiotic resistance in the Asia-Pacific region reported mean resistance rates of 17% for CLR, 18% for levofloxacin, and 44% for MTZ. There was significant heterogeneity in resistance rate across different countries (Kuo et al. 2017). The mean overall prevalence of resistance to MTZ is 44% (95% Confidence Interval 39–48) ranging from 10% in Japan to 84% in Bangladesh and 88% in Nepal.

Resistance to AMX is generally less than 1% (Mégraud 2004) and overall no significant change in the resistance has been observed (Kobayashi et al. 2007). Kuo and co-workers (2017) reported mean overall prevalence of resistance to levofloxacin of 18% (95% CI 15–22), ranging from 2% to 3% (Bhutan and Malaysia) to 66% in Bangladesh. Subgroup analysis by collection period showed that overall levofloxacin resistance increased from 2% (95% CI 0–13) before 2000 to 27% (95% CI 21–34) during in 2011–2015, with significant between-group heterogeneity. According to data from 2006 to 2015, the overall prevalence of levofloxacin (21%) was higher than those in Europe and Latin America. Resistance to levofloxacin increased over time in all included countries for which data were available. Besides, the resistance to quinolones is in the range of 20% in Europe, 15% in America, and 10% in Asia, and rapidly increasing (Liang et al. 2014).

Because levofloxacin and CLR resistance have increased worldwide such that there are only a few areas where regimens that rely on CLR or levofloxacin are still effective when used as empiric therapy and treatment strategies will need to be adapted to resistance patterns on country-by-country or region-by-region basis (Kuo et al. 2017).

2.1.1 Other Triple Therapies

As noted above, a recent high-quality meta-analysis suggests that the efficacy of both triple therapy with a PPI and AMX plus CLR or MTZ is currently low and clinical unacceptable when given for either 7 days (70% vs. 77%) or 14 days (80% vs. 84%) (Puig et al. 2016). Due to the increase to CLR resistance, levofloxacin, a broad spectrum quinolone, was substituted for CLR in triple therapy. Initial trials utilized 7 and 10 day regimens and failed to achieve even 90% treatment success. Subsequently, it was discovered that a 14-day regimen was highly successful and reliable achieves eradication rates of more than 90% in areas where the local resistance to levofloxacin is low (Savoldi et al. 2018). The worldwide use of quinolones has markedly increased such that fluoroquinolones have joined CLR in no longer being considered acceptable for empiric therapy except in the few areas where resistance is still low. Among the fourth-generation quinolones (moxifloxacin, sitafloxacin, and gemifloxacin) only sitafloxacin has proven successful in that it remains effective at a higher minimum inhibitory concentration (MIC) than other fluoroquinolones (An et al. 2018). Thus, in regions where sitafloxacin is available it is the only quinolone recommended for empiric therapy.

2.1.2 Sequential-Concomitant, Hybrid Therapies

There are a number of empirically derived 4 drug regimens using AMX, CLR, MTZ, and a PPI. They are named in relation to how the drugs are administered (e.g., sequentially, or all together). They all share the features of providing the best cure rates when given for 14 days and being rendered ineffective by the presence of dual CLR and MTZ (Graham et al. 2014). They all function as if one were giving CLR triple and MTZ triple therapy simultaneously (i.e., the MTZ will kill the CLR-resistant strains, and the CLR will kill the MNZ-resistant strains, such as that only the presence of dual resistant strains will cause this regimen fail). Since simultaneous administration of all 4 drugs always provides results equivalent or superior to the other versions, it is recommended over the other combinations using the same drugs (e.g., sequential therapy). All these regimens arose as empiric therapies in response to failing CLR triple therapy and lack of susceptibility data. It is now recognized that every patient treated receives one unneeded antibiotic (either CLR or MTZ and failures receive two unneeded antibiotics). This regimen was recommended by Maastricht V, the American College of Gastroenterology and Toronto consensus conferences without their recognition or understanding that each one million treatments would also administer approximately 15 tons of unneeded CLRor MTZ(which is actually a potential human carcinogen) (Chey et al. 2017; Fallone et al. 2016; Malfertheiner et al. 2012). This is an extremely high price to pay for failure to provide antimicrobial surveillance programs for *H. pylori* infection (Graham and Shiotani 2008).

2.1.3 Bismuth Therapies

Although bismuth triple therapy and subsequently quadruple therapy were introduced early in the history of *H. pylori*, they became never popular. The issues included intrinsic complexity, the large number of tablets, four times per day administration, side effects, lack of pharmaceutical company support, and importantly markedly negative assessments by key opinion leaders working with CLR triple therapies which were simpler and more tolerable. The combination of high dose MTZ and high dose tetracycline was associated with a high frequency of side effects such as abdominal pain, nausea, and vomiting which often resulted in poor adherence. A proprietary three-in-one capsule (Pylera) containing bismuth subcitrate potassium, MTZ and tetracycline fared no better in the US. Its introduction in Europe, however, coincided with loss of big Pharma support for triple therapy requiring key opinion leaders to re-examine their prior objections and they now found Pylera highly acceptable. Unfortunately, despite all current *H. pylori* treatment guidelines recommending 14-day bismuth quadruple therapy, Pylera is currently packaged as a 10 day regimen. The resulting reduction of cure rates in MTZ resistant infections is likely responsible for the cure rates from Europe being very heterogeneous as they represent combining a group with very high cure rates (i.e., in susceptible infections 7 days were sufficient) and lower cure rates in those with MTZ -resistant infections. In reality, 14 day MTZ triple therapy or 7 day bismuth quadruple therapy would be better tolerated choices for those with MTZ susceptible infections. The patent on the Pylera formulation expires in December 2018 and hopefully other manufacturers will offer different packaging amenable to using different durations of therapy. Another option is to prescribe generics but these are often difficult to obtain because of difficulties with the supply and availability of both bismuth or tetracycline in Europe and tetracycline in the US.

2.2 Alternate Bismuth Quadruple Therapies

Bismuth-containing quadruple therapy is recommended by the current European, US, Canadian and Chinese guideline (Chey et al. 2017; Fallone et al. 2016; Liu et al. 2018; Malfertheiner et al. 2017). Bismuth is additive adding 20–25% to the outcome of the other components of the quadruple regimen (Graham et al. 2018). Different highly successful bismuth quadruple therapies have been developed in China and are effective despite MTZ resistance. Most use a PPI twice a day, bismuth 220 mg twice a day, along with tetracycline 500 mg four times a day, and either AMX 1 g three times a day or MTZ 400 or 500 mg four times a day for 14 days (Chen et al. 2018; Dore et al. 2016; Zhang et al. 2015). Alternatively, AMX 1 g three times a day has been used to substitute tetracycline in areas, where tetracycline is difficult to obtain. These regimens have proved to be well tolerated and highly successful quadruple therapies.

2.3 Dual Therapy

Dual AMX and PPI therapy has been examined since the year 1989 and in some instances has proven to be highly effective (Unge et al. 1989; reviewed by Dore et al. 1998). The population where it seems most effective is among those with low acid secretion (e.g., significant corpus gastritis). The Food and Drug Administration (FDA) approved regimen is lansoprazole 60 mg three times per day and AMPC 1 g three times a day for 14 days with a cure rate between 60% and 70% (Dore et al. 1998; Laine et al. 1998; Unge et al. 1989). Some recent studies in Taiwan have been proven successful, whereas others from China have not (Graham et al. 2017; Yang et al. 2015; Yang et al. 2011). The major barrier to overcome seems to be the difficulty in obtaining and maintaining the intragastric pH at 6 or greater. This is almost impossible with traditional PPIs, except in the presence of significant corpus inflammation (Graham and Tansel 2018).

However, it can be achieved with the newest type of PPI, the potassium-competitive acid blocker (P-CAB) although the dose, dosing frequency, and duration remain to be established (Graham and Dore 2018). Early studies suggest that this will be possible with 14 day therapy and 1 mg AMX three times a day or 500–750 mg four times a day.

AMX is considered a time-dependent antibiotic that is rapidly absorbed into the plasma, but is excreted between 6 and 8 h after administration (Barbhaiya et al. 1979). A dosage of 500–750 mg per 6 h compared with 1000 mg twice daily is more likely to maintain higher plasma concentration of AMX. The bactericidal effect of AMX against *H. pylori* is also pH-dependent because AMX is more stable at a higher intragastric pH (Furuta et al. 2007). Moreover, *H. pylori* replicates when the intragastric pH increases to over 6, and thus become susceptible to AMX. On the contrary, the bacteria move into a non-replicative but viable state when the pH is less than 6 (Scott et al. 1998). Higher PPI dose increases antimicrobial effectiveness by maintaining a high pH level and it also improves the stability and bioavailability of AMX in gastric juice (Erah et al. 1997).

Kwack and co-workers (2016) reported the high doses of the PPI, liaprazole (80 mg per day) and AMX 3000 mg per day for 14 days as the first line therapy and the cure rate was 79.3%. A meta-analysis of dual therapy as a rescue therapy for *H. pylori* compared to other rescue therapies found eradication rates were almost same, 81.3% vs. 81.5% (Gao et al. 2016). Compliance and adverse events were also same rates compared with others, so some guidelines this regimen is recommended as a rescue treatment.

2.3.1 Vonoprazan

As noted above, vonoprazan is the new P-CAB. The inhibitory effect (pKa 9.4) is largely unaffected by ambient pH and it accumulates in parietal cells under both secreting and resting conditions (Hori et al. 2011). PPIs require 3 or more days to reach full anti-secretory effectiveness, whereas vonoprazan essentially achieves full effectiveness on the first day. Vonoprazan is currently approved in Japan for first-line *H. pylori* eradication with CLR-containing triple therapy and for second-line therapy with MTZ and AMX (Murakami et al. 2016). In the pivotal study, in CLR-susceptible infections both lansoprazole and vonoprazan containing CLR triple therapies cured almost 100% of *H. pylori* infections (Murakami et al. 2013). In contrast, with CLR-resistant strains, the resulting dual therapies (vonoprazan or lansoprazole plus AMX) were markedly different with lansoprazole- AMX curing 40% and vonoprazan-AMX curing 80% (Murakami et al. 2016). Thus, in a comparative trial in the presence of CLR resistance vonoprazan-containing triple therapy will "appear" superior to lansoprazole triple therapy (Jung et al. 2017). However, considered from the prospective of antimicrobial misuse, since 80% of those receiving CLR would have been cured if the CLR had been omitted this regimen utilizes about 3000 kg of unneeded CLR per million cases treated. Clearly, use of vonoprazan in CLR triple therapy needs to be rethought by the Japanese government. (Table 1)

2.3.2 Probiotics

Probiotic supplementation is designed to alter the microbiome and hopefully improve the outcome of *H. pylori* therapy and also reduce side effects of antibiotic therapy such as diarrhea. The interest in probiotics therapy as an adjunct to eradication therapy has resulted in an increasing number of publications and meta-analyses as discussed in detail in Chap. 14 of this book. For example, one recent study reported that the addition of a probiotic reduced the frequency of adverse events

Table 1 Drug commonly used for *H. pylori* eradication[a]

Amoxicillin
Bismuth (subcitrate of subsalicylate)
Clarithromycin (macrolides)
Metronidazole/Tinidazole
Tetracycline HCl
Fluoroquinolones (levofloxacin)
Rifabutin
Proton pump inhibitors

[a]Adapted from El-Serag et al. (2018)

from 28.2% to 12.2% (Jung et al. 2018). Oh and co-workers (2016) analyzed the microbiome of patients receiving probiotics compared to those receiving antimicrobial eradication therapy alone and found that although the microbiota were similar, as assessed by metagenomes sequencing, there was a greater proportional shift in functional gene families in those receiving antibiotics compared to those receiving the probiotic (Medilac-S®; *Streptococcus faecium* 9×10^8 and *Bacillus subtilis* 1×10^8). When used with bismuth quadruple therapy, probiotic supplementation resulted in an improvement in eradication rates (92.1% vs. 63.2%) (Shafaghi et al. 2016). Another study reported that the combination of *Bacillus mesentericus, Clostridium butyricum*, and *Streptococcus faecalis* was reported to be the optimal probiotic regime for reducing side effects and improving eradication rates when used to supplement the 14-day triple therapy (Wen et al. 2017). A systematic review analyzing 30 Randomized Control Trials involving 4302 patients reported that the addition of probiotics increased eradication rates by 12.2% analyzed as-per-protocol (PP) and 14.1% intention-to-treat (ITT) (Lau et al. 2016; McFarland et al. 2018). However, the effective strains that produce benefits of increase eradication rate and decrease of side effect have not been established. Overall, the role of probiotics is unclear. And, consensus groups have generally no recommended probiotics (Malfertheiner et al. 2017).

3 Recent Guidelines and Consensus Reports

Since the 1990s, different national and international guidelines for the management of *H. pylori*-related diseases have been published and periodically updated regarding indications for treatment, diagnostic procedures, and preferred treatment regimens (Table 2). All now agree that *H. pylori* is an important human pathogen and whenever feasible, and all with the infection should be offered curative therapy irrespective of whether they currently had clinical manifestation of the infection. This option is currently only practiced in developed countries.

A number of recent major consensus conferences have been published (Chey et al. 2017; El-Serag et al. 2018; Fallone et al. 2016; Liu et al. 2018; Mahachai et al. 2018; Malfertheiner et al. 2017; Sugano et al. 2015). The Kyoto consensus included the designation of *H. pylori* gastritis as an infectious disease with recommendation of treatment for all *H. pylori* infected subjects. This was the first to codify the apparent paradigm shift advocating treatment be no longer reserved for patients with clinical manifestations of the infection. This was followed by the Houston consensus on diagnosis of *H. pylori* infection, which updated indication for considering diagnostic testing (El-Serag et al. 2018). All of the recent consensus conferences recognized the problem of increasing resistance and failure of commonly prescribed regimens. However, most do not engage issues such as rational use of antibiotics, antimicrobial stewardship, the critical role of susceptibility testing and of supplying updated regional, local, or patient-specific susceptibility data. Rather, most guidelines provide sophisticated meta-analyses examining the outcome of different regimens done in regions with variable, often high rates of resistance, for which the prevalence and effects of resistance was often ignored (Graham et al. 2017). This approach resulted in comparisons between incomparable groups not receiving optimal regimens but rather receiving regimes including one or more of the antimicrobials to which drug resistance was induced by *H. pylori* (Chey et al. 2017; Fallone et al. 2016). Many compared regimens proven to be highly successful for treatment of adherent patients with susceptible infections that yielded unacceptably poor cure rates without comments regarding why. Overall, other than the recommendation longer no use certain combinations, most of the conclusions have been of limited value for their expressed purpose of providing up-to-date guidance.

Table 2 Recommended regimens for *Helicobacter pylori* eradication

Treatment	Drugs, dosages and duration
Susceptibility-based	**No drug allergies**
Clarithromycin triple therapy (susceptible to clarithromycin)	Amoxicillin (1 g) and clarithromycin (500 mg) plus a PPI all given twice daily for 14 days (40 mg omeprazole equivalent per dose)
Metronidazole triple therapy (susceptible to metronidazole)	Amoxicillin (1 g) and metronidazole or tinidazole (500 mg) plus a PPI all given twice daily for 14 days (40 mg omeprazole equivalent per dose)
Fluoroquinolone triple therapy (susceptible to fluoroquinolones)	Fluoroquinolone (e.g. levofloxacin 500 mg once daily), plus a PPI and amoxicillin 1 g twice daily for 14 days (40 mg omeprazole equivalent per dose)
Susceptibility-based	**Allergic to penicillin**
Susceptible to clarithromycin and metronidazole	Clarithromycin (500 mg), and metronidazole or tinidazole (500 mg) plus a PPI (40 mg omeprazole equivalent per dose) all given twice daily for 14 days
Resistant to clarithromycin and/or metronidazole	Bismuth quadruple therapy (see **susceptibility testing unavailable**)
Empiric therapies	**Susceptibility testing unavailable**
Bismuth quadruple therapy	Bismuth subsalicylate or bismuth subcitrate 2 tablets 2 or 4 times daily after meals plus tetracycline hydrochloride (500 mg) 4 times daily with meals and at bedtime plus metronidazole (400 or 500 mg) 4 times daily with meals and a PPI twice daily for 14 days.
Prepackaged bismuth quadruple therapy	PYLERA for 14 days; add a PPI b.i.d. (20 mg to 40 mg omeprazole equivalent twice a day)
New bismuth quadruple therapy (amoxicillin replaces tetracycline)	Bismuth 2 tablets 2 or 4 times daily after meals plus metronidazole (400 or 500 mg) four times daily with meals and amoxicillin 1 g three times daily along with a PPI (40 mg omeprazole equivalent or more twice a day) for 14 days.
New bismuth quadruple therapy (amoxicillin replaces metronidazole)	Bismuth 2 tablets 2 or 4 times daily after meals plus tetracycline HCl 500 mg four times daily with meals and amoxicillin 1 g three times daily along with a PPI (40 mg omeprazole equivalent or more twice a day) for 14 days.
Furazolidone quadruple therapy	Furazolidone therapies are obtained by replacing metronidazole in bismuth quadruple therapies with furazolidone 100 mg three times daily
Empiric likely effective regimens	
Rifabutin triple therapy	Rifabutin (150 mg once or twice daily), amoxicillin (1 g three times daily and omeprazole 40 mg (or an equivalent PPI) every 8 h for 14 days.
Rifabutin bismuth therapy	Add bismuth subcitrate or subsalicylate 2 tablets twice daily to above therapy
Experimental regimens	
High dose PPI-amoxicillin dual therapy	PPI (e.g. rabeprazole 40 mg, esomeprazole 40 mg) plus amoxicillin (500–750 mg) all four times daily at approximately 6 h intervals for 14 days (can use 8 h interval at night)
Vonoprazan-amoxicillin dual therapy	Vonoprazan, the potassium competitive acid blocker, 20 mg twice a day plus 500–750 mg amoxicillin every 6 h for 14 days is recommended.
Obsolete regimens	**Sequential, hybrid, concomitant therapies, empiric use of triple therapies**

Preferred PPI's: 40 mg omeprazole, 60 mg lansoprazole, 20 mg rabeprazole or esomeprazole; pantoprazole not recommended as 40 mg = 9 mg omeprazole

4 Modern Approach to Therapy

In some countries, there are constraints on which anti-*H. pylori* therapies can be used. These constraints include governmental restrictions on which regimes are approved for reimbursement and restrictions may include drugs, doses, and treatment duration. In some regions some commonly recommended antimicrobials and anti-secretory drugs are not approved for use (examples include bismuth, furazolidone, sitafloxacin, and vonoprazan) or the doses may be restricted, all of which limits the range of available therapies. Approved drugs may also not be available because of shortage or high local cost (e.g., tetracycline in the USA). Finally, the approved regimens may differ from the previously recommended regimens. For example, the bismuth quadruple formulation, Pylera, is packaged for 10 day therapy despite the recommendation, that in the presence of MTZ resistance, it can be given for 14 days. None of these restrictions are consistent with the principles of antimicrobial stewardship, which includes promoting the use of the optimal drug regimes, including drugs, dosing, duration of therapy, and routes of administration as well as take measure to ensure sustainable access to effective therapy (i.e., to prevent development of resistance) (Dyar et al. 2017).

5 Recommended Therapies

Rational antimicrobial therapy is always susceptibility-based and regulatory and government agencies should be encouraged to institute *H. pylori* resistance surveillance programs to provide up-to-date regional and local resistance prevalence reports and treatment guidelines to the physicians for better treatment of their patients. The regulatory agents should only approve highly effective therapies, which should include data about effects of resistances. Regions with different antimicrobial susceptibility patterns may need to promote different strategies.

A general strategy would be to use a susceptibility-based therapy. The order of recommended therapy with susceptible infections is based on using CLR, first because it has the lowest MICs and is generally well tolerated. Levofloxacin is used last, because of the increasing concerns regarding side-effect reflected in Black-box warnings. The recommendations should provide regimens for multi-resistant infections or for those where susceptibility testing is unavailable or refused.

The first rule is to use only proven locally effective therapies (i.e., those that reliably provide cure rates ≥90%, preferably ≥95% with patients who are adherent to the therapy). This approach relies on antimicrobials, where resistance is either rare (AMX, bismuth, tetracycline, furazolidone, rifabutin) or can be overcome (MTZ) (Fig. 1). Because of patient intolerances, drug allergies, and local experiences, the clinician should have at least two options for first-line therapy. As with any antimicrobial therapy, a thorough review of prior antibiotic use is invaluable to identify the presence of probably resistance. For example, prior use of a macrolide or quinolone makes resistance to clarithromycin and levofloxacin highly likely. The second key is patient education regarding potential and expected side-effects and the importance of completing the course of antibiotics.

5.1 Approach after Treatment Failure

Treatment failures will occur. One of the most common causes of treatment failure is when the drugs are not taken as prescribed. The problem can lie with the physician for not providing adequate instruction (to the patients, or because of intolerable side effects. With well-chosen treatment and patient education, treatment failures should be rare (e.g., <5%). Another option for failure is the presence of an unusual resistance such as to AMX. For example, if the region is one with rare CLR resistance and CLR triple therapy is the preferred first choice and is given empirically, failure suggests the patient had preexisting CLR resistance. Where available, the best approach of choosing a second choice therapy is to perform susceptibility testing. If unavailable, a second choice would utilize a different regimen. For example, if the patient was prescribed traditional bismuth quadruple therapy with bismuth, MTZ, tetracycline and a PPI, one could substitute AMX 1 g three times per day for the MTZ. Two failures require susceptibility testing.

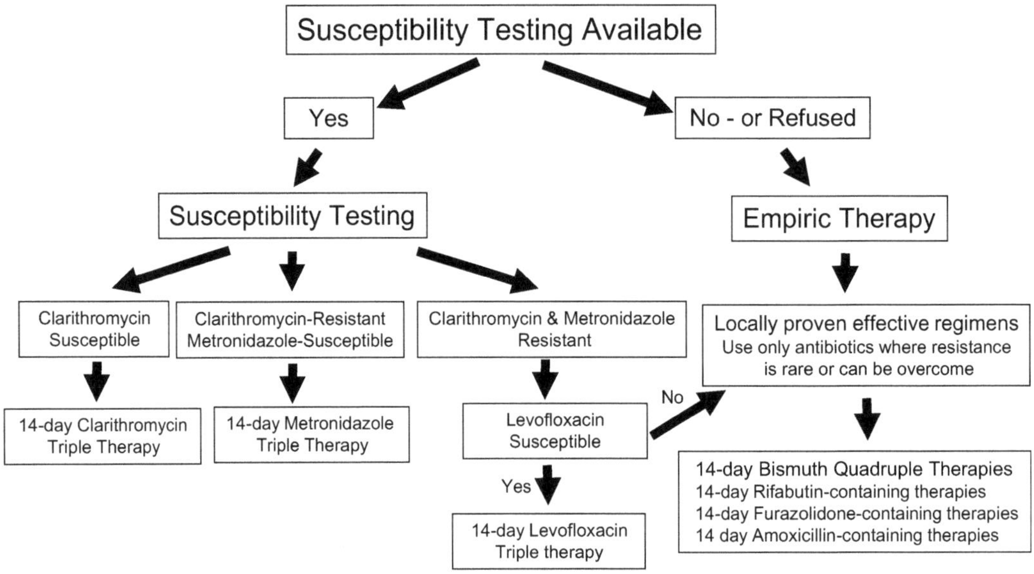

Fig. 1 Strategy for incorporation of susceptibility-based therapy against *H. pylori* into practice
The best tolerated and most effective regimens should be used first. For multidrug resistant infections or in instances when therapy must be chosen empirically only regimens proven to be highly effective should be chosen and should emphasize use of drugs for which resistance is rare

5.1.1 Vaccination

Efforts to develop an effective vaccine against *H. pylori* began in the early 1990s (Del Giudice et al. 2009; Salama et al. 2013, see also Chap. 15 of this book). Findings have shown that protection against *H pylori* infection can be achieved both prophylactically and therapeutically in animal models. However, previous trials of *H pylori* vaccine candidates have all been at an early stage, with no real efficacy reported (Corthesy et al. 2005; Del Giudice et al. 2009; Malfertheiner et al. 2008). An oral recombinant *H pylori* vaccine using urease B subunit fused with heat-labile enterotoxin B subunit was developed by Third Military Medical University and Chongqing Kang Wei Biotechnology in China. The vaccine has been assessed in phase 1 and phase 2 clinical trials for preliminary safety, immunogenicity, and optimum dose (unpublished). Zeng and co-workers (2015) reported that oral administration with the *H. pylori* vaccine provided good protection against the infection in children aged 6–15 years up to 1 year after vaccination. Although a previous study of the vaccine's protectiveness showed a mild warning of efficacy, overall protection was sustained up to 3 years. Other recent trials have failed (Zeng et al. 2015). For example, Malfertheiner et al. (2018) reported intramuscular immunization with a vaccine composed of three recombinant *H. pylori* antigens-vacuolating cytotoxin A (VacA), cytotoxin-associated antigen (CagA), and neutrophil-activating protein (NAP)-prevented infection in animal models and was well tolerated and highly immunogenic in healthy adults. However, compared with placebo, the vaccine did not confer additional protection against *H pylori* infection after challenge with a CagA-positive strain, despite increased systemic humoral responses to *H. pylori* antigens. Clearly, effective vaccination is needed if the problem of very high prevalence of *H. pylori* infection in developing countries is to be solved.

6 Conclusions

There have been several studies in the past year evaluating novel treatment options for *H. pylori.* Debraekeleer and Remaut (2018) reported the future perspective for potential *H. pylori* eradication therapies. *H. pylori* urease has been at the center of attention for many years for the development of more narrow-spectrum treatment or treatment supplements, and several potentially *in vitro* inhibitors have been found. Nevertheless, many suffer from a lack of specificity. The only marketed bacterial urease inhibitor, Acetohydroxamic acid (Lithstat), is approved only as an orphan drug for use in struvite inducing chronic urinary tract infection caused by urea-splitting pathogens. It is not advised for *H. pylori* eradication due to its many and frequent side effects. Two approved and marketed mucolytic agents, erdosteine and N-acetylcysteine (NAC), have suggested to increase *H. pylori* eradication efficiency in clinical trials when given in supplement with triple therapy (Yoon et al. 2016). Nevertheless, they have not made general use due to the high dosage required, increased cost of treatment, the additional patient discomfort, and increased risk for bleeding peptic ulcers associated with mucolytic agents.

An intervolin (anti-tumor) derivative, AS-1934, was found to have selective anti-*H. pylori* activity, including against antibiotic-resistant strains, without any effect on intestinal bacteria (Ohishi et al. 2018). There have been innumerable *in vitro* studies of agents that inhibit *H. pylori* and some animal studies but almost no human studies. Jeong and co-workers (2018) reported the efficacy of gentamicin-intercalated smectite hybrid-based treatment regimens in the murine model. Kouitcheu Mabeku et al. (2017) reported that *Bryophyllum pinnatum*, a medical plant with antioxidant and antimicrobial properties, could inhibit *H. pylori* growth and also protects gastric mucosa against reactive oxygen species. An Iran group showed the standard triple therapy with curcumin (a turmeric extract) increased *H. pylori* eradication rates and reduced endoscopic inflammation score (Judaki et al. 2017). A study of an Egyptian group suggested high eradication rates with nitazoxanide, which is a very expensive anti-infective drug against protozoa and anaerobic bacteria but also *H. pylori*. When MTZ was replaced with nitazoxanide in triple therapy, the reported eradication rates were 94.6% compared with 60.6% (Shehata et al. 2017).

The future will likely include a new generation PPI plus AMX dual therapy. As noted above, this regimen has been investigated since 1989 (Unge et al. 1989) with variable success (reviewed by Dore et al. 2016). The keys to making this regimen appear to include AMPC dosage and the ability to reliably maintain a high intragastric pH. The introduction of a new class of PPIs, the P-CABs, suggests that this may be possible (Dore et al. 2016; Graham and Dore 2018). The details of therapy have yet to be worked out in terms of doses and durations.

Acknowledgements Dr. Shiotani received a research grant and lecture fees from Takeda.
Pharmaceutical Co. Ltd. and Otsuka Pharmaceutical Co., Ltd.. Dr. Graham is in part by the Research Service Department of Veterans Affairs and by Public Health Service grant DK56338 which funds the Texas Medical Center Digestive Diseases Center.

References

An Y, Wang Y, Wu S, Wang YH, Qian X, Li Z, Fu YJ, Xie Y (2018) Fourth-generation quinolones in the treatment of *Helicobacter pylori* infection: a meta-analysis. World J Gastroenterol 24:3302–3312. https://doi.org/10.3748/wjg.v24.i29.3302

Barbhaiya R, Thin RN, Turner P, Wadsworth J (1979) Clinical pharmacological studies of amoxycillin: effect of probenecid. Br J Vener Dis 55:211–213

Bastos J, Peleteiro B, Barros R, Alves L, Severo M, de Fatima PM, Pinto H, Carvalho S, Marinho A, Guimaraes JT, Azevedo A, La Vecchia C, Barros H, Lunet N (2013) Sociodemographic determinants of prevalence and incidence of *Helicobacter pylori* infection in Portuguese adults. Helicobacter 18:413–422. https://doi.org/10.1111/hel.12061

Bruce MG, Maaroos HI (2008) Epidemiology of *Helicobacter pylori* infection. Helicobacter 13(Suppl 1):1–6. https://doi.org/10.1111/j.1523-5378.2008.00631.x

Chen Y, Blaser MJ (2012) Association between gastric *Helicobacter pylori* colonization and glycated hemoglobin levels. J Infect Dis 205:1195–1202. https://doi.org/10.1093/infdis/jis106

Chen MJ, Chen CC, Chen YN, Chen CC, Fang YJ, Lin JT, Wu MS, Liou JM, Taiwan Gastrointestinal Disease and Helicobacter Consortium (2018) Systematic review with meta-analysis: concomitant therapy vs. triple therapy for the first-line treatment of *Helicobacter pylori* infection. Am J Gastroenterol. https://doi.org/10.1038/s41395-018-0217-2

Chey WD, Leontiadis GI, Howden CW, Moss SF (2017) ACG clinical guideline: treatment of *Helicobacter pylori* infection. Am J Gastroenterol 112:212–239. https://doi.org/10.1038/ajg.2016.563

Corthesy B, Boris S, Isler P, Grangette C, Mercenier A (2005) Oral immunization of mice with lactic acid bacteria producing *Helicobacter pylori* urease B subunit partially protects against challenge with *Helicobacter felis*. J Infect Dis 192:1441–1449. https://doi.org/10.1086/444425

Debraekeleer A, Remaut H (2018) Future perspective for potential *Helicobacter pylori* eradication therapies. Future Microbiol 13:671–687. https://doi.org/10.2217/fmb-2017-0115

Del Giudice G, Malfertheiner P, Rappuoli R (2009) Development of vaccines against *Helicobacter pylori*. Expert Rev Vaccines 8:1037–1049. https://doi.org/10.1586/erv.09.62

Dore MP, Piana A, Carta M, Atzei A, Are BM, Mura I, Massarelli G, Maida A, Sepulveda AR, Graham DY, Realdi G (1998) Amoxycillin resistance is one reason for failure of amoxycillin-omeprazole treatment of *Helicobacter pylori* infection. Aliment Pharmacol Ther 12:635–639

Dore MP, Lu H, Graham DY (2016) Role of bismuth in improving *Helicobacter pylori* eradication with triple therapy gut. https://doi.org/10.1136/gutjnl-2015-311019

Dyar OJ, Huttner B, Schouten J, Pulcini C (2017) What is antimicrobial stewardship? Clin Microbiol Infect 23:793–798. https://doi.org/10.1016/j.cmi.2017.08.026

El-Serag HB, Kao JY, Kanwal F, Gilger M, LoVecchio F, Moss SF, Crowe S, Elfant A, Haas T, Hapke RJ, Graham DY (2018) Houston consensus conference on testing for *Helicobacter pylori* infection in the United States. Clin Gastroenterol Hepatol. https://doi.org/10.1016/j.cgh.2018.03.013

Erah PO, Goddard AF, Barrett DA, Shaw PN, Spiller RC (1997) The stability of amoxycillin, clarithromycin and metronidazole in gastric juice: relevance to the treatment of *Helicobacter pylori* infection. J Antimicrob Chemother 39:5–12

Fallone CA, Chiba N, van Zanten SV, Fischbach L, Gisbert JP, Hunt RH, Jones NL, Render C, Leontiadis GI, Moayyedi P, Marshall JK (2016) The Toronto consensus for the treatment of *Helicobacter pylori* infection in adults. Gastroenterology 151:51–69.e14. https://doi.org/10.1053/j.gastro.2016.04.006

Fujimoto Y, Furusyo N, Toyoda K, Takeoka H, Sawayama Y, Hayashi J (2007) Intrafamilial transmission of *Helicobacter pylori* among the population of endemic areas in Japan. Helicobacter 12:170–176. https://doi.org/10.1111/j.1523-5378.2007.00488.x

Furuta T, Shirai N, Kodaira M, Sugimoto M, Nogaki A, Kuriyama S, Iwaizumi M, Yamade M, Terakawa I, Ohshi K, Ishizaki T, Hishida A (2007) Pharmacogenomics-based tailored versus standard therapeutic regimen for eradication of *H. pylori*. Clin Pharmacol Ther 81:521–528. https://doi.org/10.1038/sj.clpt.6100043

Gao CP, Zhou Z, Wang JZ, Han SX, Li LP, Lu H (2016) Efficacy and safety of high-dose dual therapy for *Helicobacter pylori* rescue therapy: a systematic review and meta-analysis. J Dig Dis 17:811–819. https://doi.org/10.1111/1751-2980.12432

Grad YH, Lipsitch M, Aiello AE (2012) Secular trends in *Helicobacter pylori* seroprevalence in adults in the United States: evidence for sustained race/ethnic disparities. Am J Epidemiol 175:54–59. https://doi.org/10.1093/aje/kwr288

Graham DY (2015) *Helicobacter pylori* update: gastric cancer, reliable therapy, and possible benefits. Gastroenterology 148:719–731.e713. https://doi.org/10.1053/j.gastro.2015.01.040

Graham DY, Dore MP (2018) Update on the use of vonoprazan: a competitive acid blocker. Gastroenterology 154:462–466. https://doi.org/10.1053/j.gastro.2018.01.018

Graham DY, Fischbach L (2010) *Helicobacter pylori* treatment in the era of increasing antibiotic resistance. Gut 59:1143–1153. https://doi.org/10.1136/gut.2009.192757

Graham DY, Lee SY (2015) How to effectively use bismuth quadruple therapy: the good, the bad, and the ugly. Gastroenterol Clin N Am 44:537–563. https://doi.org/10.1016/j.gtc.2015.05.003

Graham DY, Shiotani A (2008) New concepts of resistance in the treatment of *Helicobacter pylori* infections. Nat Clin Pract Gastroenterol Hepatol 5:321–331. https://doi.org/10.1038/ncpgasthep1138

Graham DY, Tansel A (2018) Interchangeable use of proton pump inhibitors based on relative potency. Clin Gastroenterol Hepatol 16:800–808.e807. https://doi.org/10.1016/j.cgh.2017.09.033

Graham DY, Lee YC, Wu MS (2014) Rational *Helicobacter pylori* therapy: evidence-based medicine rather than medicine-based evidence. Clin Gastroenterol Hepatol 12:177–186.e173;. discussion e112–e173. https://doi.org/10.1016/j.cgh.2013.05.028

Graham DY, Lu H, Shiotani A (2017) Failure of optimized dual proton pump inhibitor amoxicillin therapy: what now? Saudi J Gastroenterol 23:265–267. https://doi.org/10.4103/sjg.SJG_292_17

Graham DY, Dore MP, Lu H (2018) Understanding treatment guidelines with bismuth and non-bismuth quadruple *Helicobacter pylori* eradication therapies. Expert Rev Anti-Infect Ther 16:679–687. https://doi.org/10.1080/14787210.2018.1511427

Hori Y, Matsukawa J, Takeuchi T, Nishida H, Kajino M, Inatomi N (2011) A study comparing the antisecretory effect of TAK-438, a novel potassium-competitive acid blocker, with lansoprazole in animals. J Pharmacol Exp Ther 337:797–804. https://doi.org/10.1124/jpet.111.179556

Hsu PI, Tsai FW, Kao SS, Hsu WH, Cheng JS, Peng NJ, Tsai KW, Hu HM, Wang YK, Chuah SK, Chen A, Wu DC (2017) Ten-day quadruple therapy comprising proton pump inhibitor, bismuth, tetracycline, and levofloxacin is more effective than standard levofloxacin triple therapy in the second-line treatment of *Helicobacter pylori* infection: a randomized controlled trial. Am J Gastroenterol 112:1374–1381. https://doi.org/10.1038/ajg.2017.195

Hunt RH, Xiao SD, Megraud F, Leon-Barua R, Bazzoli F, van der Merwe S, Vaz Coelho LG, Fock M, Fedail S, Cohen H, Malfertheiner P, Vakli N, Hamid S, Goh KL, Wong BC, Krabshuis J, Le Mair A, World Gastroenterology Organization (2011) *Helicobacter pylori* in developing countries. World gastroenterology organisation global guideline. J Gastrointestin Liver Dis: JGLD 20:299–304

Jeong SJ, Kim JH, Jung DH, Lee KH, Park SY, Song Y, Kang IM, Song YG (2018) Gentamicin-intercalated smectite as a new therapeutic option for *Helicobacter pylori* eradication. J Antimicrob Chemother 73:1324–1329. https://doi.org/10.1093/jac/dky011

Judaki A, Rahmani A, Feizi J, Asadollahi K, Hafezi Ahmadi MR (2017) Curcumin in combination with triple therapy regimes ameliorates oxidative stress and histopathologic changes in chronic gastritis-associated *Helicobacter Pylori* infection Arquivos de Gastroenterologia 54:177–182 doi:https://doi.org/10.1590/s0004-2803.201700000-18

Jung YS, Kim EH, Park CH (2017) Systematic review with meta-analysis: the efficacy of vonoprazan-based triple therapy on *Helicobacter pylori* eradication. Aliment Pharmacol Ther 46:106–114. https://doi.org/10.1111/apt.14130

Jung JH, Cho IK, Lee CH, Song GG, Lim JH (2018) Clinical Outcomes of Standard Triple Therapy Plus Probiotics or Concomitant Therapy for Helicobacter pylori Infection. Gut Liver 12: 165–172.

Kamada T, Haruma K, Ito M, Inoue K, Manabe N, Matsumoto H, Kusunoki H, Hata J, Yoshihara M, Sumii K, Akiyama T, Tanaka S, Shiotani A, Graham DY (2015) Time trends in *Helicobacter pylori* infection and atrophic gastritis over 40 years in Japan. Helicobacter, 20:192–198. https://doi.org/10.1111/hel.12193

Kobayashi I, Murakami K, Kato M, Kato S, Azuma T, Takahashi S, Uemura N, Katsuyama T, Fukuda Y, Haruma K, Nasu M, Fujioka T (2007) Changing antimicrobial susceptibility epidemiology of *Helicobacter pylori* strains in Japan between 2002 and 2005. J Clin Microbiol 45:4006–4010. https://doi.org/10.1128/jcm.00740-07

Kouitcheu Mabeku LB, Eyoum Bille B, Tchouangueu TF, Nguepi E, Leundji H (2017) Treatment of *Helicobacter pylori* infected mice with *Bryophyllum pinnatum*, a medicinal plant with antioxidant and antimicrobial properties, reduces bacterial load. Pharm Biol 55:603–610. https://doi.org/10.1080/13880209.2016.1266668

Kuo YT, Liou JM, EL-Omar, Wu JY, Leow AHR, Goh KL, Das R, Lu H, Lin JT, Tu YK, Yamaoka Y, Wu MS (2017) Primary antibiotic resistance in *Helicobacter pylori* in the Asia-Pacific region: a systematic review and meta-analysis Lancet Gastroenterol Hepatol 2:707–715 doi:https://doi.org/10.1016/s2468-1253(17)30219-4

Kwack W, Lim Y, Lim C, Graham DY (2016) High dose ilaprazole/amoxicillin as first-line regimen for *Helicobacter pylori* infection in Korea. Gastroenterol Res Pract 2016:1648047. https://doi.org/10.1155/2016/1648047

Kwok T, Backert S, Schwarz H, Berger J, Meyer TF (2002) Specific entry of *Helicobacter pylori* into cultured gastric epithelial cells via a zipper-like mechanism. Infect Immun 70:2108–2120

Laine L, Suchower L, Frantz J, Connors A, Neil G (1998) Twice-daily, 10-day triple therapy with omeprazole, amoxicillin, and clarithromycin for *Helicobacter pylori* eradication in duodenal ulcer disease: results of three multicenter, double-blind, United States trials. Am J Gastroenterol 93:2106–2112. https://doi.org/10.1111/j.1572-0241.1998.00602.x

Lau CS, Ward A, Chamberlain RS (2016) Probiotics improve the efficacy of standard triple therapy in the eradication of *Helicobacter pylori*: a meta-analysis. Infect Drug Resist 9:275–289. https://doi.org/10.2147/idr.s117886

Liang CM, Cheng JW, Kuo CM, Chang KC, Wu KL, Tai WC, Chiu KW, Chiou SS, Lin MT, Hu TH, Chuah SK (2014) Levofloxacin-containing second-line anti-*Helicobacter pylori* eradication in Taiwanese real-world practice. Biomed J 37:326–330. https://doi.org/10.4103/2319-4170.125650

Liu WZ, Xie Y, Lu H, Cheng H, Zeng ZR, Zhou LY, Chen Y, Wang JB, Du YQ, Lu NH (2018) Fifth Chinese National Consensus Report on the management *of Helicobacter pylori* infection. Helicobacter 23:e12475. https://doi.org/10.1111/hel.12475

Macias-Garcia F, Baston-Rey I, de la Iglesia-Garcia D, Calvino-Suarez C, Nieto-Garcia L, Dominguez-Munoz JE (2018) Bismuth-containing quadruple therapy versus concomitant quadruple therapy as first-line treatment for *Helicobacter pylori* infection in an area of high resistance to clarithromycin: a prospective, cross-sectional, comparative, open trial. Helicobacter: e12546. https://doi.org/10.1111/hel.12546

Mahachai V, Vilachone RK, Pittayanon R, Rottayanon R, Rojborwonwitaya J, Leelaksolovong S, Mannerattanaporn M, Chotivitayatarakorn P, Treeprasertsuk S, Kositchaiwat C, Pisespongsa P, Mairiang P, Rani A, Leow A, Mya SM, Lee YC, Vannarath S, Rasachak B, Chakravuth O, Aung MM, Ang TL, Sollano JD, Trong Quach D, Sansak I, Wiwattanachang O, Hamosomburana P, Syam AF, Yamaoka Y, Fock KM, Goh KL, Sugano K, Graham DY (2018) *Helicobacter pylori* management in ASEAN: The Bangkok consensus report. J Gastroenterol Hepatol 33:37–56. https://doi.org/10.1111/jgh.13911

Malfertheiner P, Schultze V, Rosenkranz B, Kaufmann SH, Ulrichs T, Novicki D, Norelli F, Contorni M, Peppoloni S, Berti D, Tornese D, Ganju J, Palla E,

Rappuoli R, Scharschmidt BF, Del Giudice G (2008) Safety and immunogenicity of an intramuscular *Helicobacter pylori* vaccine in noninfected volunteers: a phase I study. Gastroenterology 135:787–795. https://doi.org/10.1053/j.gastro.2008.05.054

Malfertheiner P, Megrad F, O'Morain CA, Atherton J, Axon AT, Bazzoli F, Gensini GF, Gisbert JP, Graham DY, Rokkas T, El-Omar EM, Kuipers EJ (2012) Management of *Helicobacter pylori* infection--the Maastricht IV/Florence Consensus Report. Gut 61:646–664. https://doi.org/10.1136/gutjnl-2012-302084

Malfertheiner P, Megraud F, O'Morain CA, Gisbert JP, Kuiper EJ, Kuiper A, Axon AT, Bazzoli F, Gasbarini A, Atherton J, Graham DY, Hunt R, Moayyedi P, Rokkas T, Rugge M, Selgrad M, Suerbaum S, Sugano K, El-Omar EM (2017) Management of *Helicobacter pylori* infection-the Maastricht V/Florence Consensus report. Gut 66:6–30. https://doi.org/10.1136/gutjnl-2016-312288

Malfertheiner P, Selgrad M, Wex T, Romi B, Borgogni E, Spensieri F, Zedda L, Ruggiero P, Pancotto L, Censini S, Palla E, Kansesa-Thasan N, Scharschmidt B, Rappuoli R, Graham DY, Schiavetti F, Del Giudice G (2018) Efficacy, immunogenicity, and safety of a parenteral vaccine against *Helicobacter pylori* in healthy volunteers challenged with a cag-positive strain: a randomised, placebo-controlled phase 1/2 study. Lancet Gastroenterol Hepatol 3:698–707. https://doi.org/10.1016/s2468-1253(18)30125-0

McFarland LV, Evans CT, Goldstein EJC (2018) Strain-specificity and disease-specificity of probiotic efficacy: a systematic review and meta-analysis. Front Med 5:124. https://doi.org/10.3389/fmed.2018.00124

Mégraud F (2004) *H pylori* antibiotic resistance: prevalence, importance, and advances in testing. Gut 53:1374–1384. https://doi.org/10.1136/gut.2003.022111

Meyer JM, Silliman NP, Wang W, Siepman NY, Sugg JE, Morris D, Zhang J, Bhattacharyya H, King EC, Hopkins RJ (2002) Risk factors for *Helicobacter pylori* resistance in the United States: the surveillance of *H. pylori* antimicrobial resistance partnership (SHARP) study, 1993–1999. Ann Intern Med 136:13–24

Murakami K, Furuta T, Ando T, Nakajima T, Inui Y, Oshima T, Tomita T, Mabe K, Sasaki M, Suganuma T, Nomura H, Satoh K, Hori S, Inoue S, Tomokane T, Kudo M, Inaba T, Take S, Ohkusa T, Yamamoto S, Mizuno S, Kamoshida T, Amagi K, Iwamoto J, Miwa H, Kodama M, Okimoto T, Kato M, Asaka M (2013) Multi-center randomized controlled study to establish the standard third-line regimen for *Helicobacter pylori* eradication in Japan. J Gastroenterol 48:1128–1135. https://doi.org/10.1007/s00535-012-0731-8

Murakami K, Sakurai Y, Shiino M, Funao N, Nishimura A, Asaka M (2016) Vonoprazan, a novel potassium-competitive acid blocker, as a component of first-line and second-line triple therapy for *Helicobacter pylori* eradication: a phase III, randomised, double-blind study. Gut. https://doi.org/10.1136/gutjnl-2015-311304

Oh B, Kim BS, Kim JW, Kim JS, Koh SJ, Kim BG, Lee KL, Chun J (2016) The effect of probiotics on gut microbiota during the *Helicobacter pylori* eradication: randomized controlled trial. Helicobacter 21:165–174. https://doi.org/10.1111/hel.12270

Ohishi T, Masuda T, Abe H, Hayashi C, Adachi H, Ohba SI, Igarashi M, Watanabe T, Mimuro H, Amalia E, Inaoka DK, Mochizuki K, Kita K, Shibasaki M, Kawada M (2018) Monotherapy with a novel intervenolin derivative, AS-1934, is an effective treatment for *Helicobacter pylori* infection. Helicobacter. https://doi.org/10.1111/hel.12470

Puig I, Baylina M, Sánchez-Delgado J, López-Gongora S, Suarez D, García-Iglesias P, Muñoz N, Gisbert JP, Dacoll C, Cohen H, Calvet X (2016) Systematic review and meta-analysis: triple therapy combining a proton-pump inhibitor, amoxicillin and metronidazole for *Helicobacter pylori* first-line treatment. J AntimicrobChemother 71:2740–53.

Ruiter R, Wunderink HF, Veenendaal RA, Visser LG, de Boer MGJ (2017) *Helicobacter pylori* resistance in the Netherlands: a growing problem? Neth J Med 75:394–398

Salama NR, Hartung ML, Muller A (2013) Life in the human stomach: persistence strategies of the bacterial pathogen *Helicobacter pylori*. Nat Rev Microbiol 11:385–399. https://doi.org/10.1038/nrmicro3016

Sanchez Ceballos F, Taxonera Samso C, Garcia Alonso C, Alba Lopez C, Sainz de Los Terreros Soler L, Diaz-Rubio M (2007) Prevalence of *Helicobacter pylori* infection in the healthy population of Madrid (Spain). Revista espanola de enfermedades digestivas: organo oficial de la Sociedad Espanola de Patologia Digestiva 99:497–501

Savoldi A, Carrara E, Graham DY, Conti M, Tacconelli E (2018) Prevalence of antibiotic resistance in *Helicobacter pylori*: a systematic review and meta-analysis in World Health Organization regions. Gastroenterology. https://doi.org/10.1053/j.gastro.2018.07.007

Scott D, Weeks D, Melchers K, Sachs G (1998) The life and death of *Helicobacter pylori*. Gut 43(Suppl 1): S56–S60

Selgrad M, Malfertheiner P (2011) Treatment of *Helicobacter pylori*. Curr Opin Gastroenterol 27:565–570. https://doi.org/10.1097/MOG.0b013e32834bb818

Shafaghi A, Pourkazemi A, Khosravani M, Fakhrie Asl S, Amir Maafi A, Atrkar Roshan Z, Abaspour Rahimabad J (2016) The effect of probiotic plus prebiotic supplementation on the tolerance and efficacy of *Helicobacter pylori* eradication quadruple therapy: a randomized prospective double blind controlled trial. Middle East J Dig Dis 8:179–188. https://doi.org/10.15171/mejdd.2016.30

Shehata MA, Talaat R, Soliman S, Elmesseri H, Soliman S, Abd-Elsalam S (2017) Randomized controlled study of a novel triple nitazoxanide (NTZ)-

containing therapeutic regimen versus the traditional regimen for eradication of *Helicobacter pylori* infection. Helicobacter 22. https://doi.org/10.1111/hel.12395

Sugano K, Tack J, Kuiper EJ, Graham DY, EL-Omar EM, Miura S, Haruma K, Asaka M, Uemura N, Malfertheiner P (2015) Kyoto global consensus report on *Helicobacter pylori* gastritis. Gut 64:1353–1367. https://doi.org/10.1136/gutjnl-2015-309252

Thung I, Aramin H, Vavinskaya V, Gupta S, Park JY, Crowe SE, Valasek MA (2016) Review article; the global emergence of *Helicobacter pylori* antibiotic resistance aliment. Pharmacol Ther 43:514–533. https://doi.org/10.1111/apt.13497

Unge P, Gad A, Gnarpe H, Olsson J (1989) Does omeprazole improve antimicrobial therapy directed towards gastric *Campylobacter pylori* in patients with antral gastritis? A pilot study Scandinavian. J Gastroenterol Suppl 167:49–54

Vale FF, Vitor JM (2010) Transmission pathway of *Helicobacter pylori*: does food play a role in rural and urban areas? Int J Food Microbiol 138:1–12. https://doi.org/10.1016/j.ijfoodmicro.2010.01.016

Wen J, Peng P, Chen P, Zeng L, Pan Q, Wei W, He J (2017) Probiotics in 14-day triple therapy for Asian pediatric patients with *Helicobacter pylori* infection: a network meta-analysis. Oncotarget 8:96409–96418. https://doi.org/10.18632/oncotarget.21633

Williams MP, Pounder RE (1999) *Helicobacter pylori*: from the benign to the malignant. Am J Gastroenterol 94:S11–S16

Yang JC, Lin CJ (2010) CYP2C19 genotypes in the pharmacokinetics/pharmacodynamics of proton pump inhibitor-based therapy of *Helicobacter pylori* infection. Expert Opin Drug Metab Toxicol 6:29–41. https://doi.org/10.1517/17425250903386251

Yang JC, Wang HL, Chern HD, Shun CT, Lin BR, Lin CJ, Wang TH (2011) Role of omeprazole dosage and cytochrome P450 2C19 genotype in patients receiving omeprazole-amoxicillin dual therapy for Helicobacter pylori eradication. Pharmacotherapy 31:227–238. https://doi.org/10.1592/phco.31.3.227

Yang JC, Lin CJ, Wang HL, Chen JD, Kao JY, Shun CT, Lu CW, Lin BR, Shieh MJ, Chang MC, Chang YT, Wei SC, Lin LC, Yeh WC, Kuo JS, Tung CC, Leong YL, Wand TH, Wong JM (2015) High-dose dual therapy is superior to standard first-line or rescue therapy for *Helicobacter pylori* infection. Clin Gastroenterol Hepatol 13:895–905.e895. https://doi.org/10.1016/j.cgh.2014.10.036

Yoon H, Lee DH, Jang ES, Kim J, Shin CM, Park YS, Hwang JH, Kim JW, Jeong SH, Kim N (2016) Effects of N-acetylcysteine on first-line sequential therapy for *Helicobacter pylori* infection: a randomized controlled pilot trial. Gut Liver 10:520–525. https://doi.org/10.5009/gnl15048

Zeng M, Mao XH, Li JX, Tong WD, Wang B, Zhang YJ, Guo G, Zhao ZJ, Li L, Wu DL, Lu DS, Tan ZM, Liang HY, Wu C, Li DH, Luo P, Zeng H, Zhang WJ, Zhang JY, Guo BT, Zhu FC, Zou QC (2015) Efficacy, safety, and immunogenicity of an oral recombinant *Helicobacter pylori* vaccine in children in China: a randomised, double-blind, placebo-controlled, phase 3 trial. Lancet 386:1457–1464. https://doi.org/10.1016/s0140-6736(15)60310-5

Zhang J, Jin HC, Zhu AK, Ying RC, Wei W, Zhang FJ (2014) Prognostic significance of plasma chemerin levels in patients with gastric cancer. Peptides 61:7–11. https://doi.org/10.1016/j.peptides.2014.08.007

Zhang YX, Zhou LY, Song ZQ, Zhang JZ, He LH, Ding Y (2015) Primary antibiotic resistance of *Helicobacter pylori* strains isolated from patients with dyspeptic symptoms in Beijing: a prospective serial study. World J Gastroenterol 21:2786–2792. https://doi.org/10.3748/wjg.v21.i9.2786

Adv Exp Med Biol - Advances in Microbiology, Infectious Diseases and Public Health (2019) 1149: 227–241
https://doi.org/10.1007/5584_2019_368

Published online: 24 April 2019

Structural Aspects of *Helicobacter pylori* Antibiotic Resistance

Giuseppe Zanotti and Laura Cendron

Abstract

Resistance to antibiotics of *Helicobacter pylori* infections is growing rapidly together with the need for more potent antimicrobials or novel strategies to recover the efficacy of the existing ones. Despite the main mechanisms according to which *H. pylori* acquires resistance are common to other microbial infections affecting humans, *H. pylori* has its own peculiarities, mostly due to the unique conditions experienced by the bacterium in the gastric niche. Possibly the most used of the antibiotics for *H. pylori* are those molecules that bind to the ribosome or to the DNA and RNA machinery, and in doing so they interfere with protein synthesis. Another important class is represented by molecules that binds to some enzyme essential for the bacterium survival, as in the case of enzymes involved in the bacterial wall biosynthesis. The mechanism used by the bacterium to fight antibiotics can be grouped in three classes: (i) mutations of some key residues in the protein that binds the inhibitor, (ii) regulation of the efflux systems or of the membrane permeability in order to reduce the uptake of the antibiotic, and (iii) other more complex indirect effects. Interestingly, the production of enzymes that degrade the antibiotics (as in the case of β-lactamases in many other bacteria) has not been clearly detected in *H. pylori*. The structural aspects of resistance players have not been object of extensive studies yet and the structure of very few *H. pylori* proteins involved in the resistance mechanisms are determined till now. Models of the proteins that play key roles in reducing antimicrobials susceptibility and their implications will be discussed in this chapter.

Keywords

Bacterial resistance · Antibiotics · Resistance mechanism · Mixed resistance · Efflux pump

G. Zanotti (✉)
Department of Biomedical Sciences, University of Padua, Padua, Italy
e-mail: giuseppe.zanotti@unipd.it

L. Cendron
Department of Biology, University of Padua, Padua, Italy
e-mail: laura.cendron@unipd.it

1 Introduction

Like *Staphylococcus aureus*, *Campylobacter* spp., *Enterococcus faecium* and few other antibiotic resistant bacteria, *H. pylori* has been categorized as a high-priority target that pose the greatest threat to human health by WHO (http://www.who.int/medicines/publications/global-priority-list-antibiotic-resistant-bacteria/en/). Resistance to feasible antibiotics is growing rapidly worldwide, making *H. pylori* infections more difficult to cure (Alba et al. 2017). The classical therapy used by most physicians consists of a cocktail of two antibiotics, mostly chlaritromycin (CLR) and metronidazole (MTZ) or amoxicillin (AMX), and a proton pump inhibitor (PPI) or, in the case of eradication failure, of a quadruple therapy, either bismuth including or not (Fallone et al. 2016). The combination of multiple antibiotics is an empirical approach deriving from the assumption that resistance does not frequently affect two antibiotics of different classes at the same time, thus avoiding the usage of antimicrobials susceptibility tests that require 10 days or more to give a response. However, growing resistance incidence makes these approaches no longer capable of achieving high cure rates (Zhang 2015).

More details on the different protocols and novel strategies currently in use are discussed in a separate book chapter (See Chap. 12). While describing the main source of failure in the treatment, often due to bacterial resistance and therapeutic inefficacy, one should keep in mind that infection reoccurrence can also be the result of reinfections, especially in populations of developing countries where inappropriate sanitary conditions could play a significant role in this sense. At the same time, resistance spreading per se is sustained by inadequate hygiene and food handling and consequent recursive reinfections (Hu et al. 2017).

The main mechanisms according to which *H. pylori* acquire antibiotics resistances are common to other bacterial infections affecting humans and can be classified into the following main categories: mutations that impair the capability of antibiotics to interfere with ribosomal activity and protein synthesis; mutations that affect DNA replication, recombination, and transcription; mutations that alter the proper redox-state of bacterial cells altering the activity of oxidoreductases and mutations that modify penicillin binding proteins, involved in peptidoglycan biosynthesis and typical target of β-lactams activity. It should be pointed out in this context that a peculiar feature of *H. pylori* is the absence or very rarely detectable β-lactamases activity within the major features identified in resistant strains. This could be the consequence of the features of the highly variable environment colonized by the bacterium, given the fact that β-lactamases coding genes are acquired by horizontal transfer and, once translated, they are secreted in the periplasmic space to hydrolyze antibiotics.

Another major role in drugs tolerance is played by outer membrane porin and efflux systems. They strongly contribute to keep the toxic agent concentrations inside the bacterial cell lower than expected and make them less able in killing bacteria (Hirata et al. 2010).

Finally, when exploring the sources of reduced susceptibility to antibiotics, other relevant aspects that should be considered in *H. pylori* infections are the capability to oscillate from rod-shaped active bacteria to dormant resting coccoidal state, in response to antimicrobials, and the ability to penetrate and colonize gastric mucosa, forming biofilms on its surface (Yonezawa et al. 2015). Such large aggregates can not only protect the bacteria from the surrounding hostile environment, helping establishing a chronic infection, but also contribute to its reduced susceptibility to antimicrobials agents, if compared to planktonic organization. Indeed, a biofilm organization where bacteria coexist as multiple species, both dead and alive, strongly interconnected by an external matrix of mixed composition, implies a different gene expression profile, the activation of the so-called quorum sensing system for cell-to-cell communication and a signaling pattern of molecules acting as auto-inducers (Attaran et al. 2017). All these components can contribute to an altered response to eradication therapies, as demonstrated by Yonezawa and co-workers (Yonezawa et al. 2015).

The type of mutations responsible for the resistance are in general well known, since several resistant strains have been sequenced. On the opposite, the structural aspects of this resistance have not been object of extensive studies and the structure of very few *H. pylori* proteins involved in the resistance mechanisms have been determined. Fortunately, their amino acid sequences are relatively similar to those of other bacteria whose 3D structures are known and reliable molecular models can be built with mid to high degree of confidence. Their structures and the implications for the resistance mechanisms will be discussed in this chapter.

2 Inhibitors of the Protein Synthesis Through Interaction with the Ribosome Machinery

2.1 Resistance to Macrolide Clarithromycin (CLR)

CLR is a classical bacteriostatic agent adopted as first option in the eradication therapy of symptomatic *H. pylori* infections, in combination with metronidazole or AMX (Gong and Yuan 2018). It belongs to second generation 14-membered-ring macrolides, composed of three structural subgroups: the lactone ring, cladinose, and desosamine sugars. It derives from erythromycin and acts on a large spectrum of infections with good pharmacokinetic properties and relative safety.

CLA and other macrolides interfere with protein synthesis through reversible binding with nanomolar affinity to the peptidyl-transferase region (domain V) of the 23S rRNA, part of the bacterial ribosome subunit 50S. Macrolide-23S rRNA interaction blocks the peptide bond formation and peptidyl tRNA translocation from the A- to P-site. Further consequences described are the premature dissociation of peptidyl tRNA with the accumulation of truncated peptides.

The structural determinants of the inhibitory mechanism have been clarified by crystallographic studies of the complex of CLA and other macrolides with 50S subunit of *Deinococcus radiodurans* ribosome (PDB: 1J5A; Schlünzen et al. 2001). CLA evidenced a common binding mode to all the tested macrolides. Established interactions with specific nucleotides in the peptidyl transferase cavity can be assigned to a multi-branched loop of domain V of the 23S rRNA. No significant conformational changes are induced by macrolides binding.

CLR widespread intensive use, combined with the high frequency of reinfections especially in developing countries, has catalyzed the emergence of resistant strains and more in general reduced susceptibility towards macrolides, with occurrence rates higher than 20% in peculiar contexts (De Francesco et al. 2010).

The most frequently described mechanisms of resistance towards CLR imply the weakening of macrolide interactions, due to mutations occurring at three main positions in 23S-rRNA gene: A2142C, A2142G, and A2143G, the latter accounting for the majority of cases (Chen et al. 2018). Less frequent mutations in the same 23S subunit have been reported to be involved in CLR resistance, such as A2115G, T2117C, G2141A, A2144T, T2182C, G2223A, T2288C and T2711C, despite the roles covered by such alterations remain elusive and sometimes even controversial.

More recently, in an effort to clarify whether other proteins and/or ribosomal subunits could be responsible of higher MIC (minimum inhibitory concentration) values vs. CLR in *H. pylori* isolates, novel candidates have been detected and characterized by low CLR doses exposure of susceptible strains and next generation sequencing (Binh et al. 2014). The authors discovered that mutations in two candidate genes, *infB* and *rpl22*, confer resistance to *H. pylori* and have a synergic effect when coexisting with 23S point mutations. In particular, a point mutation (G160A) has been found to confer higher tolerance towards CLR in translation initiation factor IF-2, also called *infB* (HP1048). IF-2/InfB prevents hydrolysis of formylmethionyl-tRNA, promote the appropriate binding to the ribosomal subunit 30S, formation of Initiation Complex (IC) by joining 30S and 50S subunits to define the 70S IC and initiation of protein synthesis. More in general, IF-2 is one of the three components that assure appropriate velocity of IC formation and translation accuracy in bacteria (Wang et al. 2015).

The structure of IF-2/InfB, a 944 amino acid multidomain protein, can be inferred by homology modeling (Fig. 1b): the core region spanning residues 370–940 shows a high degree of similarity towards classical *Thermus thermopilus* IF-2

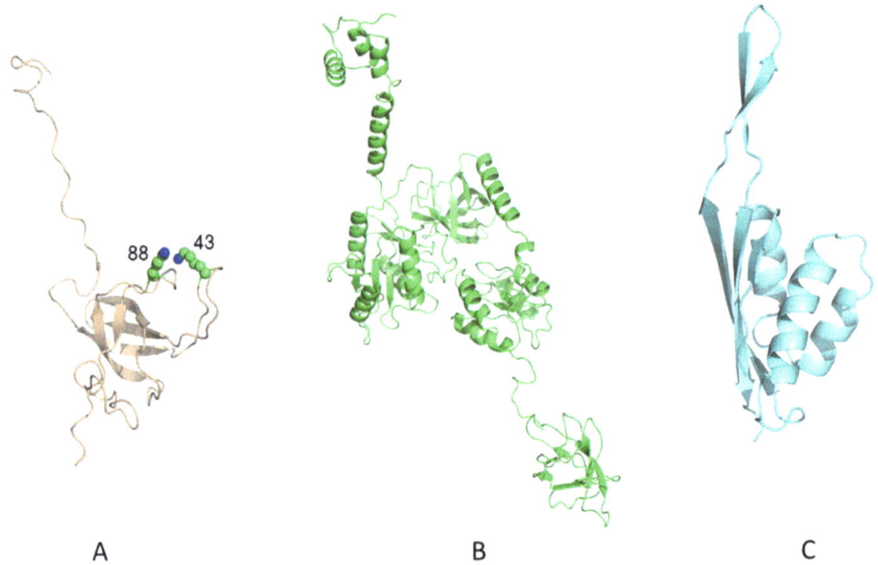

Fig. 1 **Inhibitors of the ribosome machinery**. (**a**) Cartoon view of Rpsl chain of the 30S ribosome subunit. Two mutations found in *H. pylori* as responsible for resistance to streptomycin are shown as small spheres. Both point mutations correspond to K → R replacements, a substitution that is normally not very significant (the only effect is that a positively charged side chains is replaced by a larger, positively charged one). Nevertheless, the effect is a misplacement of the bound *t*RNA in such a way to induce a codon misreading that impairs protein synthesis. (**b**) Cartoon view of the model of translation initiation factor IF-2 (InfB) from *H. pylori*. The homology modeling has been performed using Phyre 2 software (template structure PDB ID 3J4J). (**c**) Cartoon of 50S ribosomal protein L22. The model was obtained by homology with structure 3jw3 (Phyre2)

(PDB: 3J4J, Simonetti et al. 2013), *Escherichia coli* (PDB: 1ZO1; Allen et al. 2005) or even mammalian one. Such portion includes a GTPase switch domain and undergoes large conformational changes upon binding to 30S subunit and proper engagement of 50S, as well as in the following steps, where IF-2 should get repositioned and finally released from the ribosome.

The N-terminus, where the resistance-conferring mutation occurs, does not share remarkable homologies with sequences of known structures. However, modeling tools suggest a coiled-coil arrangement of the first 200 amino acids, analogously to the N-terminus region of PCSB protein from *Streptococcus pneumoniae* (PDB: 4cgk; Bartual et al. 2014). PCSB is recruited at the septum by interacting with FtsEX at the N-terminus and is involved in septal cross-wall hydrolysis and cell division.

Multiple types of deletions (three base pairs or nine base pairs deletions) were also reported at the level of ribosomal protein L22 (Rpl22, HP1314), a structural constituent of the large ribosomal subunit interacting with all domains of 23S rRNA and located close to the exit tunnel where new polypeptide chains are assembled during protein synthesis. Remarkably, *rpl22* is known to be involved in the resistance to first generation macrolides in *E. coli*: deletions at positions Met82-Lys83-Arg84 in L22 are reported to cause resistance to erythromycin in *E. coli* (Zaman et al. 2007). A model of the *H. pylori* Rpl22 protein (Fig. 1c) can be easily obtained with high degree of confidence for the full-length sequence, using the same 50S ribosomal protein L22 from *Bacillus subtilis* as a template (PDB:3jw3, with an overall identity of 36%). L22 is a 120 amino acids subdomain with a mixed alpha-beta structure and a protruding beta-hairpin that confers an elongated shape to such subunit.

The reduced strain susceptibility conferred by detected deletions of L22 subdomain can be explained more by an indirect effect on the affinity of the antibiotic with 50S ribosome subunit than an impact on direct interactions of the macrolide with L22 domain. Indeed, the observed binding and orientation of CLR and other macrolides at the entrance of the tunnel of 50S subunit do not involve direct interactions with L22 residues, with minimal distances never less than 8 Å.

Finally, many studies in the context of macrolides resistance underlined the importance of intrinsic contribution of efflux systems to CLR resistance in *H. pylori.* In particular, the multi-component HefABC (HP0605/606/607) efflux pump was observed to support the strongest contribution to CLR resistance, given its higher level of expression over the different RND (Resistance Nodulation-Division) transporters observed in CLR-resistant *H. pylori* strains. Being HefABC a multidrug efflux pump, its structural and functional features and their implication in the resistance toward chemotherapeutic agents will be described in details later in this chapter (see Sect. 6).

2.2 Resistance to Tetracyclines

Tetracyclines (TETs) are currently used in second or third-line therapies for *H. pylori* eradication, where either AMX, CLR or MTZ fail to be effective. Extensive usage of TETs in the past decades has severely limited the usefulness of this class of therapeutics nowadays. However, in some developing countries cost-effectiveness considerations still imply the usage of TETs in first-line therapies (Dunn et al. 1997).

Given the limited usage, *H. pylori* TET resistance is less frequently observed in most countries, with reported rates around 2% in several studies, but much higher peaks (>10%) in specific countries and a general tendency to increase through the time (Suzuki et al. 2010).

TETs belong to polyketides, share a four-hydrocarbon-ring structure with hydrophilic functional groups along one side, and behave as reversible bacteriostatic agents. They block *de facto* protein biosynthesis by binding site A of 30S subunit of the ribosome and preventing aminoacyl-tRNA loading during translation. Moreover, TETs prevent binding of both release factors RF-1 and 2 during the termination step, regardless of the stop codon (Brown et al. 1993).

TET binding on the small ribosomal subunit has been structurally characterized. It shows a main high affinity binding site and multiple secondary low affinity sites, whose roles have been poorly characterized so far. Multifactorial resistance mechanisms have been described in the case of TETs, some specific, such as mutations in the ribosomal subunit and ribosomal protection proteins, and others more general, often taking advantage of resistance devices toward other classes of antibiotics.

However, despite more than 60 different classes of genes encoding for TET resistance factors are known both in Gram-positive and Gram-negative bacteria, only a few have been searched and detected in *H. pylori.* The most frequent and better characterized cause of *H. pylori* resistance towards TETs is reported to be due to mutations in the *16S rRNA* gene, occurring at position 926–928 as triple-base changes (AGA to TTC). Double or single-base pair mutations insisting on the same site were also detected, conferring intermediate level of resistance to the corresponding strains as assessed by the Minimum Inhibitory Concentration (MIC) assays (Wu et al. 2005). Other reported mutations occurring at 956–958 site could be implicated in TET resistance, even if with lower frequency.

Several studies have demonstrated that TETs resistance can occur in the absence of mutations in 16S rRNA, implying other escape strategies through the accumulation of changes that may affect TET-ribosome affinity and other functions (Dailidiene et al. 2002). Within the ribosome protecting proteins (RPPs), TET (O) and TET (M) are the most extensively studied, but never characterized in *H. pylori* according to our knowledge.

tet genes products are mainly composed of two subgroups, the RPPs, that dislodge TETs by binding ribosomes and enable the protein synthesis to go ahead, and efflux pumps, that promote extrusion of toxic agents such as antibiotics. Indeed, a protein homologue of the well-known TET efflux gene *tetA* (P), HP1165, has been proved to be involved in tetracycline resistance in *H. pylori* 26,695 strain (Li and Dannelly 2006). The occurrence, role and structural properties in the context of antimicrobial resistance are discussed later in a separate paragraph (see Efflux pumps, Section 6).

2.3 Resistance to Aminoglycoside

Streptomycin (STR), an antibiotic belonging to the aminoglycosides family, is mostly used against tuberculosis and is quite effective also against *H. pylori* resistant strains (Hu et al. 2016). STR binds to 30S subunit of the bacterial ribosome, altering the correct binding of formyl-methionyl-tRNA to the ribosome and, consequently, impairing the protein synthesis. The mechanism of resistance in *H. pylori* has not been extensively studied, but it appears similar to that in *E. coli* and *Mycobacterium tuberculosis*: point mutations in the *rpsL* gene, coding for the ribosomal protein S12 (HP1197), are responsible for mismatch binding of the tRNA to the ribosome (Sharma et al. 2007; Ulger et al. 2009). Mutations in *H. pylori* have been observed at positions 43 or 88, two positions close in space. They both correspond to a lysine mutated to arginine. A molecular model of the protein has been built by homology modeling based on the structure of the orthologous protein of chloroplast ribosome from spinach (PDB ID 5X8R), that bears a surprisingly high sequence identity with our protein, 74.8%. Interestingly, both point mutations are conservative, i.e. a potentially positively charged residue, lysine, is replaced by the similar, positively charged residue arginine (Fig. 1a). This mutation does not prevent the correct functioning of the ribosome, nevertheless it prevents the binding of the antibiotic. More in depth investigations are probably necessary to better clarify this kind of resistance mechanism.

3 Inhibitors that Interfere with DNA or RNA Machinery

3.1 Resistance to Fluoroquinolones

Levofloxacin (LVFX) is used, eventually in combination with other antibiotics, to treat several bacterial infections, from pneumonia to urinary tract infections, to tuberculosis, meningitis and others. LVFX functions as a bactericide by inhibiting DNA gyrases and/or topoisomerase IV. In *H. pylori,* DNA gyrase is encoded by two genes, *gyrA* and *gyrB*, and topoisomerase IV by genes *parC* and *parE*. Mutations in the last two genes have not been observed in the bacterium, and mutations in *gyrA* gene seem to be the main reason of the resistance. Point mutations conferring resistance have been found at positions 87, 88 and 91 (Barnard and Maxwell 2001; Moore et al. 1995; Miyachi et al. 2006; Lee et al. 2008; Murakami et al. 2009, Hanafi et al. 2016; Miftahussurur et al. 2016). The structure of GyrA is not available, so the molecular model (Fig. 2a) of the product of *gyrA* gene was obtained by homology modelling from the structure of topoisomerase from *S. pneumoniae* (PDB ID 4Z2C) that presents 56% identity with the protein from *H. pylori*. The mutated residues belong to the initial part of helix 86–97 that is close to one of the regions of binding of the DNA double helix. In the complex of gyrase from *S. pneumoniae* with moxiflavicin, the inhibitor is bound to this area, suggesting that LVFX binds preventing the binding of the double-stranded DNA. Nevertheless, the mutations do not influence the functionality of the protein and do not prevent the binding of DNA. The region of the protein that binds DNA is quite extended, and to overcome the resistance a drug that binds to a different area of the protein should be tried.

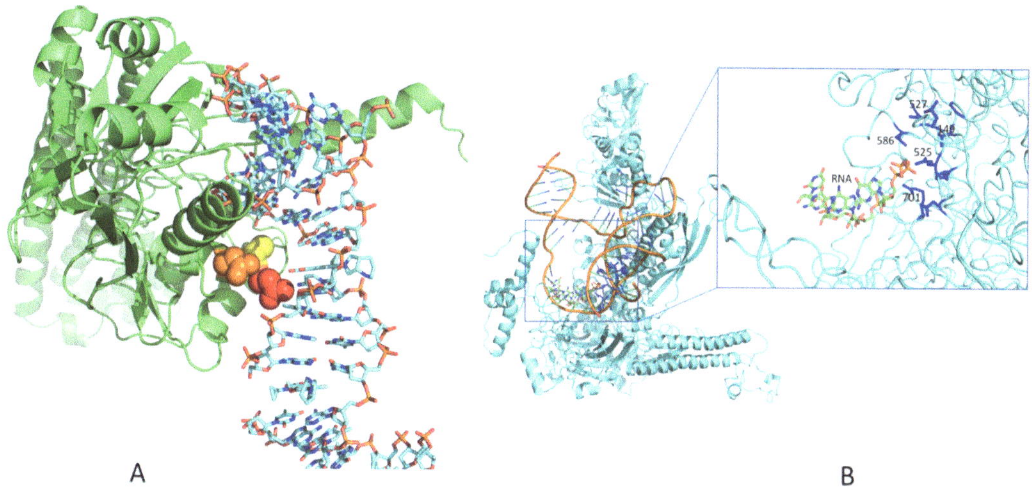

Fig. 2 Inhibitors that interfere with DNA or RNA machinery. (**a**) Cartoon view of one monomer of DNA gyrase (GyrA) model (green) with side chains of the mutated residues (87, red, 88, yellow and 91, orange) in the resistant strains of *H. pylori*. For clarity of the picture, only one monomer is shown; the second monomer is related to the first one by a twofold axis. A portion of a bound double-helix DNA is shown in cyan and brown. (**b**) Cartoon view of the model of *H. pylori* RpoB subunit (cyan). The traces of DNA promoter and RNA transcript are shown in orange, as bound to *E. coli* RopB (PDB ID 5IPM). A detail region of the protein where mutations responsible for the resistance in *H. pylori* are located is shown in the insect, where side chains of the most frequent mutated residues are shown in blue

3.2 Resistance to Rifamycinoid Antibiotics

Rifabutin (RBU), a derivative of rifamacyn S, is very effective towards Gram-positive and Gram-negative bacteria, since it inhibits transcription by binding to the β-subunit of the DNA-dependent RNA polymerase RpoB (Jin and Gross 1988). Its high efficiency at low concentrations makes it a component of the second-line therapy in drug-resistant infections (Malfertheiner et al. 2012).

The crystal structure of RpoB (HP1198) is not known, but a model can be built by homology modelling using as template 5ipl.1.C, its ortholog from *E. coli* with 47.30% of similarity (Liu et al. 2016). The entire protein complex in bacteria includes at least six polypeptide chains, in addition to the promoter DNA and the nascent RNA. In the model in Fig. 2b, only the *H. pylori* model of the β–subunit is shown, along with the DNA and RNA chains as bound to the *E. coli* complex (PDB ID 5IPL).

Mutations in RpoB observed in some resistant strains (Heep et al. 2000) are essentially confined to an area close to the active site, where the nascent RNA chain is forming. This indicates that the antibiotic binds there and that single-point mutations hinder the binding of the antibiotics, without altering or preventing the enzymatic activity of the complex. The positions of the mutations detected are illustrated in Fig. 2b. Moreover, some other resistant strains do not show mutations in RpoB, suggesting that these mutations are not the sole cause of resistance.

4 Mixed Resistances

4.1 Resistance to β-Lactams

AMX is a crucial component of the triple therapy in the attempt to eradicate the *H. pylori* infection. The general mechanism of β-lactams, a group of antibiotics containing a four-membered ring cyclic amide, is the binding to penicillin-binding proteins (PBPs); the latter are involved in peptidoglycan synthesis and its inhibition blocks the bacterial wall biosynthesis (Cho et al. 2014). The resistance to this class of antibiotics in bacteria is mostly due to the presence of β-lactamases, enzymes able to catalyze the breaking of the β-lactam ring, or eventually to a reduced membrane permeability that reduces the uptake of the antibiotic (Livermore 1995). *H. pylori* seems to differ in this contest from other bacteria, since a significant β-lactamase activity has not been very rarely detected in AMX-resistant strains. On the contrary, resistance has been associated to point-mutations to the PBPs. There are nine different PBPs, three of high molecular weight and six of low molecular weight (labelled from PBP1 to PBP9).

The crystal structure of PBP2 of *H. pylori* is available (PDB ID 5LP5, Contreras-Martel et al. 2017) and models of PBP1 and PBP3 can be quite confidently built by homology modelling (PBP1 shares 45% sequence identity with structure 2OQO; PBP3 28.1% with structure 5DF9). The overall folds of PBP2 and PBP3 proteins are quite similar (the overall root mean square deviation of the Cα atoms for the entire structure is 2.2 Å): the protein is organized in anchor, head, linker e transpeptidase domains (Fig. 3a). In the isolated PBP3 protein, the anchor is clasped against the head, whilst when PBP forms a complex with the MreC core elongation factor the anchor region shifts away from the head, exposing a hydrophobic surface that allows the protein-protein interaction. The binding of MreC to PBP represents a key event in the peptidoglycan biosynthesis and consequently in the cell wall elongation.

PBP1 is the most different among the three proteins (Fig. 3b): whilst the transpeptidase domain is quite similar (the root mean square deviation of the equivalent Cα atoms is 1.9 Å), the other domains present a different fold and a different orientation.

H. pylori resistance is primarily due to mutations in PBP1 and are localized mostly in two areas, characterized by conserved residues 402–404 and 555–557. Both are relatively close to the binding site of cephalosporin, ampicillin, aztreonam and others (Fig. 3b). The model of *H. pylori* PBP1 with superimposed the inhibitor acyl-ampicillin as bound to *E. coli* PBP1 (PDB ID 5hl9; King et al. 2017) clearly explains why the mutated residues that confer the major resistance against β-lactams are T556S, N562Y and T593A (in red in the figure). The latter are in fact located very close to the binding site of the drug and a mutation in one of these residues hamper the binding of the compound. The other residues (in blue in Fig. 3b) are more distant and possibly they have some influence on the binding, but not at a level to confer a strong resistance.

Mutations in PBP2 or PBP3 also seem to facilitate the effect of the primary mutations. Probably a suite of mutations on the three proteins altogether increase the effect and contribute to a stronger resistance (Rimbara et al. 2008).

Finally, it seems that a decreased permeability of the membrane to AMX can also contribute to the resistance, through mutations of *hopB* and *hopC* genes (Co and Schiller 2006), indicating that the resistance to AMX is a multifactorial event, not simple to contrast.

5 Resistance Due to Indirect Effects

5.1 Resistance to Nitroimidazoles

5-nitroimidazole, including MTZ, is a prototype of nitro-imidazoles used for several infections, from anaerobic bacteria to protozoa, and is also useful against *H. pylori*. It is a prodrug that must be activated, giving rise to a complex mechanism of action: a bacterial nitroreductase enzyme reduces nitroso and hydroxylamine derivatives to inhibit acid synthesis. In the presence of molecular oxygen, the latter brings to the formation of superoxides that damage bacterial DNA. Since the action of MTZ is not direct, the resistance mechanism is also complex and four possible mechanisms have been proposed: (i) increased activity of oxygen scavengers, (ii) increased activity of the enzymes, (iii) reduced activity of nitroreductases and (iv) reduced uptake of the compound. About point (i), a mutant of the ferric-uptake regulator Fur that overexpresses SodB affects the resistance to MTZ, despite the fact that there is no evidence of differences in the superoxide dismutase (SOD) activity in resistant

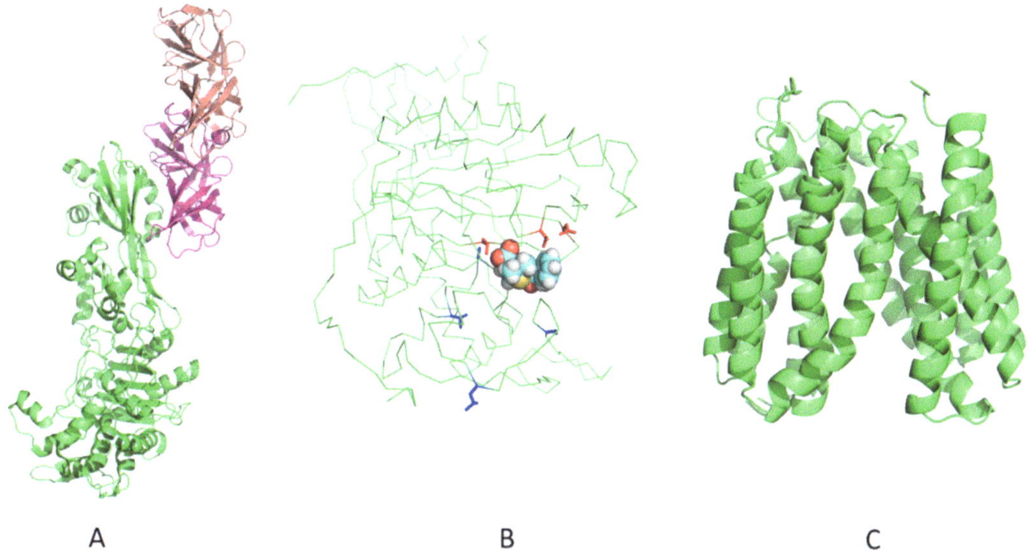

Fig. 3 Resistance due to other effects. (**a**) Cartoon structure of protein PBP2 from *H. pylori*, (coordinates from PDB ID 5LP5, Contreras-Martel et al. 2017). (**b**) Ribbon view of the C-terminal domain of the model of PBP1 of *H. pylori*, built by homology modelling with the *E. coli* protein (PDB ID 5hlb, 29% identity, King et al. 2017). The acyl-ampicillin (represented as van der Waals spheres) was positioned by superimposing the structure of the complex of the *E. coli* protein (PDB ID 5u2g) to our model and assuming that the binding position was conserved. Side chains of residues in red are those that confer stronger resistance, the blue ones are less effective. (**c**) Ribbon view of the model of HP1165, a protein homologue of the tetracycline efflux gene *tetA*, demonstrated to be involved in tetracycline resistance

strains. In addition, mechanism (ii) has not been proven, since the lack of *recA* gene, coding for a repair enzyme, does not seem to show a significant decreased resistance to MTZ. The diminished presence of an efflux pump has not been proven too. The predominant effect could be the absence of a low enough redox potential due to mutations that inactivate oxidoreductases, as RdxA (oxygen insensitive NADPH nitroreductase), FrxA (NADP:Flavin oxidoreductase) and FdxB (ferrodoxin-like enzyme), since they reduce the amount of nitroreductase present, necessary for the activation of MTZ.

6 Resistance Due to Efflux Pumps

6.1 Outer Membrane Proteins and Efflux Pumps

Together with escape mutations and drug inactivation, drugs import inside the bacterial cells and their efflux define the major mechanisms of treatment reduced susceptibility (Nikaido 1998). Import and export balance is part of the intrinsic response to antibiotics exposure, since it occurs in the absence of any genetic alterations. Moreover, efflux systems expression can be constitutive and/or induced by antibiotics acting at the transcriptional level by interacting with regulatory mechanisms (Roberts 1996; Ryan et al. 2001).

H. pylori genome codes for 32 outer membrane proteins (OMPs) and 27 drug transporters, some of which only annotated by homology searches, other better characterized and experimentally proved to act as drug exporters. The most relevant efflux systems play a significant role in multidrugs resistance by transporting a wide spectrum of structurally diverse compounds.

Efflux systems can be classified according to five main classes: (i) major facilitator superfamily (MFS), (ii) Resistance-Nodulation-Division family of transporters (RND), (iii) multi drug and toxic extrusion family (MATE), (iv) small multidrug resistance members (SMR) and (v) the ancient ATP-dependent ATP-binding cassette transporters (ABC transporters, members of a transport system superfamily). The energy costs for drugs ejection are sustained either by ATP hydrolysis, as in the case of ABC transporters, or proton gradient.

The role of specific members of any efflux-mediator families in *H. pylori* has been investigated in multiple studies, sometimes with contradictory results. What became clear is that they definitely play a major role in antibiotics resistance. Bina and co-workers first identified the presence of transporters potentially associated with drugs tolerance; they described active members belonging to at least four classes of efflux systems (Bina et al. 2000). Proofs of their significant contribution to antibiotics efflux and the consequent reduced accumulation of toxic agents in the bacterial cytoplasm came later (Dailidiene et al. 2002; van Amsterdam et al. 2005).

Together with RND transporters, the YajR homologue TetA (HP1165) has been discovered to contribute to multidrugs exposure tolerance in *H. pylori*, while ATP dependent efflux systems have never been found to be over-expressed in stressed conditions as well as in clinical isolates.

In this context, efflux pumps inhibitors received increasing attention, given their potential both as tools and therapeutic agents, as they should restore the activity of standard eradication treatments. Their roles on multidrug resistance have been investigated selecting a panel of inhibitors targeting different transporters families. A synergic effect on five of the nine tested antibiotics (chloramphenicol, TET, erythromycin, cefotaxime and ceftriaxone) was observed using carbonyl cyanide m-chlorophenyl hydrazine (CCCP) inhibitor, with reduced MIC values ranging from 19- to 4-fold, respectively. CCCP is an energy blocker which inhibits efflux pumps driven by proton gradient, and not those that are ATP dependent. These results confirmed that most likely the kind of efflux systems involved in *H. pylori* resistance do not act in a ATP hydrolysis dependent manner.

Finally, the same PPIs that are currently proposed in the first-line eradication regimen are small molecules acting as proton motive force uncoupling agents. They have been demonstrated to contribute to eradication therapy and, in particular, rabeprazole and pantoprazole positively impact on MICs of antibiotics in multidrug-resistant *H. pylori* strains. However, the mechanism of synergy and actual targets of such PPIs are far from being understood and will require further studies to be clarified (Liu et al. 2008).

6.2 RND Efflux Systems

RND efflux systems represent a relevant mechanism of resistance for multiple classes of antibiotics. In *H. pylori* cultures, they have been proven to be implicated in the resistance toward highly diverse treatments, ranging from macrolides (CLR, erythromycin and others) to penicillin G, MTZ, cefotaxime, clindamycin, novobiocin, and ethidium bromide.

The mRNA expression profile of at least four RND efflux mediators have been detected in *H. pylori*, that is HP0605–HP0607, HP0971–HP0969, HP1327–HP1329, HP1489–HP1487 (Kutsche and de Jonge 2005). Like the classical three-components RND transporters present in many Gram-negative bacteria, *H. pylori* members are characterized by a translocase, a TolC homolog and an accessory component spanning the membrane and connecting the other two components.

The best characterized in this context is the HP0605–HP0607 transporter, elsewhere identified as HefABC or acrA/B/TolC system, where *hp0605* codes for a TolC homologue (HefA), *hp0606* corresponds to the membrane fusion mediator AcrA/HefB and *hp0607* codes for the pump component on the cytoplasmic side (AcrB/HefC) (Hirata et al. 2010).

Homology modeling of the HP0605/HP0606/HP0607 members of RND transporter allows to describe the main features of such apparatus and compare it with other extensively characterized HefABC transporters such as the *E. coli* prototype. Best template for building the model of HP0607 has been identified in the AcrA protein from *E. coli* and the AcrA homologue ZneB from *Cupriavidus metallidurans,* a membrane fusion protein (Mfp) implicated in heavy metal cation efflux (De Angelis et al. 2010). A model with high confidence can be obtained (Phyre2, $\geq$ 99% confidence) despite the limited sequence identity, which do not exceed 16% over the entire protein sequence.

It conserves an elongated multidomain shape, where three tertiary structure motifs can be identified: a beta-barrel domain, predicted to localize close to the inner membrane, composed of six antiparallel beta-strands, hosting a key cysteine residue for lipid acylation and anchoring; a central lipoyl domain, composed of seven beta-strands arranged in a beta-sandwich; a coiled-coil alpha-helical hairpin that includes two helices composed of hepta-peptides repeats, deriving from the interruption of the two sides of the sandwich in the primary sequence. The beta-hairpin helices of *H. pylori* HP0606 and Zneb from *C. metallidurans* are slightly shorter that the *E. coli* ones but, analogously to any of the AcrA homologues studied till know, is connected to the central lipoyl domain by a hinge region that confers large flexibility to such coiled-coil structure and could explain its capability to induce conformational changes implied in the channel opening at the outer membrane face. Intriguingly, ZneB can bind zinc ions and such feature seems to be involved in conformational dynamics. However, the residues involved in the metal coordination at the N-terminus include an extended fragment that is not conserved in *H. pylori.*

HP0605 protein can be modeled with a relatively high degree of confidence using the crystal structure of the *C. jejuni* CmeC outer membrane channel or the multidrug efflux outer membrane protein OprM from *Pseudomonas aeruginosa*, again despite a limited sequence identity (15%; 4MT4; Su et al. 2014).

CmeC is part of a CmeABC tripartite multidrug efflux pump, homologous to HefABC apparatus. It belongs to TolC/OprM family with the typical trimeric assembly (about 500 aa per protomer) forming a long tunnel devoted to antimicrobial export, heavily charged by acidic residues on the internal side of the cavity. Each protomer has an elongated α/β-structure that contributes to form a basal membrane spanning β-barrel (4 β-strands per protomer) and a helical periplasmic domain composed of six α-helices per protomer. CmeC from *C. jejuni* and HP0606/TolC model from *H. pylori* share an extra domain decorating the channel at the mid-section. Such domain is present also in OprM from *P. aeruginosa,* while in *E. coli* and other family members it is smaller, less ordered and less pronounced.

Analogously to HP0606/AcrA, a reliable model for HP0607/AcrB protein can be built, based on the CmeB component of *C. jejuni* drugs efflux system as a template (5LQ3). A homotrimer assembles according to a typical cytoplasmic RND-type pump, with each monomer contributing with 12 transmembrane helices to the core of the structure and an elongated periplasmic protrusion defined by the association of six mixed α/β subdomains, two per monomer.

The overall trimer undergoes asymmetrical changes, depicted by the structures obtained for many AcrB homologues, since each protomer can alternatively explore different conformational states often called "access", "binding" and "extrusion" states, where the clefts present both in the transmembrane and periplasmic subdomains can alternatively switch from closed to open arrangements. This implies that such RND pump must synchronize each protomer state to go through the different steps in the drugs export mechanism (Su et al. 2017).

6.3 Tetracycline Efflux Protein

Tetracycline efflux protein (TetA), HP1165, has been demonstrated to be involved in TET resistance in *H. pylori* strain 26,695 (Li and Dannelly 2006). Structural homology searches against Protein Data Bank (PDB) evidenced 25.2% identity with the drug efflux protein YajR from *E. coli* (PDB ID 3WDO; Jiang et al. 2013).

YajR is a 49-kDa secondary active transporter, part of the highly conserved major facilitator superfamily (MFS), playing a role in substrate sensing, signaling and active export of antibiotics. MFS members follow a mechanism supported by the electrochemical potential across the cell membrane and share a canonical 12 transmembrane helices core, generated by four three-helix repeats and divided in two subdomains of six helices, each of them related by intra and inter pseudo twofold symmetries.

The two subdomains (H 1-6 and H 7-12) undergo about 40° rotation between outward and inward states during transport process, where outward conformation represent the ground state and inward the exited and protonated one. The negative inside rule drives the inward shift of YajR in the excited state while a patch of basic amino acids, enriched on the cytoplasmic side of the transporter, participates in the transport mechanism providing an energy contribution which facilitates the inward-to-outward conformational recycling.

Motif A, a highly-conserved sequence present in multiple insertion loops between TM-helices of different MSF, contribute to outward state stabilization through interactions centered at a charge-helix dipole interaction. The model (Fig. 3c) evidences a strict homology of HP1165 with YajR and other representative *E. coli* members of MSF family such as lactose permease (LacY proton/sugar symporter, PDB ID 1PV7) or multidrug transporter MdfA (PDB ID 4ZP0).

The transmembrane core is highly conserved (27% sequence similarity over 95% of the sequence, 17% identity with YajR) and all the members show a heavily charged state of cytoplasmic and periplasmic sides of the molecule, but only *E. coli* YajR and not *H. pylori* homologue HP1165 presents an extra C-terminal negative charged ferredoxin-like domain protruding as a flexible and independent appendage on the cytoplasmic side.

Evidence of antimicrobial drugs binding and export properties have been strikingly demonstrated by Heng and co-workers (Heng et al. 2015), who were able to trap *E. coli* MdfA in complex with chloramphenicol in the inward facing conformation. The aspartic acid residue buried inside the structure and located in close proximity to antibiotic binding site (Asp34) has been proved to respond to chloramphenicol binding by changing its protonation state. However, such acidic residue is not conserved in HP1165, which exposes a histidine residue (His24) in the corresponding position (same transmembrane helix and internal orientation). His24 could eventually play an analogous role given its pH responsive nature.

7 Conclusions

The mechanisms used by the Gram-negative bacterium *H. pylori* to counteract the effects of antibiotics are partially known and they share several properties in common with many other pathogenic bacteria. They present also specific features, mostly due to the peculiar niche where *H. pylori* is living, the human stomach, and possibly to the quite limited contacts the bacterium has with other bacterial species. For example, only very few examples of β-lactamases activity have been detected in *H. pylori*, and bacterium has developed its own resistance mechanism to β-lactam antibiotics.

Finally, it is important to consider that about 30% of the proteins coded by the *H. pylori* genome are essential for the survival and colonization, many of which being putative targets for the design of novel antimicrobials. Despite the efforts of many research groups all over the world, the discovery and introduction of new drugs against *H. pylori* have not been achieved till now and new approaches are required.

References

Alba C, Blanco A, Alarcón T (2017) Antibiotic resistance in *Helicobacer pylori*. Curr Opin Infect Dis 30:489–497. https://doi.org/10.1097/QCO.0000000000000396

Allen GS, Zavialov A, Gursky R, Ehrenberg M, Frank J (2005) The cryo-EM structure of a translation initiation complex from *Escherichia coli*. Cell 121:703–712. https://doi.org/10.1016/j.cell.2005.03.023

Attaran B, Falsafi T, Ghorbanmehr N (2017) Effect of biofilm formation by clinical isolates of *Helicobacter pylori* on the efflux-mediated resistance to commonly used antibiotics. World J Gastroenterol 23:1163–1170. https://doi.org/10.3748/wjg.v23.i7.116

Barnard FM, Maxwell A (2001) Interaction between DNA gyrase and quinolones: effects of alanine mutations at GyrA subunit residues Ser(83) and Asp(87). Antimicrob Agents Chemother 45:1994–2000. https://doi.org/10.1128/AAC.45.7.1994-2000.2001

Bartual SG, Straume D, Stamsas GA, Munoz IG, Alfonso C, Martinez-Ripoll M, Havarstein LS, Hermoso JA (2014) Structural basis of PcsB-mediated cell separation in *Streptococcus pneumoniae*. Nat Commun 5:3842. https://doi.org/10.1038/ncomms4842

Bina JE, Alm RA, Uria-Nickelsen M, Thomas SR, Trust TJ, Hancock RE (2000) *Helicobacter pylori* uptake and efflux: basis for intrinsic susceptibility to antibiotics in vitro. Antimicrob Agents Chemother 44:248–254

Binh TT, Shiota S, Suzuki R, Matsuda M, Trang TT, Kwon DH, Iwatani S, Yamaoka Y (2014) Discovery of novel mutations for clarithromycin resistance in *Helicobacter pylori* by using next-generation sequencing. J Antimicrob Chemother 69(7):1796–1803. https://doi.org/10.1093/jac/dku050

Brown CM, McCaughan KK, Tate WP (1993) Two regions of the *Escherichia coli* 16S ribosomal RNA are important for decoding stop signals in polypeptide chain termination. Nucleic Acids Res 21 (9):2109–2115

Chen J, Ye L, Jin L, Xu X, Xu P, Wang X, Li H (2018) Application of next-generation sequencing to characterize novel mutations in clarithromycin-susceptible *Helicobacter pylori* strains with A214G of 23S rRNA gene. Ann Clin Microbiol Antimicrob 17:10. https://doi.org/10.1186/s12941-018-0259-8

Cho H, Uehara T, Bernhardt TG (2014) Beta-lactam antibiotics induce a lethal malfunctioning of the bacterial cell wall synthesis machinery. Cell 159:1300–1311. https://doi.org/10.1016/j.cell.2014.11.017

Co EM, Schiller NL (2006) Resistance mechanisms in an in vitro-selected amoxicillin-resistant strain of *Helicobacter pylori*. Antimicrob Agents Chemother 50:4174–4176

Contreras-Martel C, Martins A, Ecobichon C, Trindade DM, Mattei PJ, Hicham S, Hardouin P, Ghachi ME, Boneca IG (2017) Molecular architecture of the PBP2-MreC core bacterial cell wall synthesis complex. Nat Commun 8:776–776. https://doi.org/10.1038/s41467-017-00783-2

Dailidiene D, Bertoli MT, Miciuleviciene J, Mukhopadhyay AK, Dailide G, Pascasio MA, Kupcinskas L, Berg DE (2002) Emergence of tetracycline resistance in *Helicobacter pylori*: multiple mutational changes in 16S ribosomal DNA and other genetic loci. Antimicrob Agents Chemother 46:3940–3946

De Angelis F, Lee JK, O'Connell JD, Miercke LJ, Verschueren KH, Srinivasan V, Bauvois C, Govaerts C, Robbins RA, Ruysschaert JM, Stroud RM, Vandenbussche G (2010) Metal-induced conformational changes in ZneB suggest an active role of membrane fusion proteins in efflux resistance systems. Proc Natl Acad Sci U S A 107:11038–11043. https://doi.org/10.1073/pnas.1003908107

De Francesco V, Giorgio F, Hassan C, Manes G, Vannella L, Panella C, Ierardi E, Zullo A (2010) Worldwide *H. pylori* antibiotic resistance: a systematic review. J Gastrointestin Liver Dis 19:409–414

Dunn BE, Cohen H, Blaser MJ (1997) *Helicobacter pylori*. Clin Microbiol Rev 10:720–741

Fallone CA, Chiba N, van Zanten SV, Fischbach L, Gisbert JP, Hunt RH, Jones NL, Render C, Leontiadis GI, Moayyedi P, Marshall JK (2016) The Toronto Consensus for the treatment of *Helicobacter pylori* infection in adults. Gastroenterology 151:51–69. https://doi.org/10.1053/j.gastro.2016.04.006

Gong Y, Yuan Y (2018) Resistance mechanisms of *Helicobacter pylori* and its dual target precise therapy. Crit Rev Microbiol 44:371–392. https://doi.org/10.1080/1040841X.2017.1418285

Hanafi A, Lee WC, Loke MF, Teh X, Shaari A, Dinarvand M, Lehours P, Mégraud F, Leow AH, Vadivelu J, Goh KL (2016) Molecular and proteomic analysis of levofloxacin and metronidazole resistant *Helicobacter pylori*. Front Microbiol 7:2015. https://doi.org/10.3389/fmicb.2016.02015

Heep M, Odenbreit S, Beck D, Decker J, Prohaska E, Rieger U, Lehn R (2000) Mutations of four distinct regions of rpoB gene can reduce the susceptibility of *Helicobacter pylori* to rifamycin. Antimicrob Agents Chemother 44:1713–1715

Heng J, Zhao Y, Liu M, Liu Y, Fan J, Wang X, Zhao Y, Zhang XC (2015) Substrate-bound structure of the *E. coli* multidrug resistance transporter MdfA. Cell Res 25:1060–1073. https://doi.org/10.1038/cr.2015.94

Hirata K, Suzuki H, Nishizawa T, Tsugawa H, Muraoka H, Saito Y, Matsuzaki J, Hibi T (2010) Contribution of efflux pumps to clarithromycin resistance in *Helicobacter pylori*. J Gastroenterol Hepatol 25 (Suppl 1):S75–S79. https://doi.org/10.1111/j.1440-1746.2009.06220.x

Hu Y, Zhang M, Lu B, Dai J (2016) *Helicobacter pylori* and antibiotic resistance, a continuing and intractable problem. Helicobacter 21:349–363. https://doi.org/10.1111/hel.12299

Hu Y, Zhu Y, Lu NH (2017) Novel therapeutic regimens for *Helicobacter pylori* in an era of increasing antibiotic resistance. Front Cell Infect Microbiol 7:168. https://doi.org/10.3389/fcimb.2017.00168

Jiang D, Zhao Y, Wang X, Fan J, Heng J, Liu X, Feng W, Kang X, Huang B, Liu J, Zhang XC (2013) Structure of the YajR transporter suggests a transport mechanism based on the conserved motif A. Proc Natl Acad Sci U S A 110:14664–14669. https://doi.org/10.1073/pnas.1308127110

Jin DJ, Gross CA (1988) Mapping and sequencing of mutations in the *Escherichia coli* rpoB gene that lead to rifampicin resistance. J Mol Biol 202:45–58

King DT, Wasney GA, Nosella M, Fong A, Strynadka NC (2017) Structural insights into inhibition of *Escherichia coli* penicillin-binding protein 1B. J Biol Chem 292(3):979–993. https://doi.org/10.1074/jbc.M116.718403

Kutsche A, de Jonge BL (2005) Compound efflux in *Helicobacter pylori*. Antimicrob Agents Chemoter 49:3009–3010. https://doi.org/10.1128/AAC.49.7.3009-3010.2005

Lee CC, Lee VW, Chan FK, Ling TK (2008) Levofloxacin resistant *Helicobacter pylori* in Hong Kong. Chemotherapy 54:50–53. https://doi.org/10.1159/000112416

Li Y, Dannelly HK (2006) Inactivation of the putative tetracycline resistance gene HP1165 in *H. pylori* led to loss of inducible tetracycline resistance. Arch Microbiol 185:255–262. https://doi.org/10.1007/s00203-006-0093-9

Liu ZQ, Zheng PY, Yang PC (2008) Efflux pump gene hefA of *Helicobacter pylori* plays an important role in m multidrug resistance. World J Gastroenterol 14:5217–5222

Liu B, Zyuo Y, Steitz TA (2016) Structure of *E. coli* σS-transcription initiation complex provide new insights into polymerase mechanism. PNAS 113:4051–4056. https://doi.org/10.1073/pnas.1520555113

Livermore DM (1995) beta-lactamases in laboratory and clinical resistance. Clin Microbiol Rev 8:557–584

Malfertheiner P, Megraud F, O'Morain CA, Atherton J, Axon AT, Bazzoli F, Gensini GF, Gisbert JP, Graham DY, Rokkas T, El-Omar EM, Kuipers EJ (2012) Management of *Helicobacter pylori* infection—the Maastricht IV/Florence Consensus report. Gut 61:646–664. https://doi.org/10.1136/gutjnl-2012-302084

Miftahussurur M, Shrestha PK, Subsomwong P et al (2016) Emerging *Helicobacter pylori* levofloxacin resistance and novel genetic mutation in Nepal. BMC Microbiol 16:256. https://doi.org/10.1186/s12866-016-0873-6

Miyachi H, Miki I, Aoyama N et al (2006) Primary levofloxacin resistance and gyrA/B mutations among *Helicobacter pylori* in Japan. Helicobacter 11:243–249. https://doi.org/10.1111/j.1523-5378.2006.00415.x

Moore RA, Beckthold B, Wong S, Kureishi A, Bryan LE (1995) Nucleotide sequence of the gyrA gene and characterization of ciprofloxacin-resistant mutants of *Helicobacter pylori*. Antimicrob Agents Chemother 39:107–111

Murakami K, Okimoto T, Kodama M, Tanahashi J, Fujioka T, Ikeda F, Muraoka H, Takigawa M, Saika T, Hasegawa M, Kobayashi I (2009) Sitafloxacin activity against *Helicobacter pylori* isolates, including those with gyrA mutations. Antimicrob Agents Chemother 53:3097–3099. https://doi.org/10.1128/AAC.01552-08

Nikaido H (1998) Antibiotic resistance caused by gram-negative multidrug efflux pumps. Clin Infect Dis 27: S32–S41

Rimbara E, Noguchi N, Kawai T, Sasatsu M (2008) Mutations in penicillin-binding proteins 1, 2 and 3 are responsible for amoxicillin resistance in *Helicobacter pylori*. J Antimicrob Chemother 61:995–998. https://doi.org/10.1093/jac/dkn051

Roberts IS (1996) The biochemistry and genetics of capsular polysaccharide production in bacteria. Annu Rev Microbiol 50:285–315. https://doi.org/10.1146/annurev.micro.50.1.285

Ryan BM, Dougherty TJ, Beaulieu D, Chuang J, Dougherty BA, Barrett JF (2001) Efflux in bacteria: what do we really know about it? Expert Opin Investig Drugs 10(8):1409–1422. https://doi.org/10.1517/13543784.10.8.1409

Schlünzen F, Zarivach R, Harms J, Bashan A, Tocilj A, Albrecht R, Yonath A, Franceschi F (2001) Structural basis for the interaction of antibiotics with the peptidyl transferase centre in eubacteria. Nature 413:814–821. https://doi.org/10.1038/35101544

Sharma D, Cukras AR, Rogers EJ, Southworth DR, Green R (2007) Mutational analysis of S12 protein and implications for the accuracy of decoding by the ribosome. J Mol Biol 374:1065–1076. https://doi.org/10.1016/j.jmb.2007.10.003

Simonetti A, Marzi S, Billas IM, Tsai A, Fabbretti A, Myasnikov AG, Roblin P, Vaiana AC, Hazemann I, Eiler D, Steitz TA, Puglisi JD, Gualerzi CO, Klaholz BP (2013) Involvement of protein IF2 N domain in ribosomal subunit joining revealed from architecture and function of the full-length initiation factor. Proc Natl Acad Sci U S A 110:15656–15661. https://doi.org/10.1073/pnas.1309578110

Su CC, Radhakrishnan A, Kumar N, Long F, Bolla JR, Lei HT, Delmar JA, Do SV, Chou TH, Rajashankar KR, Zhang Q, Yu EW (2014) Crystal structure of the *Campylobacter jejuni* CmeC outer membrane channel. Protein Sci 23:954–961. https://doi.org/10.1002/pro.2478

Su CC, Yin L, Kumar N, Dai L, Radhakrishnan A, Bolla JR, Lei HT, Chou TH, Delmar JA, Rajashankar KR, Zhang Q, Shin YK, Yu EW (2017) Structures and transport dynamics of a *Campylobacter jejuni* multidrug efflux pump. Nat Commun 8:171. https://doi.org/10.1038/s41467-017-00217-z

Suzuki H, Nishizawa T, Hibi T (2010) *Helicobacter pylori* eradication therapy. Future Microbiol 5:639–648. https://doi.org/10.2217/fmb.10.25

Ulger M, Aslan G, Emekdas G, Tezcan S, Serin MS (2009) Investigation of *rpsL* and *rrs* gene region mutations in streptomycin resistant *Mycobacterium tuberculosis* complex isolates. Mikrobiyol Bul 43:115–120

van Amsterdam K, Bart A, van der Ende A (2005) A *Helicobacter pylori* TolC efflux pump confers resistance to metronidazole. Antimicrob Agents Chemother 49:1477–1482. https://doi.org/10.1128/AAC.49.4.1477-1482.2005

Wang J, Caban K, Gonzalez RL Jr (2015) Ribosomal initiation complex-driven changes in the stability and dynamics of initiation factor 2 regulate the fidelity of translation initiation. J Mol Biol 427:1819–1834. https://doi.org/10.1016/j.jmb.2014.12.025

Wu JY, Kim JJ, Reddy R, Wang WM, Graham DY, Kwon DH (2005) Tetracycline-resistant clinical *Helicobacter pylori* isolates with and without mutations in 16S rRNA-encoding genes. Antimicrob Agents Chemother 49:578–583. https://doi.org/10.1128/AAC.49.2.578-583.2005

Yonezawa H, Osaki T, Kamiya S (2015) Biofilm formation by *Helicobacter pylori* and its involvement for antibiotic resistance. Biomed Res Int 2015:914791. https://doi.org/10.1155/2015/914791

Zaman S, Fitzpatrick M, Lindahl L et al (2007) Novel mutations in ribosomal proteins L4 and L22 that confer erythromycin resistance in *Escherichia coli*. Mol Microbiol 66:1039–1050. https://doi.org/10.1111/j.1365-2958.2007.05975.x

Zhang M (2015) High antibiotic resistance rate: a difficult issue form *Helicobacter pylori* eradication treatment. World J Gastroenterol 21:13432–13437. https://doi.org/10.3748/wjg.v21.i48.13432

Adv Exp Med Biol - Advances in Microbiology, Infectious Diseases and Public Health (2019) 1149: 243–255
https://doi.org/10.1007/5584_2019_369

Published online: 24 April 2019

Role of Probiotics in Eradication Therapy for *Helicobacter pylori* Infection

Shigeru Kamiya, Hideo Yonezawa, and Takako Osaki

Abstract

Probiotics are defined as, "Live microorganisms that, when administered in adequate amounts, confer a health benefit on the host", and have various effects including inhibitory capabilities on pathogens, stimulation of organ functions and activation of immune responses in the human. Probiotics were reported to inhibit *Helicobacter pylori* not only *in vitro,* but also *in vivo* studies. The mechanisms by which probiotics inhibit *H. pylori* infection include competition for nutrients, production of bactericidal substances, competitive inhibition of adherence and stimulation of host functions and immunity. In addition, probiotics are clinically used for eradication therapy of *H. pylori* infection, and the effects of probiotics as single treatment and combination use with other drugs including proton pump inhibitors and antibiotics against *H. pylori* are reported. It has been testified that probiotics increase the eradication rate and prevent adverse events including diarrhea, nausea, vomiting and taste disorder. In the Maastrich V/Florence Consensus Report 2017, it was stated that some probiotics may have a beneficial effect on *H. pylori* eradication and are effective in reducing side effects of eradication therapy, but more research is needed to answer several questions concerning the mechanisms of probiotics action. In addition, strain specificity, dosages and duration times of probiotics used for *H. pylori* eradication therapy need to be clarified in future studies.

Keywords

H. pylori · Probiotics · Microbiota · Eradication therapy · Adverse event

S. Kamiya (✉)
Faculty of Health Sciences, Kyorin University, Tokyo, Japan

Department of Infectious Diseases, Kyorin University School of Medicine, Tokyo, Japan
e-mail: skamiya@ks.kyorin-u.ac.jp

H. Yonezawa and T. Osaki
Department of Infectious Diseases, Kyorin University School of Medicine, Tokyo, Japan
e-mail: yonezawa@ks.kyorin-u.ac.jp; osaki@ks.kyorin-u.ac.jp

1 Definition, Classification and Properties of Probiotics

The use of antibiotic drugs in treating various microbial pathogens in humans has been proven to become less effective due to the alarming increase of antibiotics resistant strains, the side effects of antibiotics on the common gut microflora and antibiotics-associated gastroenteritis, all of which signalise a growing need for alternative treatment regimes. Probiotics are defined as "Live microorganisms that, when administered in adequate amounts, confer a health benefit on the host" (Hill et al. 2014). A similar term defines prebiotics as "non-digestible food ingredients (oligofructose, inulin *etc.*) that beneficially affect the host by selectively stimulating the growth of certain bacterial species already established in the colon and thus improve host health" (Quigley 2010; Kamiya 2011). In addition, synbiotics are the combination of probiotics and prebiotics (Pandey et al. 2015).

Many kinds of microorganisms including bacteria and fungi such as *Lactobacillus, Leuconostoc, Pediococcus, Streptococcus, Enterococcus, Bifidobacterium*, *Bacillus*, *Clostridium*, *Escherichia, Saccharomyces*, *Torulopsis* and *Aspergillus* are used for probiotics (Table 1). *Lactobacillus* is a non-spore-forming Gram-positive facultative anaerobic bacterium, and composes part of the indigenous microbiota in the oral cavity, intestine and vagina. *Bifidobacterium* is a non-spore-forming Gram-positive obligate anaerobe with a characteristic curved-rod, branched and V- or Y-shaped morphology, and produce acetic acid and lactic acid in a ratio of 3:2 (Tannock 1999).

Probiotics have various effects on human host (Thomas et al. 2010; Kamiya 2011). Inhibitory effects of probiotics on pathogenic bacteria are reported. The mechanisms include competence of nutrients, production of bactericidal substances including bacteriocin, acid and short chain fatty acids (SCFA) such as acetic, butyric and propionic acid production, competitive inhibition of adherence, stimulation of intestinal peristalsis, normalization of intestinal permeability, production of mucus and activation of host immune responses in innate, fluid and cellular immunity (Galdeano and Perdigon 2006; Hoarau et al. 2008). Considered activity mechanisms among probiotics are indicated in Table 2 (Hill et al. 2014).

Widespread mechanisms may include colonization resistance, which means resistance of the host against colonization of exogenous pathogenic bacteria, acid and SCFA production, regulation of intestinal transit, normalization of perturbed microbiota, increased turnover of enterocytes and competitive exclusion of pathogens among studied probiotics. Frequent

Table 1 Microorganisms used for probiotics

A: Lactic acid bacteria
(1) *Lactobacillus* spp.
L. acidophilus
L. brevis
L. bulgaricus
L. casei
L. dulbrueckii
L. murinus
L. plantarum
L. reuteri
L. rhamnosus
(2) *Lactococcus* spp.
L. lactis
(3) *Leuconostoc* spp.
L. mesenterioides
(4) *Pediococcus* spp.
P. acidilactici
P. cerevisiae
(5) *Streptococcus/Enterococcus* spp.
E. faecalis
E. faecium
S. thermophilus
B: *Bifidobacterium* spp.
B. animalis
B. bifidum
B. breve
B. infantis
B. longum
B. pseudolongum
B. thermophilum
C: Spore formers
(1) *Bacillus* spp.
B. cereus
B. clausii
B. licheniformis
B. mesentericus
B. subtilis
B. toyoi
(2) *Clostridium* spp.
C. butyricum
D: Enterobacteriaceae
Escherichia coli
E: Fungi
(1) *Saccharomyces* spp.
S. bourlardii
S. cerevisiae
S. frogilis
(2) *Aspergillus* spp.
A. oryzae
(3) *Torulopsis* spp.

Table 2 Considered mechanisms among probiotics

Category	Effect	Mechanism
A	Widespread (among studied probiotics)	Colonization resistance
		Acid and SCFA production
		Regulation of intestinal transit
		Normalization of perturbed microbiota
		Increased turnover of enterocytes
		Competitive exclusion of pathogens
B	Frequent (Species-level effects)	Vitamin synthesis
		Direct antagonism
		Gut barrier reinforcement
		Enzyme activity
		Bile salt metabolism
		Neutralization of carcinogens
C	Rare (Strain-level effects)	Neurological effects
		Immunological effects
		Endocrinological effects
		Production of specific bioactive substances

Modified from the report by Hill et al. (2014)

mechanisms for species-level effects may include vitamin synthesis, direct antagonism, gut barrier reinforcement, bile salt metabolism, enzymatic activity and neutralization of carcinogens. Rare mechanisms for strain-level effects may include neurological effects, immunological effects, endocrinological effects and production of specific bioactive substances.

Although it has been thought that probiotics have no adverse effects on their host, a few reports of adverse effects have been published. *Lactobacillus* sepsis and endocarditis were observed after treatment with probiotics in a patient undergoing intestinal decontamination therapy prior to liver transplantation (Patel et al. 1994). In addition, a disseminated fungemia and serious diarrhea were reported after administration of *Saccharomyces boulardii* to a one-year-old child (Pletincx et al. 1995). In a multi-center randomized, double-blind, placebo-controlled clinical trial for the patients with fulminant acute pancreatitis, it was reported that the mortality (15.7%; 24 cases/153 cases) in the probiotics group (a novel probiotics combined with 4 lactobacilli and 2 bifidobacteria) was significantly higher than that (6.2%; 9 cases/145 cases) of the placebo group (Besselink et al. 2008). However, several problems in the report were pointed out: (i) The number of cases of both organ failure and multi-organ failure were significantly higher in probiotics group than in placebo group, (ii) Safety of the probiotics used was not previously reported, and (iii) There was no description of the treatment for each of the patients. Although the number of patients suffering from adverse effects during probiotic treatment is still very low, the administration of probiotics to neonates and the immunocompromised host needs to be considered carefully.

2 *In Vitro* Effect of Probiotics on *Helicobacter pylori*

It was reported that many kinds of bacteria and yeasts including *Staphylococcus auricularis, Staphylococcus epidermidis, Enterobacter cloacae, Stenotrophomonas maltophilia, Bacillus* spp., *Lactobacillus acidophilus, Lactobacillus fermentum, Lactobacillus paracasei, Lactobacillus pentosus, Lactobacillus plantarum, Lactococcus lactis, Leuconostoc citreum, Leuconostoc lactis, Enterococcus faecium, Weissella confusa* and *Saccharomyces cerevisiae* inhibited the growth of *H. pylori* evaluated by not only liquid broth assay but agar plate method

(Lopez-Brea et al. 2008; Lin et al. 2009; Techo et al. 2018). Adhesion of *H. pylori* to gastric epithelial cells (established cell lines including MKN45) was inhibited by *Lactobacillus salivarius, L. plantarum* and *Clostridium butyricum* (Kabir et al. 1997; Takahashi et al. 2000; Zhao et al. 2018). It was also shown that culture supernatants of *Bacillus subtilis, L. plantarum, C. butyricum* and *W. confusa* inhibited the growth and morphology of *H. pylori* (Pinchuk et al. 2001; Takahashi et al. 2000; Nam et al. 2002; Zhao et al. 2018). These effects were not due to decreased pH by the cultivation of the above bacteria but the production of growth inhibitory substances (Pinchuk et al. 2001; Takahashi et al. 2000). Culture supernatants of *Lactobacillus* strains were also reported to inhibit the production of IL-8 and TNF-alpha from gastric epithelial cells including human gastric cancer AGS cell line (Zhou et al. 2008; Kim et al. 2008a; Zhao et al. 2018). It was shown that heat shock protein 60 (GroEL) of *Lactobacillus johnsonii* La1 strain binds effectively to human intestinal adenocarcinoma HT29 cells, resulting in a selective aggregation of *H. pylori* (Bergonzelli et al. 2006). The effects of culture supernatant containing conjugated linoleic acid, which inhibits the pro-inflammatory IkappaB kinase (IKK) activity of *Lactobacillus* on inflammation in cultured gastric epithelial cells infected with *H. pylori* was reported (Kim et al. 2008a). It has been recently reported that reuterin, one of the bacteriocins, produced by *Lactobacillus reuteri* inhibited the growth of *H. pylori* and downregulated virulence gene expression of *vacA* and *flaA* (Urrutia-Beca et al. 2018).

3 *In Vivo* Effect of Probiotics on *H. pylori*

There have been many reports using animal models to examine the effects of probiotics on *H. pylori* (Kabir et al. 1997; Johnson-Henry et al. 2004; Lesbros-Pantoflickova et al. 2007). It was shown that *L. salivarius* WB1004 strain decreased the number of *H. pylori* colonized to the level of less than one-hundredths in the gastric mucosa by using germfree mice, and that the titer of the anti-*H. pylori* antibodies was not detected in the gnotobiotic mice infected with *H. pylori* by administration of the probiotics (Kabir et al. 1997; Aiba et al. 1998). Similarly, in the experiment using germfree mice, *C. butyricum* M588 strain was reported to decrease the number of *H. pylori* colonized in gastric mucosa to the level of less than one-hundredth using germfree mice model (Takahashi et al. 2000). It was also shown that oral administration of *Lactobacillus casei* Shirota strain for 9 months decreased the number of *H. pylori* colonized in the gastric mucosa of C57BL/6 mice and inflammation score of the gastric mucosa was also decreased (Sgouras et al. 2004). In 5 out of 10 mice tested, *H. pylori* was eradicated by the treatment with *Lactobacillus rhamnosus* R0001 strain and *L. acidophilus* R0052 strain, and the inflammation score was also decreased in 5 out of 7 mice (Johnson-Henry et al. 2004).

4 Effect of Probiotics in Eradication Therapy for *H. pylori* Infection

There have been many clinical studies to examine the effect of probiotics on *H. pylori* infection (Goderska et al. 2018). These studies include not only the effect of single treatment with probiotics but the effect of combination of eradication therapy using proton pump inhibitor (PPI) and antimicrobial drugs.

4.1 The Effect of Single Treatment with Probiotics on *H. pylori* infection

It was reported that intake of yogurt containing *Lactobacillus gasseri* LG21 strain for 8 weeks by *H. pylori* positive healthy volunteers decreased the value of urea breath test (UBT) compared to that of yogurt only (average value; 20.9 permili vs. 26.6 permili) (Sakamoto et al. 2001). Decrease of gastric inflammation was also detected in the volunteers who had intake of the probiotics, indicating that *L. gasseri* LG21 strain is useful for decreasing of *H. pylori* colonization and gastric inflammation, but not for eradication. The effect of *Lactobacillus brevis* on *H. pylori* infection and proliferation activity of gastric epithelial cells was examined (Linsalata et al. 2004). Patients with functional dyspepsia (11 patients each) were treated with either *L. brevis* or placebo for 3 weeks, and UBT and quantitation of ornithine decarboxylase (ODC) activity and polyamine to evaluated proliferation capability of gastric mucosa were done. Probiotic treatment decreased UBT value significantly, and the ODC activity and concentration of polyamine was also decreased in the probiotics-treated patients. A similar result was reported; the intake of yogurt containing *L. acidophilus* and *Bifidobacterium* for 4 weeks by the *H. pylori* positive gastritis patients decreased UBT and inflammation score significantly which was performed 2 and 6 weeks after the termination of probiotic intake, respectively (Wang et al. 2004). Epidemiological studies to examine the correlation between re-infection with *H. pylori* and dietary and socioeconomical factors were reported (Jarosz et al. 2009). The contribution of dietary and socio-economic factors in *H. pylori* re-infection was shown. A statistically significant lower frequency of fermented dairy products ($P < 0.0001$), vegetables ($P = 0.02$), and fruit ($P = 0.008$) consumption was noted among patients with *H. pylori* re-infection as compared to those who had not been re-infected. The anti-*H. pylori* activity of *L. gasseri* LG21 strain for Thai healthy children was reported (Boonyaritichaikij et al. 2009). Four hundred and forty children aged 5–7 years old were recruited, and *H. pylori* infection was positive in 132 children. The recruited children who ate cheese with or without *L. gasseri* LG21 strain for 12 months were categorized as probiotics and placebo groups, respectively, and children who did not eat either cheese were categorized as control group. Detection of *H. pylori* was performed by stool antigen test (HpSA). In the eradication study, among 82 *H. pylori*-positive children, *H. pylori* was eradicated by probiotics in 24 children (eradication rate; Per Protocol (PP) analysis, 29.3%; Intention-to-Treat (ITT) analysis, 24.7%). The numbers of children who were eradicated during the observation period (12 months) was 0 and 1 (eradication rate 5.6% in PP analysis) in placebo ($n = 6$) and control groups ($n = 18$), respectively. In prophylaxis study, the numbers of children who were infected with *H. pylori* during the observation period were 5 (infection rate, 4.1% in PP analysis) and 8 (infection rate, 8.1% in PP analysis) in probiotics and placebo groups, respectively, indicating that intake of probiotics was not effective statistically for prevention of *H. pylori* infection. Multi-strain probiotics containing *Bifidobacterium brevis, Bifidobacterium infantis, Bifidobacterium longum, L. acidophilus, Lactobacillus delbrueckii, L. paracasei, L. plantarum* and *Streptococcus thermophilus* were used for the treatment of the adult dyspeptic patients with *H. pylori* infection (Rosania et al. 2012). The eradication rates evaluated by UBT were 32.5% (13/40 cases) and 0% (0/40 cases) in the probiotics- and placebo-treated patients, respectively. Recently, a systemic review with pooled-data analysis to evaluate the effect of probiotic monotherapy for *H. pylori* eradication has been reported (Losurdo et al. 2018). Eleven studies were selected from 1,537 records identified through database searching. Probiotics eradicated *H. pylori* in 50 out of 403 cases (eradication rate; 12.4%). Lactobacilli and *S. boulardii* eradicated 20 (eradication rate; 8.5%) and 6 (eradication rate; 9.5%) patients out of 235 and 63 patients, respectively. In the comparison of probiotics versus placebo, an OR (odds ratio) of 7.91 in favor of probiotics (95% CI:2.97–21.05, $p < 0.001$) was found, suggesting that probiotics show a minimal effect on *H. pylori* clearance.

4.2 Combination Effect of Probiotics with Other Drugs

The effect of culture supernatant of *L. acidophilus* La1 strain with PPI of omeprazole on *H. pylori* infection was examined (Michetti et al. 1999). UBT value of *H. pylori*-positive volunteers was significantly decreased at 0 and 4 weeks after the combination treatment for 2 weeks was terminated. Similarly, the effect of combination treatment of fermented mild containing *L. johnsonii* La1 strain and chlarithromycin (CLR) was examined for 53 *H. pylori*-positive volunteers (Felley et al. 2001). At 4–8 weeks after the treatment, the concentration of *H. pylori* and inflammation score were significantly decreased.

4.3 Combined Effect of Probiotics with Eradication Therapy

Triple therapy using PPI and two kinds of antimicrobial drugs is generally used for eradication of *H. pylori*. Many studies to examine the combined effect of probiotics with triple therapy on eradication rate and adverse effect caused by the triple therapy have been reported.

(a) Effect of probiotics on eradication rate

The effect of heat-inactivated culture supernatant of *L. acidophilus* on eradication therapy for *H. pylori* was examined (Canducci et al. 2000). The eradication rate in the 60 *H. pylori* positive patients treated with rabeprozole, CLR, amoxicillin (AMX) and heat-inactivated culture supernatant of *L. acidophilus* was significantly higher (88% in PP analysis and 87% in ITT analysis) than that (72% in PP analysis and 70% in ITT analysis) in 60 patients treated with only triple therapy. It was reported that the eradication rate by combined therapy of triple therapy with yogurt containing *Lactobacillus* and *Bifidobacterium* was significantly higher (91.3%; 73 patients eradicated/80 patients treated) than that by triple therapy only (78.3%; 63 patients eradicated/80 patients treated) in PP analysis, although no significant difference between two groups was observed in ITT analysis (Sheu et al. 2002). The effect of probiotic yogurt containing *L. acidophilus, L. casei, B. longum* and *S. thermophilus* on the eradication rate by triple therapy using PPI, AMX and CLR was evaluated (Kim et al. 2008b). The eradication rate in probiotics group in PP and ITT analyses was 87.5% and 72.5%, respectively. In contrast, the eradication rate by triple therapy without probiotics was 78.7% and 72.1% in PP and ITT analyses, respectively, indicating that the significant effectiveness of probiotic supplementation was observed only in PP analysis. Meta-analysis to examine the effectiveness of probiotics on eradication rate for 11 randomized controlled trial studies using *Bacillus clausii,, L. acidophilus, S. boulardii, Bifidobacterium breve, C. butyricum, L. rhamnosus, Propionibacterium freudenreichii, B. longum, Bifidobacterium animalis, L. casei* was reported (Tong et al. 2007). Among 11 studies, 3 studies indicated that eradication rate in the patients treated with triple therapy with probiotic was significantly higher than that with triple therapy only, but other 8 studies not. The eradication rate in ITT analysis was 83.6% (463 patients eradicated/554 patients tested) and 74.8% (389 patients eradicated/520 patients tested) in probiotics and control (triple therapy only) groups, respectively, and OR for effectiveness of probiotics was 1.84 (95% confidence interval (CI) =1.34–2.54), indicating that probiotics use was effective for eradication of *H. pylori*. Similarly, the effectiveness of probiotics was also observed with the OR of 1.82 (95% CI = 1.30–2.56) in PP analysis. Another meta-analysis to examine the effect of probiotics treatment on eradication of *H. pylori* for 10 randomized controlled trial (RCT) studies was reported (Sachdeva and Nagpal 2009). Various probiotics such as *Lactobacillus, Bifidobacterium, L. acidophilus, Bifidobacterium lactis, Lactobacillus bulgaris, S. thermophilus, L. casei, B. animalis, L. johnsonii, L. acidophilus, Bifidobacterium bifidum* were used. In these studies, 1, 5 and 1 RCTs were performed with quadruple, triple and single eradication therapies, and in the remaining 3 RCT eradication therapy was not performed. In ITT analysis, the eradication rates were 67.7% (337 cases eradicated/498 cased treated) and 58.1% (270 cased eradicated/465 cased treated) in probiotics and control groups, respectively, indicating the OR of 1.91 (95% CI 1.38–2.65). These studies suggest the effectiveness of probiotics treatment for eradication of *H. pylori*. The effect of addition of synbiotics to the standard triple therapy (lansoprazole+AMX + CLR) for *H. pylori* infection in children was evaluated in a randomized controlled study (Sirvan et al. 2017). *H. pylori* eradication was achieved in 88% (44/50 cases) in group 1 who received standard therapy with additional *B. lactis*-containing synbiotics and 72% (36/50 cases) in group 2 who received standard therapy without synbiotics ($p < 0.05$), indicating that the addition of synbiotics to the triple therapy is effective for eradicating *H. pylori* infection in children. The outcomes of concomitant therapy (CT) using PPI + AMX + CLR + metronidazole (MTZ) and standard triple therapy combined with probiotics (STP) were investigated (Jung et al. 2018). For probiotics, *B. subtilis* + *Streptococcus faecium* (STP-I) or *B. subtilis* + *L. casei* var. *rhamnosus* (STP-II) were used. The ITT and PP eradication rates were 83.6% (239/286 cases) and 87.1% (237/272 cases) in the STP group and 86.7% (65/75 cases) and 91.4% (64/70 cases) in CT group, respectively, showing no significant difference in the eradication rates between two groups. It was also shown that eradication rates in STP-II group were significantly higher than those in STP-1 group.

Recently, several systemic review and meta-analyses for the efficacy of probiotic supplementation therapy for *H. pylori* eradication have been reported (McFarland et al. 2016; Lü et al. 2016; Lu et al. 2016). The pooled relative risk (RR) of eradication was significantly higher in the probiotic supplementation group than in the control group (RR = 1.15, 95% CI 1.10–1.20) (Lü et al. 2016). Subgroup analysis showed that eradication rates were significantly improved in both adults and children in the probiotic supplementation group and that no regional difference between Europe and Asia were present. It was also shown that supplementation of *Lactobacillus* alone (RR = 1.24, 95% CI 1.12–1.38) or multi-strain probiotics (RR = 1.12, 95% CI 107–1.18) was effective at improving eradication rates. A total of 19 RCTs were selected for the analysis to evaluate the efficacy of multi-strain probiotics as adjunct therapy for *H. pylori* eradication (McFarland et al. 2016). It was shown that probiotics statistically are effective for *H. pylori* eradication (RR = 1.12, 95% CI 1.08–1.17), although two of six probiotic mixtures are not significantly effective as an adjunct for *H. pylori* eradication (Table 3). In contrast, a meta-analysis that probiotic supplementation does not improve eradication rate of *H. pylori* infection compared to placebo based on standard therapy was reported (Lu et al. 2016). The effect of combining probiotics, with or without placebo, with standard therapy was examined. In ITT analysis, the probiotic group was 1.21 times more likely than the placebo group to achieve eradication of *H. pylori* infection (OR = 1.21, 95% CI 0.86–1.69) and 1.84 times more likely than the standard-therapy alone group (without placebo) (OR = 1.84, 95% CI 1.51–2.25). In PP analysis, similar results were observed.

A network meta-analysis to evaluate probiotic supplementation in the eradication therapy for *H. pylori* infection in adults has recently been performed (Wang et al. 2017). In *H. pylori* eradication therapy, patients with supplementary probiotics had a higher eradication rate than

Table 3 Relative risks of *H. pylori* eradication by probiotic mixture groups compared to control groups

Probiotics group[a]	Number of studies	Subtotal Relative Risk by probiotics group[b]
La + Ba	4	1.16 (95% CI 1.05–1.28)
La + Bb	4	1.04 (95% CI 0.94–1.14)
Lh + Lr	4	1.15 (95% CI 1.06–1.25)
La + Lc + Bl	2	1.07 (95% CI 0.98–1.18)
La + Bl + Ef	4	1.17 (95% CI 1.04–1.31)
8 strains	2	1.21 (95% CI 1.07–1.36)

Modified from McFarland et al. (2016)

[a]La, *Lactobacillus;* Ba, *Bifidobacterium animalis*: Bb, *Bifidobacterium breve;* Lh, *Lactobacillus helveticus;* Lr, *Lactobacillus rhamnosus;* Lc, *Lactobacillus casei;* Bl, *Bifidobacterium longum;* Ef, *Enterococcus faecalils;* 8 strains, *L. acidophillus, L. casei* subsp. *rhamnosus, L. plantarum, L. reuteri, L. salivarius, L. sporogenes, B. infantis* and *B. longum*

[b]Overall RR: 1.12 (95% CI 1.08–1.17)

control group for ITT and PP analyses with RR of 1.17 (95% CI 1.15–1.18) and RR of 1.15 (95% CI 1.14–1.17), respectively. In pair-wise meta-analysis, *Bacillus licheniformis, C. butyricum, Enterococcus* + *B. subtilis, L. acidophilus, Lactobacillus* + *Bifidobacterium* + *Enterococcus, Lactobacillus* + *Streptococcus lactis* and *S. boulardii* showed significant efficacy in increasing the eradication rates when supplemented in 7-day triple therapy (Wang et al. 2017). A systematic review and network meta-analysis to evaluate the efficacy of probiotic-supplemented triple therapy for eradication of *H. pylori* in children has been reported (Feng et al. 2017). Compared with placebo, probiotic-supplemented triple therapy significantly increased *H. pylori* eradication rates (RR = 1.19, 95% CI 1.13–1.25). Network meta-analysis showed that *L. casei* and multi-strain of *B. infantis* and *C. butyricum* were identified the most effective for increasing *H. pylori* eradication rates. Similar network meta-analysis for the evaluation of the efficacy of probiotics in 14-day triple therapy using PPI and two antibiotics for Asian pediatric patients with *H. pylori* infection has been reported (Wen et al. 2017). Compared with placebo, probiotic supplementation significantly improved eradication rates of *H. pylori* (RR = 1.16; 95% CI 1.07–1.26). Multi-strain probiotics containing *B. infantis* and *C. butyricum* was most beneficial for the eradication rate.

(b) Effect of probiotics on adverse actions in eradication therapy for *H. pylori* infection

Adverse effects such as diarrhea, abdominal discomfort, nausea, taste disorder etc. in eradication therapy were reported. The effects of probiotics on these adverse effects in eradication therapies for *H. pylori* have been reported. The effect of *L. casei* GG (Goldstein and Gorbach) strain on eradication therapy using rabeprazole, CAM and tinidazole was investigated (Armuzzi et al. 2001). It was reported that the frequencies of nausea, taste disorder and diarrhea were 10.0%, 23.3% and 3.3% in the probiotics group and 36.6%, 50.0% and 26.6% in control group, respectively, although there was no significant difference in the eradication rates (83.3% and 80.0% in probiotics and control groups). The effect of probiotics on prevention of adverse effects in eradication therapy was reported (Cremonini et al. 2002). Co-administration of probiotics (*Lactobacillus* GG, *S. boulardii, Lactobacillus* and *Bifidobacterium*) with triple therapy decreased the occurrence of diarrhea and taste disorder. Similar effects by other probiotics including *B. clausii* and *Lactobacillus/Bifidobacterium* containing yogurt were reported to be significantly effective in decreasing the occurrence of nausea, diarrhea, upper abdominal pain, taste disorder and constipation (Nista et al. 2004).

In contrast, it was reported that co-administration of yogurt containing *L. acidophilus, L. casei, B. longum* and *S. thermophilus* increased the occurrence rate of adverse effects such as taste disorder, diarrhea and upper abdominal pain; the rate was significantly higher in probiotics group (41.1%; 69 cases/168 individuals) than in control group (26.3%; 47 cases/179 individual) (Kim et al. 2008b). Although it was suggested that *Lactobacillus* might affect the action of antibiotics to increase the adverse effect of taste disorder, the cause of the effect by probiotics was not clarified.

A meta-analysis to examine the effects of probiotics on prevention of adverse effects by eradication therapy for 7 RCT studies was reported (Tong et al. 2007). *Lactobacillus, L. acidophilus, S. boulardii, L. rhamnosus, P. freudenreichii, C. butyricum* were used as probiotics in these studies. In 4 RCT studies, the administration of probiotics significantly decreased the occurrence rate of adverse effects; 24.7% (81 cases/328 cases) in the probiotics group versus 38.4% (114 cases/297 cases) in control group. Adverse effects such as diarrhea (6.0% versus 16.1%), taste disorder (26.0% versus 46.0%), nausea (15.6% vs. 25.2%) and epigastralgia (16.3% versus 23.0%) were significantly lower in the probiotics group. Another meta-analysis to clarify the effects of probiotics on the adverse effects by eradication therapy for

6 RCT studies was reported (Sachdeva and Nagpal 2009). Probiotics of *Lactobacillus*, *Bifidobacterium*, *L. acidophilus*, *B. lactis*, *L. bulgaris*, *S. thermophilus*, *L. casei* and *L. jonsonii* were used in these studies. In ITT analysis, the occurrence rate of adverse effects in probiotics and control groups was 31.5% (118 cases/375 cases) and 45.9% (158 cases/344 cases), indicating that administration of probiotics prevents the occurrence of adverse effects by eradication therapy.

Recently, meta-analyses for the effect of probiotics on prevention of adverse events in *H. pylori* eradication therapy in adults have been reported (McFarland et al. 2016; Lü et al. 2016; Lu et al. 2016; Wang et al. 2017). It was shown that five (La + Ba, La + Bb, Lh + Lr, La + Bl + Ef, 8 strains in Table 4) out of six mixtures of strains of probiotics significantly prevented any adverse reactions (McFarland et al. 2016) (Table 4). Overall, probiotics given with triple therapy decreased adverse events by 14% (mean adverse events in probiotics was 22% compared to 36% in controls). In the meta-analysis by Lü and co-workers, the overall incidence of side effects decreased by approximately 8% in the probiotic supplementation group compared to control group (RR = 0.71, 95% CI 0.54–0.94) (Lü et al. 2016). Significant decrease in the incidences of nausea and vomiting (RR = 0.58, 95% CI 0.35–0.95) and constipation (RR = 0.47, 95% CI 0.28–0.80) was noted. In the meta-analysis by Lu and co-workers, improvement of the adverse effects of diarrhea and nausea was noted in the probiotics group compared to standard therapy alone group, although the probiotic supplementation did not improve eradication rate for *H. pylori* infection (Lu et al. 2016). Network meta-analyses for the effects of probiotics on adverse reactions in children have been reported (Feng et al. 2017; Wen et al. 2017). It was shown that probiotic-supplemented triple therapy reduced incidence of total side effects (RR = 0.49, 95% CI 0.38–0.65). As for reducing the incidence of total side effects, multi-strain of *L. acidophilus* and *L. rhamnosus*, multi-strain of *Bacillus mesentericus*, *C. butyricum* and *Streptococcus faecalis* and single-strain of *S. boulardii* were better to supplemented triple therapy (Feng et al. 2017). There were 8 probiotic regimens supplemented 14-day triple therapy reporting data of total side effects (Wen et al. 2017). *B. mesentericus* + *C. butyricum* + *S. faecalis* was the best regimen for prevention of total side effects. As for subtypes of side effects, *B. bifidum*+*B. infantis*+*L. acidophilus*+*L. bulgaricus* + *L. casei* + *L. reuteri* + *Streptococcus* and *B. bifidum* + *L. acidophilus* were most beneficial for reducing the incidence of diarrhea and nausea/vomiting, respectively.

5 Conclusions

Probiotics were reported to have *in vitro* inhibitory effects on the growth, adhesion and expression of virulence factors of *H. pylori.* By *in vivo* studies using germfree mice and Mongolian gerbils, probiotics were shown to inhibit the colonization of *H. pylori.* Various systematic review and meta-analysis indicated that probiotic supplementation to standard eradication therapy for *H. pylori* infection was effective for the increasing the eradication rate and prevention of adverse effects. Use of probiotics was evaluated in the recent guidelines for the management of *H. pylori* infection (Zagari et al. 2015; Fallone et al. 2016; Malfertheiner et al. 2017). In Italian guideline for the management of *H. pylori* infection, "Some probiotics reduce adverse effects during *H. pylori* eradication therapy" is stated (Evidence level 3a; Grade of recommendation B) (Zagari et al. 2015). In Tronto Consensus for the treatment of *H. pylori* infection in adults, the followings statements are listed in Supplemented Therapy; (i) Statement 14. In patients with *H. pylori* infection, we recommend against routinely adding probiotics to eradication therapy for the purpose of reducing adverse events (Grade: Strong recommendation; Quality of evidence: very low) (ii) Statement 15. In patients with *H. pylori* infection, we recommend against adding probiotics to eradication therapy for the purpose of increasing eradication rates. (Grade: Strong recommendation; Quality of evidence: very low)

Table 4 Relative risks for adverse events in *H. pylori* eradication therapy by probiotic mixture groups compared to control groups

Probiotics group[a]	Number of studies	Subtotal Relative Risk by probiotics group[b]
La + Ba	2	0.31 (95% CI 0.20–0.47)
La + Bb	4	0.67 (95% CI 0.50–0.88)
Lh + Lr	2	0.12 (95% CI 0.03–0.50)
La + Lc + Bl	2	1.34 (95% CI 0.98–1.85)
La + Bl + Ef	3	0.25 (95% CI 0.15–0.42)
8 strains	2	0.24 (95% CI 0.14–0.39)

Modified from McFarland et al. (2016)

[a]La, *Lactobacillus;* Ba, *Bifidobacterium animalis*: Bb, *Bifidobacterium breve;* Lh, *Lactobacillus helveticus;* Lr, *Lactobacillus rhamnosus;* Lc, *Lactobacillus casei;* Bl, *Bifidobacterium longum;* Ef, *Enterococcus faecalils;* 8 strains, *L. acidophillus, L. casei* subsp. *rhamnosus, L. plantarum, L. reuteri, L. salivarius, L. sporogenes, B. infantis* and *B. longum*

[b]Overall RR: 0.45 (95% CI 0.30–0.65)

(Fallone et al. 2016). In Maastrich V/Florence Consensus Report, the following statements are listed in Working Group 5; (i) Statement 9. Only certain probiotics have been shown to be effective in reducing gastrointestinal (GI) side effects caused by *H. pylori* eradication therapies. Specific strains should be chosen only upon the basis of a demonstrated clinical efficacy. (Grade of recommendation: strong; Level of evidence: moderate) (ii) Statement 10. Certain probiotics may have a beneficial effect on *H. pylori* eradication. (Grade of recommendation: weak; Level of evidence: very low) (Malfertheiner et al. 2017). Although recent two guidelines recommend the use of probiotics in eradication therapy for *H. pylori* infection for increasing eradication rate and preventing adverse effects, several questions remain, including the effectiveness of specific probiotic strains, dosages and duration of adjuvant probiotic therapy and more data are definitely needed.

References

Aiba Y, Suzuki N, Kabir AM, Takagi A, Koga Y (1998) Lactic acid-mediated suppression of *Helicobacter pylori* by the oral administration of *Lactobacillus salivarius* as a probiotic in a gnotobiotic murine model. Am J Gastroenterol 93:2097–2101. https://doi.org/10.1111/j.1572-0241.1998.00600.x

Armuzzi A, Cremonini F, Bartolozzi F, Canducci F, Candelli M, Ojetti V, Cammarota G, Anti M, De L, Pola P, Gasbarrini G, Gasbarrini A (2001) The effect of oral administration of *Lactobacillus* GG on antibiotic-associated gastrointestinal side-effects during *Helicobacter pylori* eradication therapy. Alimen Pharmacol Ther 15:163–169

Bergonzelli GE, Granato D, Pridmore RD, Marvin-Guy LF, Donnicola D, Corthesy-Theulaz IE (2006) GroEL of *Lactobacillus johnsonii* La1 (NCC533) is cell surface associated: potential role in interactions with the host and the gastric pathogen *Helicobacter pylori.* Infect Immun 74:425–434. https://doi.org/10.1128/IAI.74.1.425-434.2006

Besselink MG, van Santvoort HC, Buskens E, Boermeester MA, van Goor H, Timmerman HM, Nieuwenhuijs VB, Bollen TL, van Ramshorst B, Witteman BJ, Rosman C, Ploeg RJ, Brink MA, Schaapherder AF, Dejong CH, Wahab PJ, van Laarhoven CJ, van der Harst E, van Eijck CH, Cuesta MA, Akkermans LM, Gooszen HG (2008) Dutch Acute Pancreatitis Study Group: probiotic prophylaxis in predicted severe acute pancreatitis: a randomised, double-blind, placebo-controlled trial. Lancet 371:651–659. https://doi.org/10.1016/S0140-6736(08)60207-X

Boonyaritichaikij S, Kuwabara K, Nagano J, Kobayashi K, Koga Y (2009) Long-term administration of probiotics to asymptomatic pre-school children for either eradication or the prevention of *Helicobacter pylori* infection. Helicobacter 14:202–207. https://doi.org/10.1111/j.1523-5378.2009.00675.x

Canducci F, Armuzzi A, Cremonini F, Cammarota G, Bartolozzi F, Pola P, Gasbarrini G, Gasbarrini A (2000) A lyophilized inactivated culture of *Lactobacillus acidophilus* increases *Helicobacter pylori* eradication rates. Aliment Pharmacol Ther 14:1625–1629

Cremonini F, Di Caro S, Covino M, Armuzzi A, Gabrielli M, Santarelli L, Nista EC, Cammarota G, Gasbarrini G, Gasbarrini A (2002) Effect of different probiotic preparations on anti-*Helicobacter pylori* therapy-related side effects: a parallel group, triple blind, placebo-controlled study. Am J Gastroenterol 97:2744–2749. https://doi.org/10.1111/j.1572-0241.2002.07063.x

Fallone CA, Chiba N, van Zanten SV, Fischbach L, Gisbert JP, Hunt RH, Jones NL, Render C, Leontiadis GI, Moayyedi P, Marshall JK (2016) The Tronto consensus for the treatment of *Helicobacter pylori* infection in adults. Gastroenterology 151:51–69. https://doi.org/10.1053/j.gastro.2016.04.006

Feng JR, Wang F, Qiu X, McFarland LV, Chen PF, Zhou R, Liu J, Zhao Q, Li J (2017) Efficacy and safety of probiotic-supplemented triple therapy for eradication of *Helicobacter pylori* in children: a systematic review and network meta-analysis. Eur J Clin Pharmacol 73:1199–1208. https://doi.org/10.1007/s00228-017-2291-6

Felley CP, Corthesy-Theulaz I, Rivero JL, Sipponen P, Kaufmann M, Bauerfeind P, Wiesel PH, Brassart D, Pfeifer A, Blum AL, Michetti P (2001) Favourable effect of an acididied mild (LC-1) on *Helicobacter pylori* gastritis in man. Eur J Gastroenterol Hepatol 13:25–29

Galdeano CM, Perdigon G (2006) The probiotic bacterium *Lactobacillus casei* induces activation of the gut mucosal immune system through innate immunity. Clin Vaccine Immunol 13:219–226. https://doi.org/10.1128/CVI.13.2.219-226.2006

Goderska K, Agudo Pena S, Alarcon T (2018) *Helicobacter pylori* treatment: antibiotics or probiotics. Appl Microbiol Biotechnol 102:1–7. https://doi.org/10.1007/s00253-017-8535-7

Hill C, Guarner F, Reid G, Gibson GR, Merenstein DJ, Pot B, Morelli L, Canani RB, Flint HJ, Salminen S, Calder PC, Sanders ME (2014) The international scientific association for probiotics and prebiotics consensus statement on the scope and appropriate use of the term probiotic. Nat Rev Gastroenterol Hepatol 11:506–514. https://doi.org/10.1038/nrgastro.2014.66

Hoarau C, Martin L, Faugaret D, Baron C, Dauba A, Aubert-Jacauin C, Velge-Roussel F, Lebranchu Y (2008) Supernatant from *Bifidobacterium* differentially modulates transduction signaling pathways for biological functions of human dendritic cells. PLoS One 3:e2753. https://doi.org/10.1371/journal.pone.0002753

Jarosz M, Rychlik E, Siuba M, REspondek W, Ryzko-Skiba M, Sajor I, Gugala S, Blazejczyk T, Ciok J (2009) Dietary and socio-economic factors in relation to *Helicobacter pylori* re-infection. World J Gastroenterol 15:1119–1125

Johnson-Henry KC, Mitchell DJ, Avitzur Y, Galindo-Mata E, Jones NL, Sherman PM (2004) Probiotics reduce bacterial colonization and gastric inflammation in *H. pylori*-infected mice. Dig Dis Sci 49:1095–1102

Jung JH, Cho IK, Lee CH, Song GG, Lim JH (2018) Clinical outcomes of standard triple therapy plus probiotics or concomitant therapy for *Helicobacter pylori* infection. Gut Liver 12:165–172. https://doi.org/10.5009/gnl17177

Kabir AM, Aiba Y, Takagi A, Kamiya S, Miwa T, Koga Y (1997) Prevention of *Helicobacter pylori* by lactobacilli in a gnotobiotic murine model. Gut 41:49–55

Kamiya S (2011) Effect of probiotics on intestinal infections. Intestinal Res 9:171–178

Kim JM, Kim JS, Kim YJ, Oh YK, Kim IY, Chee YJ, Han JS, Jung HC (2008a) Conjugated linoleic acids produced by *Lactobacillus* dissociates IKK-γ and Hsp90 complex in *Helicobacter pylori*-infected gastric epithelial cells. Lab Investig 88:541–552. https://doi.org/10.1038/labinvest.2008.16

Kim MN, Kim N, Lee SH, Park YS, Hwang JH, Kim JW, Jeong SH, Lee DH, Kim JS, Jung HC, Song IS (2008b) The effects of probiotics on PPI-triple therapy for *Helicobacter pylori* eradication. Helicobacter 13:261–268. https://doi.org/10.1111/j.1523-5378.2008.00601.x

Lesbros-Pantoflickova D, Corthesy-Theulaz I, Blum AL (2007) *Helicobacter pylori* and probiotics. J Nutr 137 (Suppl 2):S812–S818. https://doi.org/10.1093/jn/137.3.812S

Lin WH, Lin CK, Sheu SJ, Hwang CF, Ye WT, Hwang WZ, Tsen HY (2009) Antagonistic activity of spent culture supernatants of lactic acid bacteria against *Helicobacter pylori* growth and infection in human gastric epithelial AGS cells. J Food Sci 74:225–230. https://doi.org/10.1111/j.1750-3841.2009.01194.x

Linsalata M, Russo F, Berloco P, Caruso ML, Matteo GD, Cifone MG, Simone CD, Ierardi E, Di Leo A (2004) The influence of *Lactobacillus brevis* on ornithine decarboxylase activity and polyamine profiles in *Helicobacter pylori*-infected gastric mucosa. Helicobacter 9:165–172. https://doi.org/10.1111/j.1083-4389.2004.00214.x

Lopez-Brea M, Alarcon T, Domingo D, Diaz-Reganon J (2008) Inhibitory effect of gram-negative and gram-positive microorganisms against *Helicobacter pylori*. J Antimicrob Chemother 61:139–142. https://doi.org/10.1093/jac/dkm404

Losurdo G, Cubisino R, Barone M, Principi M, Leandro G, Ierardi E, Di Leo A (2018) Probiotic monotherapy and *Helicobacter pylori* eradication: a systemic review with pooled-data analysis. World J Gastroenterol 24:139–149. https://doi.org/10.3748/wjg.v24.i1.139

Lu C, Sang J, He H, Wan X, Lin Y, Li L, Li Y, Yu C (2016) Probiotic supplementation does not improve eradication rate of *Helicobacter pylori* infection compared to placebo based on standard therapy: a meta-analysis. Sci Rep 6:23522/DOI 10.1038. https://doi.org/10.1038/srep23522

Lü M, Yu S, Deng J, Yan Q, Yang C, Xia G, Zhou X (2016) Efficacy of probiotic supplementation therapy for *Helicobacter pylori* eradication: a meta-analysis of randomized controlled trials. PLoS One 11(10):1371. https://doi.org/10.1371/journal.pone.0163743

Malferltheiner P, Megraud F, O'Morain CA, Gisbert JP, Kuipers EJ, Axon AT, Bazzoli F, Gasbarrini A, Atherton J, Graham DY, Hunt R, Moayyedi P, Rokkas T, Rugge M, Selgrad M, Suerbaum S, Sugano K, El-Omar EM, on behalf of the European Helicobacter and Microbiota Study Group and

Consensus panel (2017) Management of *Helicobacter pylori* infection – the Maastrich V/Florence Consensus Report. Gut 66:6–30. https://doi.org/10.1136/gutjnl-2016-312288

McFarland LV, Huang Y, Wang L, Malfertheiner P (2016) Systematic review and meta-analysis: multi-strain probiotics as a adjunct therapy for *Helicobacter pylori* eradication and prevention of adverse effects. United European Gastroenterol J 4:546–561. https://doi.org/10.1177/2050640615617358

Michetti P, Dorta G, Wiesel H, Brassart D, Verdu E, Herranz M, Felley C, Porta N, Rouvet M, Blum AL, Corthesy-Theulaz I (1999) Effect of whey-based culture supernatant of *Lactobacillus acidophilus (johnsonii)* La1 on *Helicobacter pylori* infection in humans. Digestion 60:203–209. https://doi.org/10.1159/000007660

Nam H, Ha M, Bae O, Lee Y (2002) Effect of *Weissella confusa* strain PL9001 on the adherence and growth of *Hellicobacter pylori*. Appl Environ Microbiol 68:4642–4645

Nista EC, Candelli M, Cremonini F, Cazzato IA, Zocco MA, Franceschi F, Cammarota G, Gasbarrini G, Gasbarrini A (2004) *Bacillus clausii* therapy to reduce side-effects of anti-*Helicobacter pylori* treatment: ranmdomized, double-blind, placebo controlled trial. Aliment Pharmacol Ther 20:1181–1188. https://doi.org/10.1111/j.1365-2036.2004.02274.x

Pandey KR, Naik SR, Vakil BV (2015) Probiotics, prebiotics and synbiotics – a review. J Food Sci Technol 52:7577–7587. https://doi.org/10.1007/s13197-015-1921-1

Patel R, Cockerill FR, Porayko MK, Osmon DR, Ilstrup DR, Keating MR (1994) Lactobacillemia in liver transplant patients. Clin Infect Dis 18:207–212

Pinchuk IV, Bressollier P, Verneuil B, Fenet B, Sorokulova IB, Megraud F, Urdaci MC (2001) In vitro anti-*Helicobacter pylori* activity of the probiotic strain *Bacillus subtilis* 3 is due to secretion of antibiotics. Antimicrob Agents Chemother 45:3156–3161. https://doi.org/10.1128/AAC.45.11.3156-3161.2001

Pletincx M, Legein J, Vandenplas Y (1995) Fungemia with *Saccharomyces boulardii* in a 1-year-old girl with protracted diarrhea. J Pediatr Gastroenterol Nutr 21:113–115

Quigley EM (2010) Prebiotics and probiotics: modifying and mining the microbiota. Pharmacol Res 61:213–218. https://doi.org/10.1016/j.phrs.2010.01.004

Rosania R, Minenna MF, Giorgio F, Facciorusso A, De Francesco V, Hassan C, Panella C, Ierardi E (2012) Probiotic multistrain treatment may eradicate *Helicobacter pylori* from the stomach of dyspeptics; a placebo-controlled pilot study. Inflamm Allergy Drug Targets 11:244–249

Sakamoto I, Igarashi M, Kimura K, Takagi A, Miwa T, Koga Y (2001) Suppresive effect of *Lactobacillus gasseri* OLL 2716 (LG21) on *Helicobacter pylori* infection in humans. J Antimicrob Chemother 47:709–710

Sgouras D, Maragkoudakis P, Petraki K, Martinez-Gonzalez-B, Eriotou E, Michopoulos S, Kalantzopoulos G, Tsakalidou E, Mentis A (2004) In vitro and in vivo inhibition of *Helicobacter pylori* by *Lactobacillus casei* Shirota. Appl Environ Microbiol 70:518–526

Sachdeva A, Nagpal J (2009) Effect of fermented milk-based probiotic preparations on *Helicobacter pylori* eradication: a systemic review and meta-analysis of randomized-controlled trials. Eur J Gatrotenterol Hepatol 21:45–53. https://doi.org/10.1097/MEG.0b013e32830d0eff

Sheu BS, Wu JJ, Lo CY, Wu HW, Chen JH, Lin YS, Lin MD (2002) Impact of supplement with *Lactobacillus*- and *Bifidobacterium* containing yogurt on triple therapy for *Helicobacter pylori* eradication. Aliment Pharmacol Ther 16:1669–1675

Sirvan BN, Usta MK, Kizikan NU, Urganci N (2017) Are synbiotics added to the standard therapy to eradicate *Helicobacter pylori* in children beneficial? A randomized controlled study. Eur J Hepatogastroenterol 7:17–22. https://doi.org/10.5005/jp-journals-10018-1205

Tannock GW (1999) Identification of lactobacilli and bifidobacteria. Curr Issues Mol Biol 1:53–64

Takahashi M, Taguchi H, Yamaguchi H, Osaki T, Kamiya S (2000) Studies of the effect of *Clostridium butyricum* on *Helicobacter pylori* in several test models including gnotobiotic mice. J Med Microbiol 49:635–642. https://doi.org/10.1099/0022-1317-49-7-635

Techo S, Visseanguan W, Vilaichone R-k, Tanasupawat S (2018) Characterization and antibacterial activity against *Helicobacter pylori* of lactic acid bacteria isolated from Thai fermented rice noodle. Probiotics Antimicrob Proteins. https://doi.org/10.1007/s12602-018-9385-z

Thomas DW, Greer FR, American Academy of Pediatrics Committee on Nutrition, American Academy of Pediatrics Section on Gastroenterology, Hepatology and Nutrition (2010) Probiotics and prebiotics in pediatrics. Pediatrics 126:1217–1231. https://doi.org/10.1542/peds.2010-2548

Tong JL, Ran ZH, Shen J, Zhang CX, Xiao SD (2007) Meta-analysis: the effect of supplementation with probiotics on eradication rates and adverse effects during *Helicobacter pylori* eradication therapy. Aliment Pharmacol Ther 25:155–168. https://doi.org/10.1111/j.1365-2036.2006.03179.x

Urrutia-Beca VH, Escamilla-Garcia E, de la Garza-Ramos MA, Tamez-Guerra P, Gomez-Flores R, Urbina-Rios CS (2018) In vitro antimicrobial activity and downregulation of virulence gene expression on *Helicobacter pylori* by reuterin. Probiotics Antimicrob Proteins 10:168–175. https://doi.org/10.1007/s12602-017-9342-2

Wang F, Feng J, Chen P, Liu X, Ma M, Zhou R, Chang Y, Liu J, Li J, Zhao Q (2017) Probiotics in *Helicobacter pylori* eradication therapy: systemic review and network meta-analysis. Clin Res Hepatol Gastroenterol 41:466–475. https://doi.org/10.1016/j.clinre.2017.04.004

Wang KY, Li SN, Liu CS, Perng DS, Su YC, Wu DC, Jan CM, Lai CH, Wang TN, Wang WM (2004) Effects of ingesting *Lactobacillus*- and *Bifidobacterium*-containing yogurt in subjects with colonized *Helicobacter pylori*. Am J Clin Nutr 80:737–741. https://doi.org/10.1093/ajcn/80.3.737

Wen J, Peng P, Chen P, Zeng L, Pan Q, Wei W, He J (2017) Probiotics in 14-day triple therapy for Asian pediatric patients with *Helicobacter pylori* infection: a network meta-analysis. Oncotarget 8:96406–96418. https://doi.org/10.18632/oncotarget.21633

Zagari RM, Ramano M, Ojetti SR, Gullini S, Annibale B, Farianti F, Ieraudi E, Maconi G, Rugge M, Calabrese C, Di Mario F, Luzza F, Pretolani S, Savio A, Gasbarrini G, Caselli M (2015) Guidelines for the management of *Helicobacter pylori* infection in Italy: the III Working Group Consensus Report 2015. Dig Liv Dis 47:903–912. https://doi.org/10.1016/j.dld.2015.06.010

Zhao K, Xie Q, Xu D, Guo Y, Tao X, Wei H, Wan C (2018) Antagonistics of *Lactobacillus plantarum* ZDY2013 against *Helicobacter pylori* SS1 and its infection *in vitro* in human gastric epithelial AGS cells. J Biosci Bioeng 126:458–463. https://doi.org/10.1016/j.jbiosc2018.04.003

Zhou C, Ma FZ, Deng XJ, Yuan H, Ma HS (2008) *Lactobacillus* inhibit interleukin-8 production induced by *Helicobacter pylori*. World J Gastroenterol 14:5090–5095

Adv Exp Med Biol - Advances in Microbiology, Infectious Diseases and Public Health (2019) 1149: 257–275
https://doi.org/10.1007/5584_2019_370

Published online: 24 April 2019

Immunity and Vaccine Development Against *Helicobacter pylori*

Anna K. Walduck and Sukanya Raghavan

Abstract

Helicobacter pylori is a highly-adapted gastro-intestinal pathogen of humans and the immunology of this chronic infection is extremely complex. Despite the availability of antibiotic therapy, the global incidence of *H. pylori* infection remains high, particularly in low to middle-income nations. Failure of therapy and the spread of antibiotic resistance among the bacteria are significant problems and provide impetus for the development of new therapies and vaccines to treat or prevent gastric ulcer, and gastric carcinoma. The expansion of knowledge on gastric conventional and regulatory T cell responses, and the role of T_H17 in chronic gastritis from studies in mouse models and patients have provided valuable insights into how gastritis is initiated and maintained. The development of human challenge models for testing candidate vaccines has meant a unique opportunity to study acute infection, but the field of vaccine development has not progressed as rapidly as anticipated. One clear lesson learned from previous studies is that we need a better understanding of the immune suppressive mechanisms *in vivo* to be able to design vaccine strategies. There is still an urgent need to identify practical surrogate markers of protection that could be deployed in future field vaccine trials. Important developments in our understanding of the chronic inflammatory response, progress and problems arising from human studies, and an outlook for the future of clinical vaccine trials will be discussed.

Keywords

Helicobacter pylori · Gastritis · Immunity · Vaccine · Regulatory T cell

A. K. Walduck (✉)
School of Science, RMIT University, Melbourne, VIC, Australia
e-mail: Anna.walduck@rmit.edu.au

S. Raghavan
Department of Microbiology and Immunology, Institute of Biomedicine, University of Gothenburg, Gothenburg, Sweden
e-mail: sukanya.raghavan@microbio.gu.se

1 Introduction

A recent meta-analysis of publications from 73 countries reported that *Helicobacter pylori* infects 44.3% of the global population (Zamani et al. 2018). It is believed that after acquiring infection in childhood, most patients are infected for many decades. What is fascinating about *H. pylori* infection, is that only around 15% of these develop symptomatic gastric disease in the form of chronic gastritis, peptic ulcer, mucosa associated lymphoma, and 1–3% develop carcinoma (Parsonnet et al. 1991; Posselt et al. 2013). Given that *H. pylori* establishes a chronic infection in such a high percentage of the population, it is perhaps more pertinent to ask the question of why such a low proportion of infected persons develop symptoms? It would appear that this long- term relationship between the bacterium and host is finely balanced.

One overarching explanation for this tight relationship lies in the long period of co- evolution of *H. pylori* with the human host (Moodley et al. 2012). The studies that document the migrations of human populations with their *H. pylori* and have elegantly illustrated this (Falush et al. 2003). Consistent with the success in the wide distribution of this pathogen, and hand in hand with idea of co-evolution, is the proposal that *H. pylori* was a member of the gastric microbiota until relatively recently (Otero et al. 2014). It has been reasoned that improvements in hygiene, and widespread use of antibiotics in the decades since they became widely available in the 1940s mean that *H. pylori* is now carried by less people. The status of a flora organism may go some way to explain the minimal response in many *H. pylori* infected individuals. Indeed, some workers have described *H. pylori* as a pathobiont (Arnold et al. 2017).

The human host also shows evidence of co-evolution insofar that although an immune response is generated after infection, the bacterium is not cleared. Indeed, the nature of the natural immune response to *H. pylori* infection appears to be aimed not at clearance, but rather at the suppression of overt inflammation, and therefore symptomatic disease. The current status of our understanding of this suppressive milieu will be discussed here, with a focus on the knowledge gained from analysis of clinical material.

The recommendations of the 2017 consensus report on clinical management of *H. pylori* infections include what has been called a paradigm shift in the clinical view- that *H. pylori* disease is an infectious disease that should be treated in all patients were infection is detected, not just in symptomatic patients (Malfertheiner et al. 2017). This shift provides additional impetus for development of better therapies, and for prevention and vaccine development.

Given the extent of *H. pylori* disease worldwide, and problems with antibiotic resistance and treatment failure, an efficacious vaccine is a desirable aim. Vaccine development although initially promising, has struggled to progress in recent years and lacks investment. The current status of vaccine clinical trials will be discussed in the context of our understanding of the natural immune response and mechanistic studies in animals. Finally, we discuss a view to the way forward for vaccine development against this challenging pathogen.

2 The Natural Immune Response to *H. pylori* – A Tight Balance of Chronic Inflammation and Suppression That Ultimately Fails to Control Infection

From an immunological standpoint, chronic infection with *H. pylori* is a complex equilibrium between the inflammatory and suppressive aspects of immunity. A combination of bacterial and host factors are now known to orchestrate this.

Histologically, the chronic-active gastritis seen in the "snap-shot" from biopsies from symptomatic adults infected with *H. pylori* is characterized by strong infiltrates of monocytes, polymorphs, and lymphocytes. Lymphoid follicles containing primed B cells form, and infection persists despite the production of *H. pylori* specific IgA, and IgG (Dixon 2001). More detailed analysis of gastric infiltrates has revealed that in addition to macrophages and neutrophils, $CD4^+$ and $CD8^+$ and B lymphocytes, are present (Bamford et al. 1998a; Muñoz et al. 2007).

Early studies established that *H. pylori* gastritis is T_H1 biased with significant amounts of IL-8 and interferon-γ (IFN-γ) detected in infected humans (Bamford et al. 1998b; Lindholm et al. 1998) and mice (Eaton et al. 2001). Studies in $CD4^+$T cell depleted mice revealed that $CD4^+$ T cells were required for the development of *H. pylori* gastritis (Eaton et al. 2001). Seemingly at odds with these pro-inflammatory responses, is the presence of a $CD4^+$ subset of regulatory T cells (Treg). Treg are MHCII restricted and have the ability to reduce the effector functions of activated T cells, B cells, dendritic cells (DC) and natural killer cells (NK) (Sakaguchi et al. 1995).

The role of Tregs in suppressing inflammation and promoting chronic infection has been investigated in depth, and *Helicobacter*-specific Tregs were demonstrated in both humans (Lundgren et al. 2003) and mice, (Raghavan et al. 2004). In functional studies, Tregs were depleted from infected C57BL/6 mice using anti-CD25 antibody, resulting in increased levels of gastritis, increased antibody titres, and bacterial numbers (Rad et al. 2006). A study published in the same year reported however that depletion of Tregs from infected BALB/c mice did not affect either gastritis or colonization, but antibody IgG1 titres were increased (Kaparakis et al. 2006). Further, the effect was the same when the mice were infected with *H. pylori*, or the close relative *H. felis* (Kaparakis et al. 2006). The T_H2 bias of the BALB/c background probably explains the difference seen in the effects of depleting $CD25^+$ Tregs on *H. pylori* induced inflammation, and this highlights the fact that the cytokine milieu also has an overriding effect on Treg function, an observation that has also been made for other "non- responsive" mouse strains (Nedrud et al. 2012). Another aspect that Treg depletion studies have brought to light is the potential for autoimmune reactions, the use of anti-CD25 antibody to deplete Tregs did not increase the incidence of autoimmune gastritis in wild-type C57BL/6 or BALB/c *H. pylori* infected mice, but did in 1E4-TCR mice that are predisposed to auto-immune gastritis (Kaparakis et al. 2006). As discussed, there is potential for bias in results due to the genetic background of the mice used, however if Tregs are to be investigated as therapeutic targets for ulcer and tumours, the development of autoreactivity should be included in study design.

2.1 First Contact- the Intensity of the Initial Inflammatory Response Is Limited by Host Factors

The gastric mucous layer has a complex structure and is composed predominantly of gel-forming mucins MUC5AC and MUC6 which forms a protective layer above the epithelium (reviewed in (McGuckin et al. 2011)). This barrier is not only physical, and a signalling role for Muc1 has been demonstrated. Studies using *H. pylori* infected $Muc1^{-/-}$ knockout mice show reduced inflammation compared to wild-type littermates (McGuckin et al. 2007; Guang et al. 2012) and inhibition of the NLRP3 inflammasome (Ng and Sutton 2016). Studies in AGS cells with Muc1 knock-down or overexpression, have resulted in the conclusion that Muc1 acts to prevent activation of the pro-inflammatory transcription factor NF-κB (Guang et al. 2010), and that it also interacts with both CagA and β-catenin, resulting in attenuation of IL-8 production (Guang et al. 2012). A number of antibacterial epithelial factors also inhibit *H. pylori* in the inner gastric gel layer including MUC6, defensins (HBD2, LL37), galectin 3, and Trefoil factors (TFFs) also play roles in preventing access (reviewed in (Dhar et al. 2016)). The protective effects of the mucous layer, in combination with lipopolysaccharide (LPS) that is less pyrogenic and immunoreactive than that of other Gram-negative pathogens (Moran 2007), in part explain the comparatively mild inflammation triggered even in acute infections. Recently developed *in vitro* models employing organoid cultures of polarized, mucin secreting, human gastric epithelia have also provided new potential for investigations on the interactions between *H. pylori* and the epithelium (Schlaermann et al. 2016; Sigal et al. 2017). These models also present exciting possibilities to investigate the interactions between immune cell populations and the gastric epithelium.

2.2 *H. pylori* Drives Dendritic Cells (DC) to a Tolerogenic Phenotype

Activation of the gastric epithelium brings DCs into play. Dendritic cells patrol mucosal surfaces to provide early recognition of pathogens. Intra vital microscopy has shown DCs in the upper area of the stomach mucosa in mice (Kao et al. 2010), and cells with DC-like morphology were also detected by electron microscopy in biopsies from heavily infected humans (Necchi et al.

2009). In the latter study, projections were seen to contact *H. pylori,* consistent with earlier reports of intraluminal sampling of other pathogens by DCs in the intestine (Rescigno et al. 2001). This observation indicates that despite the establishment of long- term Treg populations, DCs continue to patrol and present antigen, potentially contributing to the chronic-active nature of *H. pylori* gastritis.

Another feature that may prevent excessive inflammation overall is the focal nature of *H. pylori* infection. Local damage to the mucous or epithelial barrier facilitates focal colonisation. Morey and co-workers have described a possible specific mechanism, in that the actions of *H. pylori* cholesterol-α-glucosyltransferase effectively deplete cholesterol in the epithelium, effectively interfering with inflammatory signalling and thus permitting inflammation only in adjacent areas (Morey et al. 2018). Foci of inflammation are reported in human biopsies (Cherdantseva et al. 2014) and colonisation is not distributed uniformly throughout the stomach. This is also one reason for the consensus recommendation of a minimum of two biopsies from the antrum and two from the middle of the gastric body for clinical diagnosis (Malfertheiner et al. 2017).

Despite the actions of a number of host factors which appear to dampen responses, an *H. pylori*-specific immune response does develop. Expression of epithelial Toll-like receptors (TLRs) is increased (Lagunes-Servin et al. 2013, Pachathundikandi et al. 2015), and cytokines and chemokines are secreted which attract DCs and neutrophils. Dendritic cells co-localized with lymphocytes secreting IL-23 and IL-17 were reported in antral biopsies from *H. pylori* infected patients (Khamri et al. 2010), and CagA-positive *H. pylori* were more effective in stimulating DC to produce IL-23, promoting IL-17 secretion by autologous $CD4^+$ T cells (Fig. 1). Other studies have implicated bacterial factors in driving a DC response that is skewed towards Treg responses. This response was originally reported to be independent of CagA or Vac A in murine DC *in vitro* (Kao et al. 2010). In a later study however, vacA and γ-glutamyl transpeptidase (GGT) both independently promoted generation of tolerizing DCs in a mouse model (Oertli et al. 2013). In the case of CagA, the mechanism of this effect appears to be that it impairs maturation of human DCs, and therefore promotes development of tolerogenic DCs (Kaebisch et al. 2013). In addition, a recent study using peripheral blood mononuclear cells (PBMC) suggested that *H. pylori* heat shock protein 60 (HpHSP60) acts on macrophages to trigger secretion of IL-10 and TGF-β to promote production of Tregs (Hsu et al. 2018) (Fig. 1).

2.3 Reduced Inflammatory Responses in *H. pylori* Positive Children

H. pylori infection is thought to be acquired in early childhood in the majority of cases, however symptoms frequently do not manifest until decades later. An understanding of the progression of inflammatory responses in young children will enhance our understanding of how the gastritis is largely suppressed, and perhaps how best to target a prophylactic vaccine. Studies in a cohort of Chilean children showed that in an area of high prevalence approx. 60% of children had positive stool antigen detected by ELISA by 2 years of age (O'Ryan et al. 2015). Of these children, 22% were persistently infected (several sequential positive stool antigen tests) by the age of 5, and 11% were only intermittently infected. Significantly, only 2% of children in the study had clinical and pathological findings after follow-up by a gastroenterologist. The latter observation is consistent with the findings of a number of studies that have all reported that low levels of gastritis in children (Muñoz et al. 2007; Harris et al. 2008). These reports suggest that the Treg response is established soon after initial infection, and indeed increased proportions of Tregs were reported in children compared to (symptomatic) adults in both Chile and Brazil (Harris et al. 2008; Freire de Melo et al. 2012). Results from studies with neonatal mice are consistent with this, and neonatal infection resulted in reduced inflammation that was dependent on Tregs (Arnold et al. 2011). While reduced gastric inflammatory response minimizes symptoms it does not however

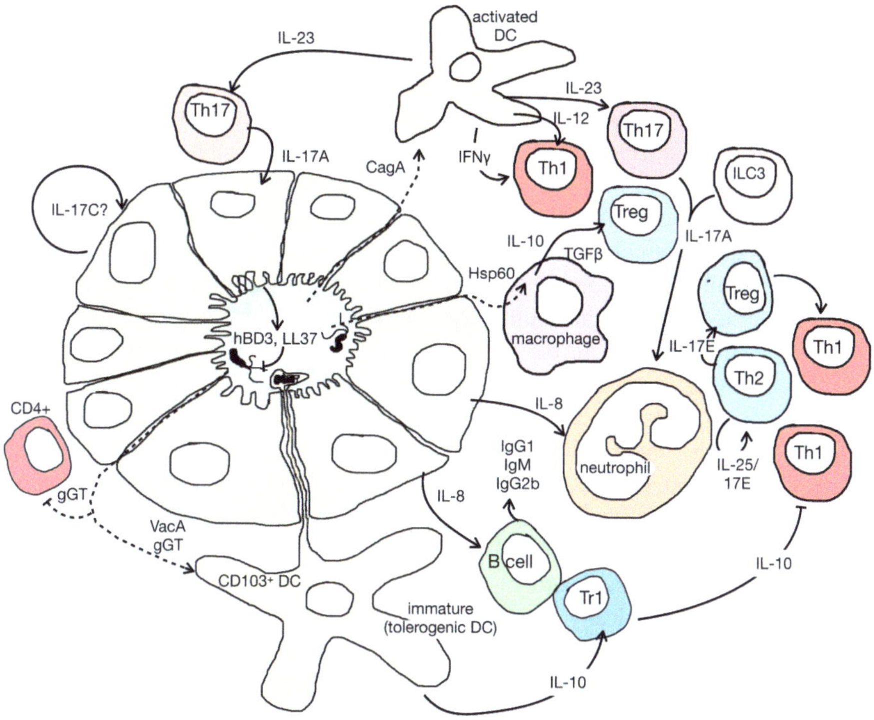

Fig. 1 Complex control of chronic inflammation in *H. pylori* infection is mediated by a balance of T_H1/ T_H2/T_H17 and Treg responses. Bacterial factors (dotted lines) VacA and GGT prevent maturation in dendritic cells (DC), promoting Treg responses, GGT also interferes with cell- cycle progression in CD4 + T cells. HpHsp60 drives macrophages to secreted TGF-β and IL-10, promoting induction of Treg. CagA positive strains stimulate IL-8 production by epithelial cells which attracts neutrophils and B cells (solid arrows), CagA can also activate DC to secreted IL-23 and IL-12 promoting pro-inflammatory T_H1 and T_H17 T cell populations (red shades). T_H17 (and possibly ILC3), secrete IL-17A which induces anti-microbial peptide (eg. HBD3, LL-37) production by epithelial cells, contributing to mucosal defence. Anti-inflammatory T cell populations (blue shades) Treg suppress the actions of T_H1 and T_H17 cells by contact dependent means, Tr1 secrete IL-10 which also suppresses T_H1

suppress the growth of *H. pylori,* and children have similar levels of colonization as adults (Freire de Melo et al. 2012), whereas neonatally infected mice had significantly higher levels of colonization (Arnold et al. 2011).

2.4 Chronic- Active Gastritis Is Driven by the Actions of $T_H1/17$ and Treg

The central role of $CD4^+$ T cells in *H. pylori* gastritis is still not completely understood and a strong focus has been the study of the role of T_H1/ T_H2/T_H17 cells and the cytokines that regulate these populations. Early studies showed that *H. pylori* specific $CD4^+$T cell clones can be isolated from the stomach of chronically infected subjects, and that they secreted IFN-γ and tumour necrosis factor-α (TNF-α) but not IL-4 or IL-5 (D'Elios et al. 1997). Since then, levels of inflammatory cytokines IFN-γ, TNF-α, IL-1, IL-6, IL-7, IL-8, IL-17-A, IL-18, and IL-23, as well as anti-inflammatory cytokines- transforming growth factor (TGF-β) and IL-10, have all been reported to be present at increased levels in the mucosa of *H. pylori* infected compared to healthy patients (Di Tommaso et al. 1995; Luzza et al. 2000; Mizuno et al. 2005; Caruso et al. 2008; Bagheri et al. 2017).

It appears that individual patient differences in the expression levels of these cytokines determine the severity of inflammation and disease outcomes. Indeed, polymorphisms in the promotor regions of many cytokines, are associated with increased cancer risk (reviewed in (Rivas-Ortiz et al. 2017)). In a recent comparison of paired antral and corpus biopsies, the frequencies of IL-17A and IFN-γ^+ cells were increased in *H. pylori* infected patients, particularly in those with gastric ulcers (Adamsson et al. 2017). Of note, was that although *H. pylori* were detected in both antrum and corpus, increased cytokine expression and the inflammation were only seen in the antrum. Studies in the mouse model have shown that IL1-7A secreted by T_H17 cells drives IL-8 production by epithelial cells, recruits neutrophils, and B cells (Algood et al. 2009), thus contributing to mucosal defence, but also to chronic active gastritis (reviewed by (Chamoun et al. 2018)) (Fig. 1).

The link between proinflammatory and regulatory responses is intrinsically linked by T_H17 plasticity (Morrison et al. 2013), and the decision of antigen stimulated cells to differentiate in to T_H17 or Tregs is dependent on the cytokine-driven activation of the transcription factors RORγt (T_H17) or FOXP3$^+$ (Treg) (Omenetti and Pizarro 2015). The actions of IL-17 are mediated by binding to multimeric receptors (IL-17RA-E) that are expressed on a variety of cell types including epithelial cells, lymphocytes, and myeloid cells, as such IL-17R signalling acts as a link between innate and adaptive immunity (Gaffen 2009).

Most studies on the role of IL-17 in *H. pylori* pathogenesis to date have focussed on IL-17A, and the related IL-17F, and studies using IL17 and IL-17R deficient mice, and depletion of IL-17 with antibodies have revealed an import role in the control on gastritis (Delyria et al. 2009) and on vaccine-induced protection (Velin et al. 2009; Flach et al. 2012). Mice that are deficient in IL-17RA $^{-/-}$ do not control *H. pylori* numbers as well as wild-type mice, and this is linked to reduced recruitment of neutrophils, but enhanced B cell recruitment (Algood et al. 2009). In contrast to IL-17A which is secreted chiefly by T cells, other members of the IL-17 family (IL-17B-E) are secreted by a wide range of cell types including the gastric epithelia, and investigation of these in the future is likely to bring new understanding to the control of inflammation in the gastric mucosa. A recent study showed that levels of IL-17C were increased in biopsies of *H. pylori* infected patients, and that expression was localized to epithelial, and chromogranin A positive endocrine cells (Tanaka et al. 2017). The mechanistic role of IL-17C remains unresolved but its receptor IL-17RE is expressed in stomach glandular tissue, suggesting a role in epithelial responses to infection.

In a model of experimental colitis, IL-17E (also known as IL-25) was reported to inhibit secretion of both IL-23 and IL-12 (Caruso et al. 2009), and is considered to promote T_H2 responses, and to enhance the function of Treg (Tang et al. 2015). IL-17E signals through a multimeric IL-17RA/RB receptor and Horvath and co-workers reported that despite some reduction in T_H2 responses, colonization and inflammation was similar in IL-17RB$^{-/-}$ and wild-type mice, indicating that neither IL-17B or E are critical for control of natural infection (Horvath et al. 2013). It would be of interest to test whether this also applies in a vaccination model.

2.5 Treg Sub-populations and the Control of Gastritis

Almost all studies on Tregs and *H. pylori* infection have focussed on FoxP3$^+$ Tregs (Lundgren et al. 2004; Enarsson et al. 2006; Harris et al. 2008; Aebischer et al. 2008). This population is CD25^{hi}, CD4$^+$, FoxP3$^+$, and are antigen-specific. A few studies have investigated the suppressive capacity of Tregs isolated from the gastric mucosa of patients (Rad et al. 2006; Enarsson et al. 2006) and mice (Raghavan et al. 2004), demonstrating the specificity for *H. pylori* antigens. Depletion of Tregs was also shown to increase severity of gastritis in mice (Raghavan et al. 2003; Rad et al. 2006).

The presence of Tregs is not sufficient to prevent disease, however, and two studies have reported that Tregs are found in both symptomatic and asymptomatic *H. pylori* infected individuals (Lundgren et al. 2004; Robinson et al. 2008). Further, a study in patients with gastric ulcers revealed that these patients had lower frequencies of $CD4^{+}/CD25^{+}$, and lower expression of FOXP3, compared to infected donors with no ulcer (Robinson et al. 2008).

FoxP3^{-}negative Tregs (Tr1) have been detected in *H. pylori* infected mice (Blanchard et al. 2004; Nedrud et al. 2012). Tr1 cells are not antigen-specific and mediate suppressive activity via IL-10 secretion and induction of cytotoxic killing (Maynard et al. 2007) (Fig. 1). An interesting additional aspect has been raised by a recent study that reported that frequencies of Tr1 cells were increased in asymptomatic *H. pylori* infected patients, and further that the Tr1 cells from patients with GC also had reduced suppressive function (Song et al. 2018). This also raises the question of whether the combined actions of antigen specific FOXP3^{+} Treg and FoxP3^{-}Tr1 act in concert in *H. pylori* disease. The study by Song and co-workers unfortunately did not report on the relative frequencies of FoxP3^{+} Treg in the same patients (Song et al. 2018). More detailed studies of Treg populations isolated from non-ulcer dyspepsia, ulcer and carcinoma lesions, using approaches such as single cell expression profiling are required to provide clarify the roles of Treg subpopulations in the pathogenesis of *H. pylori* disease.

2.6 Unconventional Lymphocytes in *H. pylori* Infections

Mucosal-associated invariant T cells (MAIT) are a class of innate-like T cells that have been found in many organs and are involved in anti-microbial defences. Investigations in the mucosa and blood of *H. pylori* infected patients revealed that significant numbers of MAIT cells were present in the gastric mucosa (Booth et al. 2015; D'Souza et al. 2018), there was however no difference in the frequency of MAIT cells in the gastric mucosa of infected and healthy patients, the numbers in the blood were significantly decreased in *H. pylori* negative patients (Booth et al. 2015). These unconventional T cell populations may have as yet unknown roles in regulating inflammation.

Innate lymphoid cells (ILC) also represent an understudied area with regard to *H. pylori* infection. ILCs are of particular interest because they are distributed at epithelial surfaces and appear to act as intermediates between innate and acquired immune cells (Moro and Koyasu 2015). The cytokine secretion patterns of ILCs allow them to be classified into types that correspond generally to T_H1 (ILC1), T_H2 (ILC2) and T_H17 (ILC3). One study implicated ILC2s to be increased in *H. pylori* infected patients, and mice (Li et al. 2017). Similarly, ILC3 are a potential source of IL-17 which may drive gastritis (Fig. 1). Because there are to date no animal models that are deficient in ILC, the specifics of their roles remain to be investigated.

2.7 Acute Infection, Spontaneous Clearance and Re-Infection in Adults

The development of *H. pylori* challenge models in adult volunteers has permitted study of acute infection in adults and provided new insight in to the early infection process. In the first study using this model, volunteers were infected with a single dose of the Baylor strain (CagA-negative) (Graham et al. 2004). While some developed mild/moderate dyspepsia, all resolved by the end of the study developed inflammation with the histological appearance of "acute chronic" gastritis by 2 weeks post infection. Immunological studies on these patients were unfortunately limited, and no data on induction of Tregs were reported. Only a small number of studies have been performed with the infection model, but the data that are available resulted in a number of findings, some of which are unexpected, including apparent clearance of the infection in a proportion of those challenged (discussed later).

The participants in the challenge volunteer trials described above were all previously *H. pylori*-negative. A report from a small study on re-infection in two *H. pylori* positive individuals provides evidence for a lack of protection induced by natural infection, and both volunteers could be repeatedly re-infected after eradication (Stenström et al. 2016).

Interestingly an epidemiology study with 10-year follow-up in Sorbo San Basile in a rural area in Italy with high infection rates for *H. pylori* reported reversion to sero-negativity in only 3% of participants (Luzza et al. 2014). It seems that clearance is unusual in adults with infections that were presumably established in childhood. *H. pylori* negative adults may however possess immune mechanisms that are at least partially effective at preventing the bacterium from establishing.

These studies in acutely and chronically infected adults have revealed two interesting points: that acutely infected adults develop chronic gastritis and Treg responses are established within a short period, and further that spontaneous clearance in adults acquiring infection for the first time is relatively frequent. Further investigation of this phenomenon may reveal useful information to guide vaccine development. The successful reinfection after eradication in the study by Stenström and co-workers was cited as a potential ground for the lack of viability of a therapeutic vaccine (Stenström et al. 2016). Analysis of the mucosae of vaccinated mice has however shown that vaccination induces a different "program" in the mucosa from infection (Mueller et al. 2003; Walduck et al. 2004). Vaccinated human volunteers, also had distinct gene expression signatures (Aebischer et al. 2008).

3 Progress Towards a Vaccine Against *H. pylori*

3.1 The Need for a Vaccine and Its Potential

The current *H. pylori* eradication treatment regime using antibiotics combined with proton pump inhibitors have been improved in recent years and is discussed in detail elsewhere in this book (Chapter 13: Treatment of *Helicobacter pylori* Infection). Indeed, currently the most cost-effective method of preventing *H. pylori* induced peptic ulcers (gastric and duodenal) is antibiotic treatment. However, since peptic ulcers and cancers can be regarded as two mutually exclusive diseases, vaccination ideally should target those individuals that do not have a previous history of peptic ulcers. Thus, it would be useful to screen *H. pylori* infected individuals particularly in regions of the world where both *H. pylori* infection and gastric cancer are highly prevalent, and to select subjects for vaccination that show pre-malignant changes in their gastric mucosa. The concept of a vaccine to prevent inflammation and cancer is well established for the prevention of cervical cancer caused by the human papilloma virus (HPV). The HPV vaccine, has been successful in dramatically reducing the prevalence of infections (Machalek et al. 2018), and are projected to reduce the number of cases of cervical cancer and mortality by over 30% by 2035 (Hall et al. 2018). Gastric cancer (GC) is responsible for 9% of cancer deaths world-wide (around 80% of which are *H. pylori* associated), and a vaccine that would prevent *H. pylori*-induced GC could be expected to have a significant impact. Ideally, a highly efficacious vaccine should be based on our understanding of the *H. pylori* bacterium and its interplay with the human host in order to eradicate the infection and protect against reinfection.

Mathematical modelling studies by Rupnow and co-workers (Rupnow et al. 2000, 2001) have been helpful in understanding the potential impact of introducing a prophylactic *H. pylori* vaccine for children in developed versus low- and middle-income countries. They report that vaccines against *H. pylori* infection could be cost-effective from a long term perspective in reducing both the prevalence of the infection, and incidence of GC in the US and Japan (Rupnow et al. 2001, 2009). Since the prevalence of the infection in developed countries is decreasing naturally without intervention, a vaccination scheme running for 10 years was predicted to almost eradicate the infection at the population level (Rupnow et al. 2009). In low and middle-income countries however the scenario is different; the high prevalence rate of *H. pylori* infection means that it would require a longer time (>10 years of continuous vaccination) and also a wider vaccine reach for it to effectively reduce the prevalence of *H. pylori* infection in the population (Rupnow et al. 2001). These are some aspects that should be considered when discussing mass vaccination schemes, which have been effective in reducing the incidence of other enteric infectious diseases.

Vaccination using a number of different strategies, and using both mucosal and parenteral route have been tested in mouse models, and in general significant reductions in colonization, but not sterilizing immunity have been observed in most models. These pre-clinical studies are too numerous to be discussed in detail here (for recent reviews see Raghavan and Quiding-Järbrink 2016; Sutton and Boag 2018).

3.2 Current Status of Clinical Trials of Candidate *H. pylori* Vaccines

After extensive pre-clinical studies that have explored a range of approaches to the design of a *H. pylori* vaccine, three have progressed to further clinical trials. These are:

1. Mucosal vaccination with *H. pylori* whole cell vaccine antigens and active *Escherichia coli* heat-labile toxin (LT) or with an LT mutant, LT_{R192G} as adjuvant (Michetti et al. 1999; Kotloff et al. 2001).
2. Oral vaccination with live attenuated *Salmonella* carriers expressing *H. pylori* urease (DiPetrillo et al. 1999; Angelakopoulos and Hohmann 2000; Bumann et al. 2001; Metzger et al. 2004; Aebischer et al. 2008) and finally;
3. Parenteral immunization with recombinantly produced *H. pylori* antigens and alum adjuvant (Malfertheiner et al. 2008, 2018).

Unfortunately, all three approaches have failed to provide a strong candidate for further clinical development. This has been in part due to setbacks experienced during the trials that were not expected, despite the designs being based on extensive pre-clinical data as discussed below.

The *H. pylori* vaccine containing the enterotoxin LT as an adjuvant, as might be anticipated resulted in side-effects such as diarrhoea in a large proportion of the volunteers (Michetti et al. 1999). Furthermore, even though both adjuvants, LT or LT_{R192G} were able to induce strong specific serum IgG or mucosal IgA antibody against the co-administered antigens, *H. pylori* colonisation was either not studied (Kotloff et al. 2001), or not affected by vaccination (Michetti et al. 1999). Thus, the long-term efficacy of the vaccine on *H. pylori* colonization will not be known due to the difficulties in studying colonization levels in the *H. pylori* infected vaccinated subjects.

Vaccination with *Salmonella* carrier expressing urease, or another candidate antigen (Hp0231), presented other challenges than the enterotoxicity of the LT or LT_{R192G} adjuvant. The *S.* Typhimurium Ty21a-based vaccine, although safe and successful in inducing strong immune responses to the *Salmonella* carrier indicating uptake and presentation to the immune system, unexpectedly resulted in only weak immune responses to the urease antigen and decreased the bacterial load only in a few volunteers (Metzger et al. 2004; Aebischer et al. 2008). The mechanisms leading to the weak immune responses to urease, particularly antibody response has not been fully explored and could possibly relate to low *in vivo* expression of urease by the *Salmonella* carrying the plasmid. In addition, another aspect to consider regarding the *Salmonella* carrier vaccination approach is that pre-existing immunity to *Salmonella* might prevent uptake during booster immunizations (Metzger et al. 2004). The two aforementioned approaches discussed above have focused on vaccination via the mucosal route.

Malfertheiner and colleagues have instead investigated the potential of a systemic route of immunization since there are known adjuvants that are registered for human use. Thus, parenteral administration of a mixture of putative protective antigens, CagA, VacA and NapA, together with an alum adjuvant (Malfertheiner et al. 2008) led to strong circulating humoral and cellular immune responses in *H. pylori* negative healthy volunteers, which was very encouraging. A follow up clinical study with the same intramuscular immunizations with CagA, VacA and NapA, together with an alum adjuvant was carried out combined with a challenge with live *H. pylori* $CagA^+$ Baylor strain 300. The results of the study were recently reported and, compared with placebo, intramuscular immunization with alum and *H. pylori* antigens, CagA, VacA and NapA did not confer additional protection against *H. pylori* infection, in spite of the strong antigen-specific antibody and T cell responses induced (Malfertheiner et al. 2018). The study also reports that a spontaneous clearance of *H. pylori* infection in the healthy adult population occurs, a phenomenon that was also reported in a proportion of the subjects in the previous challenge study. (Aebischer et al. 2008). Since at least some adults seem to spontaneously eradicate the *H. pylori*, it has to be concluded that adults are not ideal subjects for testing vaccine-efficacy as colonization cannot be used as a primary endpoint of the trial. In the discussion of their findings, Malfertheiner and colleagues suggest that the next vaccine trial with a candidate *H. pylori* vaccine should be carried out in children that become naturally infected (Malfertheiner et al. 2018).

Indeed, vaccination of *H. pylori*-negative children and follow up of acquisition of infection was explored in a phase III trial performed between 2005 and 2007 in Nanjing, Jiangsu Province, China (Zeng et al. 2015). No prior data pre-clinical studies or phase I/II trials had been published before the phase III trial was reported. The vaccine consisted of a fusion protein of *H. pylori* urease and the non-toxic *E. coli* heat labile toxin B subunit (LTB). The LTB subunit presumably functioned as an adjuvant in the vaccine, since strong enhancement of *H. pylori* specific IgG and IgA responses against the urease antigen was reported in the subjects. The immune responses were also associated with a 72% efficacy at follow up year 1, and 65% at year three. Indicating that (1) the urease-LTB fusion protein is safe and immunogenic and (2) it could provide protection against acquisition against *H. pylori*. However, since the study was performed more than 10 years ago with no follow up, the future of this vaccine remains unclear.

3.3 Ongoing Clinical Trials Listed on Public Clinical Trials Database

At the time of writing this book chapter there was one trial recruiting for a vaccine against *H. pylori* infection (www.clinicaltrials.gov). This vaccine consists of GGT and a non-toxic derivative of cholera toxin as an adjuvant. The enzyme GGT secreted by *H. pylori* has a unique function in that it can inhibit T cell proliferation by inducing cell cycle arrest in the G1 phase of T cells (Gerhard et al. 2005) (Fig. 1). Thus, the premise of the vaccine is different from those previously tested, and is based on the principle that by inducing neutralizing antibodies through vaccination blocking GGT activity, T cell inhibition can be overcome leading to a productive T cell response to *H. pylori* antigens and protection against infection. Pre-clinical work has shown that therapeutic vaccination with GGT and adjuvant, induces neutralizing antibodies to GGT and reduction in the number of the bacteria in the stomach of mice (Anderl et al. 2012). In the ongoing phase 1a/b trial (Clinical Trials Identifier: NCT03270800), safety is the primary end point and the secondary end point is the analysis of neutralizing antibodies to GGT in both in *H. pylori* positive and *H. pylori* negative subjects. Although protection against *H. pylori* infection will not be evaluated in this trial, it still holds promise for an effective vaccine against *H. pylori* infection due to the unique mechanism of targeting GGT.

3.4 What Have We Learned from the Clinical Trials That Can Help Us to Make a Better Vaccine?

There are important factors that need to be taken into consideration when designing a vaccine against *H. pylori*, namely, the choice of antigen, the route of immunization, and the safety of an adjuvant. An optimal combination of these three factors will be crucial for the success of a *H. pylori* vaccine in the future.

Antigens Vaccines based on *Salmonella* carriers expressing urease are not currently being pursued, due to the weak antibody and T cell responses generated to the heterologous antigen. In this regard, whole-cell vaccines might be more consistent in enhancing immune responses to *H. pylori* antigens, as long they can be combined with a non-toxic mucosal adjuvant since the *H. pylori* pathogen associated molecular patterns (PAMPs) are very poor at activating the immune system. Currently two independent studies have reported growth of *H. pylori* in fermenter scale and inactivation using formalin to prepare the whole-cell vaccine (Summerton et al. 2010; Holmgren et al. 2018). *H. pylori* whole cell vaccines seem to be effective in enhancing immune responses (Kotloff et al. 2001), but there is uncertainty as to whether the immune responses are strain specific. In this regard, a selection of highly immunogenic antigens produced recombinantly has the advantage of a streamlined production with minimal batch-to-batch variation (Satin et al. 2000; Malfertheiner et al. 2008).

Route The mucosal route of vaccination should be considered since the infection is acquired via the mucosal route and would need mucosal homing $CD4^+$ T cells expressing $\alpha4\beta7$ to be induced for protection (Michetti et al. 2000). In addition, since *H. pylori* is prevalent in low- and middle-income countries which are also target population for the vaccine, a drinkable needle-free preparation will be much safer to introduce for mass-vaccinations.

Adjuvants Safe and non-toxic mucosal adjuvants are available such as the non-toxic double-mutant *E. coli* heat-labile toxin (dmLT) which has shown a good safety profile in human volunteers in a bacterial vaccine directed against enterotoxigenic *E. coli* (ETEC) (Lundgren et al. 2014). It is also possible that multiple rounds of immunizations might be necessary as reported in the trial by Zeng and co-workers (Zeng et al. 2015).

The implementation of a prophylactic vaccine in children is going to require further careful work before it will have wide application. There is however potential for application of a prophylactic vaccine to benefit other patient groups. Administration of prophylactic vaccine to *H. pylori*-positive individuals at the same time as, or after treatment with antibiotics would be an excellent strategy to promote cure, and to prevent re-infection. Clinical studies in *H. pylori*-positive subjects have shown that eradication therapy frequently has an unacceptably low success rate (Graham et al. 2007), so a prophylactic vaccine could have impact as an adjunct therapy. Further, as discussed above studies in the human challenge model have shown that some volunteers spontaneously eradicate the infection, making it difficult to assess protection in vaccinated subjects. Thus, the next trial of a candidate *H. pylori* vaccine should recruit uninfected individuals (or those treated with antibiotics) and then follow the rate of infection or reinfection over a 3–5 year period, much in the same way as the previous field trial (Zeng et al. 2015). The *H. pylori* challenge model trials have unfortunately raised more questions than they have been able to answer regarding the protective efficacy of vaccine candidates, and so in our opinion field trials are going to be the only possible approach. As mentioned previously, the target for such a vaccine would initially be principally children, and possibly adults identified to be at the risk for developing GC, and possibly also patients with recurrent *H. pylori*-induced peptic ulcers.

3.5 Is a Vaccine Actually an Achievable Goal?

Given the challenges discussed above, a number of aspects will have to be addressed for a successful vaccine.

3.5.1 Vaccines Should Not Induce or Exacerbate Inflammation

Because there is no inflammation in the stomach in the absence of infection, a prophylactic vaccine can be expected to be quite safe to administer. However, when the subjects eventually acquire *H. pylori* naturally through contaminated food or water, there is a risk that in the vaccinated individuals, chronic inflammation may be activated. In the pre-clinical models, vaccinated mice receive a very high dose of bacteria to achieve consistent colonization in the sham-treated mice (usually in excess of 10^8 colony forming units (CFU)). This often leads to what is termed "post-immunization gastritis" reported in several studies (eg. (Sutton et al. 2001; Raghavan et al. 2002a, b; Velin et al. 2005; Becher et al. 2010). In reality, the dose of *H. pylori* bacteria that humans will be exposed to might well be considerably lower, and might therefore not induce significant inflammation and permit eradication by the vaccine-induced memory responses with minimal inflammation. Evidence from the studies in volunteers suggests that vaccinees do not develop post-immunization gastritis since even with the high dose of *H. pylori* bacteria administered in the challenge studies (dose of 10^5-10^6 CFU), the subjects did not have a higher pathology score of inflammation compared unvaccinated subjects (Aebischer et al. 2008; Malfertheiner et al. 2018).

The major challenge for the development of a prophylactic vaccine against *H. pylori* is the induction of long-lasting immunity. This is particularly important in the context of protection against GC since it develops much later in life. A vaccine should also be able to prevent infection/reinfections with *H. pylori* in the elderly as this group are thought to have increased susceptibility to acquire *H. pylori* infection in elderly care facilities (Regev et al. 1999) as they become immunocompromised with age. There are indications from pre-clinical studies in other infections that mucosal vaccination can generate long-lasting tissue resident memory $CD4^+$ T cells that act as a first-line of defence upon second encounter with the pathogen (Schenkel and Masopust 2014). Thus, $CD4^+$ tissue resident memory cells could potentially be induced by a *H. pylori* vaccine and even in the absence of the antigen and could provide long-term protection against infection.

3.5.2 Therapeutic Vaccines Must Overcome Suppression

We know from pre-clinical studies in mouse models that therapeutic vaccination is able to reduce colonisation, but is unable to eradicate an ongoing infection (Koesling et al. 2001; Raghavan et al. 2002a, b; Sjökvist Ottsjö et al. 2015). This is possibly due to the suppressive effect of an ongoing chronic infection on the vaccine-induced response. Indeed, we have shown that the T cell response and cytokine secretion after vaccination is lower after infection of the mice compared to the response in naïve mice (Sjökvist Ottsjö et al. 2013; Holmgren et al. 2018). A lower immune response after vaccination was also observed in *H. pylori*-positive compared to *H. pylori*-negative subjects in a clinical trial of a *H. pylori* whole cell vaccine and LT_{R192G} adjuvant (Kotloff et al. 2001). Together this suggests that an ongoing *H. pylori* infection can suppress the immune responses to a *H. pylori* vaccine. To avoid the complication of still carrying the *H. pylori* bacteria for an extended period of time and a suppressed antigen-specific immune response, the vaccination might be combined with antibiotic therapy for protection against reinfection. Follow-up of acquisition of infection in vaccinees could perhaps be combined with screening for gastric cancer (Ohata et al. 2004).

3.6 Mechanisms of Protection- What Are Desirable Responses?

3.6.1 Surrogates of Protection for Human Trials

Confirming eradication or colonisation levels with biopsies in vaccine trials is invasive and expensive. Identifying surrogates of protection remains a problem for screening of candidate prophylactic vaccines. Irrespective of the type of vaccine, $CD4^+$T cell responses are clearly essential for protection as elucidated in pre-clinical models and recently in clinical studies (Kotloff et al. 2001; Aebischer et al. 2008). Since $CD4^+$T cells cannot directly access or target the bacteria, the secretion of cytokines by the $CD4^+$T cells contributes to the anti-bacterial response seen after vaccination. Several studies in both mice and humans have suggested that IFN-γ and/or IL-17A are important effector cytokines but not T_H2 related cytokines. The IFN-γ and IL-17A secreting cells have distinct functions. Secretion of IFN-γ by the $CD4^+$T cells in the stomach leads to macrophage activation, making the environment unfavourable for colonization (Fig. 1). While the secretion of IL-17A by $CD4^+$T cells activated the epithelial cells to continue to secrete IL-8 which sustains the chronic recruitment of neutrophils to the stomach (Luzza et al. 2000) (Fig. 1). Very few clinical studies have systematically measured T cell responses in subjects after vaccination and thus we have only limited knowledge regarding the activation of cellular immune responses after vaccination. Since it is technically challenging to measure the immune cells in gastric biopsies, blood is probably the samples of choice to measure the frequency of circulating and mucosal homing $\alpha4\beta7^+CD4^+$ T cells secreting IFN-γ and IL-17A and possibly also TNF-α in response to *H. pylori* antigens (Fig. 1). These advances will greatly improve our understanding of vaccine-induced responses to *H. pylori* in subjects which is greatly needed.

The desirable biomarkers for protection will however depend on the vaccine and the type of response that it can induce. For instance, because the vaccine trials based on the vaccine GGT are based on a different strategy, analysing the neutralizing IgG antibody responses to GGT will actually be the marker for an effective vaccination schedule optimizing doses and frequency (Anderl et al. 2012).

3.7 Strategies for Design of the Next Generation of Candidate Vaccines.

3.7.1 Multi Epitope Vaccines

Single antigen vaccine strategies are less likely to induce broadly effective protection, and the rationale for testing multi-antigen or multi- epitope subunit vaccines is clear. There are two major challenges facing the design of a universal *H. pylori* vaccine for global application: firstly, the variation in strains of *H. pylori* in different regions means that vaccines containing the well-studied virulence factor antigens such as CagA and VacA would have to include a broad range of antigens to cover all types; and secondly subunit vaccines will have to be carefully designed to account the spectrum of HLA types. The second problem might be overcome by the whole cell vaccine approach, but it is not clear whether even whole cell vaccines could induce protection against all strains.

One approach to address this problem is described by Ali and co-workers, who have used a comparative genomics and pathogenomics approach to compare the genome sequences and proteomes of 39 global representative strains (Ali et al. 2015). The study selected conserved regions of the genome and used computational approaches to attempt to identify universal proteins that might be selected as targets for vaccines of therapeutics. A total of 28 proteins were identified, none of which have been previously tested in pre-clinical models (Ali et al. 2015). While the rationale for this approach is sound, and given the challenges that have arisen from the currently tested antigens to date, perhaps a completely new approach is required. To our knowledge no follow-up studies have demonstrated proof-of-principle of this approach.

4 Conclusions

Is an H. pylori vaccine an achievable aim? Decades of research by several groups worldwide on effects of different approaches to vaccination on immune responses to *H. pylori* in mice and human subjects has led to several important findings regarding the host response and correlates to protection. *H. pylori* bacteria have evolved to adapt to the hostile stomach environment and can express specific antigens that can dampen the host mucosal immune response. This could explain why in spite of a vigorous systemic immune response generated in all infected individuals, the infection is rarely spontaneously eradicated and can persist for decades in the human host. Several candidate vaccines have been evaluated in phase I clinical trials, and one study has advanced to Phase II/III clinical trials and has given us valuable insights into vaccine design and delivery. Yet, we need to consider for future clinical trials that: (1) the adult challenge model is not optimal for evaluating the efficacy of a *H. pylori* vaccine, (2) as a primary endpoint, it is difficult to accurately assess level of colonization after vaccination in *H. pylori* infected individuals and (3) evaluation of long term re-infection rates is probably the most effective method to assess the protective efficacy of a *H. pylori* vaccine.

References

Adamsson J, Ottsjö LS, Lundin SB, Svennerholm A-M, Raghavan S (2017) Gastric expression of IL-17A and IFNγ in *Helicobacter pylori* infected individuals is related to symptoms. Cytokine 99:30–34. https://doi.org/10.1016/j.cyto.2017.06.013

Aebischer T, Bumann D, Epple HJ, Metzger W, Schneider T, Cherepnev G, Walduck AK, Kunkel D, Moos V, Loddenkemper C, Jiadze I, Panasyuk M, Stolte M, Graham DY, Zeitz M, Meyer TF (2008) Correlation of T cell response and bacterial clearance in human volunteers challenged with *Helicobacter pylori* revealed by randomised controlled vaccination with Ty21a-based *Salmonella* vaccines. Gut 57:1065–1072. https://doi.org/10.1136/gut.2007.145839

Algood HMS, Allen SS, Washington MK, Peek RM, Miller GG, Cover TL (2009) Regulation of gastric B cell recruitment is dependent on IL-17 receptor A signaling in a model of chronic bacterial infection. J Immunol 183:5837–5846. https://doi.org/10.4049/jimmunol.0901206

Ali A, Naz A, Soares SC, Bakhtiar M, Tiwari S, Hassan SS, Hanan F, Ramos R, Pereira U, Barh D, Figueiredo HCP, Ussery DW, Miyoshi A, Silva A, Azevedo V (2015) Pan-genome analysis of human gastric pathogen *H. pylori*: comparative genomics and pathogenomics approaches to identify regions associated with pathogenicity and prediction of potential core therapeutic targets. Biomed Res Int 2015:139580. https://doi.org/10.1155/2015/139580

Anderl F, Bolz C, Busch DH, Gerhard M (2012) Therapeutic efficacy of a *Helicobacter pylori* vaccine dependent on antibodies and T cells. Abstract HP-83. Helsingoer, Denmark

Angelakopoulos H, Hohmann EL (2000) Pilot study of phoP/phoQ-deleted *Salmonella enterica* serovar typhimurium expressing *Helicobacter pylori* urease in adult volunteers. Infect Immun 68:2135–2141

Arnold IC, Lee JY, Amieva MR, Roers A, Flavell RA, Sparwasser T, Müller A (2011) Tolerance rather than immunity protects from *Helicobacter pylori*–induced gastric Preneoplasia. Gastroenterology 140:199–209.e8. https://doi.org/10.1053/j.gastro.2010.06.047

Arnold IC, Zhang X, Urban S, Artola-Borán M, Manz MG, Ottemann KM, Müller A (2017) NLRP3 controls the development of gastrointestinal CD11b+ dendritic cells in the steady state and during chronic bacterial infection. CELREP 21:3860–3872. https://doi.org/10.1016/j.celrep.2017.12.015

Bagheri N, Shirzad H, Elahi S, Azadegan-Dehkordi F, Rahimian G, Shafigh M, Rashidii R, Sarafnejad A, Rafieian-Kopaei M, Faridani R, Tahmasbi K, Kheiri S, Razavi A (2017) Downregulated regulatory T cell function is associated with increased peptic ulcer in *Helicobacter pylori*-infection. Microb Pathog 110:165–175. https://doi.org/10.1016/j.micpath.2017.06.040

Bamford KB, Fan X, Crowe SE, Leary JF, Gourley WK, Luthra GK, Brooks EG, Graham DY, Reyes VE, Ernst PB (1998a) Lymphocytes in the human gastric mucosa during *Helicobacter pylori* have a T helper cell 1 phenotype. Gastroenterology 114:482–492

Bamford KB, Fan X, Crowe SE, Leary JF, Gourley WK, Luthra GK, Brooks EG, Graham DY, Reyes VE, Ernst PB (1998b) Lymphocytes in the human gastric mucosa during *Helicobacter pylori* have a T helper cell 1 phenotype. Gastroenterology 114:482–492

Becher D, Deutscher ME, Simpfendorfer KR, Wijburg OL, Pederson JS, Lew AM, Strugnell RA, Walduck AK (2010) Local recall responses in the stomach involving reduced regulation and expanded help mediate vaccine-induced protection against *Helicobacter pylori* in mice. Eur J Immunol 40:2778–2790. https://doi.org/10.1002/eji.200940219

Blanchard TG, Eisenberg JC, Matsumoto Y (2004) Clearance of *Helicobacter pylori* infection through immunization: the site of T cell activation contributes to vaccine efficacy. Vaccine 22:888–897. https://doi.org/10.1016/j.vaccine.2003.11.035

Booth JS, Salerno-Goncalves R, Blanchard TG, Patil SA, Kader HA, Safta AM, Morningstar LM, Czinn SJ, Greenwald BD, Sztein MB (2015) Mucosal-associated invariant T cells in the human gastric mucosa and blood: role in *Helicobacter pylori* infection. Front Immunol 6:466. https://doi.org/10.3389/fimmu.2015.00466

Bumann D, Metzger WG, Mansouri E, Palme O, Wendland M, Hurwitz R, Haas G, Aebischer T, Specht Von B-U, Meyer TF (2001) Safety and immunogenicity of live recombinant *Salmonella enterica* serovar Typhi Ty21a expressing urease A and B from *Helicobacter pylori* in human volunteers. Vaccine 20:845–852

Caruso R, Fina D, Paoluzi OA, Del Vecchio Blanco G, Stolfi C, Rizzo A, Caprioli F, Sarra M, Andrei F, Fantini MC, Macdonald TT, Pallone F, Monteleone G (2008) IL-23-mediated regulation of IL-17 production in *Helicobacter pylori*-infected gastric mucosa. Eur J Immunol 38:470–478. https://doi.org/10.1002/eji.200737635

Caruso R, Sarra M, Stolfi C, Rizzo A, Fina D, Fantini MC, Pallone F, Macdonald TT, Monteleone G (2009) Interleukin-25 inhibits interleukin-12 production and Th1 cell-driven inflammation in the gut. Gastroenterology 136:2270–2279. https://doi.org/10.1053/j.gastro.2009.02.049

Chamoun MN, Blumenthal A, Sullivan MJ, Schembri MA, Ulett GC (2018) Bacterial pathogenesis and interleukin-17: interconnecting mechanisms of immune regulation, host genetics, and microbial virulence that influence severity of infection. Crit Rev Microbiol 19:1–22. https://doi.org/10.1080/1040841X.2018.1426556

Cherdantseva LA, Potapova OV, Sharkova TV, Belyaeva YY, Shkurupiy VA (2014) Association of *Helicobacter pylori* and iNOS production by macrophages and lymphocytes in the gastric mucosa in chronic gastritis. J Immunol Res 2014:762514–4. doi: https://doi.org/10.1155/2014/762514, 1

D'Elios MM, Manghetti M, De Carli M, Costa F, Baldari CT, Burroni D, Telford JL, Romagnani S, Del Prete G (1997) T helper 1 effector cells specific for *Helicobacter pylori* in the gastric antrum of patients with peptic ulcer disease. J Immunol 158:962–967

D'Souza C, Pediongco T, Wang H, Scheerlinck J-PY, Kostenko L, Esterbauer R, Stent AW, Eckle SBG, Meehan BS, Strugnell RA, Cao H, Liu L, Mak JYW, Lovrecz G, Lu L, Fairlie DP, Rossjohn J, McCluskey J, Every AL, Chen Z, Corbett AJ (2018) Mucosal-associated invariant T cells augment immunopathology and gastritis in chronic *Helicobacter pylori* infection. J Immunol 200:1901–1916. https://doi.org/10.4049/jimmunol.1701512

Delyria ES, Redline RW, Blanchard TG (2009) Vaccination of mice against H pylori induces a strong Th-17 response and immunity that is neutrophil dependent. Gastroenterology 136:247–256. https://doi.org/10.1053/j.gastro.2008.09.017

Dhar P, Ng GZ, Sutton P (2016) How host regulation of *Helicobacter pylori*-induced gastritis protects against peptic ulcer disease and gastric cancer. Am J Physiol Gastrointest Liver Physiol 311:G514–G520. https://doi.org/10.1152/ajpgi.00146.2016

Di Tommaso A, Xiang Z, Bugnoli M, Pileri P, Figura N, Bayeli PF, Rappuoli R, Abrignani S, De Magistris MT (1995) *Helicobacter pylori*-specific CD4+ T-cell clones from peripheral blood and gastric biopsies. Infect Immun 63:1102–1106

DiPetrillo MD, Tibbetts T, Kleanthous H, Killeen KP, Hohmann EL (1999) Safety and immunogenicity of phoP/phoQ-deleted *Salmonella typhi* expressing *Helicobacter pylori* urease in adult volunteers. Vaccine 18:449–459

Dixon MF (2001) Pathology of gastritis and peptic ulceration. In: Mobley HL, Mendz GL, Hazell SL (eds) *Helicobacter pylori* physiology and genetics. ASM Press, Washington, DC

Eaton KA, Mefford M, Thevenot T (2001) The role of T cell subsets and cytokines in the pathogenesis of *Helicobacter pylori* gastritis in mice. J Immunol 166:7456–7461

Enarsson K, Lundgren A, Kindlund B, Hermansson M, Roncador G, Banham AH, Lundin BS, Quiding-Järbrink M (2006) Function and recruitment of mucosal regulatory T cells in human chronic *Helicobacter pylori* infection and gastric adenocarcinoma. Clin Immunol 121:358–368. https://doi.org/10.1016/j.clim.2006.07.002

Falush D, Wirth T, Linz B, Pritchard JK, Stephens M, Kidd M, Blaser MJ, Graham DY, Vacher S, Perez-Perez GI, Yamaoka Y, Mégraud F, Otto K, Reichard U, Katzowitsch E, Wang X, Achtman M, Suerbaum S (2003) Traces of human migrations in *Helicobacter pylori* populations. Science 299:1582–1585. https://doi.org/10.1126/science.1080857

Flach C-FC, Mozer MM, Sundquist MM, Holmgren JJ, Raghavan SS (2012) Mucosal vaccination increases local chemokine production attracting immune cells to the stomach mucosa of *Helicobacter pylori* infected mice. Vaccine 30:1636–1643. https://doi.org/10.1016/j.vaccine.2011.12.111

Freire de Melo F, Rocha AMC, Rocha GA, Pedroso SHSP, de Assis Batista S, Fonseca de Castro LP, Carvalho SD, Bittencourt PFS, de Oliveira CA, Corrêa-Oliveira R, Magalhães Queiroz DM (2012) A regulatory instead of an IL-17 T response predominates in *Helicobacter pylori*-associated gastritis in children. Microbes Infect 14:341–347. https://doi.org/10.1016/j.micinf.2011.11.008

Gaffen SL (2009) Structure and signalling in the IL-17 receptor family. Nat Rev Immunol 9:556–567. https://doi.org/10.1038/nri2586

Gerhard M, Schmees C, Voland P, Endres N, Sander M, Reindl W, Rad R, Oelsner M, Decker T, Mempel M, Hengst L, Prinz C (2005) A secreted low-molecular-weight protein from *Helicobacter pylori* induces cell-cycle arrest of T cells. Gastroenterology 128:1327–1339

Graham DY, Opekun AR, Osato MS, El-Zimaity HMT, Lee CK, Yamaoka Y, Qureshi WA, Cadoz M, Monath TP (2004) Challenge model for *Helicobacter pylori* infection in human volunteers. Gut 53:1235–1243. https://doi.org/10.1136/gut.2003.037499

Graham DY, Lu H, Yamaoka Y (2007) A report card to grade *Helicobacter pylori* therapy. Helicobacter 12:275–278. https://doi.org/10.1111/j.1523-5378.2007.00518.x

Guang W, Ding H, Czinn SJ, Kim KC, Blanchard TG, Lillehoj EP (2010) Muc1 cell surface mucin attenuates epithelial inflammation in response to a common mucosal pathogen. J Biol Chem 285:20547–20557. https://doi.org/10.1074/jbc.M110.121319

Guang W, Twaddell WS, Lillehoj EP (2012) Molecular interactions between MUC1 epithelial mucin, β-catenin, and CagA proteins. Front Immunol 3:105. https://doi.org/10.3389/fimmu.2012.00105

Hall MT, Simms KT, Lew J-B, Smith MA, Saville M, Canfell K (2018) Projected future impact of HPV vaccination and primary HPV screening on cervical cancer rates from 2017-2035: example from Australia. PLoS One 13:e0185332. https://doi.org/10.1371/journal.pone.0185332

Harris PR, Wright SW, Serrano C, Riera F, Duarte I, Torres J, Peña A, Rollán A, Viviani P, Guiraldes E, Schmitz JM, Lorenz RG, Novak L, Smythies LE, Smith PD (2008) *Helicobacter pylori* gastritis in children is associated with a regulatory T-cell response. Gastroenterology 134:491–499. https://doi.org/10.1053/j.gastro.2007.11.006

Holmgren J, Nordqvist S, Blomquist M, Jeverstam F, Lebens M, Raghavan S (2018) Preclinical immunogenicity and protective efficacy of an oral *Helicobacter pylori* inactivated whole cell vaccine and multiple mutant cholera toxin: a novel and non-toxic mucosal adjuvant. Vaccine 36:1–8. https://doi.org/10.1016/j.vaccine.2018.07.073

Horvath DJ, Radin JN, Cho SH, Washington MK, Algood HMS (2013) The interleukin-17 receptor B subunit is essential for the Th2 response to *Helicobacter pylori*, but not for control of bacterial burden. PLoS One 8:e60363. https://doi.org/10.1371/journal.pone.0060363

Hsu W-T, Ho S-Y, Jian T-Y, Huang H-N, Lin Y-L, Chen C-H, Lin T-H, Wu M-S, Wu C-J, Chan Y-L, Liao K-W (2018) *Helicobacter pylori*-derived heat shock protein 60 increases the induction of regulatory T-cells associated with persistent infection. Microb Pathog 119:152–161. https://doi.org/10.1016/j.micpath.2018.04.016

Kaebisch R, Mejias-Luque R, Prinz C, Gerhard M (2013) *Helicobacter pylori* Cytotoxin-associated gene A impairs human dendritic cell maturation and function through IL-10-mediated activation of STAT3. J Immunol 192:316–323. https://doi.org/10.4049/jimmunol.1302476

Kao JY, Zhang M, Miller MJ, Mills JC, Wang B, Liu M, Eaton KA, Zou W, Berndt BE, Cole TS, Takeuchi T, Owyang SY, Luther J (2010) *Helicobacter pylori* immune escape is mediated by dendritic cell-induced Treg skewing and Th17 suppression in mice. Gastroenterology 138:1046–1054. https://doi.org/10.1053/j.gastro.2009.11.043

Kaparakis MM, Laurie KLK, Wijburg OO, Pedersen JJ, Pearse MM, van Driel IRI, Gleeson PAP, Strugnell RAR (2006) CD4+ CD25+ regulatory T cells modulate the T-cell and antibody responses in helicobacter-infected BALB/c mice. Infect Immun 74:3519–3529. https://doi.org/10.1128/IAI.01314-05

Khamri W, Walker MM, Clark P, Atherton JC, Thursz MR, Bamford KB, Lechler RI, Lombardi G (2010) *Helicobacter pylori* stimulates dendritic cells to induce interleukin-17 expression from CD4+ T lymphocytes. Infect Immun 78:845–853. https://doi.org/10.1128/IAI.00524-09

Koesling J, Lucas B, Develioglou L, Aebischer T, Meyer TF (2001) Vaccination of mice with live recombinant *Salmonella* typhimurium aroA against *H. pylori*: parameters associated with prophylactic and therapeutic vaccine efficacy. Vaccine 20:413–420

Kotloff KL, Sztein MB, Wasserman SS, Losonsky GA, DiLorenzo SC, Walker RI (2001) Safety and immunogenicity of Oral inactivated whole-cell *Helicobacter pylori* vaccine with adjuvant among volunteers with or without subclinical infection. Infect Immun 69:3581–3590

Lagunes-Servin H, Torres J, Maldonado-Bernal C, Pérez-Rodríguez M, Huerta-Yépez S, Madrazo de la Garza A, Muñoz-Pérez L, Flores-Luna L, Ramón-García G, Camorlinga-Ponce M (2013) Toll-like receptors and cytokines are upregulated during *Helicobacter pylori* infection in children. Helicobacter 18:423–432. https://doi.org/10.1111/hel.12067

Li R, Jiang X-X, Zhang L-F, Liu X-M, Hu T-Z, Xia X-J, Li M, Xu C-X (2017) Group 2 innate lymphoid cells are involved in skewed type 2 immunity of gastric diseases induced by *Helicobacter pylori* infection. Mediat Inflamm 2017:4927964. https://doi.org/10.1155/2017/4927964

Lindholm C, Quiding-Jarbrink M, Lönroth H, Hamlet A, Svennerholm A-M (1998) Local cytokine response in *Helicobacter pylori*-infected subjects. Infect Immun 66:5964–5971

Lundgren A, Suri-Payer E, Enarsson K, Svennerholm A-M, Lundin BS (2003) *Helicobacter pylori*-specific CD4+ CD25high regulatory T cells suppress memory T-cell responses to *H. pylori* in infected individuals. Infect Immun 71:1755–1762. https://doi.org/10.1128/IAI.71.4.1755–1762.2003

Lundgren A, Stromberg E, Sjöling Å, Lindholm C, Enarsson K, Edebo A, Johnsson E, Suri-Payer E, Larsson P, Rudin A, Svennerholm A-M, Lundin BS (2004) Mucosal FOXP3-expressing CD4+ CD25high regulatory T cells in *Helicobacter pylori*-infected patients. Infect Immun 73:523–531. https://doi.org/10.1128/IAI.73.1.523-531.2005

Lundgren A, Bourgeois L, Carlin N, Clements J, Gustafsson B, Hartford M, Holmgren J, Petzold M, Walker R, Svennerholm A-M (2014) Safety and immunogenicity of an improved oral inactivated multivalent enterotoxigenic *Escherichia coli* (ETEC) vaccine administered alone and together with dmLT adjuvant in a double-blind, randomized, placebo-controlled phase I study. Vaccine 32:7077–7084. https://doi.org/10.1016/j.vaccine.2014.10.069

Luzza F, Parrello T, Monteleone G, Sebkova L, Romano M, Zarrilli R, Imeneo M, Pallone F (2000) Up-regulation of IL-17 is associated with bioactive IL-8 expression in *Helicobacter pylori*-infected human gastric mucosa. J Immunol 165:5332–5337

Luzza F, Suraci E, Larussa T, Leone I, Imeneo M (2014) High exposure, spontaneous clearance, and low incidence of active *Helicobacter pylori* infection: the Sorbo San Basile study. Helicobacter 19:296–305. https://doi.org/10.1111/hel.12133

Machalek DA, Garland SM, Brotherton JML, Bateson D, McNamee K, Stewart M, Rachel Skinner S, Liu B, Cornall AM, Kaldor JM, Tabrizi SN (2018) Very low prevalence of vaccine human papillomavirus types among 18- to 35-year old Australian women 9 years following implementation of vaccination. J Infect Dis 217:1590–1600. https://doi.org/10.1093/infdis/jiy075

Malfertheiner P, Schultze V, Rosenkranz B, Kaufmann SHE, Ulrichs T, Novicki D, Norelli F, Contorni M, Peppoloni S, Berti D, Tornese D, Ganju J, Palla E, Rappuoli R, Scharschmidt BF, Del Giudice G (2008) Safety and immunogenicity of an intramuscular *Helicobacter pylori* vaccine in noninfected volunteers: a phase I study. Gastroenterology 135:787–795. https://doi.org/10.1053/j.gastro.2008.05.054

Malfertheiner P, Megraud F, O'Morain CA, Gisbert JP, Kuipers EJ, Axon AT, Bazzoli F, Gasbarrini A, Atherton J, Graham DY, Hunt R, Moayyedi P, Rokkas T, Rugge M, Selgrad M, Suerbaum S, Sugano K, El-Omar EM, European Helicobacter and Microbiota Study Group and Consensus panel (2017) Management of *Helicobacter pylori* infection-the Maastricht V/Florence Consensus Report, pp 6–30

Malfertheiner P, Selgrad M, Wex T, Romi B, Borgogni E, Spensieri F, Zedda L, Ruggiero P, Pancotto L, Censini S, Palla E, Kanesa-Thasan N, Scharschmidt B, Rappuoli R, Graham DY, Schiavetti F, Del Giudice G (2018) Efficacy, immunogenicity, and safety of a parenteral vaccine against *Helicobacter pylori* in healthy volunteers challenged with a Cag-positive strain: a randomised, placebo-controlled phase 1/2 study. Lancet Gastroenterol Hepatol 3:698–707. https://doi.org/10.1016/S2468-1253(18)30125-0

Maynard CL, Harrington LE, Janowski KM, Oliver JR, Zindl CL, Rudensky AY, Weaver CT (2007) Regulatory T cells expressing interleukin 10 develop from Foxp3+ and Foxp3- precursor cells in the absence of interleukin 10. Nat Immunol 8:931–941. https://doi.org/10.1038/ni1504

McGuckin MA, Every AL, Skene CD, Lindén SK, Chionh Y-T, Swierczak A, McAuley J, Harbour S, Kaparakis M, Ferrero R, Sutton P (2007) Muc1 mucin limits both *Helicobacter pylori* colonization of the murine gastric mucosa and associated gastritis. Gastroenterology 133:1210–1218. https://doi.org/10.1053/j.gastro.2007.07.003

McGuckin MA, Lindén SK, Sutton P, Florin TH (2011) Mucin dynamics and enteric pathogens. Nat Rev Micro 9:265–278. https://doi.org/10.1038/nrmicro2538

Metzger WG, Mansouri E, Kronawitter M, Diescher S, Soerensen M, Hurwitz R, Bumann D, Aebischer T, Specht Von B-U, Meyer TF (2004) Impact of vector-priming on the immunogenicity of a live recombinant *Salmonella enterica* serovar *typhi* Ty21a vaccine expressing urease A and B from *Helicobacter pylori* in human volunteers. Vaccine 22:2273–2277. https://doi.org/10.1016/j.vaccine.2003.11.020

Michetti P, Kreiss C, Kotloff KL, Porta N, Blanco JL, Bachmann D, Herranz M, Saldinger PF, Corthésy-Theulaz I, Losonsky G, Nichols R, Simon J, Stolte M, Ackerman S, Monath TP, Blum AL (1999) Oral immunization with urease and *Escherichia coli* heat-labile enterotoxin is safe and immunogenic in *Helicobacter pylori*-infected adults. Gastroenterology 116:804–812

Michetti M, Kelly CP, Kraehenbuhl JP, Bouzourene H, Michetti P (2000) Gastric mucosal alpha(4)beta(7)-integrin-positive CD4 T lymphocytes and immune protection against helicobacter infection in mice. Gastroenterology 119:109–118

Mizuno T, Ando T, Nobata K, Tsuzuki T, Maeda O, Watanabe O, Minami M, Ina K, Kusugami K, Peek RM, Goto H (2005) Interleukin-17 levels in *Helicobacter pylori*-infected gastric mucosa and pathologic sequelae of colonization. World J Gastroenterol 11:6305–6311

Moodley Y, Linz B, Bond RP, Nieuwoudt M, Soodyall H, Schlebusch CM, Bernhöft S, Hale J, Suerbaum S, Mugisha L, van der Merwe SW, Achtman M (2012) Age of the association between *Helicobacter pylori* and man. PLoS Pathog 8:e1002693. https://doi.org/10.1371/journal.ppat.1002693

Moran AP (2007) Lipopolysaccharide in bacterial chronic infection: insights from *Helicobacter pylori* lipopolysaccharide and lipid A. Int J Med Microbiol 297:307–319. https://doi.org/10.1016/j.ijmm.2007.03.008

Morey P, Pfannkuch L, Pang E, Boccellato F, Sigal M, Imai-Matsushima A, Dyer V, Koch M, Mollenkopf H-J, Schlaermann P, Meyer TF (2018) *Helicobacter pylori* depletes cholesterol in gastric glands to prevent interferon gamma signaling and escape the inflammatory response. Gastroenterology 154:1391–1404.e9. https://doi.org/10.1053/j.gastro.2017.12.008

Moro K, Koyasu S (2015) Innate lymphoid cells, possible interaction with microbiota. Semin Immunopathol 37:27–37. https://doi.org/10.1007/s00281-014-0470-4

Morrison PJ, Bending D, Fouser LA, Wright JF, Stockinger B, Cooke A, Kullberg MC (2013) Th17-cell plasticity in *Helicobacter hepaticus*-induced intestinal inflammation. Mucosal Immunol 6:1143–1156. https://doi.org/10.1038/mi.2013.11

Mueller A, O'Rourke J, Chu P, Kim CC, Sutton P, Lee A, Falkow S (2003) Protective immunity against helicobacter is characterized by a unique transcriptional signature. Proc Natl Acad Sci U S A 100:12289–12294. https://doi.org/10.1073/pnas.1635231100

Muñoz L, Camorlinga M, Hernández R, Giono S, Ramón G, Muñoz O, Torres J (2007) Immune and proliferative cellular responses to *Helicobacter pylori* infection in the gastric mucosa of Mexican children. Helicobacter 12:224–230. https://doi.org/10.1111/j.1523-5378.2007.00493.x

Necchi V, Manca R, Ricci V, Solcia E (2009) Evidence for transepithelial dendritic cells in human *H. pylori* active gastritis. Helicobacter 14:208–222. https://doi.org/10.1111/j.1523-5378.2009.00679.x

Nedrud JG, Czinn SJ, Ding H, Zagorski BM, Redline RW, Twaddell W, Blanchard TG (2012) Lack of genetic influence on the innate inflammatory response to helicobacter infection of the gastric mucosa. Front Immunol 3:181. https://doi.org/10.3389/fimmu.2012.00181

Ng GZ, Sutton P (2016) The MUC1 mucin specifically inhibits activation of the NLRP3 inflammasome. Genes Immun 17:203–206. https://doi.org/10.1038/gene.2016.10

O'Ryan ML, Lucero Y, Rabello M, Mamani N, Salinas AM, Peña A, Torres-Torreti JP, Mejías A, Ramilo O, Suarez N, Reynolds HE, Orellana A, Lagomarcino AJ (2015) Persistent and transient *Helicobacter pylori* infections in early childhood. Clin Infect Dis 61:211–218. https://doi.org/10.1093/cid/civ256

Oertli M, Noben M, Engler DB, Semper RP, Reuter S, Maxeiner J, Gerhard M, Taube C, Müller A (2013) *Helicobacter pylori* {gamma}-glutamyl transpeptidase and vacuolating cytotoxin promote gastric persistence and immune tolerance. PNAS 110:3047–3052. https://doi.org/10.1073/pnas.1211248110

Ohata H, Kitauchi S, Yoshimura N, Mugitani K, Iwane M, Nakamura H, Yoshikawa A, Yanaoka K, Arii K, Tamai H, Shimizu Y, Takeshita T, Mohara O, Ichinose M (2004) Progression of chronic atrophic gastritis associated with *Helicobacter pylori* infection increases risk of gastric cancer. Int J Cancer 109:138–143. https://doi.org/10.1002/ijc.11680

Omenetti S, Pizarro TT (2015) The Treg/Th17 Axis: a dynamic balance regulated by the gut microbiome. Front Immunol 6:639. https://doi.org/10.3389/fimmu.2015.00639

Otero LL, Ruiz VE, Perez-Perez GI (2014) *Helicobacter pylori*: the balance between a role as colonizer and pathogen. Best Pract Res Clin Gastroenterol 28:1017–1029. https://doi.org/10.1016/j.bpg.2014.09.003

Pachathundikandi SK, Lind J, Tegtmeyer N, El-Omar EM, Backert S (2015) Interplay of the gastric pathogen *Helicobacter pylori* with toll-like receptors. Biomed Res Int 2015:192420. https://doi.org/10.1155/2015/192420

Parsonnet J, Friedman GD, Vandersteen DP, Chang Y, Vogelman JH, Orentreich N, Sibley RK (1991) *Helicobacter pylori* infection and the risk of gastric carcinoma. N Engl J Med 325:1127–1131. https://doi.org/10.1056/NEJM199110173251603

Posselt G, Backert S, Wessler S (2013) The functional interplay of *Helicobacter pylori* factors with gastric epithelial cells induces a multi-step process in pathogenesis. Cell Commun Signal 11:77. https://doi.org/10.1186/1478-811X-11-77

Rad R, Brenner L, Bauer S, Schwendy S, Layland L, da Costa CP, Reindl W, Dossumbekova A, Friedrich M, Saur D, Wagner H, Schmid RM, Prinz C (2006) CD25+/Foxp3+ T cells regulate gastric inflammation and *Helicobacter pylori* colonization in vivo. Gastroenterology 131:525–537. https://doi.org/10.1053/j.gastro.2006.05.001

Raghavan S, Quiding-Järbrink M (2016) Vaccination against *Helicobacter pylori*. In: Backert S, Yamaoka Y (eds) *Helicobacter pylori* research. Springer, Tokyo

Raghavan S, Hjulström M, Holmgren J, Svennerholm A-M (2002a) Protection against experimental *Helicobacter pylori* infection after immunization with inactivated *H. pylori* whole-cell vaccines. Infect Immun 70:6383–6388

Raghavan S, Svennerholm A-M, Holmgren J (2002b) Effects of oral vaccination and immunomodulation by cholera toxin on experimental *Helicobacter pylori* infection, reinfection, and gastritis. Infect Immun 70:4621–4627

Raghavan S, Fredriksson M, Svennerholm A-M, Holmgren J, Suri-Payer E (2003) Absence of CD4+CD25+regulatory T cells is associated with a loss of regulation leading to increased pathology in *Helicobacter pylori*-infected mice. Clin Exp Immunol 132:393–400. https://doi.org/10.1046/j.1365-2249.2003.02177.x

Raghavan S, Suri-Payer E, Holmgren J (2004) Antigen-specific in vitro suppression of murine *Helicobacter pylori*-reactive immunopathological T cells by CD4+CD25+ regulatory T cells. Scand J Immunol 60:82–88. https://doi.org/10.1111/j.0300-9475.2004.01447.x

Regev A, Fraser GM, Braun M, Maoz E, Leibovici L, Niv Y (1999) Seroprevalence of *Helicobacter pylori* and length of stay in a nursing home. Helicobacter 4:89–93

Rescigno M, Urbano M, Valzasina B, Francolini M, Rotta G, Bonasio R, Granucci F, Kraehenbuhl JP, Ricciardi-Castagnoli P (2001) Dendritic cells express tight junction proteins and penetrate gut epithelial monolayers to sample bacteria. Nat Immunol 2:361–367. https://doi.org/10.1038/86373

Rivas-Ortiz CI, López-Vidal Y, Arredondo-Hernandez LJR, Castillo-Rojas G (2017) Genetic alterations in gastric cancer associated with *Helicobacter pylori* infection. Front Med 4:1–12. https://doi.org/10.3389/fmed.2017.00047

Robinson K, Kenefeck R, Pidgeon EL, Shakib S, Patel S, Polson RJ, Zaitoun AM, Atherton JC (2008) *Helicobacter pylori*-induced peptic ulcer disease is associated with inadequate regulatory T cell responses. Gut 57:1375–1385. https://doi.org/10.1136/gut.2007.137539

Rupnow MF, Shachter RD, Owens DK, Parsonnet J (2000) A dynamic transmission model for predicting trends in *Helicobacter pylori* and associated diseases in the United States. Emerg Infect Dis 6:228–237. https://doi.org/10.3201/eid0603.000302

Rupnow MFT, Shachter RD, Owens DK, Parsonnet J (2001) Quantifying the population impact of a prophylactic *Helicobacter pylori* vaccine. Vaccine 20:879–885. https://doi.org/10.1016/S0264-410X(01)00401-7

Rupnow MFT, Chang AH, Shachter RD, Owens DK, Parsonnet J (2009) Cost-effectiveness of a potential prophylactic *Helicobacter pylori* vaccine in the United States. J Infect Dis 200:1311–1317. https://doi.org/10.1086/605845

Sakaguchi S, Sakaguchi N, Asano M, Itoh M, Toda M (1995) Immunologic self-tolerance maintained by activated T cells expressing IL-2 receptor alpha-chains (CD25). Breakdown of a single mechanism of self-tolerance causes various autoimmune diseases. J Immunol 155:1151–1164

Satin B, Del Giudice G, Bianca Della V, Dusi S, Laudanna C, Tonello F, Kelleher D, Rappuoli R, Montecucco C, Rossi F (2000) The neutrophil-activating protein (HP-NAP) of *Helicobacter pylori* is a protective antigen and a major virulence factor. J Exp Med 191:1467–1476

Schenkel JM, Masopust D (2014) Tissue-resident memory T cells. Immunity 41:886–897. https://doi.org/10.1016/j.immuni.2014.12.007

Schlaermann P, Toelle B, Berger H, Schmidt SC, Glanemann M, Ordemann J, Bartfeld S, Mollenkopf HJ, Meyer TF (2016) A novel human gastric primary cell culture system for modelling *Helicobacter pylori* infection in vitro. Gut 65:202–213. https://doi.org/10.1136/gutjnl-2014-307949

Sigal M, Logan CY, Kapalczynska M, Mollenkopf H-J, Berger H, Wiedenmann B, Nusse R, Amieva MR, Meyer TF (2017) Stromal R-spondin orchestrates gastric epithelial stem cells and gland homeostasis. Nature 548:451–455. https://doi.org/10.1038/nature23642

Sjökvist Ottsjö L, Flach C-F, Clements J, Holmgren J, Raghavan S (2013) A double mutant heat-labile toxin from Escherichia coli, LT(R192G/L211A), is an effective mucosal adjuvant for vaccination against *Helicobacter pylori* infection. Infect Immun 81:1532–1540. https://doi.org/10.1128/IAI.01407-12

Sjökvist Ottsjö L, Flach C-F, Nilsson S, Malefyt RDW, Walduck AK, Raghavan S (2015) Defining the roles of IFN-γ and IL-17A in inflammation and protection against *Helicobacter pylori* infection. PLoS One 10: e0131444. https://doi.org/10.1371/journal.pone.0131444

Song Z, Zhang T, Li G, Tang Y, Luo Y, Yu G (2018) Tr1 responses are elevated in asymptomatic *H. pylori*-infected individuals and are functionally impaired in *H. pylori*-gastric cancer patients. Exp Cell Res 367:251–256. https://doi.org/10.1016/j.yexcr.2018.04.002

Stenström B, Windsor HM, Fulurija A, Benghezal M, Kumarasinghe MP, Kimura K, Tay CY, Viiala CH, Ee HC, Lu W, Schoep TD, Webberley KM, Marshall BJ (2016) *Helicobacter pylori* overcomes natural immunity in repeated infections. Clin Case Rep 4:1026–1033. https://doi.org/10.1002/ccr3.687

Summerton NA, Welch RW, Bondoc L, Yang H-H, Pleune B, Ramachandran N, Harris AM, Bland D, Jackson WJ, Park S, Clements JD, Nabors GS (2010) Toward the development of a stable, freeze-dried formulation of *Helicobacter pylori* killed whole cell vaccine adjuvanted with a novel mutant of *Escherichia coli* heat-labile toxin. Vaccine 28:1404–1411. https://doi.org/10.1016/j.vaccine.2009.10.147

Sutton P, Boag JM (2018) Status of vaccine research and development for *Helicobacter pylori*. Vaccine. https://doi.org/10.1016/j.vaccine.2018.01.001

Sutton P, Danon SJ, Walker M, Thompson LJ, Wilson J, Kosaka T, Lee A (2001) Post-immunisation gastritis and helicobacter infection in the mouse: a long-term study. Gut 49:467–473

Tanaka S, Nagashima H, Cruz M, Uchida T, Uotani T, Jiménez Abreu JA, Mahachai V, Vilaichone R-K, Ratanachu-Ek T, Tshering L, Graham DY, Yamaoka Y (2017) Interleukin-17C in human *Helicobacter pylori* gastritis. Infect Immun 85. https://doi.org/10.1128/IAI.00389-17

Tang J, Zhou X, Liu J, Meng Q, Han Y, Wang Z, Fan H, Liu Z (2015) IL-25 promotes the function of CD4+CD25+ T regulatory cells and prolongs skin-graft survival in murine models. Int Immunopharmacol 28:931–937. https://doi.org/10.1016/j.intimp.2015.03.036

Velin D, Bachmann D, Bouzourene H, Michetti P (2005) Mast cells are critical mediators of vaccine-induced helicobacter clearance in the mouse model. Gastroenterology 129:142–155

Velin D, Favre L, Bernasconi E, Bachmann D, Pythoud C, Saiji E, Bouzourene H, Michetti P (2009) Interleukin-17 is a critical mediator of vaccine-induced reduction of helicobacter infection in the mouse model. Gastroenterology 136:2237–2246.e1. https://doi.org/10.1053/j.gastro.2009.02.077

Walduck AK, Schmitt A, Lucas B, Aebischer T, Meyer TF (2004) Transcription profiling analysis of the mechanisms of vaccine-induced protection against *H. pylori*. FASEB J 18:1955–1957. https://doi.org/10.1096/fj.04-2321fje

Zamani M, Ebrahimtabar F, Zamani V, Miller WH, Alizadeh-Navaei R, Shokri-Shirvani J, Derakhshan MH (2018) Systematic review with meta-analysis: the worldwide prevalence of *Helicobacter pylori* infection. Aliment Pharmacol Ther 47:868–876. https://doi.org/10.1111/apt.14561

Zeng M, Mao X-H, Li J-X, Tong W-D, Wang B, Zhang Y-J, Guo G, Zhao Z-J, Li L, Wu D-L, Lu D-S, Tan Z-M, Liang H-Y, Wu C, Li D-H, Luo P, Zeng H, Zhang W-J, Zhang J-Y, Guo B-T, Zhu F-C, Zou Q-M (2015) Efficacy, safety, and immunogenicity of an oral recombinant *Helicobacter pylori* vaccine in children in China: a randomised, double-blind, placebo-controlled, phase 3 trial. Lancet 386:1457–1464. https://doi.org/10.1016/S0140-6736(15)60310-5

Adv Exp Med Biol - Advances in Microbiology, Infectious Diseases and Public Health (2019) 1149: 277–284
https://doi.org/10.1007/978-3-030-21916-1

Index

MIX
Papier aus verantwortungsvollen Quellen
Paper from responsible sources
FSC® C105338

If you have any concerns about our products,
you can contact us on
ProductSafety@springernature.com

In case Publisher is established outside the EU,
the EU authorized representative is:
Springer Nature Customer Service Center GmbH
Europaplatz 3, 69115 Heidelberg, Germany

Printed by Libri Plureos GmbH
in Hamburg, Germany

Cinderella
of
Wall Street

by

LYDIA FISHER

GALYDA MEDIA, LLC

Published by Galyda Media, LLC, P.O. Box 11831, Chicago, IL 60611.
Printed at Grace Printing, 2050 W. Devon Ave. Chicago, IL 60659.

Disclaimer: This book is a memoir, based on the author's reflections, recollections, and perceptions of the financial industry as experienced by the author. While autobiographical, the book is supplemented with fictional vignettes. The individuals referenced in the book are inspired by people from the author's life. Names of some individuals have been changed or identifying features modified to preserve anonymity and privacy. In some cases, characters were created. To the extent that any financial institution is referenced or alluded to, it is meant to illustrate concepts rather than to single out any one entity or financial sector. The book is not intended to be and should not be taken as an accurate authoritative factual history. The author and Publisher make no representation or warranty of any sort as to accuracy of fact, data, events or completeness of the contents of this book. Under no circumstances will the author and Publisher have any liability to any entity or person for any loss or damages of any sort resulting from information contained herein. The book is not intended to render any advice, financial or otherwise. If any advice is required, services of a financial or any other professional should be sought.

The Publisher does not have any control over and does not assume any responsibility for third party websites or their content.

Grateful acknowledgement is made to the following for permission to reprint: William Tyler Page; "The American's Creed"; http://www.ushistory.org/documents/creed.htm; Independence Hall Association, on the web at ushistory.org; copyright 2009.

All inquiries should be sent to Galyda Media, LLC, P.O. Box 11831, Chicago, Illinois, 60611.

First Edition, September 2009.

Library of Congress Cataloging-in-Publication Data:

ISBN #: 978-0-615-30049-8